A CENTURY OF CALCULUS
PART II

A CENTURY OF CALCULUS
PART II

The Raymond W. Brink
Selected Mathematical Papers

Reprinted from the

AMERICAN MATHEMATICAL MONTHLY
(Volumes 76–98)

MATHEMATICS MAGAZINE
(Volumes 41–64)

COLLEGE MATHEMATICS JOURNAL
(Volumes 1–22)

Selected and arranged by an editorial committee consisting of

Tom M. Apostol, Chairman
California Institute of Technology

Dale H. Mugler
University of Akron

David R. Scott
University of Puget Sound

Andrew Sterrett, Jr.
Denison University

Ann E. Watkins
California State University, Northridge

Published and distributed by The Mathematical Association of America

NOV 0 2 1992

Library of Congress Catalog Number 92-060987
ISBN 0-883385-206-3
Copyright © 1992 by
The Mathematical Association of America (Incorporated)
PRINTED IN THE UNITED STATES OF AMERICA

PREFACE

In 1969 the Mathematical Association of America published SELECTED PAPERS IN CALCULUS, the first of a set of volumes that contained articles from issues of the Association's journals. This book is a sequel to the 1969 volume and contains selected papers on calculus reprinted from three MAA publications: The American Mathematical Monthly, the Mathematics Magazine, and the College Mathematics Journal (formerly the Two-Year College Mathematics Journal).

The editorial committee read more than 500 articles, of which 150 are reproduced here. The selection procedure was similar to that described in the Preface of the 1969 volume. The papers are arranged into categories treated as chapters of the book, following more or less the same plan as in the earlier volume. Some of the chapters are further divided into subcategories.

In each subcategory the selections from the *Monthly* appear first, followed by those from the other journals in the order mentioned above. Some exceptions to this arrangement were made when it seemed preferable to group together papers on similar topics. Most subcategories are followed by a bibliography listing further related papers. Some of the bibliographic entries are accompanied by brief editorial comments describing in more detail the contents of the paper.

The committee attempted to keep the volume to a reasonable size, while including a broad spectrum of ideas and maintaining some sort of balance among the various categories. Many of the papers contain material that can be used directly in the classroom. Others provide insights, background, or source material for special projects. The inclusion of any paper or bibliographic entry is not to be considered in any way as an endorsement by the editorial committee of the point of view expressed by its author.

It is the committee's hope that there will be much of interest in this volume, both for the beginning teacher and the more experienced veteran, as well as for students.

THE EDITORIAL COMMITTEE

CONTENTS

3. FUNCTIONS

4. LIMITS AND CONTINUITY

5. DIFFERENTIATION

6. MEAN VALUE THEOREM FOR DERIVATIVES, INDETERMINATE FORMS

7. TAYLOR POLYNOMIALS, BERNOULLI POLYNOMIALS AND SUMS OF POWERS OF INTEGERS

(b) BERNOULLI POLYNOMIALS AND SUMS OF POWERS OF INTEGERS

(no papers reproduced)

8. MAXIMA AND MINIMA

9. INTEGRATION

(e) MULTIPLE INTEGRALS AND LINE INTEGRALS

10. NUMERICAL, GRAPHICAL, AND MECHANICAL METHODS AND APPROXIMATIONS (INCLUDING USE OF COMPUTERS)

11. INFINITE SEQUENCES AND SERIES

12. SPECIAL NUMBERS

13. THE LIGHT TOUCH

1

HISTORY

WHO GAVE YOU THE EPSILON?
CAUCHY AND THE ORIGINS OF RIGOROUS CALCULUS

JUDITH V. GRABINER

424 West 7th Street, Claremont, California 91711

Student: The car has a speed of 50 miles an hour. What does that mean?

Teacher: Given any $\varepsilon > 0$, there exists a δ such that if $|t_2 - t_1| < \delta$, then $\left| \dfrac{s_2 - s_1}{t_2 - t_1} - 50 \right| < \varepsilon$.

Student: How in the world did anybody ever think of such an answer?

*   *   *   *   *   *   *   *   *   *   *   *   *

Perhaps this exchange will remind us that the rigorous basis for the calculus is not at all intuitive—in fact, quite the contrary. The calculus is a subject dealing with speeds and distances, with tangents and areas—not inequalities. When Newton and Leibniz invented the calculus in the late seventeenth century, they did not use delta-epsilon proofs. It took a hundred and fifty years to develop them. This means that it was probably very hard, and it is no wonder that a modern student finds the rigorous basis of the calculus difficult. How, then, did the calculus get a rigorous basis in terms of the algebra of inequalities?

Delta-epsilon proofs are first found in the works of Augustin-Louis Cauchy (1789–1867). This is not always recognized, since Cauchy gave a purely verbal definition of limit, which at first glance does not resemble modern definitions: "When the successively attributed values of the same variable indefinitely approach a fixed value, so that finally they differ from it by as little as desired, the last is called the *limit* of all the others" [1]. Cauchy also gave a purely verbal definition of the derivative of $f(x)$ as the limit, when it exists, of the quotient of differences $(f(x + h) - f(x))/h$ when h goes to zero, a statement much like those that had already been made by Newton, Leibniz, d'Alembert, Maclaurin, and Euler. But what is significant is that Cauchy translated such verbal statements into the precise language of inequalities when he needed them in his proofs. For instance, for the derivative [2]:

(1)   Let δ, ε be two very small numbers; the first is chosen so that for all numerical [i.e., absolute] values of h less than δ, and for any value of x included [in the interval of definition], the ratio $(f(x + h) - f(x))/h$ will always be greater than $f'(x) - \varepsilon$ and less than $f'(x) + \varepsilon$.

Judith V. Grabiner has taught the history of science since 1972 at California State University, Dominguez Hills, where she is Professor of History. After getting a B. S. in Mathematics from the University of Chicago, she received her M. A. and Ph. D. (1966) at Harvard University. She is Book Review Editor of *Historia Mathematica*, Chairman of the Southern California Section of the Mathematical Association of America, and the author of *The Origins of Cauchy's Rigorous Calculus* (M. I. T. Press, 1981). In 1982–1983 she was Visiting Professor of History at U. C. L. A.

This paper is a revised version of a talk given at various mathematics colloquia, at the Summer Meeting of the Mathematical Association of America in Ann Arbor, Michigan, in 1980, and at the New York Academy of Sciences in March, 1982. Some of the research was supported by the National Science Foundation under Grant No. SOC 7907844.

This one example will be enough to indicate how Cauchy did the calculus, because the question to be answered in the present paper is not, "how is a rigorous delta-epsilon proof constructed?" As Cauchy's intellectual heirs we all know this. The central question is, how and why was Cauchy able to put the calculus on a rigorous basis, when his predecessors were not? The answers to this historical question cannot be found by reflecting on the logical relations between the concepts, but by looking in detail at the past and seeing how the existing state of affairs in fact developed from that past. Thus we will examine the mathematical situation in the seventeenth and eighteenth centuries—the background against which we can appreciate Cauchy's innovation. We will describe the powerful techniques of the calculus of this earlier period and the relatively unimpressive views put forth to justify them. We will then discuss how a sense of urgency about rigorizing analysis gradually developed in the eighteenth century. Most important, we will explain the development of the mathematical techniques necessary for the new rigor from the work of men like Euler, d'Alembert, Poisson, and especially Lagrange. Finally, we will show how these mathematical results, though often developed for purposes far removed from establishing foundations for the calculus, were used by Cauchy in constructing his new rigorous analysis.

The Practice of Analysis: From Newton to Euler. In the late seventeenth century, Newton and Leibniz, almost simultaneously, independently invented the calculus. This invention involved three things. First, they invented the general concepts of differential quotient and integral (these are Leibniz's terms; Newton called the concepts "fluxion" and "fluent"). Second, they devised a notation for these concepts which made the calculus an algorithm: the methods not only worked, but were easy to use. Their notations had great heuristic power, and we still use Leibniz's dy/dx and $\int y\,dx$, and Newton's $\dot{x}$, today. Third, both men realized that the basic processes of finding tangents and areas, that is, differentiating and integrating, are mutually inverse—what we now call the Fundamental Theorem of Calculus.

Once the calculus had been invented, mathematicians possessed an extremely powerful set of methods for solving problems in geometry, in physics, and in pure analysis. But what was the nature of the basic concepts? For Leibniz, the differential quotient was a ratio of infinitesimal differences, and the integral was a sum of infinitesimals. For Newton, the derivative, or fluxion, was described as a rate of change; the integral, or fluent, was its inverse. In fact, throughout the eighteenth century, the integral was generally thought of as the inverse of the differential. One might imagine asking Leibniz exactly what an infinitesimal was, or Newton what a rate of change might be. Newton's answer, the best of the eighteenth century, is instructive. Consider a ratio of finite quantities (in modern notation, $(f(x + h) - f(x))/h$ as h goes to zero). The ratio eventually becomes what Newton called an "ultimate ratio." Ultimate ratios are "limits to which the ratios of quantities decreasing without limit do always converge, and to which they approach nearer than by any given difference, but never go beyond, nor ever reach until the quantities vanish" [3]. Except for "reaching" the limit when the quantities vanish, we can translate Newton's words into our algebraic language. Newton himself, however, did not do this, nor did most of his followers in the eighteenth century. Moreover, "never go beyond" does not allow a variable to oscillate about its limit. Thus, though Newton's is an intuitively pleasing picture, as it stands it was not and could not be used for proofs about limits. The definition sounds good, but it was not understood or applied in algebraic terms.

But most eighteenth-century mathematicians would object, "Why worry about foundations?" In the eighteenth century, the calculus, intuitively understood and algorithmically executed, was applied to a wide range of problems. For instance, the partial differential equation for vibrating strings was solved; the equations of motion for the solar system were solved; the Laplace transform and the calculus of variations and the gamma function were invented and applied; all of mechanics was worked out in the language of the calculus. These were great achievements on the part of eighteenth-century mathematicians. Who would be greatly concerned about foundations when such important problems could be successfully treated by the calculus? Results were what counted.

This point will be better appreciated by looking at an example which illustrates both the "uncritical" approach to concepts of the eighteenth century and the immense power of eighteenth-century techniques, from the work of the great master of such techniques: Leonhard Euler. The problem is to find the sum of the series

$$1/1 + 1/4 + 1/9 + \cdots + 1/k^2 + \cdots.$$

It clearly has a finite sum since it is bounded above by the series

$$1 + 1/1 \cdot 2 + 1/2 \cdot 3 + 1/3 \cdot 4 + \cdots + 1/(k-1) \cdot k + \cdots,$$

whose sum was known to be 2; Johann Bernoulli had found this sum by treating $1/1 \cdot 2 + 1/2 \cdot 3 + 1/3 \cdot 4 + \cdots$ as the difference between the series $1/1 + 1/2 + 1/3 + \cdots$ and the series $1/2 + 1/3 + 1/4 + \cdots$, and observing that this difference telescopes [4].

Euler's summation of $\sum_{k=1}^{\infty} 1/k^2$ makes use of a lemma from the theory of equations: given a polynomial equation whose constant term is one, the coefficient of the linear term is the product of the reciprocals of the roots with the signs changed. This result was both discovered and demonstrated by considering the equation $(x-a)(x-b) = 0$, having roots a and b. Multiplying and then dividing out ab, we obtain

$$(1/ab)x^2 - (1/a + 1/b)x + 1 = 0;$$

the result is now obvious, as is the extension to equations of higher degree.

Euler's solution then considers the equation $\sin x = 0$.

Expanding this as an infinite series, Euler obtained

$$x - x^3/3! + x^5/5! - \cdots = 0.$$

Dividing by x yields

$$1 - x^2/3! + x^4/5! - \cdots = 0.$$

Finally, substituting $x^2 = u$ produces

$$1 - u/3! + u^2/5! - \cdots = 0.$$

But Euler thought that power series could be manipulated just like polynomials. Thus, we now have a polynomial equation in u, whose constant term is one. Applying the lemma to it, the coefficient of the linear term with the sign changed is $1/3! = 1/6$. The roots of the equation in u are the roots of $\sin x = 0$ with the substitution $u = x^2$, namely $\pi^2, 4\pi^2, 9\pi^2, \ldots$. Thus the lemma implies

$$1/6 = 1/\pi^2 + 1/4\pi^2 + 1/9\pi^2 + \cdots.$$

Multiplying by π^2 yields the sum of the original series [5]:

$$1/1 + 1/4 + 1/9 + \cdots + 1/k^2 + \cdots = \pi^2/6.$$

Though it is easy to criticize eighteenth-century arguments like this for their lack of rigor, it is also unfair. Foundations, precise specifications of the conditions under which such manipulations with infinites or infinitesimals were admissible, were not very important to men like Euler, because without such specifications they made important new discoveries, whose results in cases like this could readily be verified. When the foundations of the calculus were discussed in the eighteenth century, they were treated as secondary. Discussions of foundations appeared in the introductions to books, in popularizations, and in philosophical writings, and were not—as they are now and have been since Cauchy's time—the subject of articles in research-oriented journals.

Thus, where we once had one question to answer, we now have two. The first remains, where do Cauchy's rigorous techniques come from? Second, one must now ask, why rigorize the calculus in the first place? If few mathematicians were very interested in foundations in the eighteenth century [6], then when, and why, were attitudes changed?

Of course, to establish rigor, it is necessary—though not sufficient—to think rigor is significant. But more important, to establish rigor, it is necessary (though also not sufficient) to have a set of techniques in existence which are suitable for that purpose. In particular, if the calculus is to be made rigorous by being reduced to the algebra of inequalities, one must have both the algebra of inequalities, and facts about the concepts of the calculus that can be expressed in terms of the algebra of inequalities.

In the early nineteenth century, three conditions held for the first time: Rigor was considered important; there was a well-developed algebra of inequalities; and, certain properties were known about the basic concepts of analysis—limits, convergence, continuity, derivatives, integrals—properties which could be expressed in the language of inequalities if desired. Cauchy, followed by Riemann and Weierstrass, gave the calculus a rigorous basis, using the already-existing algebra of inequalities, and built a logically-connected structure of theorems about the concepts of the calculus. It is our task to explain how these three conditions—the developed algebra of inequalities, the importance of rigor, the appropriate properties of the concepts of the calculus—came to be.

The Algebra of Inequalities. Today, the algebra of inequalities is studied in calculus courses because of its use as a basis for the calculus, but why should it have been studied in the eighteenth century when this application was unknown? In the eighteenth century, inequalities were important in the study of a major class of results: approximations. For example, consider an equation such as $(x + 1)^\mu = a$, for μ not an integer. Usually a cannot be found exactly, but it can be approximated by an infinite series. In general, given some number n of terms of such an approximating series, eighteenth-century mathematicians sought to compute an upper bound on the error in the approximation—that is, the difference between the sum of the series and the nth partial sum. This computation was a problem in the algebra of inequalities. Jean d'Alembert solved it for the important case of the binomial series; given the number of terms of the series n, and assuming implicitly that the series converges to its sum, he could find the bounds on the error—that is, on the remainder of the series after the nth term—by bounding the series above and below with convergent geometric progressions [7]. Similarly, Joseph-Louis Lagrange invented a new approximation method using continued fractions and, by extremely intricate inequality-calculations, gave necessary and sufficient conditions for a given iteration of the approximation to be closer to the result than the previous iteration [8]. Lagrange also derived the Lagrange remainder of the Taylor series [9], using an inequality which bounded the remainder above and below by the maximum and minimum values of the nth derivative and then applying the intermediate-value theorem for continuous functions. Thus through such eighteenth-century work [10], there was by the end of the eighteenth century a developed algebra of inequalities, and people used to working with it. Given an n, these people are used to finding an error—that is, an epsilon.

Changing Attitudes toward Rigor. Mathematicians were much more interested in finding rigorous foundations for the calculus in 1800 than they had been a hundred years before. There are many reasons for this: no one enough by itself, but apparently sufficient when acting together. Of course one might think that eighteenth-century mathematicians were always making errors because of the lack of an explicitly-formulated rigorous foundation. But this did not occur. They were usually right, and for two reasons. One is that if one deals with real variables, functions of one variable, series which are power series, and functions arising from physical problems, errors will not occur too often. A second reason is that mathematicians like Euler and Laplace had a deep insight into the basic properties of the concepts of the calculus, and were able to choose fruitful methods and evade pitfalls. The only "error" they committed was to use methods that shocked mathematicians of later ages who had grown up with the rigor of the nineteenth century.

What then were the reasons for the deepened interest in rigor? One set of reasons was philosophical. In 1734, the British philosopher Bishop Berkeley had attacked the calculus on the ground that it was not rigorous. In *The Analyst, or a Discourse Addressed to an Infidel Mathematician*, he said that mathematicians had no business attacking the unreasonableness of religion, given the way they themselves reasoned. He ridiculed fluxions—"velocities of evanescent increments"—calling the evanescent increments "ghosts of departed quantities" [11]. Even more to the point, he correctly criticized a number of specific arguments from the writings of his mathematical contemporaries. For instance, he attacked the process of finding the fluxion (our derivative) by reviewing the steps of the process: if we consider $y = x^2$, taking the ratio of the differences $((x + h)^2 - x^2)/h$, then simplifying to $2x + h$, then letting h vanish, we obtain $2x$. But is h zero? If it is, we cannot meaningfully divide by it; if it is not zero, we have no right to throw it away. As Berkeley put it, the quantity we have called h "might have signified either an increment or nothing. But then, which of these soever you make it signify, you must argue consistently with such its signification" [12].

Since an adequate response to Berkeley's objections would have involved recognizing that an equation involving limits is a shorthand expression for a sequence of inequalities—a subtle and difficult idea—no eighteenth-century analyst gave a fully adequate answer to Berkeley. However, many tried. Maclaurin, d'Alembert, Lagrange, Lazare Carnot, and possibly Euler, all knew about Berkeley's work, and all wrote something about foundations. So Berkeley did call attention to the question. However, except for Maclaurin, no leading mathematician spent much time on the question because of Berkeley's work, and even Maclaurin's influence lay in other fields.

Another factor contributing to the new interest in rigor was that there was a limit to the number of results that could be reached by eighteenth-century methods. Near the end of the century, some leading mathematicians had begun to feel that this limit was at hand. D'Alembert and Lagrange indicate this in their correspondence, with Lagrange calling higher mathematics "decadent" [13]. The philosopher Diderot went so far as to claim that the mathematicians of the eighteenth century had "erected the pillars of Hercules" beyond which it was impossible to go [14]. Thus, there was a perceived need to consolidate the gains of the past century.

Another "factor" was Lagrange, who became increasingly interested in foundations, and, through his activities, interested other mathematicians. In the eighteenth century, scientific academies offered prizes for solving major outstanding problems. In 1784, Lagrange and his colleagues posed the problem of foundations of the calculus as the Berlin Academy's prize problem. Nobody solved it to Lagrange's satisfaction, but two of the entries in the competition were later expanded into full-length books, the first on the Continent, on foundations: Simon L'Huilier's *Exposition élémentaire des principes des calculs supérieurs*, Berlin, 1787, and Lazare Carnot's *Réflexions sur la métaphysique du calcul infinitésimal*, Paris, 1797. Thus Lagrange clearly helped revive interest in the problem.

Lagrange's interest stemmed in part from his respect for the power and generality of algebra; he wanted to gain for the calculus the certainty he believed algebra to possess. But there was another factor increasing interest in foundations, not only for Lagrange, but for many other mathematicians by the end of the eighteenth century: the need to teach. Teaching forces one's attention to basic questions. Yet before the mideighteenth century, mathematicians had often made their living by being attached to royal courts. But royal courts declined; the number of mathematicians increased; and mathematics began to look useful. First in military schools and later on at the Ecole Polytechnique in Paris, another line of work became available: teaching mathematics to students of science and engineering. The Ecole Polytechnique was founded by the French revolutionary government to train scientists, who, the government believed, might prove useful to a modern state. And it was as a lecturer in analysis at the Ecole Polytechnique that

Lagrange wrote his two major works on the calculus which treated foundations; similarly, it was 40 years earlier, teaching the calculus at the Military Academy at Turin, that Lagrange had first set out to work on the problem of foundations. Because teaching forces one to ask basic questions about the nature of the most important concepts, the change in the economic circumstances of mathematicians—the need to teach—provided a catalyst for the crystallization of the foundations of the calculus out of the historical and mathematical background. In fact, even well into the nineteenth century, much of foundations was born in the teaching situation; Weierstrass's foundations come from his lectures at Berlin; Dedekind first thought of the problem of continuity while teaching at Zurich; Dini and Landau turned to foundations while teaching analysis; and, most important for our present purposes, so did Cauchy. Cauchy's foundations of analysis appear in the books based on his lectures at the Ecole Polytechnique; his book of 1821 was the first example of the great French tradition of *Cours d'analyse*.

The Concepts of the Calculus. Arising from algebra, the algebra of inequalities was now there for the calculus to be reduced to; the desire to make the calculus rigorous had arisen through consolidation, through philosophy, through teaching, through Lagrange. Now let us turn to the mathematical substance of eighteenth-century analysis, to see what was known about the concepts of the calculus before Cauchy, and what he had to work out for himself, in order to define, and prove theorems about, limit, convergence, continuity, derivatives, and integrals.

First, consider the concept of limit. As we have already pointed out, since Newton the limit had been thought of as a *bound* which could be approached closer and closer, though not surpassed. By 1800, with the work of L'Huilier and Lacroix on alternating series, the restriction that the limit be one-sided had been abandoned. Cauchy systematically translated this refined limit-concept into the algebra of inequalities, and used it in proofs once it had been so translated; thus he gave reality to the oft-repeated eighteenth-century statement that the calculus could be based on limits.

For example, consider the concept of convergence. Maclaurin had said already that the sum of a series was the limit of the partial sums. For Cauchy, this meant something precise. It meant that, given an ε, one could find n such that, for more than n terms, the sum of the infinite series is within ε of the nth partial sum. That is the reverse of the error-estimating procedure that d'Alembert had used. From his definition of a series having a sum, Cauchy could prove that a geometric progression with radius less in absolute value than 1 converged to its usual sum. As we have said, d'Alembert had shown that the binomial series for, say, $(1 + x)^{p/q}$ could be bounded above and below by convergent geometric progressions. Cauchy assumed that if a series of positive terms is bounded above, term-by-term, by a convergent geometric progression, then it converges; he then used such comparisons to prove a number of tests for convergence: the root test, the ratio test, the logarithm test. The treatment is quite elegant [15]. Taking a technique used a few times by men like d'Alembert and Lagrange on an ad hoc basis in approximations, and using the definition of the sum of a series based on the limit-concept, Cauchy created the first rigorous theory of convergence.

Let us now turn to the concept of continuity. Cauchy gave essentially the modern definition of continuous function, saying that the function $f(x)$ is continuous on a given interval if for each x in that interval "the numerical [i.e., absolute] value of the difference $f(x + \alpha) - f(x)$ decreases indefinitely with α" [16]. He used this definition in proving the intermediate-value theorem for continuous functions [17]. The proof proceeds by examining a function $f(x)$ on an interval, say $[b, c]$, where $f(b)$ is negative, $f(c)$ positive, and dividing the interval $[b, c]$ into m parts of width $h = (c - b)/m$. Cauchy considered the sign of the function at the points $f(b), f(b + h),\ldots,$ $f(b + (m - 1)h), f(c)$; unless one of the values of f is zero, there are two values of x differing by h such that f is negative at one, positive at the other. Repeating this process for new intervals of width $(c - b)/m, (c - b)/m^2,\ldots,$ gives an increasing sequence of values of x: $b, b_1, b_2,\ldots$ for which f is negative, and a decreasing sequence of values of x: $c, c_1, c_2,\ldots$ for which f is positive, and such that the difference between b_k and c_k goes to zero. Cauchy asserted that these two sequences must have a common limit a. He then argued that since $f(x)$ is continuous, the sequence of negative values $f(b_k)$ and of positive values $f(c_k)$ both converge toward the common limit $f(a)$, which must therefore be zero.

Cauchy's proof involves an already existing technique, which Lagrange had applied in approximating real roots of polynomial equations. If a polynomial was negative for one value of the variable, positive for another, there was a root in between, and the difference between those two values of the variable bounded the error made in taking either as an approximation to the root [18]. Thus again we have the algebra of inequalities providing a technique which Cauchy transformed from a tool of approximation to a tool of rigor.

It is worth remarking at this point that Cauchy, in his treatment both of convergence and of continuity, implicitly assumed various forms of the completeness property for the real numbers. For instance, he treated as obvious that a series of positive terms, bounded above by a convergent geometric progression, converges: also, his proof of the intermediate-value theorem assumes that a bounded monotone sequence has a limit. While Cauchy was the first systematically to exploit inequality proof techniques to prove theorems in analysis, he did not identify all the implicit assumptions about the real numbers that such inequality techniques involve. Similarly, as the reader may have already noticed, Cauchy's definition of continuous function does not distinguish between what we now call point-wise and uniform continuity; also, in treating series of functions, Cauchy did not distinguish between pointwise and uniform convergence. The verbal formulations like "for all" that are involved in choosing deltas did not distinguish between "for any epsilon and for all x" and "for any x, given any epsilon" [19]. Nor was it at all clear in the 1820's how much depended on this distinction, since proofs about continuity and convergence were in themselves so novel. We shall see the same confusion between uniform and point-wise convergence as we turn now to Cauchy's theory of the derivative.

Again we begin with an approximation. Lagrange gave the following inequality about the derivative:

$$(2) \qquad\qquad f(x + h) = f(x) + hf'(x) + hV,$$

where V goes to 0 with h. He interpreted this to mean that, given any D, one can find h sufficiently small so that V is between $-D$ and $+D$ [20]. Clearly this is equivalent to (1) above, Cauchy's delta-epsilon characterization of the derivative. But how did Lagrange obtain this result? The answer is surprising; for Lagrange, formula (2) was a consequence of Taylor's theorem. Lagrange believed that any function (that is, any analytic expression, whether finite or infinite, involving the variable) had a unique power-series expansion (except possibly at a finite number of isolated points). This is because he believed that there was an "algebra of infinite series," an algebra exemplified by work of Euler such as the example we gave above. And Lagrange said that the way to make the calculus rigorous was to reduce it to algebra. Although there is no "algebra" of infinite series that gives power-series expansions without any consideration of convergence and limits, this assumption led Lagrange to define $f'(x)$ without reference to limits, as the coefficient of the linear term in h in the Taylor series expansion for $f(x + h)$. Following Euler, Lagrange then said that, for any power series in h, one could take h sufficiently small so that any given term of the series exceeded the sum of all the rest of the terms following it; this approximation, said Lagrange, is assumed in applications of the calculus to geometry and mechanics [21]. Applying this approximation to the linear term in the Taylor series produces (2), which I call the Lagrange property of the derivative. (Like Cauchy's (1), the inequality-translation Lagrange gives for (2) assumes that, given any D, one finds h sufficiently small so $|V| \leq D$ with no mention whatever of x.)

Not only did Lagrange state property (2) and the associated inequalities, he used them as a basis for a number of proofs about derivatives: for instance, to prove that a function with positive derivative on an interval is increasing there, to prove the mean-value theorem for derivatives, and to obtain the Lagrange remainder for the Taylor series. (Details may be found in the works cited in [22].) Lagrange also applied his results to characterize the properties of maxima and minima, and orders of contact between curves.

With a few modifications, Lagrange's proofs are valid—provided that property (2) can be justified. Cauchy borrowed and simplified what are in effect Lagrange's inequality proofs about derivatives, with a few improvements, basing them on his own (1). But Cauchy made these proofs legitimate because Cauchy defined the derivative precisely to satisfy the relevant inequalities. Once again, the key properties come from an approximation. For Lagrange, the derivative was *exactly*—no epsilons needed—the coefficient of the linear term in the Taylor series; formula (2), and the corresponding inequality that $f(x + h) - f(x)$ lies between $h(f'(x) \pm D)$, were approximations. Cauchy brought Lagrange's inequality properties and proofs together with a definition of derivative devised to make those techniques rigorously founded [22].

The last of the concepts we shall consider, the integral, followed an analogous development. In the eighteenth century, the integral was usually thought of as the inverse of the differential. But sometimes the inverse could not be computed exactly, so men like Euler remarked that the integral could be approximated as closely as one liked by a sum. Of course, the geometric picture of an area being approximated by rectangles, or the Leibnizian definition of the integral as a sum, suggests this immediately. But what is important for our purposes is that much work was done on approximating the values of definite integrals in the eighteenth century, including considerations of how small the subintervals used in the sums should be when the function oscillates to a greater or lesser extent. For instance, Euler treated sums of the form $\sum_{k=0}^{n} f(x_k)(x_{k+1} - x_k)$ as approximations to the integral $\int_{x_0}^{x_n} f(x) \, dx$ [23].

In 1820, S.-D. Poisson, who was interested in complex integration and therefore more concerned than most people about the existence and behavior of integrals, asked the following question. If the integral F is defined as the antiderivative of f, and if $b - a = nh$, can it be proved that $F(b) - F(a) = \int_a^b f(x) \, dx$ is the limit of the sum

$$S = hf(a) + hf(a + h) + \cdots + hf(a + (n - 1)h)$$

as h gets small? (S is an approximating sum of the eighteenth-century sort.) Poisson called this result "the fundamental proposition of the theory of definite integrals." He proved it by using another inequality-result: the Taylor series with remainder. First, he wrote $F(b) - F(a)$ as the telescoping sum

$$(3) \quad F(a + h) - F(a) + F(a + 2h) - F(a + h) + \cdots + F(b) - F(a + (n - 1)h).$$

Then, for each of the terms of the form $F(a + kh) - F(a + (k - 1)h)$, Taylor's series with remainder gives, since by definition $F' = f$,

$$F(a + kh) - F(a + (k - 1)h) = hf(a + (k - 1)h) + R_k h^{1+w}$$

where $w > 0$, for some R_k. Thus the telescoping sum (3) becomes

$$hf(a) + hf(a + h) + \cdots + hf(a + (n - 1)h) + (R_1 + \cdots + R_n)h^{1+w}.$$

So $F(b) - F(a)$ and the sum S differ by $(R_1 + \cdots + R_n)h^{1+w}$. Letting R be the maximum value for the R_k,

$$(R_1 + \cdots + R_n)h^{1+w} \le n \cdot R(h^{1+w}) = R \cdot nh \cdot h^w = R(b - a)h^w.$$

Therefore, if h is taken sufficiently small, $F(b) - F(a)$ differs from S by less than any given quantity [24].

Poisson's was the first attempt to prove the equivalence of the antiderivative and limit-of-sums conceptions of the integral. However, besides the implicit assumptions of the existence of antiderivatives and bounded first derivatives for f on the given interval, the proof assumes that the subintervals on which the sum is taken are all equal. Should the result not hold for unequal divisions also? Poisson thought so, and justified it by saying, "If the integral is represented by the area of a curve, this area will be the same, if we divide the difference... into an infinite number of equal parts, or an infinite number of unequal parts following any law" [25]. This, however, is an assertion, not a proof. And Cauchy saw that a proof was needed.

Cauchy did not like formalistic arguments in supposedly rigorous subjects, saying that most algebraic formulas hold "only under certain conditions, and for certain values of the quantities they contain" [26]. In particular, one could not assume that what worked for finite expressions automatically worked for infinite ones. Thus, Cauchy showed that the sum of the series $1/1 + 1/4 + 1/9 + \cdots$ was $\pi^2/6$ by actually calculating the difference between the nth partial sum and $\pi^2/6$ and showing that it was arbitrarily small [27]. Similarly, just because there was an operation called taking a derivative did not mean that the inverse of that operation always produced a result. The existence of the definite integral had to be proved. And how does one prove existence in the 1820's? One constructs the mathematical object in question by using an eighteenth-century approximation that converges to it. Cauchy defined the integral as the limit of Euler-style sums $\Sigma f(x_k)(x_{k+1} - x_k)$ for $x_{k+1} - x_k$ sufficiently small. Assuming explicitly that $f(x)$ was continuous on the given interval (and implicitly that it was uniformly continuous), Cauchy was able to show that all sums of that form approach a fixed value, called by definition the integral of the function on that interval. This is an extremely hard proof [28]. Finally, borrowing from Lagrange the mean-value theorem for integrals, Cauchy proved the Fundamental Theorem of Calculus [29].

Conclusion. Here are all the pieces of the puzzle we originally set out to solve. Algebraic approximations produced the algebra of inequalities; eighteenth-century approximations in the calculus produced the useful properties of the concepts of analysis: d'Alembert's error-bounds for series, Lagrange's inequalities about derivatives, Euler's approximations to integrals. There was a new interest in foundations. All that was needed was a sufficiently great genius to build the new foundation.

Two men came close. In 1816, Carl Friedrich Gauss gave a rigorous treatment of the convergence of the hypergeometric series, using the technique of comparing a series with convergent geometric progressions; however, Gauss did not give a general foundation for all of analysis. Bernhard Bolzano, whose work was little known until the 1860's, echoing Lagrange's call to reduce the calculus to algebra, gave in 1817 a definition of continuous function like Cauchy's and then proved—by a different technique from Cauchy's—the intermediate-value theorem [30]. But it was Cauchy who gave rigorous definitions and proofs for all the basic concepts; it was he who realized the far-reaching power of the inequality-based limit concept; and it was he who gave us—except for a few implicit assumptions about uniformity and about completeness—the modern rigorous approach to calculus.

Mathematicians are used to taking the rigorous foundations of the calculus as a completed whole. What I have tried to do as a historian is to reveal what went into making up that great achievement. This needs to be done, because completed wholes by their nature do not reveal the separate strands that go into weaving them—especially when the strands have been considerably transformed. In Cauchy's work, though, one trace indeed was left of the origin of rigorous calculus in approximations—the letter epsilon. The ε corresponds to the initial letter in the word "erreur" (or "error"), and Cauchy in fact used ε for "error" in some of his work on probability [31]. It is both amusing and historically appropriate that the "ε," once used to designate the "error" in approximations, has become transformed into the characteristic symbol of precision and rigor in the calculus. As Cauchy transformed the algebra of inequalities from a tool of approximation to a tool of rigor, so he transformed the calculus from a powerful method of generating results to the rigorous subject we know today.

References

1. A. -L. Cauchy, Cours d'analyse, Paris, 1821; in Oeuvres complètes d'Augustin Cauchy, series 2, vol. 3, Paris, Gauthier-Villars, 1899, p. 19.

2. A. -L. Cauchy, Résumé des leçons données à l'école royale polytechnique sur le calcul infinitésimal, Paris, 1823; in Oeuvres, series 2, vol. 4, p. 44. Cauchy used i for the increment; otherwise the notation is his.

3. Isaac Newton, Mathematical Principles of Natural Philosophy, 3rd ed., 1726, tr. A. Motte, revised by Florian Cajori, University of California Press, Berkeley, 1934, Scholium to Lemma XI, p. 39.

4. Johann Bernoulli, Opera Omnia, IV, 8; section entitled "De seriebus varia, Corollarium III," cited by D. J. Struik, A Source Book in Mathematics, 1200–1800, Harvard, Cambridge, 1969, p. 321.

5. Boyer, History of Mathematics, p. 487; Euler's paper is in Comm. Acad. Sci. Petrop., 7, 1734–5, pp. 123–34; in Leonhard Euler, Opera omnia, series 1, vol. 14, pp. 73–86.

6. J. V. Grabiner, The Origins of Cauchy's Rigorous Calculus, M. I. T. Press, Cambridge and London, 1981, chapter 4.

7. J. d'Alembert, Réflexions sur les suites et sur les racines imaginaires, in Opuscules mathématiques, vol. 5, Briasson, Paris, 1768, pp. 171–215; see especially pp. 175–178.

8. J. -L. Lagrange, Traité de la résolution des équations numériques de tous les degrés, 2nd ed., Courcier, Paris, 1808; in Oeuvres de Lagrange, Gauthier-Villars, Paris, 1867–1892, vol. 8, pp. 162–163.

9. Lagrange, Théorie des fonctions analytiques, 2nd ed., Paris, 1813, in Oeuvres, vol. 9, pp. 80–85; compare Lagrange, Leçons sur le calcul des fonctions, Paris, 1806, in Oeuvres, vol. 10, pp. 91–95.

10. Grabiner, Origins of Cauchy's Rigorous Calculus, pp. 56–68; compare H. Goldstine, A History of Numerical Analysis from the 16th through the 19th Century, Springer-Verlag, New York, Heidelberg, Berlin, 1977, chapters 2–4.

11. George Berkeley, The Analyst, section 35.

12. Analyst, section 15. Berkeley used the function x^n where we have used x^2, and a Newtonian notation, lower-case o, for the increment.

13. Letter from Lagrange to d'Alembert, 24 February 1772, in Oeuvres de Lagrange, vol. 13, p. 229.

14. D. Diderot, De l'interprétation de la nature, in Oeuvres philosophiques, ed., P. Vernière, Garnier, Paris, 1961, pp. 180–181.

15. Cauchy, Cours d'analyse, Oeuvres, series 2, vol. 3; for real-valued series, see especially pp. 114–138.

16. Cauchy, op. cit., p. 43. So did Bolzano; see below, and note 30.

17. Cauchy, op. cit., pp. 378–380. For an English translation of this proof, see Grabiner, Origins, pp. 167–168. For clarity, I have substituted $b, b_1, b_2, \ldots$ and $c, c_1, c_2, \ldots$ for Cauchy's $x_0, x_1, x_2, \ldots$ and $X, X', X'', \ldots$ in the present version.

18. Lagrange, Equations numériques, sections 2 and 6, in Oeuvres, vol. 8; also in Lagrange, Leçons élémentaires sur les mathématiques données à l'école normale en 1795, Séances des Ecoles Normales, Paris, 1794–1795; in Oeuvres, vol. 7, pp. 181–288; this method is on pp. 260–261.

19. I. Grattan-Guinness, Development of the Foundations of Mathematical Analysis from Euler to Riemann, M. I. T. Press, Cambridge and London, 1970, p. 123, puts it well: "Uniform convergence was tucked away in the word "always," with no reference to the variable at all."

20. Lagrange, Leçons sur le calcul des fonctions, Oeuvres 10, p. 87; compare Lagrange, Théorie des fonctions analytiques, Oeuvres 9, p. 77. I have substituted h for the i Lagrange used for the increment.

21. Lagrange, Théorie des fonctions analytiques, Oeuvres 9, p. 29. Compare Leçons sur le calcul des fonctions, Oeuvres 10, p. 101. For Euler, see his Institutiones calculi differentialis, St. Petersburg, 1755; in Opera, series 1, vol. 10, section 122.

22. Grabiner, Origins of Cauchy's Rigorous Calculus, chapter 5; also J. V. Grabiner, The origins of Cauchy's theory of the derivative, Hist. Math., 5, 1978, pp. 379–409.

23. The notation is modernized. For Euler, see Institutiones calculi integralis, St. Petersburg. 1768–1770, 3 vols; in Opera, series 1, vol. 11, p. 184. Eighteenth-century summations approximating integrals are treated in A. P. Iushkevich, O vozniknoveniya poiyatiya ob opredelennom integrale Koshi, Trudy Instituta Istorii Estestvoznaniya, Akademia Nauk SSSR, vol. 1, 1947, pp. 373–411.

24. S. D. Poisson, Suite du mémoire sur les intégrales définies, Journ. de l'Ecole polytechnique, Cah. 18, 11, 1820, pp. 295–341, 319–323. I have substituted h, w for Poisson's α, k, and have used R_1 for his R_0.

25. Poisson, op. cit., pp. 329–330.

26. Cauchy, Cours d'analyse, Introduction, Oeuvres, Series 2, vol. 3, p. iii.

27. Cauchy, Cours d'analyse, Note VIII, Oeuvres, series 2, vol. 3, pp. 456–457.

28. Cauchy, Calcul infinitésimal, Oeuvres, series 2, vol. 4, 122–25; in Grabiner, Origins of Cauchy's Rigorous Calculus, pp. 171–175 in English translation.

29. Cauchy, op. cit., pp. 151–152.

30. B. Bolzano, Rein analytischer Beweis des Lehrsatzes dass zwischen je zwey Werthen, die ein entgegenge-setztes Resultat gewaehren, wenigstens eine reele Wurzel der Gleichung liege, Prague, 1817. English version, S. B. Russ, A translation of Bolzano's paper on the intermediate value theorem, Hist. Math., 7, 1980, pp. 156–185. The contention by Grattan-Guinness, Foundations, p. 54, that Cauchy took his program of rigorizing analysis, definition of continuity, Cauchy criterion, and proof of the intermediate-value theorem, from Bolzano's paper without acknowledgement is not, in my opinion, valid; the similarities are better explained by common prior influences, especially that of Lagrange. For a documented argument to this effect, see J. V. Grabiner, Cauchy and Bolzano: Tradition and transformation in the history of mathematics, to appear in E. Mendelsohn, Transformation and Tradition in the Sciences, Cambridge University Press, Cambridge, forthcoming; see also Grabiner, Origins of Cauchy's Rigorous Calculus, pp. 69–75, 102–105, 94–96, 52–53.

31. Cauchy, Sur la plus grande erreur à craindre dans un resultat moyen, et sur le système de facteurs qui rend cette plus grande erreur un minimum, Comptes rendus 37, 1853; in Oeuvres, series 1, vol. 12, pp. 114–124.

An Application of Geography to Mathematics: History of the Integral of the Secant

V. FREDERICK RICKEY
Bowling Green State University
Bowling Green, OH 43403

PHILIP M. TUCHINSKY
Ford Motor Company
Engineering Computing Center
Dearborn, MI 48121

Every student of the integral calculus has done battle with the formula

$$\int \sec \theta \, d\theta = \ln|\sec \theta + \tan \theta| + c. \tag{1}$$

This formula can be checked by differentiation or "derived" by using the substitution $u = \sec \theta + \tan \theta$, but these ad hoc methods do not make the formula any more understandable. Experience has taught us that this troublesome integral can be motivated by presenting its history. Perhaps our title seems twisted, but the tale to follow will show that this integral should be presented not as an application of mathematics to geography, but rather as an application of geography to mathematics.

The secant integral arose from cartography and navigation, and its evaluation was a central question of mid-seventeenth century mathematics. The first formula, discovered in 1645 before the work of Newton and Leibniz, was

$$\int \sec \theta \, d\theta = \ln\left|\tan\left(\frac{\theta}{2} + \frac{\pi}{4}\right)\right| + c, \tag{2}$$

which is a trigonometric variant of (1). This was discovered, not through any mathematician's cleverness, but by a serendipitous historical accident when mathematicians and cartographers sought to understand the Mercator map projection. To see how this happened, we must first discuss sailing and early maps so that we can explain why Mercator invented his famous map projection.

From the time of Ptolemy (c. 150 A.D.) maps were drawn on rectangular grids with one degree of latitude equal in length to one degree of longitude. When restricted to a small area, like the Mediterranean, they were accurate enough for sailors. But in the age of exploration, the Atlantic presented vast distances and higher latitudes, and so the navigational errors due to using the "plain charts" became apparent.

The magnetic compass was in widespread use after the thirteenth century, so directions were conveniently given by distance and compass bearing. Lines of fixed compass direction were called **rhumb** lines by sailors, and in 1624 Willebrord Snell dubbed them **loxodromes**. To plan a journey one laid a straightedge on a map between origin and destination, then read off the compass bearing to follow. But rhumb lines are spirals on the globe and curves on a plain chart —facts sailors had difficulty understanding. They needed a chart where the loxodromes were represented as straight lines.

It was Gerardus Mercator (1512–1594) who solved this problem by designing a map where the lines of latitude were more widely spaced when located further from the equator. On his famous world map of 1569 ([1], p. 46), Mercator wrote:

> In making this representation of the world we had...to spread on a plane the surface of the sphere in such a way that the positions of places shall correspond on all sides with each other both in so far as true direction and distance are concerned and as concerns correct longitudes and latitudes... . With this intention we have had to employ a new proportion and a new arrangement of the meridians with reference to the parallels. ... It is for these reasons that we have progressively increased the degrees of latitude towards each pole in proportion to the lengthening of the parallels with reference to the equator.

Mercator wished to map the sphere onto the plane so that both angles and distances are preserved, but he realized this was impossible. He opted for a conformal map (one which preserves angles) because, as we shall see, it guaranteed that loxodromes would appear on the map as straight lines.

Unfortunately, Mercator did not explain how he "progressively increased" the distances between parallels of latitude. Thomas Harriot (c. 1560–1621) gave a mathematical explanation in the late 1580's, but neither published his results nor influenced later work (see [6], [11]-[15]). In his *Certaine Errors in Navigation...* [22] of 1599, Edward Wright (1561–1615) finally gave a mathematical method for constructing an accurate Mercator map. The Mercator map has its meridians of longitude placed vertically and spaced equally. The parallels of latitude are horizontal and unequally spaced. Wright's great achievement was to show that the parallel at latitude θ should be stretched by a factor of $\sec\theta$ when drawn on the map. Let us see why.

FIGURE 1 represents a wedge of the earth, where AB is on the equator, C is the center of the earth, and T is the north pole. The parallel at latitude θ is a circle, with center P, that includes arc MN between the meridians AT and BT. Thus BC and NP are parallel and so angle $PNC = \theta$. The "triangles" ABC and MNP are similar figures, so

$$\frac{AB}{MN} = \frac{BC}{NP} = \frac{NC}{NP} = \sec\theta,$$

or $AB = MN\sec\theta$. Thus when MN is placed on the map it must be stretched horizontally by a factor $\sec\theta$. (This argument is not the one used by Wright [22]. His argument is two dimensional and shows that $BC = NP\sec\theta$.)

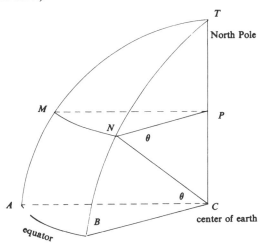

FIGURE 1.

Suppose we can construct a map where angles are preserved, i.e., where the globe-to-map function is conformal. Then a loxodrome, which makes the same angle with each meridian, will appear on this map as a curve which cuts all the map's meridians (a family of parallel straight lines) at the same angle. Since a curve that cuts a family of parallel straight lines at a fixed angle is a straight line, loxodromes on the globe will appear straight on the map. Conversely, if loxodromes are mapped to straight lines, the globe-to-map function must be conformal.

In order for angles to be preserved, the map must be stretched not only horizontally, but also vertically, by $\sec\theta$; this, however, requires an argument by infinitesimals. Let $D(\theta)$ be the distance *on the map* from the equator to the parallel of latitude θ, and let dD be the infinitesimal change in D resulting from an infinitesimal change $d\theta$ in θ. If we stretch vertically by $\sec\theta$, i.e., if

$$dD = \sec\theta\, d\theta$$

then an infinitesimal region on the globe becomes a similar region on the map, and so angles are preserved. Conversely, if the map is to be conformal the vertical multiplier must be $\sec\theta$.

Finally, "by perpetuall addition of the Secantes," to quote Wright, we see that the distance on the map from the equator to the parallel at latitude θ is

$$D(\theta) = \int_0^\theta \sec\theta\, d\theta.$$

Of course Wright did not express himself as we have here. He said ([2], pp. 312–313):

> the parts of the meridian at euery poynt of latitude must needs increase with the same proportion wherewith the Secantes or hypotenusae of the arke, intercepted betweene those pointes of latitude and the aequinoctiall [equator] do increase. ... For...by perpetuall addition of the Secantes answerable to the latitudes of each point or parallel vnto the summe compounded of all former secantes,...we may make a table which shall shew the sections and points of latitude in the meridians of the nautical planisphaere: by which sections, the parallels are to be drawne.

Wright published a table of "meridional parts" which was obtained by taking $d\theta = 1'$ and then computing the Riemann sums for latitudes below 75°. Thus the methods of constructing Mercator's "true chart" became available to cartographers.

Wright also offered an interesting physical model. Consider a cylinder tangent to the earth's equator and imagine the earth to "swal [swell] like a bladder." Then identify points on the earth with the points on the cylinder that they come into contact with. Finally unroll the cylinder; it will be a Mercator map. This model has often been misinterpreted as the cylindrical projection (where a light source at the earth's center projects the unswollen sphere onto its tangent cylinder), but this projection is not conformal.

We have established half of our result, namely that the distance on the map from the equator to the parallel at latitude θ is given by the integral of the secant. It remains to show that it is also given by $\ln|\tan(\frac{\theta}{2} + \frac{\pi}{4})|$.

In 1614 John Napier (1550–1617) published his work on logarithms. Wright's authorized English translation, *A Description of the Admirable Table of Logarithms*, was published in 1616. This contained a table of logarithms of sines, something much needed by astronomers. In 1620 Edmund Gunter (1581–1626) published a table of common logarithms of tangents in his *Canon triangulorum*. In the next twenty years numerous tables of logarithmic tangents were published and so were widely available. (Not even a table of secants was available in Mercator's day.)

In the 1640's Henry Bond (c. 1600–1678), who advertised himself as a "teacher of navigation, survey and other parts of the mathematics," compared Wright's table of meridional parts with a log-tan table and discovered a close agreement. This serendipitous accident led him to conjecture that $D(\theta) = \ln|\tan(\frac{\theta}{2} + \frac{\pi}{4})|$. He published this conjecture in 1645 in Norwood's *Epitome of Navigation*. Mainly through the correspondence of John Collins this conjecture became widely known. In fact, it became one of the outstanding open problems of the mid-seventeenth century, and was attempted by such eminent mathematicians as Collins, N. Mercator (no relation), Wilson, Oughtred and John Wallis. It is interesting to note that young Newton was aware of it in 1665 [18], [21].

The "Learned and Industrious *Nicolaus Mercator*" in the very first volume of the *Philosophical Transactions* of the Royal Society of London was "willing to lay a *Wager* against any one or more persons that have a mind to engage... *Whether the Artificial* [logarithmic] *Tangent-line be the true Meridian-line*, yea or no?" ([9], pp. 217–218). Nicolaus Mercator is not, as the story is often told, wagering that he knows more about logarithms than his contemporaries; rather, he is offering a prize for the solution of an open problem.

The first to prove the conjecture was, to quote Edmund Halley, "the excellent Mr. *James Gregory* in his *Exercitationes Geometricae*, published *Anno* 1668, which he did, not without a long train of Consequences and Complication of Proportions, whereby the evidence of the Demonstration is in a great measure lost, and the Reader wearied before he attain it" ([7], p. 203). Judging by Turnbull's modern elucidation [19] of Gregory's proof, one would have to agree with Halley. At any rate, Gregory's proof could not be presented to today's calculus students, and so we omit it here.

Isaac Barrow (1630–1677) in his *Geometrical Lectures* (Lect. XII, App. I) gave the first "intelligible" proof of the result, but it was couched in the geometric idiom of the day. It is especially noteworthy in that it is the earliest use of partial fractions in integration. Thus we reproduce it here in modern garb:

$$\int \sec\theta \, d\theta = \int \frac{1}{\cos\theta} \, d\theta$$

$$= \int \frac{\cos\theta}{\cos^2\theta} \, d\theta$$

$$= \int \frac{\cos\theta}{1 - \sin^2\theta} \, d\theta$$

$$= \int \frac{\cos\theta}{(1 - \sin\theta)(1 + \sin\theta)} \, d\theta$$

$$= \frac{1}{2} \int \frac{\cos\theta}{1 - \sin\theta} + \frac{\cos\theta}{1 + \sin\theta} \, d\theta$$

$$= \frac{1}{2} [-\ln|1 - \sin\theta| + \ln|1 + \sin\theta|] + c$$

$$= \frac{1}{2} \ln \left| \frac{1 + \sin\theta}{1 - \sin\theta} \right| + c$$

$$= \frac{1}{2} \ln \left| \frac{1 + \sin\theta}{1 - \sin\theta} \cdot \frac{1 + \sin\theta}{1 + \sin\theta} \right| + c$$

$$= \frac{1}{2} \ln \left| \frac{(1 + \sin\theta)^2}{1 - \sin^2\theta} \right| + c$$

$$= \frac{1}{2} \ln \left| \frac{(1 + \sin\theta)^2}{(\cos\theta)^2} \right| + c$$

$$= \ln \left| \frac{1 + \sin\theta}{\cos\theta} \right| + c$$

$$= \ln|\sec\theta + \tan\theta| + c.$$

We became interested in this topic after noting one line of historical comment in Spivak's excellent *Calculus* (p. 326). As we ferreted out the details and shared them with our students, we found an ideal soapbox for discussing the nature of mathematics, the process of mathematical discovery, and the role that mathematics plays in the world. We found this so useful in the classroom that we have prepared a more detailed version for our students [17].

References

The following works contain interesting information pertaining to this paper. The best concise source of information about the individuals mentioned in this paper is the excellent *Dictionary of Scientific Biography*, edited by C. C. Gillespie.

[1]   Anonymous, Gerard Mercator's Map of the World (1569), Supplement no. 2 to Imago Mundi, 1961.

[2]   Florian Cajori, On an integration ante-dating the integral calculus, Bibliotheca Mathematica, 3rd series, 14(1915) 312–319.

[3]   H. S. Carslaw, The story of Mercator's map. A chapter in the history of mathematics, Math. Gaz., 12(1924) 1–7.

[4]   Georgina Dawson, Edward Wright, mathematician and hydrographer, Amer. Neptune, 37(1977) 174–178.

[5]   Jacques Delevsky, L'invention de la projection de Mercator et les enseignements de son histoire, Isis, 34(1942) 110–117.

[6]   Frank George, Hariot's meridional parts, J. Inst. Navigation, London, 21(1968) 82–83.

[7]   E. Halley, An easie demonstration of the analogy of the logarithmick tangents to the meridian line or sum of secants: with various methods for computing the same to the utmost exactness, Philos. Trans., Roy. Soc. London, 19(1695–97) 202–214.

[8]   Johannes Keuning, The history of geographical map projections until 1600, Imago Mundi, 12(1955) 1–24.

[9]   Nicolaus Mercator, Certain problems touching some points of navigation, Philos. Trans., Roy. Soc. London, 1(1666) 215–218.

[10] E. J. S. Parsons and W. F. Morris, Edward Wright and his work, Imago Mundi, 3(1939) 61–71.

[11] Jon V. Pepper, Hariot's calculation of the meridional parts as logarithmic tangents, Archive for History of Exact Science, 4(1967) 359–413.

[12] _____, A note on Hariot's method of obtaining meridional parts, J. Inst. Navigation, London, 20(1967) 347–349.

[13] _____, The study of Thomas Hariot's manuscripts, II: Hariot's unpublished papers, History of Science, 6 (1967) 17–40.

[14] _____, Hariot's earlier work on mathematical navigation: theory and practice. With an appendix, 'The early development of the Mercator chart,' in Thomas Hariot: Renaissance Scientist, Clarendon Press, Oxford, 1974, John W. Shirley, editor, pp. 54–90.

[15] D. H. Sadler and Eva G. R. Taylor, The doctrine of nauticall triangles compendious. Part I–Thomas Hariot's manuscript (by Taylor). Part II–Calculating the meridional parts (by Sadler), J. Inst. Navigation, London, 6(1953) 131–147.

[16] Eva G. R. Taylor, The Haven-Finding Art, Hollis and Carter, London, 1971.

[17] P. M. Tuchinsky, Mercator's World Map and the Calculus, Modules and Monographs in Undergraduate Mathematics and its Applications (UMAP) Project, Education Development Center, Newton, Mass., 1978.

[18] H. W. Turnbull, editor, The Correspondence of Isaac Newton, Cambridge Univ. Press, 1959–1960, vol. 1, pp. 13–16, and vol. 2, pp. 99–100.

[19] H. W. Turnbull, James Gregory Tercentenary Memorial Volume, G. Bell & Sons, London, 1939, pp. 463–464.

[20] D. W. Waters, The Art of Navigation in England in Elizabethan and Early Stuart Times, Yale Univ. Press, New Haven, 1958.

[21] D. T. Whiteside, editor, The Mathematical Papers of Isaac Newton, vol. 1, Cambridge Univ. Press, 1967, pp. 466–467, 473–475.

[22] Edward Wright, Certaine Errors in Navigation, Arising either of the ordinaire erroneous making or vsing of the sea Chart, Compasse, Crosse staffe, and Tables of declination of the Sunne, and fixed Starres detected and corrected, Valentine Sims, London, 1599. Available on microfilm as part of Early English Books 1475–1640, reels 539 and 1018 (these two copies from 1599 have slightly different title pages). The preface and table of meridional parts have been reproduced as "Origin of meridional parts," International Hydrographic Review, 8(1931) 84–97.

The Changing Concept of Change:
The Derivative from Fermat to Weierstrass

First the derivative was used, then discovered, explored and developed, and only then, defined.

JUDITH V. GRABINER

Department of History
University of California, Los Angeles
Los Angeles, CA 90024

Some years ago while teaching the history of mathematics, I asked my students to read a discussion of maxima and minima by the seventeenth-century mathematician, Pierre Fermat. To start the discussion, I asked them, "Would you please define a relative maximum?" They told me it was a place where the derivative was zero. "If that's so," I asked, "then what is the definition of a relative minimum?" They told me, *that's* a place where the derivative is zero. "Well, in that case," I asked, "what is the difference between a maximum and a minimum?" They replied that in the case of a maximum, the second derivative is negative.

What can we learn from this apparent victory of calculus over common sense?

I used to think that this story showed that these students did not understand the calculus, but I have come to think the opposite: they understood it very well. The students' answers are a tribute to the power of the calculus in general, and the power of the concept of derivative in particular. Once one has been initiated into the calculus, it is hard to remember what it was like *not* to know what a derivative is and how to use it, and to realize that people like Fermat once had to cope with finding maxima and minima without knowing about derivatives at all.

Historically speaking, there were four steps in the development of today's concept of the derivative, which I list here in chronological order. The derivative was first *used*; it was then *discovered*; it was then *explored and developed*; and it was finally *defined*. That is, examples of what we now recognize as derivatives first were used on an ad hoc basis in solving particular problems; then the general concept lying behind these uses was identified (as part of the invention of the calculus); then many properties of the derivative were explained and developed in applications both to mathematics and to physics; and finally, a rigorous definition was given and the concept of derivative was embedded in a rigorous theory. I will describe the steps, and give one detailed mathematical example from each. We will then reflect on what it all means—for the teacher, for the historian, and for the mathematician.

The seventeenth-century background

Our story begins shortly after European mathematicians had become familiar once more with Greek mathematics, learned Islamic algebra, synthesized the two traditions, and struck out on their own. François Vieta invented symbolic algebra in 1591; Descartes and Fermat independently invented analytic geometry in the 1630's. Analytic geometry meant, first, that curves could be represented by equations; conversely, it meant also that every equation determined a curve. The Greeks and Muslims had studied curves, but not that many—principally the circle and the conic sections plus a few more defined as loci. Many problems had been solved for these, including

17

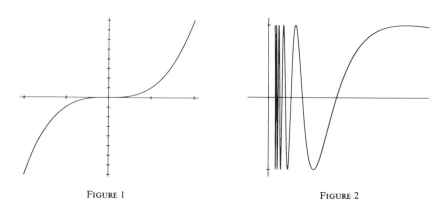

FIGURE 1   FIGURE 2

finding their tangents and areas. But since any equation could now produce a new curve, students of the geometry of curves in the early seventeenth century were suddenly confronted with an explosion of curves to consider. With these new curves, the old Greek methods of synthetic geometry were no longer sufficient. The Greeks, of course, had known how to find the tangents to circles, conic sections, and some more sophisticated curves such as the spiral of Archimedes, using the methods of synthetic geometry. But how could one describe the properties of the tangent at an arbitrary point on a curve defined by a ninety-sixth degree polynomial? The Greeks had defined a tangent as a line which touches a curve without cutting it, and usually expected it to have only one point in common with the curve. How then was the tangent to be defined at the point $(0,0)$ for a curve like $y = x^3$ (FIGURE 1), or to a point on a curve with many turning points (FIGURE 2)?

The same new curves presented new problems to the student of areas and arc lengths. The Greeks had also studied a few cases of what they called "isoperimetric" problems. For example, they asked: of all plane figures with the same perimeter, which one has the greatest area? The circle, of course, but the Greeks had no general method for solving all such problems. Seventeenth-century mathematicians hoped that the new symbolic algebra might somehow help solve all problems of maxima and minima.

Thus, though a major part of the agenda for seventeenth-century mathematicians—tangents, areas, extrema—came from the Greeks, the subject matter had been vastly extended, and the solutions would come from using the new tools: symbolic algebra and analytic geometry.

Finding maxima, minima, and tangents

We turn to the first of our four steps in the history of the derivative: its *use*, and also illustrate some of the general statements we have made. We shall look at Pierre Fermat's method of finding maxima and minima, which dates from the 1630's [**8**]. Fermat illustrated his method first in solving a simple problem, whose solution was well known: *Given a line, to divide it into two parts so that the product of the parts will be a maximum.* Let the length of the line be designated B and the first part A (FIGURE 3). Then the second part is $B - A$ and the product of the two parts is

$$A(B - A) = AB - A^2. \tag{1}$$

Fermat had read in the writings of the Greek mathematician Pappus of Alexandria that a problem which has, in general, two solutions will have only one solution in the case of a maximum. This remark led him to his method of finding maxima and minima. Suppose in the problem just stated there is a second solution. For this solution, let the first part of the line be designated as $A + E$; the second part is then $B - (A + E) = B - A - E$. Multiplying the two parts together, we obtain

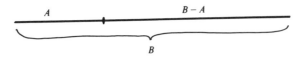

FIGURE 3

for the product

$$BA + BE - A^2 - AE - EA - E^2 = AB - A^2 - 2AE + BE - E^2. \qquad (2)$$

Following Pappus' principle for the maximum, instead of two solutions, there is only one. So we set the two products (1) and (2) "sort of" equal; that is, we formulate what Fermat called the pseudo-equality:

$$AB - A^2 = AB - A^2 - 2AE + BE - E^2.$$

Simplifying, we obtain

$$2AE + E^2 = BE$$

and

$$2A + E = B.$$

Now Fermat said, with no justification and no ceremony, "suppress E." Thus he obtained

$$A = B/2,$$

which indeed gives the maximum sought. He concluded, "We can hardly expect a more general method." And, of course, he was right.

Notice that Fermat did not call E infinitely small, or vanishing, or a limit; he did not explain why he could first divide by E (treating it as nonzero) and then throw it out (treating it as zero). Furthermore, he did not explain what he was doing as a special case of a more general concept, be it derivative, rate of change, or even slope of tangent. He did not even understand the relationship between his maximum-minimum method and the way one found tangents; in fact he followed his treatment of maxima and minima by saying that the same method—that is, adding E, doing the algebra, then suppressing E—could be used to find tangents [8, p. 223].

Though the considerations that led Fermat to his method may seem surprising to us, he did devise a method of finding extrema that worked, and it gave results that were far from trivial. For instance, Fermat applied his method to optics. Assuming that a ray of light which goes from one medium to another always takes the quickest path (what we now call the Fermat least-time principle), he used his method to compute the path taking minimal time. Thus he showed that his least-time principle yields Snell's law of refraction [7] [12, pp. 387–390].

Though Fermat did not publish his method of maxima and minima, it became well known through correspondence and was widely used. After mathematicians had become familiar with a variety of examples, a pattern emerged from the solutions by Fermat's method to maximum-minimum problems. In 1659, Johann Hudde gave a general verbal formulation of this pattern [3, p. 186], which, in modern notation, states that, *given a polynomial of the form*

$$y = \sum_{k=0}^{n} a_k x^k,$$

there is a maximum or minimum when

$$\sum_{k=1}^{n} k a_k x^{k-1} = 0.$$

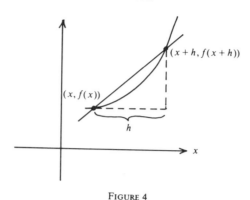

Of even greater interest than the problem of extrema in the seventeenth century was the finding of tangents. Here the tangent was usually thought of as a secant for which the two points came closer and closer together until they coincided. Precisely what it meant for a secant to "become" a tangent was never completely explained. Nevertheless, methods based on this approach worked. Given the equation of a curve

$$y = f(x),$$

Fermat, Descartes, John Wallis, Isaac Barrow, and many other seventeenth-century mathematicians were able to find the tangent. The method involves considering, and computing, the slope of the secant,

$$\frac{f(x+h) - f(x)}{h},$$

doing the algebra required by the formula for $f(x + h)$ in the numerator, then dividing by h. The diagram in FIGURE 4 then suggests that when the quantity h vanishes, the secant becomes the tangent, so that neglecting h in the expression for the slope of the secant gives the slope of the tangent. Again, a general pattern for the equations of slopes of tangents soon became apparent, and a rule analogous to Hudde's rule for maxima and minima was stated by several people, including René Sluse, Hudde, and Christiaan Huygens [3, pp. 185–186].

By the year 1660, both the computational and the geometric relationships between the problem of extrema and the problem of tangents were clearly understood; that is, a maximum was found by computing the slope of the tangent, according to the rule, and asking when it was zero. While in 1660 there was not yet a general concept of derivative, there was a general method for solving one type of geometric problem. However, the relationship of the tangent to other geometric concepts—area, for instance—was not understood, and there was no completely satisfactory definition of tangent. Nevertheless, there was a wealth of methods for solving problems that we now solve by using the calculus, and in retrospect, it would seem to be possible to generalize those methods. Thus in this context it is natural to ask, how did the derivative as we know it come to be?

It is sometimes said that the idea of the derivative was motivated chiefly by physics. Newton, after all, invented both the calculus and a great deal of the physics of motion. Indeed, already in the Middle Ages, physicists, following Aristotle who had made "change" the central concept in his physics, logically analyzed and classified the different ways a variable could change. In particular, something could change uniformly or nonuniformly; if nonuniformly, it could change uniformly-nonuniformly or nonuniformly-nonuniformly, etc. [3, pp. 73–74]. These medieval classifications of variation helped to lead Galileo in 1638, without benefit of calculus, to his successful treatment of uniformly accelerated motion. Motion, then, could be studied scientifically. Were such studies the origin and purpose of the calculus? The answer is no. However plausible this suggestion may sound, and however important physics was in the later development of the calculus, physical questions were in fact neither the immediate motivation nor the first application of the calculus. Certainly they prepared people's thoughts for some of the properties of the derivative, and for the introduction into mathematics of the concept of change. But the immediate motivation for the general concept of derivative—as opposed to specific examples like speed or slope of tangent—did not come from physics. The first problems to be solved, as well as the first applications, occurred in mathematics, especially geometry (see [1, chapter 7]; see also [3; chapters 4–5], and, for Newton, [17]). The concept of derivative then developed gradually, together with the ideas of extrema, tangent, area, limit, continuity, and function, and it interacted with these ideas in some unexpected ways.

Tangents, areas, and rates of change

In the latter third of the seventeenth century, Newton and Leibniz, each independently, invented the calculus. By "inventing the calculus" I mean that they did three things. First, they took the wealth of methods that already existed for finding tangents, extrema, and areas, and they subsumed all these methods under the heading of two general concepts, the concepts which we now call **derivative** and **integral**. Second, Newton and Leibniz each worked out a notation which made it easy, almost automatic, to use these general concepts. (We still use Newton's $\dot{x}$ and we still use Leibniz's dy/dx and $\int y\,dx$.) Third, Newton and Leibniz each gave an argument to prove what we now call the Fundamental Theorem of Calculus: the derivative and the integral are mutually inverse. Newton called our "derivative" a *fluxion*—a rate of flux or change; Leibniz saw the derivative as a ratio of infinitesimal differences and called it the *differential quotient*. But whatever terms were used, the concept of derivative was now embedded in a general subject—the calculus—and its relationship to the other basic concept, which Leibniz called the integral, was now understood. Thus we have reached the stage I have called *discovery*.

Let us look at an early Newtonian version of the Fundamental Theorem [13, sections 54–5, p. 23]. This will illustrate how Newton presented the calculus in 1669, and also illustrate both the strengths and weaknesses of the understanding of the derivative in this period.

Consider with Newton a curve under which the area up to the point $D = (x, y)$ is given by z (see FIGURE 5). His argument is general: "Assume any relation betwixt x and z that you please;" he then proceeded to find y. The example he used is

$$z = \frac{n}{m+n} ax^{(m+n)/n};$$

however, it will be sufficient to use $z = x^3$ to illustrate his argument.

In the diagram in FIGURE 5, the auxiliary line bd is chosen so that $Bb = o$, where o is not zero. Newton then specified that $BK = v$ should be chosen so that area $BbHK$ = area $BbdD$. Thus ov = area $BbdD$. Now, as x increases to $x + o$, the change in the area z is given by

$$z(x + o) - z(x) = x^3 + 3x^2o + 3xo^2 + o^3 - x^3 = 3x^2o + 3xo^2 + o^3,$$

which, by the definition of v, is equal to ov. Now since $3x^2o + 3xo^2 + o^3 = ov$, dividing by o produces $3x^2 + 3ox + o^2 = v$. Now, said Newton, "If we suppose Bb to be diminished infinitely and to vanish, or o to be nothing, v and y in that case will be equal and the terms which are multiplied by o will vanish: so that there will remain..."

$$3x^2 = y.$$

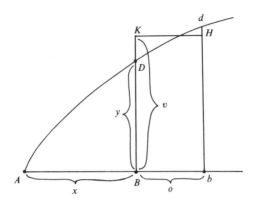

FIGURE 5

What has he shown? Since $(z(x + o) - z(x))/o$ is the rate at which the area z changes, that rate is given by the ordinate y. Moreover, we recognize that $3x^2$ would be the slope of the tangent to the curve $z = x^3$. Newton went on to say that the argument can be reversed; thus the converse holds too. We see that derivatives are fundamentally involved in areas as well as tangents, so the concept of derivative helps us to see that these two problems are mutually inverse. Leibniz gave analogous arguments on this same point (see, e.g. [**16**, pp. 282–284]).

Newton and Leibniz did not, of course, have the last word on the concept of derivative. Though each man had the most useful properties of the concept, there were still many unanswered questions. In particular, what, exactly, is a differential quotient? Some disciples of Leibniz, notably Johann Bernoulli and his pupil the Marquis de l'Hospital, said a differential quotient was a ratio of infinitesimals; after all, that is the way it was calculated. But infinitesimals, as seventeenth-century mathematicians were well aware, do not obey the Archimedean axiom. Since the Archimedean axiom was the basis for the Greek theory of ratios, which was, in turn, the basis of arithmetic, algebra, and geometry for seventeenth-century mathematicians, non-Archimedean objects were viewed with some suspicion. Again, what is a fluxion? Though it can be understood intuitively as a velocity, the proofs Newton gave in his 1671 *Method of Fluxions* all involved an "indefinitely small quantity o," [**14**, pp. 32–33] which raises many of the same problems that the o which "vanishes" raised in the Newtonian example of 1669 we saw above. In particular, what is the status of that little o? Is it zero? If so, how can we divide by it? If it is not zero, aren't we making an error when we throw it away? These questions had already been posed in Newton's and Leibniz's time. To avoid such problems, Newton said in 1687 that quantities defined in the way that $3x^2$ was defined in our example were the *limit* of the ratio of vanishing increments. This sounds good, but Newton's understanding of the term "limit" was not ours. Newton in his *Principia* (1687) described limits as "ultimate ratios"—that is, the value of the ratio of those vanishing quantities just when they are vanishing. He said, "Those ultimate ratios with which quantities vanish are not truly the ratios of ultimate quantities, but limits towards which the ratios of quantities decreasing without limit do always converge; and to which they approach nearer than by any given difference, but never go beyond, nor in effect attain to, till the quantities are diminished in infinitum" [**15**, Book I, Scholium to Lemma XI, p. 39].

Notice the phrase "but never go beyond"—so a variable cannot oscillate about its limit. By "limit" Newton seems to have had in mind "bound," and mathematicians of his time often cite the particular example of the circle as the limit of inscribed polygons. Also, Newton said, "nor... attain to, till the quantities are diminished in infinitum." This raises a central issue: it was often asked whether a variable quantity ever actually reached its limit. If it did not, wasn't there an error? Newton did not help clarify this when he stated as a theorem that "Quantities and the ratios of quantities which in any finite time converge continually to equality, and before the end of that time approach nearer to each other than by any given difference, become ultimately equal" [15, Book I, Lemma I, p. 29]. What does "become ultimately equal" mean? It was not really clear in the eighteenth century, let alone the seventeenth.

In 1734, George Berkeley, Bishop of Cloyne, attacked the calculus on precisely this point. Scientists, he said, attack religion for being unreasonable; well, let them improve their own reasoning first. A quantity is either zero or not; there is nothing in between. And Berkeley characterized the mathematicians of his time as men "rather accustomed to compute, than to think" [2].

Perhaps Berkeley was right, but most mathematicians were not greatly concerned. The concepts of differential quotient and integral, concepts made more effective by Leibniz's notation and by the Fundamental Theorem, had enormous power. For eighteenth-century mathematicians, especially those on the Continent where the greatest achievements occurred, it was enough that the concepts of the calculus were understood sufficiently well to be applied to solve a large number of problems, both in mathematics and in physics. So, we come to our third stage: *exploration and development*.

Differential equations, Taylor series, and functions

Newton had stated his three laws of motion in words, and derived his physics from those laws by means of synthetic geometry [15]. Newton's second law stated: "*The change of motion* [*our 'momentum'*] *is proportional to the motive force impressed, and is made in the direction of the* [*straight*] *line in which that force is impressed*" [15, p. 13]. Once translated into the language of the calculus, this law provided physicists with an instrument of physical discovery of tremendous power—because of the power of the concept of the derivative.

To illustrate, if F is force and x distance (so $m\dot{x}$ is momentum and, for constant mass, $m\ddot{x}$ the rate of change of momentum), then Newton's second law takes the form $F = m\ddot{x}$. Hooke's law of elasticity (when an elastic body is distorted the restoring force is proportional to the distance [in the opposite direction] of the distortion) takes the algebraic form $F = -kx$. By equating these expressions for force, Euler in 1739 could easily both state and solve the differential equation $m\ddot{x} + kx = 0$ which describes the motion of a vibrating spring [10, p. 482]. It was mathematically surprising, and physically interesting, that the solution to that differential equation involves sines and cosines.

An analogous, but considerably more sophisticated problem, was the statement and solution of the partial differential equation for the vibrating string. In modern notation, this is

$$\frac{\partial^2 y}{\partial t^2} = \frac{T \partial^2 y}{\mu \partial x^2},$$

where T is the tension in the string and μ is its mass per unit length. The question of how the solutions to this partial differential equation behaved was investigated by such men as d'Alembert, Daniel Bernoulli, and Leonhard Euler, and led to extensive discussions about the nature of continuity, and to an expansion of the notion of function from formulas to more general dependence relations [10, pp. 502–514], [16, pp. 367–368]. Discussions surrounding the problem of the vibrating string illustrate the unexpected ways that discoveries in mathematics and physics can interact ([16, pp. 351–368] has good selections from the original papers). Numerous other examples could be cited, from the use of infinite-series approximations in celestial mechanics to the dynamics of rigid bodies, to show that by the mid-eighteenth century the differential equation had become the most useful mathematical tool in the history of physics.

Another useful tool was the Taylor series, developed in part to help solve differential equations. In 1715, Brook Taylor, arguing from the properties of finite differences, wrote an equation expressing what we would write as $f(x + h)$ in terms of $f(x)$ and its quotients of differences of various orders. He then let the differences get small, passed to the limit, and gave the formula that still bears his name: the Taylor series. (Actually, James Gregory and Newton had anticipated this discovery, but Taylor's work was more directly influential.) The importance of this property of derivatives was soon recognized, notably by Colin Maclaurin (who has a special case of it named after him), by Euler, and by Joseph-Louis Lagrange. In their hands, the Taylor series became a powerful tool in studying functions and in approximating the solution of equations.

But beyond this, the study of Taylor series provided new insights into the nature of the derivative. In 1755, Euler, in his study of power series, had said that for any power series,

$$a + bx + cx^2 + dx^3 + \cdots,$$

one could find x sufficiently small so that if one broke off the series after some particular term—say x^2—the x^2 term would exceed, in absolute value, the sum of the entire remainder of the series [6, section 122]. Though Euler did not prove this—he must have thought it obvious since he usually worked with series with finite coefficients—he applied it to great advantage. For instance, he could use it to analyze the nature of maxima and minima. Consider, for definiteness, the case of maxima. If $f(x)$ is a relative maximum, then by definition, for small h,

$$f(x - h) < f(x) \quad \text{and} \quad f(x + h) < f(x).$$

Taylor's theorem gives, for these inequalities,

$$f(x - h) = f(x) - h\frac{df(x)}{dx} + h^2\frac{d^2f(x)}{dx^2} - \cdots < f(x) \tag{3}$$

$$f(x + h) = f(x) + h\frac{df(x)}{dx} + h^2\frac{d^2f(x)}{dx^2} + \cdots < f(x). \tag{4}$$

Now if h is so small that $h\,df(x)/dx$ dominates the rest of the terms, the only way that both of the inequalities (3) and (4) can be satisfied is for $df(x)/dx$ to be zero. Thus the differential quotient is zero for a relative maximum. Furthermore, Euler argued, since h^2 is always positive, if $d^2f(x)/dx^2 \neq 0$, the only way both inequalities can be satisfied is for $d^2f(x)/dx^2$ to be negative. This is because the h^2 term dominates the rest of the series—unless $d^2f(x)/dx^2$ is itself zero, in which case we must go on and think about even higher-order differential quotients. This analysis, first given and demonstrated geometrically by Maclaurin, was worked out in full analytic detail by Euler [6, sections 253–254], [9, pp. 117–118]. It is typical of Euler's ability to choose computations that produce insight into fundamental concepts. It assumes, of course, that the function in question has a Taylor series, an assumption which Euler made without proof for many functions; it assumes also that the function is uniquely the sum of its Taylor series, which Euler took for granted. Nevertheless, this analysis is a beautiful example of the exploration and development of the concept of the differential quotient of first, second, and nth orders—a development which completely solves the problem of characterizing maxima and minima, a problem which goes back to the Greeks.

Lagrange and the derivative as a function

Though Euler did a good job analyzing maxima and minima, he brought little further understanding of the nature of the differential quotient. The new importance given to Taylor series meant that one had to be concerned not only about first and second differential quotients, but about differential quotients of any order.

The first person to take these questions seriously was Lagrange. In the 1770's, Lagrange was impressed with what Euler had been able to achieve by Taylor-series manipulations with differential quotients, but Lagrange soon became concerned about the logical inadequacy of all the existing justifications for the calculus. In particular, Lagrange wrote in 1797 that the Newtonian limit-concept was not clear enough to be the foundation for a branch of mathematics. Moreover, in not allowing variables to surpass their limits, Lagrange thought the limit-concept too restrictive. Instead, he said, the calculus should be reduced to algebra, a subject whose foundations in the eighteenth century were generally thought to be sound [11, pp. 15–16].

The algebra Lagrange had in mind was what he called the algebra of infinite series, because Lagrange was convinced that infinite series were part of algebra. Just as arithmetic deals with infinite decimal fractions without ceasing to be arithmetic, Lagrange thought, so algebra deals with infinite algebraic expressions without ceasing to be algebra. Lagrange believed that expanding $f(x + h)$ into a power series in h was always an algebraic process. It is obviously algebraic when one turns $1/(1 - x)$ into a power series by dividing. And Euler had found, by manipulating formulas, infinite power-series expansions for functions like $\sin x, \cos x, e^x$. If functions like those have power-series expansions, perhaps everything could be reduced to algebra. Euler, in his book *Introduction to the analysis of the infinite* (*Introductio in analysin infinitorum*, 1748), had studied infinite series, infinite products, and infinite continued fractions by what he thought of as purely algebraic methods. For instance, he converted infinite series into infinite products by treating a series as a very long polynomial. Euler thought that this work was purely algebraic, and—what is crucial here—Lagrange also thought Euler's methods were purely algebraic. So Lagrange tried to make the calculus rigorous by reducing it to the algebra of infinite series.

Lagrange stated in 1797, and thought he had proved, that any function (that is, any analytic expression, finite or infinite) had a power-series expansion:

$$f(x + h) = f(x) + p(x)h + q(x)h^2 + r(x)h^3 + \cdots, \tag{5}$$

except, possibly, for a finite number of isolated values of x. He then defined a new function, the coefficient of the linear term in h which is $p(x)$ in the expansion shown in (5)) and called it the **first derived function** of $f(x)$. Lagrange's term "derived function" (*fonction dérivée*) is the origin of our term "derivative." Lagrange introduced a new notation, $f'(x)$, for that function. He defined $f''(x)$ to be the first derived function of $f'(x)$, and so on, recursively. Finally, using these definitions, he proved that, in the expansion (5) above, $q(x) = f''(x)/2, r(x) = f'''(x)/6$, and so on [11, chapter 2].

What was new about Lagrange's definition? The concept of *function*—whether simply an algebraic expression (possibly infinite) or, more generally, any dependence relation—helps free the concept of derivative from the earlier ill-defined notions. Newton's explanation of a fluxion as a rate of change appeared to involve the concept of motion in mathematics; moreover, a fluxion seemed to be a different kind of object than the flowing quantity whose fluxion it was. For Leibniz, the differential quotient had been the quotient of vanishingly small differences; the second differential quotient, of even smaller differences. Bishop Berkeley, in his attack on the calculus, had made fun of these earlier concepts, calling vanishing increments "ghosts of departed quantities" [2, section 35]. But since, for Lagrange, the derivative was a function, it was now the same sort of object as the original function. The second derivative is precisely the same sort of object as the first derivative; even the nth derivative is simply another function, defined as the coefficient of h in the Taylor series for $f^{(n-1)}(x + h)$. Lagrange's notation $f'(x)$ was designed precisely to make this point.

We cannot fully accept Lagrange's definition of the derivative, since it assumes that every differentiable function is the sum of a Taylor series and thus has infinitely many derivatives. Nevertheless, that definition led Lagrange to a number of important properties of the derivative. He used his definition together with Euler's criterion for using truncated power series in approximations to give a most useful characterization of the derivative of a function [9, p. 116, pp.

P. Fermat   R. Descartes   I. Newton   G.W. Leibniz

1637–38   1669   1684

Dates refer to these mathematician's major works which

118–121]:

$$f(x+h)=f(x)+hf'(x)+hH, \text{ where } H \text{ goes to zero with } h.$$

(I call this the *Lagrange property of the derivative*.) Lagrange interpreted the phrase "H goes to zero with h" in terms of inequalities. That is, he wrote that,

$$\text{Given } D, h \text{ can be chosen so that } f(x+h)-f(x) \quad (6)$$
$$\text{lies between } h(f'(x)-D) \text{ and } h(f'(x)+D).$$

Formula (6) is recognizably close to the modern delta-epsilon definition of the derivative.

Lagrange used inequality (6) to prove theorems. For instance, he proved that a function with positive derivative on an interval is increasing there, and used that theorem to derive the Lagrange remainder of the Taylor series [9, pp. 122–127], [11, pp. 78–85]. Furthermore, he said, considerations like inequality (6) are what make possible applications of the differential calculus to a whole range of problems in mechanics, in geometry, and, as we have described, the problem of maxima and minima (which Lagrange solved using the Taylor series remainder which bears his name [11, pp. 233–237]).

In Lagrange's 1797 work, then, the derivative is defined by its position in the Taylor series—a strange definition to us. But the derivative is also *described* as satisfying what we recognize as the appropriate delta-epsilon inequality, and Lagrange applied this inequality and its nth-order analogue, the Lagrange remainder, to solve problems about tangents, orders of contact between curves, and extrema. Here the derivative was clearly a function, rather than a ratio or a speed.

Still, it is a lot to assume that a function has a Taylor series if one wants to define only *one* derivative. Further, Lagrange was wrong about the algebra of infinite series. As Cauchy pointed out in 1821, the algebra of finite quantities cannot automatically be extended to infinite processes. And, as Cauchy also pointed out, manipulating Taylor series is not foolproof. For instance, e^{-1/x^2} has a zero Taylor series about $x=0$, but the function is not identically zero. For these reasons, Cauchy rejected Lagrange's definition of derivative and substituted his own.

Definitions, rigor, and proofs

Now we come to the last stage in our chronological list: *definition*. In 1823, Cauchy defined the derivative of $f(x)$ as the limit, when it exists, of the quotient of differences $(f(x+h)-f(x))/h$ as h goes to zero [4, pp. 22–23]. But Cauchy understood "limit" differently than had his predecessors. Cauchy entirely avoided the question of whether a variable ever reached its limit; he just didn't discuss it. Also, knowing an absolute value when he saw one, Cauchy followed Simon l'Huilier and S.-F. Lacroix in abandoning the restriction that variables never surpass their limits.

L. Euler	J.-L. Lagrange	A.-L. Cauchy	K. Weierstrass
1755	1797	1823	1861

contributed to the evolution of the concept of the derivative.

Finally, though Cauchy, like Newton and d'Alembert before him, gave his definition of limit in words, Cauchy's understanding of limit (most of the time, at least) was algebraic. By this, I mean that when Cauchy needed a limit property in a proof, he used the algebraic inequality-characterization of limit. Cauchy's proof of the mean value theorem for derivatives illustrates this. First he proved a theorem which states: *if $f(x)$ is continuous on $[x, x + a]$, then*

$$\min_{[x, x+a]} f'(x) \leqslant \frac{f(x+a) - f(x)}{a} \leqslant \max_{[x, x+a]} f'(x). \tag{7}$$

The first step in his proof is [**4**, p. 44]:

> Let δ, ε be two very small numbers; the first is chosen so that for all [absolute] values of h less than δ, and for any value of x [on the given interval], the ratio $(f(x+h) - f(x))/h$ will always be greater than $f'(x) - \varepsilon$ and less than $f'(x) + \varepsilon$.

(The notation in this quote is Cauchy's, except that I have substituted h for the i he used for the increment.) Assuming the intermediate-value theorem for continuous functions, which Cauchy had proved in 1821, the mean-value theorem is an easy corollary of (7) [**4**, pp. 44–45], [**9**, pp. 168–170].

Cauchy took the inequality-characterization of the derivative from Lagrange (possibly via an 1806 paper of A.-M. Ampère [**9**, pp. 127–132]). But Cauchy made that characterization into a definition of derivative. Cauchy also took from Lagrange the name derivative and the notation $f'(x)$, emphasizing the functional nature of the derivative. And, as I have shown in detail elsewhere [**9**, chapter 5], Cauchy adapted and improved Lagrange's inequality proof-methods to prove results like the mean-value theorem, proof-methods now justified by Cauchy's definition of derivative.

But of course, with the new and more rigorous definition, Cauchy went far beyond Lagrange. For instance, using his concept of limit to define the integral as the limit of sums, Cauchy made a good first approximation to a real proof of the Fundamental Theorem of Calculus [**9**, pp. 171–175], [**4**, pp. 122–125, 151–152]. And it was Cauchy who not only raised the question, but gave the first proof, of the existence of a solution to a differential equation [**9**, pp. 158–159].

After Cauchy, the calculus itself was viewed differently. It was seen as a rigorous subject, with good definitions and with theorems whose proofs were based on those definitions, rather than merely as a set of powerful methods. Not only did Cauchy's new rigor establish the earlier results on a firm foundation, but it also provided a framework for a wealth of new results, some of which could not even be formulated before Cauchy's work.

Of course, Cauchy did not himself solve all the problems occasioned by his work. In particular, Cauchy's definition of the derivative suffers from one deficiency of which he was unaware. Given an ε, he chose a δ which he assumed would work for any x. That is, he assumed that the quotient of differences converged uniformly to its limit. It was not until the 1840's that G. G. Stokes, V. Seidel, K. Weierstrass, and Cauchy himself worked out the distinction between convergence and uniform convergence. After all, in order to make this distinction, one first needs a clear and algebraic understanding of what a limit is—the understanding Cauchy himself had provided.

In the 1850's, Karl Weierstrass began to lecture at the University of Berlin. In his lectures, Weierstrass made algebraic inequalities replace words in theorems in analysis, and used his own clear distinction between pointwise and uniform convergence along with Cauchy's delta-epsilon techniques to present a systematic and thoroughly rigorous treatment of the calculus. Though Weierstrass did not publish his lectures, his students—H. A. Schwartz, G. Mittag-Leffler, E. Heine, S. Pincherle, Sonya Kowalevsky, Georg Cantor, to name a few—disseminated Weierstrassian rigor to the mathematical centers of Europe. Thus although our modern delta-epsilon definition of derivative cannot be quoted from the *works* of Weierstrass, it is in fact the *work* of Weierstrass [3, pp. 284–287]. The rigorous understanding brought to the concept of the derivative by Weierstrass is signaled by his publication in 1872 of an example of an everywhere continuous, nowhere differentiable function. This is a far cry from merely acknowledging that derivatives might not always exist, and the example shows a complete mastery of the concepts of derivative, limit, and existence of limit [3, p. 285].

Historical development versus textbook exposition

The span of time from Fermat to Weierstrass is over two hundred years. How did the concept of derivative develop? Fermat implicitly used it; Newton and Liebniz discovered it; Taylor, Euler, Maclaurin developed it; Lagrange named and characterized it; and only at the end of this long period of development did Cauchy and Weierstrass define it. This is certainly a complete reversal of the usual order of textbook exposition in mathematics, where one starts with a definition, then explores some results, and only then suggests applications.

This point is important for the teacher of mathematics: the historical order of development of the derivative is the reverse of the usual order of textbook exposition. Knowing the history helps us as we teach about derivatives. We should put ourselves where mathematicians were before Fermat, and where our beginning students are now—back on the other side, before we had any concept of derivative, and also before we knew the many uses of derivatives. Seeing the historical origins of a concept helps motivate the concept, which we—along with Newton and Leibniz—want for the problems it helps to solve. Knowing the historical order also helps to motivate the rigorous definition—which we, like Cauchy and Weierstrass, want in order to justify the uses of the derivative, and to show precisely when derivatives exist and when they do not. We need to remember that the rigorous definition is often the end, rather than the beginning, of a subject.

The real historical development of mathematics—the order of discovery—reveals the creative mathematician at work, and it is creation that makes doing mathematics so exciting. The order of exposition, on the other hand, is what gives mathematics its characteristic logical structure and its incomparable deductive certainty. Unfortunately, once the classic exposition has been given, the order of discovery is often forgotten. The task of the historian is to recapture the order of discovery: not as we think it might have been, not as we think it should have been, but as it really was. And this is the purpose of the story we have just told of the derivative from Fermat to Weierstrass.

This article is based on a talk delivered at the Conference on the History of Modern Mathematics, Indiana Region of the Mathematical Association of America, Ball State University, April 1982; earlier versions were presented at the Southern California Section of the M. A. A. and at various mathematics colloquia. I thank the MATHEMATICS MAGAZINE referees for their helpful suggestions.

References

[1] Margaret Baron, Origins of the Infinitesimal Calculus, Pergamon, Oxford, 1969.

[2] George Berkeley, The Analyst, or a Discourse Addressed to an Infidel Mathematician, 1734. In A. A. Luce and T. R. Jessop, eds., The Works of George Berkeley, Nelson, London, 1951 (some excerpts appear in [16, pp. 333–338]).

[3] Carl Boyer, History of the Calculus and Its Conceptual Development, Dover, New York, 1959.

[4] A.-L. Cauchy, Résumé des leçons données à l'école royale polytechnique sur le calcul infinitésimal, Paris, 1823. In Oeuvres complètes d'Augustin Cauchy, Gauthier-Villars, Paris, 1882- , series 2, vol. 4.

[5] Pierre Dugac, Fondements d'analyse, in J. Dieudonné, Abrégé d'histoire des mathématiques, 1700–1900, 2 vols., Hermann, Paris, 1978.

[6] Leonhard Euler, Institutiones calculi differentialis, St. Petersburg, 1755. In Operia omnia, Teubner, Leipzig, Berlin, and Zurich, 1911- , series 1, vol. 10.

[7] Pierre Fermat, Analysis ad refractiones, 1661. In Oeuvres de Fermat, ed., C. Henry and P. Tannery, 4 vols., Paris, 1891–1912; Supplement, ed. C. de Waard, Paris, 1922, vol. 1, pp. 170–172.

[8] _____, Methodus ad disquirendam maximam et minimum et de tangentibus linearum curvarum, Oeuvres, vol. 1, pp. 133–136. Excerpted in English in [16, pp. 222–225].

[9] Judith V. Grabiner, The Origins of Cauchy's Rigorous Calculus, M. I. T. Press, Cambridge and London, 1981.

[10] Morris Kline, Mathematical Thought from Ancient to Modern Times, Oxford, New York, 1972.

[11] J.-L. Lagrange, Théorie des fonctions analytiques, Paris, 2nd edition, 1813. In Oeuvres de Lagrange, ed. M. Serret, Gauthier-Villars, Paris, 1867–1892, vol. 9.

[12] Michael S. Mahoney, The Mathematical Career of Pierre de Fermat, 1601–1665, Princeton University Press, Princeton, 1973.

[13] Isaac Newton, Of Analysis by Equations of an Infinite Number of Terms [1669], in D. T. Whiteside, ed., Mathematical Works of Isaac Newton, Johnson, New York and London, 1964, vol. 1, pp. 3–25.

[14] _____, Method of Fluxions [1671], in D. T. Whiteside, ed., Mathematical Works of Isaac Newton, vol. 1, pp. 29–139.

[15] _____, Mathematical Principles of Natural Philosophy, tr. A. Motte, ed. F. Cajori, University of California Press, Berkeley, 1934.

[16] D. J. Struik, Source Book in Mathematics, 1200–1800, Harvard University Press, Cambridge, MA, 1969.

[17] D. T. Whiteside, ed., The Mathematical Papers of Isaac Newton, Cambridge University Press, 1967–1982.

BIBLIOGRAPHIC ENTRIES: HISTORY

1. *Monthly* Vol. 81, No. 10, pp. 1095–1096. Martin G. Beumer, The definite integral symbol.

Anecdotal comment that *integral* comes from *integer* and *Aal* (German for *eel*).

2. TYCMJ Vol. 1, No. 1, pp. 60–86. Carl B. Boyer, The history of THE CALCULUS.

Covers contributions to calculus from the time of the Babylonians to the end of the 19th century. A reprint of Chapter 7 of the NCTM 31st Yearbook (1969), *Historical Topics for the Classroom*.

3. CMJ Vol. 18, No. 5, pp. 362–389. V. Frederick Rickey, Isaac Newton: Man, Myth, and Mathematics.

Biography emphasizing Newton's mathematical works and readings, especially Descartes' *Geometry*. (This paper won a Polya award.)

2

PEDAGOGY

CALCULUS AS AN EXPERIMENTAL SCIENCE*

R. P. Boas, Jr., Northwestern University

I hope that my title was not too misleading. I am not going to suggest that calculus should somehow be based on experiment, but rather that calculus should be presented to the student in the same spirit as the experimental sciences. The point that I hope to make is, briefly, that proofs are to mathematics what experiments are to physics (or chemistry, or biology), and that our teaching can profit by the analogy.

Let us first of all be clear about what calculus is. There are two big ideas, the derivative and the integral. Geometrically, these are the slope of a curve and the area under a curve. Of course they frequently, even usually, appear in non-geometrical forms: a derivative might represent a mass density and an integral might represent work, for instance. However, translating to and from geometry should not bother anyone who has ever done something like drawing a vector diagram of forces. The predecessors of Newton and Leibnitz knew perfectly well how to determine tangents and areas, but they had to approach each problem from first principles. The great contribution of Newton and Leibnitz was precisely to make the procedures for finding tangents, areas, etc. into a calculus, that is, a systematic way of calculating—a collection of algorithms, to use the currently fashionable word. Moreover, they didn't really understand—in the modern sense—why the algorithms worked. Perhaps it will make the point clearer if I use a very elementary example. If you write two numbers with Roman numerals, and want to multiply them, you can work out the product if

---

* Vice-presidential address given at the meeting of the American Association for the Advancement of Science, December 28, 1970, as part of a Symposium on Mathematics in the Undergraduate Program in the Sciences, jointly sponsored by CUPM. The opinions expressed are those of the author, and are not to be taken as representing CUPM policy.

This article is also appearing in the Two Year College Mathematics Journal. Since Professor Boas is a prominent analyst and has recently served as chairman of the Committee on the Undergraduate Program in Mathematics, the editors feel that the mathematical community at large would be especially interested to know what he has to say about the teaching of elementary calculus. We have therefore departed from normal procedures, and duplicated publication of this article. *The Editors.*

you understand the commutative, associative, and distributive laws of multiplication and addition for integers, but it takes time. Presumably the Romans used some other method, perhaps some kind of abacus (a simple digital computer). A more convenient way is to use Arabic numerals and the rules for manipulating them—a kind of calculus, in fact. In either case we have substituted rules or procedures for thinking about what is actually going on. Note incidentally that one can often use a calculus successfully without fully understanding why it works. (Does a digital computer understand arithmetic?)

Once invented, differential and integral calculus were very successful at solving certain kinds of geometrical problems, and hence physical problems that can be represented geometrically, and hence problems in physics, chemistry, economics, etc., even when they are not represented geometrically. Consequently calculus became the standard language for talking about the subjects in which it was most successful, it has remained so to this day, and seems likely to continue for some time into the future. This is why every scientist has to study calculus, although he often wonders at first why he should have to. Another way of saying much the same thing is that calculus is used because it facilitates the study of models of observed phenomena. If a biological process, for instance, can be modeled as a differential equation, calculus can take over and predict properties of the process without using any biological thought, and the biologist can then compare the prediction with experiment—in this way he may save considerable time and thought. It is hard to get this idea across to the beginning student, especially when he doesn't know any biology yet.

Everybody admits, I suppose, that the sciences other than mathematics are based on experiment. Things that can be checked by experiment are accepted: things that disagree with experiment are not. However, I am not aware of any physics or chemistry or biology course that repeats all the classical experiments, or even any of those that are particularly difficult or time-consuming. At least in the science courses I took (of course, this was a long time ago) we were told that certain things had been established experimentally, and maybe (not always) what the experiment was like. Inspection of some current textbooks suggests that things haven't changed much in 40 years.

Now I do not know any experimental scientists who seem to feel uncomfortable about this state of affairs, although for all I know they may worry about it in secret. Nor do they seem to worry about the necessity of sometimes giving oversimplified or even mildly fallacious reasons why the experiment comes out as it does. For example, why does an airplane stay up? Elementary texts give theoretical reasons that do not seem very convincing; the real theoretical reasons are clearly too sophisticated for elementary courses; it presumably would be possible also to rely on experimental measurements of the flow around a wing, but few, if any, physics courses bring wind-tunnels into the classroom. Similarly, first-year physics courses usually teach Newtonian mechanics, rather than relativistic mechanics or quantum mechanics. In calling attention to this, I do not intend to criticize the current teaching of the experimental sciences; in fact, I do not see what else could be done, and indeed I want to use the experimental scientists' approach as a model.

Mathematics is not at all an experimental science, but there is a rather exact parallel between mathematics and the experimental sciences. In mathematics we believe things, not because we did an experiment, but because we proved them. At least—and this continues the analogy—we believe things because *somebody* proved them; we have not necessarily studied the proof ourselves. Thus the proof is to mathematics as the experiment is to physics, chemistry, biology, and so on. Perhaps we are better off than the experimental scientist in one respect—we know that we *could* read the proof if we tried, whereas the experiment may be too difficult or too expensive for the experimental scientist to hope to repeat for himself or his class. Of course we want our students to believe what we say because we have done something that really carries conviction. What shall we do? The answer that most mathematicians believe in, or profess to believe in, or act as if they believed in, is that they ought to present their students with formal proofs of everything that they tell them. The effect of this is to make calculus a chapter in the theory of functions of a real variable. There are several reasons for this attitude. There are mathematicians who didn't understand calculus themselves to begin with, but now do; and, filled with missionary zeal, wish to spread the light. There are those who feel that it is intellectually dishonest not to tell everything that they know. There are those who feel that anything less than full explanations cheats the student. And there are those who understand the proofs but can't solve the problems: theory is always easier than technique. In any case, only just so much time is available. In order to make the best use of it, I claim that the teacher of calculus would do well to follow the lead of the experimental scientist: let him give proofs when they are easy and justify unexpected things; let him omit tedious or difficult proofs, especially those of plausible things. Let him give easy proofs under simplified assumptions rather than complicated proofs under general hypotheses. Let him by all means always give correct statements, but not necessarily the most general ones that he knows.

Let us see how these principles apply to some topics in calculus. (1) One of the unexpected results is the formula for the derivative of a product. Most beginners will guess it wrong. The proof is easy and completely convincing. One should by all means give it. (2) A function with a positive derivative is increasing. This looks, but isn't, tautological; the point at issue, generating a global property from a local one, is rather subtle. The proof is not illuminating, and might well be skipped. (3) It is certainly necessary to define the definite integral, but to prove that the integral of a continuous function exists is both technically demanding and time-consuming. This seems to be a clear case for "it can be proved." (4) The uniqueness theorem for solutions of a second-order linear differential equation is only too plausible—don't the initial position and velocity determine the motion? The proof is time-consuming, but the facts are easy to state precisely and meaningfully. (5) Assuming that Fourier series get into the calculus (they usually don't), it would be difficult, time-consuming, and unconvincing to *prove* any really useful convergence theorem. On the other hand, should Fourier series be left out just because we cannot prove a satisfying theorem about them? It is easy enough to state one, and there is no excuse for stating an incorrect one.

I am going to be accused by my colleagues of advocating a cookbook approach to calculus. This I deny. There was once a really cookbook approach to calculus, in which the student had to listen to incomprehensible nonsense until he developed a sound intuition (if he ever did). The approach that some of my colleagues favor makes the student listen to incomprehensible sense instead. I think the experimental scientists do better. Let me illustrate the difference with an example. A cookbook approach to maximum and minimum problems leads the student to approach all problems by setting a derivative equal to zero and testing by the sign of the second derivative. This traditional procedure can, in fact, lead to mistakes. A rigorous approach demands a long series of preliminary theorems about maxima, mean values and derivatives. What I prefer is something like this: observe that if there is a maximum where there is a derivative, the derivative must be zero; then the maximum must be at one of the (usually small number of) points where the derivative is zero or doesn't exist, or else at an endpoint. A small amount of computation will usually decide; and we avoid the second-derivative test, which in spite of its theoretical elegance is usually quite impractical. It seems to me that an approach of this kind is very much in the spirit in which experimental sciences are usually presented; and in practice it seems to give the students more capability with calculus, and sooner, than the theorem-proving approach that has been so popular.

THE PROBLEM OF LEARNING TO TEACH

I. THE TEACHING OF PROBLEM SOLVING — BY P. R. HALMOS

The best way to learn is to do; the worst way to teach is to talk.

About the latter: did you ever notice that some of the best teachers of the world are the worst lecturers? (I can prove that, but I'd rather not lose quite so many friends.) And, the other way around, did you ever notice that good lecturers are not necessarily good teachers? A good lecture is usually systematic, complete, precise — and dull; it is a bad teaching instrument. When given by such legendary outstanding speakers as Emil Artin and John von Neumann, even a lecture can be a useful tool — their charisma and enthusiasm come through enough to inspire the listener to go forth and do something — it looks like such fun. For most ordinary mortals, however, who are not so bad at lecturing as Wiener was — nor so stimulating!— and not so good as Artin — and not so dramatic! — the lecture is an instrument of last resort for good teaching.

My test for what makes a good teacher is very simple: it is the pragmatic one of judging the performance by the product. If a teacher of graduate students consistently produces Ph. D.'s who are mathematicians and who create high-quality new mathematics, he is a good teacher. If a teacher of calculus consistently produces seniors who turn into outstanding graduate students of mathematics, or into leading engineers, biologists, or economists, he is a good teacher. If a teacher of third-grade "new math" (or old) consistently produces outstanding calculus students, or grocery store check-out clerks, or carpenters, or automobile mechanics, he is a good teacher.

For a student of mathematics to hear someone talk about mathematics does hardly any more good than for a student of swimming to hear someone talk about swimming. You can't learn swimming technique by having someone tell you where to put your arms and legs; and you can't learn to solve problems by having someone tell you to complete the square or to substitute sin u for y.

Can one learn mathematics by reading it? I am inclined to say no. Reading has an edge over listening because reading is more active — but not much. Reading with pencil and paper on the side is very much better — it is a big step in the right direction. The very best way to read a book, however, with, to be sure, pencil and paper on the side, is to keep the pencil busy on the paper and throw the book away.

Having stated this extreme position, I'll rescind it immediately. I know that it is extreme, and I don't really mean it -- but I wanted to be very emphatic about not going along with the view that learning means going to lectures and reading books. If we had longer lives, and bigger brains, and enough dedicated expert teachers to have a student/teacher ratio of 1/1, I'd stick with the extreme view — but we don't. Books and lectures don't do a good job of transplanting the facts and techniques of

Talks given at the Annual Meeting in San Francisco, January 17, 1974, at a joint AMS-MAA Panel discussion.

34

the past into the bloodstream of the scientist of the future — but we must put up with a second best job in order to save time and money. But, and this is the text of my sermon today, if we rely on lectures and books only, we are doing our students, and their students, a grave disservice.

What mathematics is really all about is solving concrete problems. Hilbert once said (but I can't remember where) that the best way to understand a theory is to find, and then to study, a prototypal concrete example of that theory, a root example that illustrates everything that can happen. The biggest fault of many students, even good ones, is that although they might be able to spout correct statements of theorems, and remember correct proofs, they cannot give examples, construct counterexamples, and solve special problems. I have seen many students who could state something they called the spectral theorem for Hermitian operators on Hilbert space but who had no idea how to diagonalize a 3×3 real symmetric matrix. That's bad — that's bad learning, probably caused, at least in part, by bad teaching. The full-time professional mathematician and the occasional user of mathematics, and the whole spectrum of the scientific community in between — they all need to solve problems, mathematical problems, and our job is to teach them how to do it, or, rather, to teach their future teachers how to teach them to do it.

I like to start every course I teach with a problem. The last time I taught the introductory course in set theory, my first sentence was the definition of algebraic numbers, and the second was a question: are there any numbers that are not algebraic? The last time I taught the introductory course in real function theory, my first sentence was a question: is there a non-decreasing continuous function that maps the unit interval into the unit interval so that length of its graph is equal to 2? For almost every course one can find a small set of questions such as these — questions that can be stated with the minimum of technical language, that are sufficiently striking to capture interest, that do not have trivial answers, and that manage to embody, in their answers, all the important ideas of the subject. The existence of such questions is what one means when one says that mathematics is really all about solving problems, and my emphasis on problem solving (as opposed to lecture attending and book reading) is motivated by them.

A famous dictum of Pólya's about problem solving is that if you can't solve a problem, then there is an easier problem that you can't solve — find it! If you can teach that dictum to your students, teach it so that they can teach it to theirs, you have solved the problem of creating teachers of problem solving. The hardest part of answering questions is to ask them; our job as teachers and teachers of teachers is to teach how to ask questions. It's easy to teach an engineer to use a differential equations cook book; what's hard is to teach him (and his teacher) what to do when the answer is not in the cook book. In that case, again, the chief problem is likely to be "what is the problem?". Find the right question to ask, and you're a long way toward solving the problem you're working on.

What then is the secret — what is the best way to learn to solve problems? The answer is implied by the sentence I started with: solve problems. The method I advocate is sometimes known as the "Moore method," because R. L. Moore developed and used it at the University of Texas. It is a method of teaching, a method of creating the problem-solving attitude in a student, that is a mixture of what Socrates taught us and the fiercely competitive spirit of the Olympic games.

The way a bad lecturer can be a good teacher, in the sense of producing good students, is the way a grain of sand can produce pearl-manufacturing oysters. A smooth lecture and a book entitled "Freshman algebra for girls" may be pleasant; a good teacher challenges, asks, annoys, irritates, and maintains high standards — all that is generally not pleasant. A good teacher may not be a popular teacher (except perhaps with his *ex*-students), because some students don't like to be challenged, asked, annoyed, and irritated — but he produces pearls (instead of casting them in the proverbial manner).

Let me tell you about the time I taught a course in linear algebra to juniors. The first hour I handed to each student a few sheets of paper on which were dittoed the precise statements of fifty theorems. That's all — just the statements of the theorems. There was no introduction, there were no definitions, there were no explanations, and, certainly, there were no proofs.

The rest of the first hour I told the class a little about the Moore method. I told them to give up reading linear algebra (for that semester only!), and to give up consulting with each other (for that semester only). I told them that the course was in their hands. The course was those fifty theorems; when they understood them, when they could explain them, when they could buttress them with the necessary examples and counterexamples, and, of course, when they could prove them, then they would have finished the course.

They stared at me. They didn't believe me. They thought I was just lazy and trying to get out of work. They were sure that they'd never learn anything that way.

All this didn't take as much as a half hour. I finished the hour by giving them the basic definitions that they needed to understand the first half dozen or so theorems, and, wishing them well, I left them to their own devices.

The second hour, and each succeeding hour, I called on Smith to prove Theorem 1, Kovacs to prove Theorem 2, and so on. I encouraged Kovacs and Herrero and all to watch Smith like hawks, and to pounce on him if he went wrong. I myself listened as carefully as I could, and, while I tried not to be sadistic, I too pounced when I felt I needed to. I pointed out gaps, I kept saying that I didn't understand, I asked questions about side issues, I asked for, and sometimes supplied, counterexamples, I told about the history of the subject when I had a chance, and I pointed out connections with other parts of mathematics. In addition I took five minutes or so of most hours to introduce the new definitions needed. Altogether I probably talked 20 minutes out of each of the 50-minute academic hours that we were together. That's a lot — but it's a lot less than 50 (or 55) out of 50.

It worked like a charm. By the second week they were proving theorems and finding errors in the proofs of others, and obviously taking pleasure in the process. Several of them had the grace to come to me and confess that they were skeptical at first, but they had been converted. Most of them said that they spent more time on that course than on their other courses that semester, and learned more from it.

What I just now described is like the "Moore method" as R. L. Moore used it, but it's a much modified Moore method. I am sure that hundreds of modifications could be devised, to suit the temperaments of different teachers and the needs of different subjects. The details don't matter. What matters is to make students ask and answer questions.

Many times when I've used the Moore method, my colleagues commented to me, perhaps a semester or two later, that they could often recognize those students in their classes who had been exposed to a "Moore class" by those students' attitude and behavior. The distinguishing characteristics were greater mathematical maturity than that of the others (the research attitude), and greater inclination and ability to ask penetrating questions.

The "research attitude" is a tremendous help to all teachers, and students, and creators, and users of mathematics. To illustrate, for instance, how it is a help to me when I teach elementary calculus (to a class that's too large to use the Moore method on), I must first of all boast to you about my wonderful memory. Wonderfully bad, that is. If I don't teach calculus, say, for a semester or two, I forget it. I forget the theorems, the problems, the formulas, the techniques. As a result, when I prepare next week's lecture, which I do by glancing at the prescribed syllabus, or, if there is none, at the table of contents of the text, but never at the text itself, I start almost from scratch — I do research in calculus. The result is that I have more fun than if I had it all by rote, that time after time I am genuinely surprised and pleased by some student's re-discovery of what Leibniz probably knew when he was a teenager, and that my fun, surprise, pleasure, and enthusiasm is felt by the class, and is taken as an accolade by each discoverer.

To teach the research attitude, every teacher should do research and should have had training in doing research. I am not saying that everyone who teaches trigonometry should spend half his time proving abstruse theorems about categorical teratology and joining the publish-or-perish race. What I am saying is that everyone who teaches, even if what he teaches is high-school algebra, would be a better teacher if he thought about the implications of the subject outside the subject, if he read about the connections of the subject with other subjects, if he tried to work out the problems that those implications and connections suggest — if, in other words, he did research in and around high-school algebra. That's the only way to keep the research attitude, the question-asking attitude, alive in himself, and thus to keep it in a condition suitable for transmitting it to others.

Here it is, summed up, in a few nut shells:

The best way to learn is to do — to ask, and to do.

The best way to teach is to make students ask, and do. Don't preach facts — stimulate acts.

The best way to teach teachers is to make them ask and do what they, in turn, will make their students ask and do.

Good luck, and happy teaching, to us all.

DEPARTMENT OF MATHEMATICS, INDIANA UNIVERSITY, BLOOMINGTON, IN 47401.

II. THE PROBLEM OF LEARNING TO TEACH — BY E. E. MOISE

It was a real pleasure to listen to Professor Halmos's talk. Seldom have I heard so much to agree with, and so much to applaud. He has given us a beautiful description of our task as teachers. And the description implied — as it had to — a wholesale rejection of the naive empiricism and naive behaviorism which have become an endemic plague in much of the educational world.

In the present state of our knowledge, teaching is an art. In mathematics, at least, attempts to turn it into a science have been retrogressive, in every case that I know of. Even when mathematics is taught poorly or only passably, we take for granted that students will have an opportunity to react to it in different ways, and to learn it at different levels, according to their own talents, temperaments, and motivations. At least, we *used* to take this for granted, until various people found ways to put a stop to it. I believe that the ultimate caricature of good mathematical teaching is linear error-free programming. Under this scheme, instead of taking care to ensure that every student is provided with the most stimulating challenges that he can react to successfully, people use their best efforts to create a situation in which nobody is faced with any challenge at all.

Certain ways of using "modules" have the same vice in a milder form. Some schools are now using a scheme under which courses are split up into small parts (the "modules"), with a standard test for each of them. When a student had passed the test on one module, he is ready to move on to the next. In some schools, at least, the rules prescribe that a student's grade at the end of the year is based on the number of modules that he has completed. Since the tests are of such a sort that almost any student can pass them eventually, the moral conveyed by all this is that an A-student is one who acquires a C-knowledge of mathematics at high speed. I suppose it is possible for a student in such a program to analyze ideas in depth, and to spend lots of time working on hard problems. But to behave in such a way, the student would have to resist the suggestions conveyed to him by the people who are receiving pay on the ground that they are promoting the student's intellectual development.

One of the difficulties with the pseudoscientific "learning theorists" is that they concentrate their attention on those aspects of the learning process that are capable of being meticulously observed and measured. Such a proceeding is not valid, or even safe; we simply don't know enough about learning processes to do anything predicated on the notion that our knowledge is complete. Curiously, there is empirical evidence against the validity of the empiricists' conception of learning.

In the early 1960's, Dr. Lyn Carlsmith (Harvard Educational Review, vol. 34 (1964), pp. 3–21) found a group of 20 male students, in the Harvard College class of 1964, whose fathers had gone overseas when their sons were no more than six months old, and had not returned until at least two years later. She then took a carefully matched control group of 20 male students whose fathers had not been absent in their early childhood. The SAT test was given to both groups. This test is in two parts, mathematical and verbal. Ordinarily, the difference $M - V$ of the mathematical and verbal scores M and V is positive for boys and negative for girls. The control group conformed to this expectation: in 18 cases out of 20, $M - V$ was positive. But in the "father-absent" group, $M - V$ was positive in only 7 cases out of 20. For a smaller group of 18 doctors' sons, similarly matched, the results were even more striking: in the control group, $M - V$ was positive in 7 cases out of 9, while in the father-absent group, $M - V$ was positive in only *one* case out of 9.

Further study of larger samples confirmed all this. Apparently, $M - V$ diminishes sharply as the duration of the father's absence increases; and the absence of the father in the *first six months* of a boy's life makes a significant difference in the relation between his SAT scores twenty years later.

These results are hard to reconcile with two views now widely held, namely, (1) intellectual capacities that seem to be purely cognitive really are, and (2) these capacities are acquired in ways that are readily accessible to empirical study. I believe that both these notions are not just inexact but very wide of the mark. It would be interesting to know just what it is that fathers teach their baby boys, and how the fathers go about it.

Obviously this study left important questions unanswered. For example, did the absence of the father inhibit the growth of mathematical faculties, or promote the growth of verbal ability, or both? (There is the prior question whether the "mathematical" part of the SAT test measures the sort of ability that produces a mathematician.) The study reminds us, however, of something that we should have known all along, that some of the most important learning processes go on when nobody is looking, and that they go on in ways that are very hard to keep track of. It is simplistic to suppose that people remember what they are told, and understand the things that are explained to them clearly. More commonly, people remember what interests them, and understand the things that they enjoy understanding. Thus intellectual development is linked with development of personality, and the refinement and enlargement of esthetic perceptions is a vital part of intellectual growth. This sort of growth does not lend itself to mechanization.

The processes by which people learn to teach are equally obscure. Some years ago — or so the story goes — a class in Social Relations at Harvard played an elaborate prank on their section man. The section man was in the habit of pacing back and forth while talking. In a secret caucus, the class agreed on a imaginary line, down the middle of the classroom. When the teacher moved to the left of the line, the class became eager and alert. When he stepped to the right of the line, the class became apathetic. When the class had taught the teacher to stay to the left of the center line, they gradually moved the line, until at the end of two or three weeks they had the teacher boxed into a corner. He had no idea of what was going on.

This was different, in a way, from the usual process under which students teach teachers to teach. But I think that the main difference is the students knew what they were doing. Ordinarily, I believe the process is unconscious for everybody.

This brings us, at last, to the question that I was supposed to be discussing at the outset: granted that teaching is an art, learned by experience, what can we do to help people to learn it? It seems to me that beginning teachers can probably get a great deal of help from policies which could easily be carried out in most departments.

One of the greatest troubles, I believe, in the initial teaching experience, is that the learning of teaching is virtually solitary. At the places that I know about, senior faculty members visit each teaching fellow's classes about once a semester. Even for purposes of evaluation, these procedures are perfunctory, and their value as teacher training is nil. It is hard to think of another art that people are expected to learn in such a way, with no significant help in the form of knowledgeable criticism.

This suggests that we should try to turn the learning of teaching into a group activity. I propose the following scheme. Beginning teachers would be organized in groups of about five, with *identical* teaching assignments, preferably a single course. They would share an office, so that they could conveniently discuss the problems that they all faced. Schedules would be arranged so that they could visit one another's classes. They would all meet, at least once a week, in a sort of "teaching seminar," to discuss what was going on. Each would have full responsibility for his own section, pacing the course to suit himself, subject only to the loose constraints imposed by the place that the course was supposed to fill in the curriculum. Each would make up his own assignments, and write his own hour tests and final examination. If a highly skilled senior faculty member formed part of the group, or met with them as an advisor, this would no doubt be helpful; but I believe that the senior man ought to be an advisor and not a boss. Classroom visits by peers would be much more frequent than visits by the advisor.

I see reason to hope that this sort of consultative effort would improve and vastly accelerate the process by which teachers learn by experience. Some features of it may need further explanation.

(1) I believe that there are such things as pedagogical principles. But even if we agree on what these are, they are hard to demonstrate, or even to convey, by abstract statements, and the art of putting them into practice takes quite a while to learn. I think that discussions of pedagogic questions are of immediate practical utility in proportion to their specificity. Hence an arrangement under which beginning teachers

would discuss not the general problems of education but rather, at a given moment, the problem of teaching a particular topic at a particular stage in a particular course. Under these conditions, I think that general ideas will emerge, in such forms that their meanings will be clear and the extent of their validity will be evident. This is why I think it vital for the teaching assignments to be identical; we need a situation in which the people discussing teaching problems have the same problems on their minds.

(2) If the group has a supervisor who tells everybody exactly what to do, he will almost certainly be telling some of them the wrong things. It is no part of any teacher's job to duplicate the performance of any other teacher however skilled. Teaching is an interpersonal relation, and optimal styles depend on the personalities of individuals. Such styles do and should change in response to class reactions.

(3) Moreover, if all important decisions are made by some higher authority, the beginning teacher will be less likely to come to grips, in his own mind, with the sort of problem that he will have to solve for himself in future years when the boss is gone. Hence the proposal that beginning teachers have full responsibility for their own courses, at a time when they have the benefits of consultation and criticism.

It seems likely that this scheme would amply repay the effort that it would require. Obviously, the only way to find out is to try it. I believe, however, that it involves at least one important hazard and has at least two important limitations.

First, it may be that working under observation, even by peers, will make people over-cautious, in an attempt to avoid the possibility of looking foolish. Probably this danger can be minimized if people are clearly aware of it. It seems especially important for the advisor to be aware of it, and for him to be of a gentle disposition.

Second, the whole scheme, in the form described, deals with fairly traditional teaching, in which the general content and method of the course are taken as given. This means that the skills acquired are only the beginning of professional maturation. The best courses that I know of were of the teacher's own design, and in some cases they were improvisations, whose outcome was not known even to the teacher at the outset. I believe, however, that fairly conventional teaching is a natural first step in professional development. This is a limit to the problems that one man can think about in one semester.

Finally, I don't think that we ought to feel complacent about our present lack of an adequate theory of teaching. If we had such a theory, we would be better off, and I think that one of the tasks of the coming generation is to create one. I have no idea of the form that such a theory might take. Perhaps its most likely inventors are people each of whom has a sophisticated and creative grasp both of mathematics and of psychology.

DEPARTMENT OF MATHEMATICS, QUEENS COLLEGE, FLUSHING, N. Y. 11367.

III. THE PROMOTION OF PARTICIPATION — BY GEORGE PIRANIAN

I address this to teachers of graduate and undergraduate students, to teachers in junior colleges, and perhaps to high-school teachers. Teachers in elementary schools already know what I have to say.

My colleagues have discussed ways to stimulate classroom participation. Paul Halmos has talked about participation by students, and Ed Moise proposes to inject life into the teachers. I shall try to reinforce their message with a story, and I'll mention a few relevant technicalities.

In 1967–68, The University of Michigan let me teach a section of the honors course in calculus. Because of a long period without freshman contacts, the prospect filled me with fear; but the students were a lovable lot, and we soon developed effective cooperation.

We had a solid book. Unfortunately, the author had taken himself a bit too seriously, and consequently the text was on the dreary side. To compensate, I regularly assigned special problems. For example, I asked the students to prove or disprove that if a real-valued function on the line is continuous at a point, then it is continuous throughout some neighborhood of that point. The subsequent classroom discussion of such a problem could chew up an entire period. But the course ran well, and I was so pleased that at the end I asked one of the girls to grade papers for me during her sophomore year.

In June, Addison-Wesley sent me a copy of Joseph Kitchen's *Calculus of One Variable*. Because the book looked lively, I thought we should try it, and to show my affection for the grader, I wrote to the publisher and requested that he send a copy to her home.

In September, when Lisa came to my office, I asked her opinion, and she said "It's just like the book we used, except that the Piranian problems are already in it." The students bought the book, and I looked forward to a great year.

After one week, I felt apprehensive, and soon I sensed the cold shadow of failure. Despite the excellent text and the bright students, the class sat glued to the runway. And then it happened that Kitchen skipped a point I consider important, and this forced me to devise a special problem for the occasion. The consequence was dramatic. With a roar of the engines and a slight shudder of the fuselage, we took off for the white clouds in the blue sky.

The moral is simple: no matter how sound, complete, and clear my text or lecture notes may be, the students should know that I'm developing the course especially for them, and that I'm turning myself inside-out in their behalf. For example, I must not assign homework by opening the text to page 93 and saying "for next time, try problems 3, 7, 10, 16, 19, and let me see, 21; class dismissed." We'll come back to this in a few minutes.

I must not give the impression of a man hired to teach as many students as possible and wired to do it with maximum industrial efficiency. I must indulge in extensive participation; the best way to achieve this is to recognize that this year's students require a new course, and that regardless of the cost, my section deserves special treatment. You can't teach with the left hand; you can't teach with the right hand. Like playing volley-ball, swimming, or racing a small sailboat, the job requires both hands, both arms, and the muscles of the legs and the torso.

The job takes more. I can preach an eloquent sermon on the gospel according to Darboux and Riemann, or give a spirited performance on Cantor sets, or use both hands and feet in a glorious axiomatic fugue, and yet reap substantial failure. No man can please all the people all the time, and no style of teaching is effective for all students. Therefore, successful teaching requires cooperation from the class.

You and I would know how to live, if we were young again. Meanwhile, multitudes of boys and girls suffer from awkwardness, uncertainty, and hesitation. Ask a dozen of your students with how many of their classmates they are acquainted, and you'll be astonished to learn about the bleakness and academic isolation in which some of them exist. A few years ago, I hit upon an unobtrusive way of sending a bit of mature wisdom across the generation gap. Early each term, I distribute a dittoed sheet listing the Ann Arbor addresses and telephone numbers of the entire class. This may encourage collaboration on homework; but it does not produce the miserable situation in which Archibald copies Merthiolate's paper fifteen minutes before it is due. Half of the class may meet for a great jam session. Leaders emerge, and the strong give guidance to the weak. I should share my salary with four or five students. They do some of my most difficult work, and I receive credit for their success. The kids learn to communicate, and when the homework is done, they may be so full of social steam that they go jogging together. If a few hundred of us were to trot from the Hilton to the Fisherman's Wharf, San Francisco would notice our physical condition and our social cohesion.

I've come back to homework. I do not know how to present mathematical ideas so effectively that students can take possession of them simply by sitting at my feet and smelling my socks. Let me change to a slightly less offensive metaphor: after grazing in my lush pastures, the students must ruminate; they must dedicate substantial time to the chewing of the cud. That's why we need homework.

Suppose now that our calculus text has a set of problems on integration by parts, a set on masses and centroids, a set on cylindrical and spherical coordinates. In each set, the problems range from the trivial to bread-and-butter drill, and they may end with a few important stinkers.

This is a reasonable arrangement of the text. A natural way of running the homework show is to assign problems from Set 23 today, problems from Set 24 tomorrow, and so forth. This is efficient for the teacher, for the students, and for the grader, and it is consistent with the principle of orderly progress. Nevertheless, the practice is a manifestation of pedagogic brutality. The poor boy who can barely manage

Problem 7 never gets the benefit of Problems 10, 16, and 19, except during a discussion that he endures passively because in his inexperienced view it comes too late to be of any use.

A more effective assignment for tomorrow might look like this:

Problem 18 in Set 22,

Problems 15 and 16 in Set 23,

Problems 9 and 12 in Set 24,

Problems 1 and 4 in Set 25.

Under this plan, the difficult problem comes after a week of experience with easier exercises in the same topic. The student profits from repeated exposure, and the teacher has several opportunities for clarifying the basic principles and demonstrating the necessary technique. A tough piece of meat calls for slow cooking, and a difficult idea requires thought on several consecutive days. Use the scheme of staggered assignments, tell the students that you've carefully programmed the homework for maximum effectiveness, and make certain that you're telling the truth.

Staggered homework is a small technicality; but it makes a difference. Also, it illustrates the dictum that genius is the capacity for taking trouble.

I urge the mathematical community to strengthen its pedagogical effort, not by buying new gadgets, not by creating new committees of experts, but by intensification of personal effort. Let each man assume the responsibility for teaching with greater vitality. If this reduces his rate of publication by thirty percent, so much the better. It will be good news for libraries, and it will help save Mathematical Reviews.

In the deliberations among the elders, the first question about a man should be how well he teaches, the second question, how good his publications are — never, how numerous.

I do not say this because we should create more mathematicians; there are enough of us. Nor am I concerned with the problem of generating stronger enrollment in mathematics classes to prevent economic dislocation of superannuated fuddy-duddies. There's one commodity that the world needs above everything else, and for which we'll never develop a satisfactory substitute. We need good men and women. As teachers, we have the desperately urgent task of communicating to the young some of the intellectual values of civilized mankind. We have the task of inspiring students to rise to the highest level of excellence that they can attain. Our survival depends on our collective success. I apologize for ending on such a serious note; but we face a problem of the utmost importance.

DEPARTMENT OF MATHEMATICS, UNIVERSITY OF MICHIGAN, ANN ARBOR, MI 48104.

A MOCK SYMPOSIUM FOR YOUR CALCULUS CLASS

DENNIS WILDFOGEL

Mathematics Program, Stockton State College, Pomona, NJ 08240

Dr. Gregory Campe, Professor of Organic Chemistry at Frostbite Falls (Minn.) State College. Dr. Linda Gillespe, Professor of Environmental Science at the Susan B. Anthony University for Women. Dr. Daniel Hain, Professor of Surfing at Could's Hole Oceanographic Institute. An unknowing colleague of mine at Stockton State College looked at the impressive roster of thirty-five participants in the First Stockton Symposium on Mathematical Modeling in the Life Sciences and exclaimed, "You got all those people to come here?" What he didn't know was that the people on the roster were the students in my Calculus For Life Scientists II class, given phony titles and affiliations at fictitious institutes.

Calculus For Life Scientists is a two-semester, introductory level calculus course aimed primarily at students majoring in environmental science, marine science, and biology. The course emphasizes mathematical modeling. At the "Symposium," which acts as a capstone to the students' year-long encounter with calculus, teams of students make presentations about a modeling problem on which they have been working for about two weeks. The benefits of doing these projects are numerous: (1) the projects give the students a holistic view of the use of higher mathematics in their own disciplines; (2) most projects require them to learn some techniques they would not ordinarily encounter in a one-year calculus course; (3) they learn about the benefits and difficulties of working on a team under a pressing deadline; (4) they experience the gamut of emotions associated with the problem-solving process, from the pleasure of initial idea generation, through the frustration of the intermediate stages, to the triumph of the completed project; (5) they gain experience in making oral presentations and written reports; (6) they see how the need for mathematical techniques grows out of realistic problems.

I have conducted such a symposium four times now. Students consistently rate it as one of their most rewarding and useful academic experiences. Several other faculty members have successfully adapted the symposium idea for use in their own classes.

Here's how it works. About two and a half weeks before the end of the term, I divide the class of thirty to forty students into teams of about six students each. Each team is given a fairly difficult modeling problem and is responsible for making both an oral presentation and a written report about the results of their investigation of that problem. Each team meets during class time for the remainder of the term while I give hints and feedback on their work. The teams invariably find it necessary to schedule meetings outside of class time, too. I always offer enough suggestions so that each team will develop a good model by the time of the symposium.

The key aspect of running a symposium like this is the selection of appropriate problems for the teams. Each problem must be sufficiently challenging to occupy a team of six students for two weeks and yet still be within reach of their capabilities. There are a few texts ([1]-[3]) which I have used repeatedly as sources for problems. In a few instances I have been able to adapt material from books or journals in other fields, and I have made up several problems on my own. Colleagues in the life sciences have provided a great deal of assistance. Below is a partial list of titles of problems I have used.

Competition of two species for limited resources
Biogeography: a species equilibrium model
The maximum brightness of Venus
Operating strategies for publicly owned commuter bus systems
A box model for airshed pollutant capacity estimation
The effects of natural selection on gene frequency
Passive transport of chemical substances through a thick section of tissue
Determination of the shape of subterranean deposits by use of gravitational anomalies

Parasitic relationships which are not harmful to the host
Ventilation systems and the accumulation of toxic pollutants
A model for the clinical detection of diabetes
A rare example of a closed ecological system
The chemical kinetics of bimolecular reactions
Excretion of a drug
An optimal inventory policy model for an import wholesaler

The composition of the student teams is important. I make sure that each student works on a project in an area of his or her own interest. I have tried dividing the class into groups homogeneous or heterogeneous according to ability. The homogeneous groups make it easier to tailor the difficulty of the problem to the appropriate level; however, it is difficult to keep the least capable groups from becoming discouraged. In heterogeneous groupings, the less successful students can learn from the better ones; however, too much of the burden then falls on the better students in each group. The best compromise I have found so far is to have two "all-star" teams of the best students, and to have heterogeneous teams composed of the remaining students.

The symposium itself occupies the last three days of the term. Each team makes a half-hour presentation. The entire event, always attended by several other faculty members, is done up in tongue-in-cheek style. I circulate in advance to all mathematics and science faculty members a Roster of Participants and a Schedule of the Symposium, making it look as much like a real scholarly meeting as possible. (I always fool at least one new faculty member!)

At the beginning of each session, one of my mathematics or science colleagues makes a humorous presentation, e.g., a "double talk" address that sounded like a commentary on the specific models to be presented, or a short discourse on the three-body problem while juggling three balls. Once Miss America visited, and another time I sang a song about calculus which I composed. The student teams get into the act, dressing up in suits and ties or lab coats, calling each other "Doctor" or "Professor," and occasionally putting on brief skits. The merriment makes it enjoyable without detracting from the serious work to be done and serves to alleviate some of the anxiety the students have about making oral presentations. Afterwards, each student receives a copy of the "Proceedings of the Symposium" containing the written reports of the several teams and a few memorable photographs of the symposium.

The symposium creates an opportunity for students to understand the way mathematics is actually used in their own fields and to understand both their own potential as users of mathematics and the difficulties inherent in the modeling process. It is thus a valuable experience in their mathematical education.

References

1. M. Braun, Differential Equations and Their Applications: An Introduction to Applied Mathematics, 2nd ed., Springer-Verlag, New York, 1978.

2. D. P. Maki and M. Thompson, Mathematical Models and Applications: With Emphasis on the Social, Life, and Management Sciences, Prentice-Hall, Englewood Cliffs, N. J., 1973.

3. E. O. Wilson and W. H. Bossert, A Primer of Population Biology, Sinauer Associates, Stamford, Conn., 1971.

Calculus by Mistake

LOUISE S. GRINSTEIN

LOUISE S. GRINSTEIN is Professor of Mathematics at Kingsborough Community College of the City University of New York, where she has been teaching since 1966. Her Ph.D. in mathematics education was received from Columbia University in 1965. In addition to pedagogical experience, Professor Grinstein has worked in industry as a computer programmer and systems analyst. She has also contributed articles to various professional journals.

The analysis of mistakes is important in the teaching of mathematical concepts. Too often, beginning students accept without question any proof presented in print or on the blackboard. Fostering a critical approach to mathematics is a necessary and important part of the teacher's function. Some examples of mistaken reasoning in calculus are presented and briefly discussed in this article. The "proofs" are given initially without comment. (The superscripts which appear in the examples refer to the Notes and References at the end of the article.)

Why? What's Wrong?

(1)[1] *The derivative of any function is zero.* The derivative of any function is defined at a value of x. The function is constant at that value of x. Hence the derivative is zero because the derivative of a constant is zero.

(2)[2] *2 = 1.* Let

$$x^2 = (x)(x), (\text{i.e., } xx\text{'s}) = x + x + \cdots + x,$$

a total of x addends;

$$\frac{d(x^2)}{dx} = \frac{d(x + x + \cdots + x)}{dx}$$

$$2x = 1 + 1 + \cdots + 1$$

$$2x = x \text{ and therefore } 2 = 1.$$

(3)[3] *No point on the hyperbola $x^2 - y^2 = a^2$ is closest to the origin.* (An analogous theorem is: *No point on the circle $x^2 + y^2 = r^2$ is closest to or farthest from a given point inside the circle.*[4]) Let $L = \sqrt{x^2 + y^2}$ be the distance of the point from the origin. Substitute the value of y:

$$L = \sqrt{2x^2 - a^2}$$

and

$$dL/dx = 2x/\sqrt{2x^2 - a^2}.$$

Let $dL/dx = 0$. Then $x = 0$ and y is imaginary. Let dL/dx be undefined. Then $x = +a/\sqrt{2}$ and y is imaginary. Therefore, there are no closest points. (Note: it is

47

obvious, however, that the points $(\pm a, 0)$ are nearer the origin than any other point.

(4)[5] $0 = 1$. A student integrates $\int \sin x \cos x \, dx$ by letting $u = \sin x$. He obtains $\frac{1}{2} \sin^2 x + C$. He notices that the same integral equals $= -\frac{1}{2} \cos^2 x + C$ by letting $u = \cos x$. Equating these answers gives

$$\tfrac{1}{2} \sin^2 x = -\tfrac{1}{2} \cos^2 x \text{ or } \sin^2 x + \cos^2 x = 0.$$

From trigonometry, however, $\sin^2 x + \cos^2 x = 1$. Thus, $0 = 1$.

(5)[6] $0 = 1$. Evaluate $\int dx/x$ by parts. Let $u = 1/x$ and $dv = 1 \, dx$. Thus $\int dx/x = 1 + \int dx/x$ or $0 = 1$. (Analogous results may be obtained by evaluating[7] $\int \log x \, dx/x$ or[8] $\int \cot x \, dx$ by parts).

(6)[9] $(x^2 + 1)/(x^2 - 1) \equiv 2/(x^2 - 1)$. To break up $(x^2 + 1)/(x^2 - 1)$ into partial fractions, a student assumes that

$$(x^2 + 1)/(x^2 - 1) = A/(x - 1) + B/(x + 1).$$

Then $x^2 + 1 = A(x + 1) + B(x - 1)$. Setting $x = 1$ yields $A = 1$ while for $x = -1$, $B = -1$. Thus, $(x^2 + 1)/(x^2 - 1) = 1/(x - 1) - 1/(x + 1) = 2/(x^2 - 1)$.

(7)[10] *The length of an arch of the cycloid is zero.* The parametric equations of the cycloid are

$$x = t + \sin t; \qquad y = 1 + \cos t.$$

An arch is covered by values of t running from 0 to 2π. Let s denote length of arc, then $s = \int_0^{2\pi} s' \, dt$.

$$s' = \sqrt{(1 + \cos t)^2 + (-\sin t)^2}$$

$$= \sqrt{1 + 2\cos t + \cos^2 t + \sin^2 t}$$

$$= \sqrt{2 + 2\cos t} = 2\cos \tfrac{1}{2}t.$$

Thus $s = \int_0^{2\pi} 2\cos \tfrac{1}{2}t \, dt = 4 \sin \tfrac{1}{2}t \,]_0^{2\pi} = 0$

(8)[11] $\pi = 0$. Evaluate $\int_0^{2\pi} f(\theta) \cos \theta \, d\theta$, where $f(\theta)$ is any function of θ. Let $\sin \theta = t$, then $\cos \theta \, d\theta = dt$; $f\{\sin^{-1} t\} = g(t)$; $\theta = 0$ implies $t = 0$, $\theta = \pi$ implies $t = 0$. Thus $\int_0^\pi f(\theta) \cos \theta \, d\theta = \int_0^0 g(t) \, dt = 0$. Now, consider the special case where $f(\theta) = \cos \theta$. It follows that

$$\int_0^\pi \cos^2 \theta \, d\theta = 0.$$

However,

$$\int_0^\pi \cos^2 \theta \, d\theta = \tfrac{1}{2} \int_0^\pi (1 + \cos 2\theta) \, d\theta$$

$$= \left[\tfrac{1}{2}\theta + \tfrac{1}{4} \sin 2\theta\right]_0^\pi = \pi/2.$$

Therefore $\pi = 0$.

(9)[12] $\pi = 0$. Consider $\int_{-1}^{1} dx/(1 + x^2)$. Let $x = 1/t$, then $dx = -dt/t^2$. Thus

$$\int_{-1}^{1} dx/(1 + x^2) = -\int_{-1}^{1} \frac{dt/t^2}{1 + 1/t^2}$$

$$= -\int_{-1}^{1} dt/(1 + t^2)$$

and $\int_{-1}^{1} dx/(1 + x^2) = 0$. By use of the standard formula

$$\int_{-1}^{1} dx/(1 + x^2) = \tan^{-1} x \Big]_{-1}^{1} = \pi/2.$$

Therefore $\pi = 0$.

(10)[13] $2 = 1$. Let $f(x)$ be any given function. Then:

$$\int_{1}^{2} f(x)\, dx = \int_{0}^{2} f(x)\, dx - \int_{0}^{1} f(x)\, dx$$

Letting $x = 2y$ in the first integral on the right:

$$\int_{0}^{2} f(x)\, dx = 2\int_{0}^{1} f(2y)\, dy$$

$$= 2\int_{0}^{1} f(2x)\, dx.$$

Take $f(x)$ such that $f(2x) \equiv \tfrac{1}{2} f(x)$ for all values of x. Then

$$\int_{1}^{2} f(x)\, dx = 2\int_{0}^{1} \tfrac{1}{2} f(x)\, dx - \int_{0}^{1} f(x)\, dx$$

$$= 0.$$

Now $f(2x) \equiv \tfrac{1}{2} f(x)$ is satisfied by $f(x) = 1/x$. Thus, $\int_{1}^{2} dx/x = 0$, so that $\log 2 = 0$ or $2 = 1$.

(11)[14] *An integral that is simultaneously zero and undefined.*

$$\int_{-\infty}^{\infty} 4x^3\, dx/(1 + x^4) = \lim_{t \to \infty} \int_{-t}^{t} 4x^3\, dx/(1 + x^4)$$

$$= \lim_{t \to \infty} \log(1 + x^4)\Big]_{-t}^{t}$$

$$= \lim_{t \to \infty} \left[\log(1 + t^4) - \log(1 + t^4)\right] = 0.$$

However, $\int_{-\infty}^{\infty} 4x^3\, dx/(1 + x^4) = \int_{-\infty}^{0} 4x^3\, dx/(1 + x^4) + \int_{0}^{\infty} 4x^3\, dx/(1 + x^4)$. Now $\int_{0}^{\infty} 4x^3\, dx/(1 + x^4) = \lim_{t \to \infty} \log(1 + x^4)]_{0}^{t} = \lim_{t \to \infty} \log(1 + t^4)$, which does not exist, thus implying that the original integral does not exist.

(12)[15] $0 < 0$. Take $f(x) = (1 + e^{1/x})^{-1}$ and $g(x) = e^{1/x}(x + xe^{1/x})^{-2}$. Then $g(x) \geq 0$ and $f'(x) = g(x)$. Now: $0 \leq \int_{-1}^{1} g(x)\, dx = f(1) - f(-1) = (1 - e)/(1 + e) < 0$. Thus, $0 < 0$.

(13)[16] $0 = 1$. To evaluate $\lim_{x \to 0}(3x^2 - 1)/(x - 1)$, apply l'Hôpital's rule:

$$\lim_{x \to 0} (3x^2 - 1)/(x - 1) = \lim_{x \to 0} 6x/1 = 0.$$

From the original function, however, it can be seen that as x approaches zero, the function approaches one. Thus, $0 = 1$.

(14)[17] $2 = 1$. It is known that

$$\log 2 = 1 - \tfrac{1}{2} + \tfrac{1}{3} - \tfrac{1}{4} + \tfrac{1}{5} - \cdots .$$

Let

$$S = 1 + \tfrac{1}{2} + \tfrac{1}{3} + \tfrac{1}{4} + \tfrac{1}{5} + \cdots .$$

Thus

$$\log 2 + S = 1 - \tfrac{1}{2} + \tfrac{1}{3} - \tfrac{1}{4} + \cdots + 2(\tfrac{1}{2} + \tfrac{1}{4} + \tfrac{1}{6} + \cdots)$$
$$= 1 + \tfrac{1}{2} + \tfrac{1}{3} + \tfrac{1}{4} + \tfrac{1}{5} + \cdots$$
$$= S.$$

Therefore $\log 2 = 0$ and $2 = 1$.

(15)[18] *A series that converges and diverges simultaneously.* Given the series

$$\tfrac{1}{2} + \tfrac{1}{5} + \tfrac{1}{8} + \tfrac{1}{11} + \cdots , \tag{1}$$

each term is less than the corresponding term of the series

$$1 + \tfrac{1}{2} + \tfrac{1}{4} + \tfrac{1}{8} + \cdots . \tag{2}$$

Series (2) converges because it is a geometric series with common ratio $\tfrac{1}{2}$ and thus (1) converges. Each term of (1) is greater than the corresponding term of the series

$$\tfrac{1}{3} + \tfrac{1}{6} + \tfrac{1}{9} + \tfrac{1}{12} + \cdots .$$

Series (3) diverges because it is a nonzero multiple of the harmonic series. Thus (1) diverges.

(16)[19] $0 = 1 = \tfrac{1}{2}$. Let $S = 1 - 1 + 1 - 1 + \cdots$. Then

(a) $S = (1 - 1) + (1 - 1) + (1 - 1) \cdots = 0$,
(b) $S = 1 - (1 - 1) - (1 - 1) - (1 - 1) \cdots = 1$,
(c) $S = 1 - (1 - 1 + 1 - 1 + \cdots) = 1 - S$ and $S = \tfrac{1}{2}$.

(17)[20] *A sum of positive terms can be negative.* Let

$$S = 1 + 2 + 4 + 8 + \cdots ,$$
$$2S = 2 + 4 + 8 + 16 + \cdots .$$

Thus $2S = S - 1$ and $S = -1$.

(18)[21] $24 = 0$. To find the volume cut from the sphere $x^2 + y^2 + z^2 = 9$ by the cylinder $x^2 + y^2 = 3x$, use cylindrical coordinates. The element of volume is thus a column of cross-section dr by $r\,d\theta$ and of height $z = \sqrt{9 - r^2}$. Considering symmetry with respect to the xy plane only, the volume integral becomes

$$V_1 = 2 \int_{-\pi/2}^{\pi/2} \int_0^{3\cos\theta} (9 - r^2)^{1/2} r \, dr \, d\theta$$

$$= 18\pi.$$

Considering symmetry with respect to both the xy- and xz-planes, the volume integral becomes:

$$V_2 = 4 \int_0^{\pi/2} \int_0^{3\cos\theta} (9 - r^2)^{1/2} r \, dr \, d\theta$$

$$= 18\pi + 24.$$

Thus, $24 = 0$.

(19)[22] $\log 4 = 6/5$. To compute the area bounded by the curves $y = x/(x^2 + 1)$ and $5y = x$ by double integration:

$$A_1 = 2 \int_0^2 \int_{x/5}^{x/(x^2+1)} dy \, dx = \log 5 - 4/5.$$

Integration with respect to x first, however, yields:

$$A_2 = 2 \lim_{\epsilon \to 0} \int_\epsilon^{2/5} \int_{\frac{1}{2y} - \frac{\sqrt{1-4y^2}}{2y}}^{5y} dx \, dy$$

$$= 2/5 - \log(4/5).$$

Thus, $\log 5 - 4/5 = 2/5 - \log(4/5)$ or $\log 4 = 6/5$.

Clues

(1) The derivative of a function is defined at a fixed value of x and the function is constant at that value of x. If the function varies as x varies, it is not constant for all x and therefore its derivative need not be zero.

(2) The definition of multiplication as repeated addition is valid only for integers. Thus, the function is not continuous. Differentiation is meaningless for this definition.

(3) The domain is x is $\{x : |x| \geq a\}$. At the endpoints $x = \pm a$, L is minimum without the necessity of L' being zero or undefined.

(4) The constants of integration have been ignored.

(5) Same reason as #4.

(6) The original assumption leads to the erroneous identity:

$$x^2 + 1 \equiv A(x + 1) + B(x - 1).$$

This is impossible because one side is quadratic and the other linear.

(7) It is to be noted that:

$$(s')^2 = 4\cos^2 \tfrac{1}{2}t$$

implies

$$s' = 2\cos\tfrac{1}{2}t \quad \text{for } 0 \leq t \leq \pi$$
$$= -2\cos\tfrac{1}{2}t \quad \text{for } \pi \leq t \leq 2\pi.$$

(8) It is to be noted that:

$$\cos^2\theta = 1 - t^2$$

implies

$$\cos\theta = +\sqrt{1 - t^2} \text{ for } 0 \leq \theta \leq \tfrac{1}{2}\pi$$
$$= -\sqrt{1 - t^2} \text{ for } \tfrac{1}{2}\pi \leq \theta \leq \pi.$$

(9) The substitution $x = 1/t$ is meaningless for $t = 0$. Thus, the new form of the integral $\int_{-1}^{1} f(t)\, dt$ contains a discontinuity which has been ignored.

(10) The integral $\int_0^1 f(x)\, dx$ does not exist when $f(x) \equiv 1/x$.

(11) The quantities t and $-t$ must approach ∞ and $-\infty$ independently of each other.

(12) the fact that the integrand $g(x)$ has a discontinuity at $x = 0$ has been ignored.

(13) The original function is not an indeterminate form. Therefore l'Hôpital's rule does not apply.

(14) The result is based on the invalid assumption that S exists.

(15) The next terms in the given series are $1/14, 1/17, 1/20, \ldots$ which are greater, term by term, than the next terms in the geometric series $1/16, 1/32, 1/64, \ldots$.

(16) In (a), grouping gives only the even-numbered partial sums of the original series. In (b), grouping gives only the odd-numbered partial sums of the original series. The error in (c) is the same as that in #14.

(17) The same reason as #14.

(18) The answer $18\pi + 24$ is correct. The discrepancy in answers arises in the following:

$$\int_0^{3\cos\theta} (9 - r^2)^{1/2} r\, dr\, d\theta = \left[-(9 - r^2)^{3/2}/3 \right]_0^{3\cos\theta}$$
$$= -(9 - 9\cos^2\theta)^{3/2}/3 + 9$$
$$= 9 - 9(1 - \cos^2\theta)^{3/2} = 9 - 9\sin^3\theta.$$

The expression $(1 - \cos^2\theta)^{3/2}$ is a nonnegative quantity. Its "equivalent", $\sin^3\theta$, is positive in quadrants I and II but negative in quadrants III and IV. V_2 gives the correct result, since the limits from 0 to $\tfrac{1}{2}\pi$ do not go outside of the first quadrant.

(19) It is to be noted that $y = x/(x^2 + 1)$ has a maximum at $x = 1$, $y = \tfrac{1}{2}$. A_2 neglects the portion of area bounded by $y = x/(x^2 + 1)$ and $y = 2/5$ as well as that between $y = x/(x^2 + 1)$ and $y = -2/5$. Thus, the correct answer is $A_1 = \log 5 - 4/5$.

The "whys" presented here are not an exhaustive list. In addition to the entertainment value, these examples should provide a springboard for deeper understanding and appreciation of calculus concepts. In this way, they should help bring an increased awareness of desired precision in mathematical reasoning.

NOTES AND REFERENCES
(Sources also explain errors.)

1. M. Kline, Calculus: An Intuitive and Physical Approach, Wiley, New York, 1967, Part I, 34, Instructor's Manual, 4.
2. Quickie #459, Math. Mag., 42, 203, 225. Also, Math. Mag., 43, 173–4.
3. Problem #2679, School Science and Mathematics, 59, 748.
4. M. Kline, op. cit., Part I, 222, Instructor's Manual, 39. Also, E. A. Maxwell, Fallacies in Mathematics, Cambridge University Press, 1963, 55.
5. M. Kline, op. cit., Part I, 247, Instructor's Manual, 44. Also, R. A. Bonic et al., Freshman Calculus, Heath & Co., Massachusetts, 1971, 185–6.
6. M. Kline, op. cit., Part I, 392, Instructor's Manual, 70. Also, E. A. Maxwell, op. cit., 65.
7. M. Kline, op. cit., Part I, 392, Instructor's Manual, 70. Also, comment to Falsie #21, Math. Mag., 37, 360.
8. Falsie #21, Math. Mag. 37, 62.
9. M. Kline, op. cit., Part I, 399, Instructor's Manual, 72.
10. E. A. Maxwell, op. cit., 69.
11. E. A. Maxwell, op. cit., 66.
12. Problem #3251, School Science and Mathematics, 70, 472.
13. E. A. Maxwell, op. cit., 65.
14. R. A. Kurtz, Calculus Supplement: An Outline with Solved Problems, Benjamin, New York, 1970, 113.
15. Problem #35, The MATYC Journal, 7, #2, 35.
16. M. Kline, op. cit., Part I, 381, Instructor's Manual, 66.
17. Problem #1321, School Science and Mathematics, 34, 435.
18. M. Kline, op. cit., Part II, 125, Instructor's Manual, 119.
19. M. Kline, op. cit., Part II, 119, Instructor's Manual, 119.
20. Ibid.
21. Problem #E2, Amer. Math. Monthly, 40, 112.
22. Falsie #16, Math. Mag., 33, 237.

Testing Understanding and Understanding Testing

Jean Pedersen
Peter Ross

Jean Pedersen is a member of the Mathematics Department at the University of Santa Clara. For several years she has been involved in many MAA activities. She has served in the WAM program, the Visiting Lectureship Program, and on the CTUM committee. She is currently an associate editor of Mathematics Magazine and the Governor of the Northern California Section of the MAA. Her principal research interests are in polyhedral geometry, combinatorics and mathematics education, in which fields she has published numerous articles and several books. Her most recent book Fear No More: An Adult Approach to Mathematics (Addison-Wesley, 1983) is co-authored with Peter Hilton. This book is the first of a series of 3 volumes designed to display a new approach to mathematical exposition.

Peter Ross interrupted his academic program (B.S. at M.I.T., M.A. and Ph.D. at the University of California at Berkeley) twice, the first time to teach as a Peace Corps volunteer in secondary schools in India. The second interruption consisted of his working for three years as a teacher-writer on a new, "new math" program, the Comprehensive School Mathematics Project. Now in the Mathematics Department at the University of Santa Clara, he has taught at four other universities and his activities included writing self-paced calculus materials for the University of California at Santa Barbara. As his background might suggest, his mathematical interests are diverse and eclectic.

In this article, we try to deal with the Do-I-have-to-memorize-this-for-the-next-test? syndrome. The problem which we consider—one which is certainly faced by most of our colleagues—is that of teaching our students the *meaning* of mathematical concepts, so that they will be able to use those concepts in new situations. In particular, we do *not* want them simply to memorize theorems or formulas, and mindlessly execute sets of routine exercises.

One way to change the students' behavior (and, we hope, their attitude toward learning mathematics) is to include on tests—and, of course, in class—some questions that require deeper understanding of concepts than formula memorization or mechanical symbol manipulation.

Our objective here is to share some examples appropriate for a beginning course in calculus and analytic geometry. Each example involves a graph of some sort, either in the statement of the problem or in the solution. The use of an appropriate graphical representation often casts a bright light on mathematical concepts that other means leave in darkness.

On several problems (e.g., The Mean Value Theorem and Newton's method) it may be advisable to tell students they may use a piece of folded paper or a ruler as a straightedge. Students are often surprised at being asked to do ordinary constructions in a class that isn't called 'plane geometry.'

The Mean Value Theorem

(i)   Complete the following statement: The mean value theorem implies that if $f'(x)$ exists for $a \leqslant x \leqslant b$, then there exists at least one number c between a and b such that _____.

(ii)   How many such numbers c are there for the function shown below? Mark on the x-axis approximately where each such number c is.

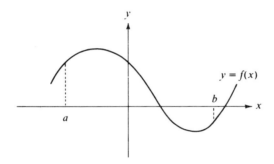

Newton's Method

1.   Suppose you are using Newton's method to find a root of $f(x) = 0$ for the function f shown below. If x_1 indicates where your first approximation is, label where (approximately) your second approximation x_2 and third approximation x_3 will be. Your completed illustration should indicate *how* you located x_2 and x_3.

2.   If Newton's method is applied to $f(x) = 3x + 4$, and x_1 is chosen to be 15, what will x_7 be? (Hint: Think geometrically.)

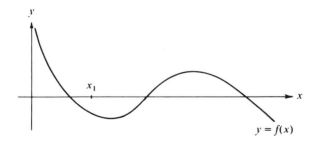

Properties of the Definite Integral

1. Assume $f(x)$ is continuous for $g \leqslant x \leqslant c$ as shown.
    (i)   Write an equation relating the 3 quantities

$$\int_a^b f(x)\,dx, \qquad \int_a^c f(x)\,dx, \qquad \int_b^c f(x)\,dx.$$

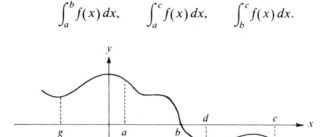

(ii)   Is the equation you wrote for part (i) still true
    (A)   if b is replaced by d?
    (B)   if b is replaced by g?
2. The graph of $f(x)$, shown below, consists of two straight line segments and two quarter circles. Find the value of $\int_0^{16} f(x)\,dx$.

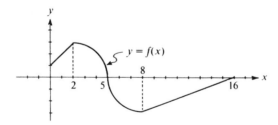

The Trapezoidal Rule for Approximating $\int_a^b f(x)\,dx$

Suppose that $f(x) > 0$ and $f''(x) > 0$ for all x between a and b, where $a < b$. What can be said about a trapezoidal approximation T for the integral $\int_a^b f(x)\,dx$?

(i) $T < \int_a^b f(x)\,dx$

(ii) $T > \int_a^b f(x)\,dx$

(iii) Can't say which is larger.

This question could be specialized by using particular data. For example, let $f(x) = x^2$. A possible follow-up question would be to ask students to find other functions satisfying the hypotheses and to test the *size* of the error for various subdivisions (on a computer, if possible). The error in any numerical approximation is an important concept to emphasize, especially since the availability of computers has made the computations much easier to obtain. Nevertheless, it is important to know just how long it is going to take the computer to arrive at a sufficiently close approximation.

Graphs of Functions with the Same Derivative

Sketched below is the graph of a function f. Suppose another function g has the following properties: $g(-1) = -2$, $g'(x) = f'(x)$ for all x. Sketch the graph of g, using the same axes.

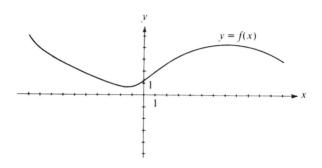

The Significance of the Signs of $f(x)$, $f'(x)$, $f''(x)$

Indicate which of the graphs below could depict a function f that has the following three properties.

$$f(c) > 0, \quad f'(c) < 0, \quad \text{and} \quad f''(c) > 0,$$

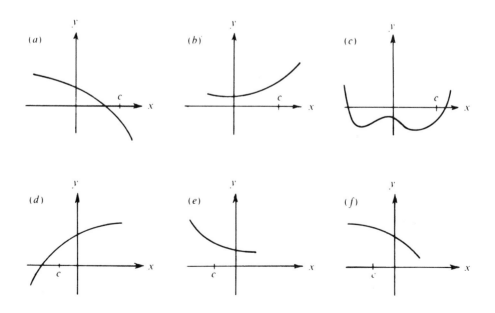

Estimation Involving Functions Defined by Their Graphs

This graph shows the average weights of boys and girls as functions of their ages.

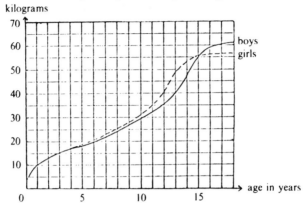

Use the graph to answer the questions below.
 (i)   Estimate the age at which teenage boys and girls weigh the same.
 (ii) Who grows faster (with respect to weight): 14-year-old girls or 14-year-old
       boys?
 (iii) Estimate the *average* rate of change of boys' weights between the ages of 10
       and 15.
 (iv) Estimate the *instantaneous* rate of change of boys' weights at the instant
       when they become 10 years old.
 (v)   Suppose that sociologists define puberty* as the age at which the fastest rate
       of change of weight occurs. Estimate:
       (A) puberty for girls, and (B) puberty for boys.
The above graph is taken from [3], which contains other graphs of a similar nature.

Slopes of Lines and Curves

 1.   The five numbers below are slopes of the five lines shown:

$$-\frac{1}{4}, \qquad \frac{2}{5}, \qquad 0, \qquad 3, \qquad -\frac{21}{6}.$$

Write the appropriate slope of each line in the blank space provided.

line	slope
l_1	
l_2	
l_3	
l_4	
l_5	

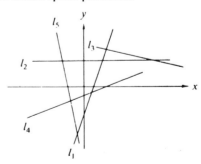

---

*As a point of interest, the Random House Dictionary states that "... in common law, (*puberty is*)
presumed to be 14 years in the male and 12 years in the female."

2. Line L is tangent to the curve $y = f(x)$ at the point $(3, 5)$.

    (i)   Find $f(3)$

    (ii)   Find $f'(3)$

    (iii) Find $f(0)$.

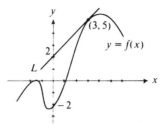

For problems similar to this one see Sections 1.2 and 1.3 of [2].

The Quadratic Formula

Suppose that $f(x) = ax^2 + bx + c$, where the numbers a, b, c satisfy $a < 0$ and $b^2 - 4ac = 0$. Which of the six parabolas shown could be the graph of $f(x)$?

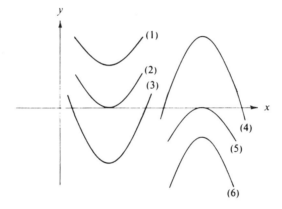

This question forces students to connect the quadratic formula with a geometrical interpretation of the graph of a quadratic function.

Graphing a Familiar Formula

If you were to measure the circumference C and the radius r of many circles:

    (i)   What would the graph of C as a function of r look like?

    (ii)   What would be its slope and intercepts?

Graphs of Distance (or Time) Versus Speed

Suppose you plan to drive at a constant speed from Philadelphia to Boston by way of New York City. Here are four variables that will depend on the speed you choose:

    y_1: your distance from Philadelphia after 1 hour of travel

    y_2: the time you require to go 100 miles

    y_3: your distance from New York City after 1 hour of travel

    y_4: the distance you still have to go after 1 hour of travel

Below are four graphs. Each of the four variables y_1, y_2, y_3, y_4 is graphed as a function of the speed in exactly one of the graphs. Match the variable with the appropriate graph; that is, identify the vertical axes in each of the graphs as one of the variables y_1, y_2, y_3, y_4.

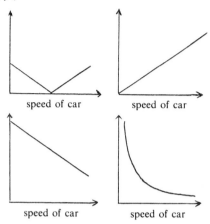

Remarks. (1) In case the reader experienced any difficulty with this problem perhaps it is due to the choice of speed as the independent variable. Once again, we are reminded of the confusion our calculus students may have with problems that are "routine" (to us!). (2) This problem is a modification of one used in a diagnostic quiz for an introductory statistics course at the University of California, Berkeley [1].

The Graphs of sin x and cos x

When the graphs of $y = \sin x$ and $y = \cos x$ are drawn on the same axes (i.e., in the same coordinate system), how many times do they intersect in the interval $0 \leqslant x \leqslant 100\pi$? (Had we asked for the interval $0 \leqslant x \leqslant 6\pi$, some students might not have thought of using periodicity. Asking a 'harder' problem sometimes suggests to students that they must do some analytical thinking.)

Graphing New Functions Related to Known Functions

Below is sketched the graph of $f(x)$. Sketch the graphs of the following functions:

(i) $y = f(x) - 2$

(ii) $y = 2f(x)$

(iii) $y = f(x - 2)$

(iv) $y = f(x + 2)$

(v) $y = f(2x)$.

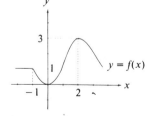

Discussing problems like this helps students see the generality of these concepts, and thus enables them to graph other functions such as $a \sin(bx + c)$, $a \ln(bx + c)$, $ae^{bt/c}$—provided they are familiar with the graphs of $\sin x$, $\ln x$, e^t, respectively.

Postscript

Students should be forewarned that conceptual test questions of this nature are coming. Such test questions should seem reasonable if they reflect classroom instruction—as, for example, when class time is spent on questions which require deeper understanding of the material covered in previous classes. This teaching technique not only provides a good review, but also encourages students to think more about the *meaning* of the new material as it is presented. We are convinced that time spent on these types of graphical problems is often more valuable than time spent on non-pictorial explanations, however lucid the latter may seem to the instructor.

REFERENCES

1. David Freedman, Robert Pisani, and Roger Purves, Instructor's Manual for Statistics, W. W. Norton and Company, 1978, p. 118.
2. Larry J. Goldstein, David C. Lay, and David I. Schneider, Calculus and Its Applications, Second Edition, Prentice-Hall, 1980.
3. EDC/Project CALC, Module I, Integration, second edition, July, 1975, p. 111.
4. Hugh Burkhardt, Vern Treilibs, Kaye Stacey, and Malcolm Swan, Beginning to Tackle Real Problems, Shell Centre for Mathematical Education, University of Nottingham, 1980.

The authors most sincerely wish to thank Peter Hilton for the very precise, perceptive, and pertinent comments he gave us during the preparation of this paper.

BIBLIOGRAPHIC ENTRIES: PEDAGOGY

1. *Monthly* Vol. 78, No. 7, pp. 789–791. Neil Davidson, The small group-discussion method as applied in calculus instruction.

Describes a pilot study with a freshman class.

2. *Monthly* Vol. 80, No. 2, pp. 195–201. M. W. Ham, The lecture method in mathematics: a student's view.

3. *Monthly* Vol. 80, No. 3, pp. 302–307. E. A. Bender, Teaching applicable mathematics.

4. *Monthly* Vol. 80, No. 8, pp. 937–942. Jerry Silver and Bert Waits. Multiple-choice examinations in mathematics, not valid for everyone.

5. *Monthly* Vol. 83, No. 5, pp. 370–375. Kathleen Sullivan, The teaching of elementary calculus using the nonstandard analysis approach.

6. *Monthly* Vol. 83, No. 5, pp. 375–378. Margaret S. Menzin, Use of canned computer programs in freshman calculus.

7. *Monthly* Vol. 94, No. 8, pp. 776–785. Report of the CUPM panel on calculus articulation.

8. *Monthly* Vol. 96, No. 4, pp. 350–354. Martha B. Burton, The effect of prior calculus experience on "Introductory" college calculus.

9. TYCMJ Vol. 6, No. 4, pp. 29–35. Samuel Goldberg, A precalculus unit on area under curves.

10. CMJ Vol. 21, No. 1, pp. 2–19. George D. Gopen and David A. Smith. What's an assignment like you doing in a course like this?: Writing to learn mathematics.

3

FUNCTIONS
(a)

CONCEPTS

ON THE NOTION OF "FUNCTION"

G. J. MINTY, Indiana University

I recently became suspicious that my students in Differential Equations did not fully appreciate the existence- and uniqueness-theorems for the usual initial-value problem for $y' = f(x, y)$ because they didn't understand what f is. I asked them to define "function," and they (at least, some of them) gave me the standard "set of ordered pairs" definition. Pushing the point further, I asked on a quiz: Which of the following can be interpreted as differential equations of the form $f(y', y, x) \equiv 0$:

$$(1) \int_0^{y'} e^{-(xt)^2} dt = 0; \quad (2) \int_0^1 \{[y'(x)]^2 + y(x)\} dx = 0; \quad (3) \ y'(y(x)) = 0?$$

The answers I received were completely random. All this makes me feel like a grade-school teacher asking his pupils "What is an integer?" and getting back Peano's axioms, and then discovering his pupils can't count to ten.

I conclude that the "set of ordered pairs" definition is doing our students even less good than the old "a number which jumps up and down while another number is jumping up and down" that I learned as an undergraduate. Certainly the ultimate test of whether the students have properly absorbed the concept is whether they can recognize a function when they see one.

It is my feeling that calculus texts and teachers ought to be doing something which they are now doing very inadequately: giving the student *many different* ways to visualize the concept (e.g.: the graph; the collection of strings tying points of the domain to points of the range; the idea that if $f(x)$ is $x^2 + e^x$ then f is $(\cdot)^2 + e^{(\cdot)}$; the "slot-machine" into which, when one inserts a number and turns the crank, one gets out another determined entirely by the first; etc.), reserving the "ordered pairs" definition until after the student has assimilated these mental pictures and made his peace with them. The two virtues of the "ordered pairs" definition are its precision—which the student cannot appreciate until he sees how it ties together all these foggy mental pictures—and the demonstration that "function" can be defined within the framework of set-theory—which could well be postponed to a later course in axiomatic set-theory, and *certainly* could be postponed until the student can see for himself the set of ordered pairs associated with $f(y', y, x) = \int_0^{y'} e^{-(xt)^2} dt$.

BIBLIOGRAPHIC ENTRIES: CONCEPTS

1. *Monthly* Vol. 81, No. 4, pp. 390–393. David Shelupsky, A proof of the binomial theorem.

> Proof for real exponents based on solution of the Cauchy functional equation $f(a + b) = f(a) + f(b)$.

2. TYCMJ Vol. 9, No. 4, pp. 205–209. Stephen J. Milles and Henry J. Schultz, Some functional equations for the calculus student.

> Discusses solutions of the Cauchy functional equation and related equations.

3. CMJ Vol. 20, No. 4, pp. 282–300. Israel Kleiner, Evolution of the function concept: A brief survey.

> (This paper won a Polya award)

(b)

TRIGONOMETRIC FUNCTIONS

On the Differentiation Formula for sin θ

Donald Hartig

Mathematics Department, California Polytechnic State University, San Luis Obispo, CA 93407

Very few textbooks on elementary calculus can lay claim to a "rigorous" proof of the fact that the sine function differentiates to the cosine function. See the remarks in Peter Ungar's insightful review [1] of three such texts. In the figure below I offer what should be regarded as another plausibility argument in support of the fact that

$$\frac{d}{d\theta}\sin\theta = \cos\theta.$$

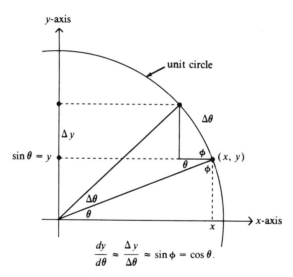

$$\frac{dy}{d\theta} \approx \frac{\Delta y}{\Delta\theta} \approx \sin\phi = \cos\theta.$$

It is as rigorous as the usual "proofs" and has the redeeming feature that the companion formula:

$$\frac{d}{d\theta}\cos\theta = -\sin\theta,$$

can be motivated in the same way:

REFERENCES

1. Peter Ungar, Review of *Calculus and Analytic Geometry* by Al Shenk, *Calculus with Analytic Geometry* by M. A. Munem and D. J. Foulis, and *Calculus with Analytic Geometry* by Howard Anton, this MONTHLY, 93 (1986) 221–230.

π and the Limit of $(\sin \alpha)/\alpha$

LEONARD GILLMAN

Department of Mathematics, University of Texas, Austin, TX 78712

1. Introduction. The formulas $C = 2\pi r$ and $A = \pi r^2$ are part of the vocabulary of our students and of many ordinary citizens, but very few know the ideas underlying them—not even my recent honors calculus class, who told me they were handed the formulas in high school without explanation. As for the presence of π in both formulas, some thoughtful students may consider it a miracle; but probably most people regard π, with its mysterious string of decimals, as just one more example of the depressing magic that math teachers dish out for them to memorize.

Devoting even one calculus or precalculus period to the fundamental ideas can accomplish several things: come to grips in a nonthreatening way with the completeness of the real numbers, afford a preview of the integral in a familiar context, clarify the basis for the trigonometric functions, show the theorem on the limit of $(\sin \alpha)/\alpha$ in its natural setting as essentially just the definition of the circumference of a circle, and of course elucidate the famous formulas themselves (which the students are going to remember long after they have forgotten all their calculus).

What follows is an expanded account of what I presented to the class (and later to an MAA section meeting). It was a pleasure to see their eyes light up at the punch line.

2. π and $(\sin \alpha)/\alpha$. The perimeter of any circumscribed polygon about a circle is greater than that of any inscribed polygon. (This is a good exercise for the students, though they may need hints.) Obviously, the area inside any circumscribed polygon exceeds that inside an inscribed polygon. It is convenient to restrict our attention to the inscribed regular polygons P_m, where m, the number of sides, is a power of 2. (I had my students draw the picture for $m = 4, 8, 16$, and 32.) We define C, the circumference of the circle, to be the least upper bound of the perimeters of the P_m, and A, the area inside the circle, to be the least upper bound of the areas they enclose.

Polygonal lengths on a circle of radius r are equal to r times the corresponding lengths for the unit circle. So, then, is their least upper bound. (Here and later we need the theorem that lub $cx = c$ lub x ($c \geq 0$); I would show students this proof.) The circumference of a circle is therefore proportional to the radius. The constant of proportionality—i.e., the circumference of the unit circle—is called 2π; this defines π. Thus, $C = 2\pi r$.

On the unit circle, a pair of consecutive vertices of P_m marks off an arc of length $2\pi/m$—at the present stage, this is an axiom rather than a theorem—and this number is defined to be the radian measure of the corresponding central angle. On a circle of radius r, the angle cuts off an arc of length $2\pi r/m$.

FIGURE 1 shows the sector associated with a side AB of a polygon P_m, and its two half-sectors corresponding to sides AD and DB of P_{2m}. The central half-angle

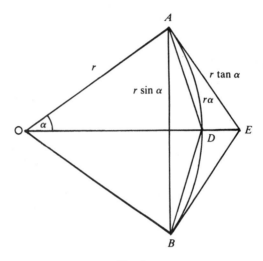

FIG. 1.

α (or α_{2m}) has arc $r\alpha$, and the half-side of P_m is $r \sin \alpha$. Since the ratio of side to arc is the same as perimeter to circumference,

$$\frac{\sin \alpha}{\alpha} = \frac{P_m}{C} \to 1. \tag{1}$$

The area inside triangle OAD is $(1/2)r^2 \sin \alpha$; and the area inside P_{2m} is $C/(r\alpha)$ times that area. Therefore,

$$\text{area } P_{2m} = \frac{1}{2}Cr\frac{\sin \alpha}{\alpha}.$$

Taking least upper bounds yields

$$A = \frac{1}{2}Cr. \tag{2}$$

(This is what so pleased the students.) With $C = 2\pi r$, $A = \pi r^2$.

3. Remarks on the proof. A variant derivation bypasses $(\sin \alpha)/\alpha$, using the more elementary fact that $\cos \alpha \to 1$ as $\alpha \to 0$. To verify this latter, assume for simplicity that $0 < \alpha < \pi/2$; then

$$0 < 1 - \cos \alpha < 1 - \cos^2 \alpha = \sin^2 \alpha < \sin \alpha < \alpha \tag{3}$$

—this last from FIGURE 1. By the squeeze theorem, $1 - \cos \alpha \to 0$, whence $\cos \alpha \to 1$, as $\alpha \to 0$. (Note too for later reference that $\sin \alpha \to 0$.)

In FIGURE 1, the area enclosed by triangle OAB is $(1/2)r^2 \sin 2\alpha$, and the length of AB, a side of P_m, is $2r \sin \alpha$. The area and perimeter of P_m are in the same ratio, which is $(r/2)\cos \alpha$. Taking limits, we get $A/C = r/2$. □

A heuristic form of these arguments is the well-known "cut-and-unroll" proof shown in FIGURE 2.

At the other extreme, [1] uses integration by parts to transform the (improper) integral for the length of arc of a quarter-circle to $(2/r)$ times the integral for the area.

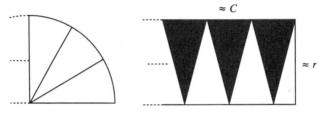

FIG. 2.

4. $\lim(1 - \cos \alpha)/\alpha = 0$. Note that this result is independent of the units used in measuring α. (Ask your students what causes this difference from the case of $(\sin \alpha)/\alpha$.) Nor do we need to know that $(\sin \alpha)/\alpha$ has a limit, but only that it is bounded—say by k. In evaluating $\lim(1 - \cos \alpha)/\alpha$, I think the following direct manipulation of inequalities is more instructive than the usual trick of multiplying and dividing by $1 + \cos \alpha$. From (3),

$$0 < \frac{1 - \cos \alpha}{\alpha} < \frac{\sin \alpha}{\alpha} \sin \alpha < k \sin \alpha \to 0,$$

and the result follows from the squeeze theorem.

5. Definitions of circumference and area. It is instructive to check that the various natural definitions of circumference and area are all equivalent. Let Q_m be the circumscribed polygon whose points of tangency are the vertices of P_m. The following result is basic.

THEOREM. *There are P_m and Q_m with arbitrarily close perimeters and enclosing arbitrarily close areas.*

Proof. This proof is a good one to show students. In FIGURE 1, the half-sides of Q_m and P_m are $r \tan \alpha$ and $r \sin \alpha$; multiplying by $C/(r\alpha)$ then gives the perimeters. The areas inside triangles OAE and OAD are $(1/2)r^2 \tan \alpha$ and $(1/2)r^2 \sin \alpha$, and multiplying by $C/(r\alpha)$ gives the areas inside Q_m and P_{2m}. These facts yield the interesting pair of formulas

$$\text{perim } Q_m - \text{perim } P_m = C \frac{\tan \alpha - \sin \alpha}{\alpha},$$

and, using (2),

$$\text{area } Q_m - \text{area } P_{2m} = A \frac{\tan \alpha - \sin \alpha}{\alpha}.$$

It is easy to see (e.g., from (3)) that these quantities approach 0 as $\alpha \to 0$.   □

It follows from the theorem that C is the *unique* number greater than all inscribed and less than all circumscribed polygonal perimeters, and is the least upper bound of the former and the greatest lower bound of the latter—and these statements hold both, for all polygons and for the P_m and Q_m; finally, the corresponding statements hold for A.

6. Underlying concepts. The development of the trigonometric functions rests on several concepts: the definition of the length of an arbitrary arc as the least upper bound of polygonal paths; the fact that congruent arcs then have the same length; the fact that arc length is additive (needed in deriving the formula for $\cos(\alpha + \beta)$); and the existence of arcs of arbitrary length, so that $\sin \alpha$ and $\cos \alpha$ really are defined for all real α [2, pp. 198–199]. Everyone will have one's own idea about whether, when, and how to bring them in. Most texts say nothing at all.

In our discussion, the length of arc on the unit circle determined by a side of P_m is defined—as C/m; therefore $\sin \alpha$ and $\cos \alpha$ are defined for $\alpha = 2\pi/4, 2\pi/8, 2\pi/16, \ldots$. Trivially, the sequence of perimeters $P_4, P_8, P_{16}, \ldots$ is increasing (which is why we stick to powers of 2); the limit notation in (1) is therefore intuitively clear. But is it reasonable to evaluate the limit of $(\sin \alpha)/\alpha$ (and the other limits) for α ranging merely over a sequence? Well, at least that's a more substantial domain than the empty set; moreover, we get the result free of charge in the course of deriving the formula $A = \pi r^2$. In any case, we are still permitted to show students the standard argument based on the areas of the right triangles and the sector.

By the way, if we now define the length along the unit circle determined by k consecutive sides of P_m to be kC/m—or, alternatively, if we decree additivity—then all dyadic rational multiples of C appear as arc lengths. This dense set of values, obtained effortlessly, gives the student something to hang onto. (One can then get the lengths of all other arcs from this set by means of least upper bounds.)

Acknowledgements. This article is an elaboration of material that was at one time part of a joint venture with R. H. McDowell, whom I wish to thank for permission to publish the material separately. I also wish to thank the referee for several helpful suggestions.

REFERENCES

1. E. F. Assmus, Jr., Pi, this MONTHLY, 92 (1985) 213–214.
2. W. F. Eberlein, The circular function(s), *Math. Mag.*, 39 (1966) 197–201.

Graphs and Derivatives of the Inverse Trig Functions

Daniel A. Moran, Michigan State University, East Lansing, MI 48824

In a calculus course the differentiation formulas for the inverse trig functions are derived by implicit differentiation (at least for two or three of the functions). To avoid tedious repetition, the formulas for the others are merely stated, and their proofs omitted or left as an exercise.

The approach outlined below gives half of the differentiation formulas as immediate consequences of the others. After the inverse functions are defined, it is established that $f^{-1}(x)$ and $\cot^{-1}(x)$ are always complementary when f is sine, tangent or secant. Along the way, there is an opportunity to use graphics (computer-driven or otherwise) and strengthen the students' grasp of the elementary geometry of reflections and translations. And the whole process takes less classroom time than the conventional method!

The archetypical demonstration:

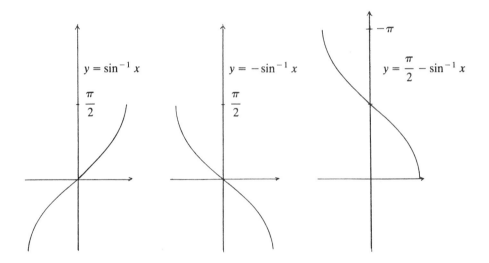

In the figure, the first graph is reflected in the horizontal axis to produce the second; the latter is then translated $\pi/2$ units upward to yield the third (which is evidently congruent to the graph of $y = \cos^{-1} x$). This establishes that $\sin^{-1} x + \cos^{-1} x = \pi/2$. We can now differentiate to discover that $D_x \cos^{-1} x = -D_x \sin^{-1} x$.

The demonstrations for $\tan^{-1}$ and $\sec^{-1}$ require practically no change from the above.

69

Trigonometric Identities through Calculus

Herb Silverman, College of Charleston, Charleston, SC 29424

Too much time is spent on trigonometric identities before calculus and too little time after. We routinely apply methods of calculus to re-solve precalculus problems concerned with curve sketching, areas, and volumes. But while trigonometric formulas are used to find derivatives and integrals of trigonometric expressions, we seldom show students that derivatives and integrals may also be used to verify trigonometric formulas.

For differentiable functions f, g, the identity $f(x) = g(x)$ is equivalent to the (sometimes more easily established) identity $f'(x) = g'(x)$ with $f(a) = g(a)$ for some value a. A differentiation of $f(x) = \sin^2 x + \cos^2 x$ yields $f'(x) = 2 \sin x \cos x - 2 \sin x \cos x = 0$, from which follows the identity that no one can forget. A student who differentiates both sides of $\sin 2x = 2 \sin x \cos x$ verifies that $\cos 2x = \cos^2 x - \sin^2 x$.

The reader may rightly be concerned with potential circular reasoning since the derivatives of trigonometric functions depend on knowing various trigonometric identities. However this note is not an attempt to derive trigonometric formulas from first principles, but rather is a reminder that the connection between mathematical formulas *is* often circular and we can use new formulas to construct older ones. For instance, the derivative of $\sin x$ is usually found from the identity

$$\sin x - \sin y = 2 \sin\left(\frac{x - y}{2}\right) \cos\left(\frac{x + y}{2}\right), \tag{1}$$

which is equivalent to

$$2 \sin x \cos y = \sin(x + y) + \sin(x - y). \tag{2}$$

Taking partials with respect to x and y in (2) gives the identities $2 \cos x \cos y = \cos(x + y) + \cos(x - y)$ and $-2 \sin x \sin y = \cos(x + y) - \cos(x - y)$; taking the partial with respect to x (or y) in (1) leads to the formula for the cosine of a sum, whose partials produce the sine of a sum, etc.

A student recently asked me where she had made her mistake in evaluating $\int \tan x \sec^2 x\, dx$. Her answer of $\sec^2 x/2 + C$ from $u = \sec x$ did not agree with the book's answer of $\tan^2 x/2 + C$ from $u = \tan x$. She was pleased to learn not only that her answer was correct but that she had discovered a new "proof" of the identity $\tan^2 x + 1 = \sec^2 x$.

One final illustration: While working on a paper, I had the vague feeling that a needed expression $\sin^{-1}(2x/(1 + x^2))$ could be put into a nicer form. After several false starts I differentiated to find that

$$\frac{d}{dx}\left(\sin^{-1}\left(\frac{2x}{1 + x^2}\right)\right) = \frac{2}{1 + x^2},$$

70

from which followed the identity $\sin^{-1}(2x/(1+x^2)) = 2\tan^{-1}x$. We leave for the reader the task of coming up with a rationale for making the substitution $u = 2x/(1+x^2)$ in order to show that

$$\int \frac{dx}{1+x^2} = \frac{1}{2}\int \frac{du}{\sqrt{1-u^2}} = \frac{1}{2}\sin^{-1}\left(\frac{2x}{1+x^2}\right) + C.$$

(For related material see the paper by Rosenthal [Lattices of trigonometric identities, *College Mathematics Journal* 20 (1989) 232–234.])

BIBLIOGRAPHIC ENTRIES: TRIGONOMETRIC FUNCTIONS

1. *Monthly* Vol. 89, No. 4, pp. 225–230. Wolfgang Walter, Old and new approaches to Euler's trigonometric expansions.

History of the infinite product for the sine function.

2. TYCMJ Vol. 10, No. 4, pp. 276–277. Norman Schaumberger, An alternate approach to the derivatives of the trigonometric functions.

This paper and the next avoid use of the limit of $(\sin h)/h$ as $h \to 0$.

3. TYCMJ Vol. 13, No. 4, pp. 274–275. Norman Schaumberger, The derivative of arc tan x.

A geometric derivation.

4. CMJ Vol. 15, No. 1, pp. 143–145. Norman Schaumberger, The derivatives of sin x and cos x.

Geometric argument squeezing area of a triangle between areas of two sectors.

5. CMJ Vol. 17, No. 3, pp. 244–246. Norman Schaumberger, The derivatives of arcsec x, arctan x, and tan x.

6. CMJ Vol. 18, No. 2, pp. 139–141. Barry A. Cipra, The derivatives of the sine and cosine functions.

7. CMJ Vol. 20, No. 3, pp. 232–234. William E. Rosenthal, Lattices of trigonometric identities.

8. CMJ Vol. 21, No. 2, pp. 90–99. William B. Gearhart and Harris S. Shultz, The function sin x/x.

(This paper won a Polya award.)

(c)

LOGARITHMIC FUNCTIONS

THE LOGARITHMIC MEAN

B. C. CARLSON, Iowa State University

Let the *logarithmic mean* of the positive numbers x and y be defined by

(1)
$$L(x,y) = \frac{x-y}{\log x - \log y}, \quad x \neq y,$$

$$L(x,x) = x.$$

Note that L is symmetric and homogeneous in x and y and continuous at $x = y$. It is not widely known that L separates the arithmetic and geometric means:

(2)
$$(xy)^{\frac{1}{2}} \leq L(x,y) \leq \frac{x+y}{2},$$

with strict inequalities if $x \neq y$. Division by y shows that (2) is equivalent to well-known inequalities in the single variable $w = x/y$, but the beauty of (2) comes from its symmetry in two variables. The right-hand inequality is due to Ostle and Terwilliger [1], and several proofs are cited by Mitrinović [2]. In both sources the symmetry is somewhat slighted by retaining the unnecessary condition $x \geq y$. The left-hand inequality was stated by Carlson [3, Eq. (3.1)], who obtained (2) by specializing some rather general integral inequalities to the case of the representation

(3)
$$\frac{1}{L(x,y)} = \int_0^1 \frac{du}{ux + (1-u)y}.$$

In the present note we first prove and sharpen (2) by an elementary method which treats x and y symmetrically.

THEOREM 1. *If the positive numbers x and y are unequal, then*

(4)
$$(xy)^{\frac{1}{2}} < (xy)^{\frac{1}{4}}\frac{\sqrt{x}+\sqrt{y}}{2} < L(x,y) < \left(\frac{\sqrt{x}+\sqrt{y}}{2}\right)^2 < \frac{x+y}{2}.$$

Proof. If $t > 0$ the inequality of the arithmetic and geometric means implies that

$$t^2 + t(x+y) + \left(\frac{x+y}{2}\right)^2 > t^2 + t(x+y) + xy > t^2 + 2t(xy)^{\frac{1}{2}} + xy.$$

Thus

$$\int_0^\infty \frac{dt}{\left(t + \frac{x+y}{2}\right)^2} < \int_0^\infty \frac{dt}{(t+x)(t+y)} < \int_0^\infty \frac{dt}{(t+\sqrt{xy})^2}.$$

Evaluating the middle integral by the method of partial fractions, we find

$$\frac{2}{x+y} < \frac{1}{x-y} \lim_{R \to \infty} \left[\log(t+y) - \log(t+x)\right]_0^R < \frac{1}{\sqrt{xy}},$$

which implies (2). We now sharpen (2) by replacing x by $\sqrt{x}$ and y by $\sqrt{y}$:

$$(xy)^{\frac{1}{4}} < \frac{2(\sqrt{x} - \sqrt{y})}{\log x - \log y} < \frac{\sqrt{x} + \sqrt{y}}{2}.$$

Multiplication by $(\sqrt{x} + \sqrt{y})/2$ proves the two inner inequalities in (4). The two outer ones follow from the inequality of the arithmetic and geometric means.

The process by which (2) was sharpened can be repeated to obtain (8). Instead of taking this route we prove a more general inequality first. For any real $t \neq 0$ and any positive x and y, we define

$$G_t(x, y) = t(xy)^{t/2} \frac{x-y}{x^t - y^t}, \quad A_t(x, y) = t \frac{x^t + y^t}{2} \frac{x-y}{x^t - y^t}, \quad x \neq y,$$

(5)

$$G_t(x, x) = A_t(x, x) = x.$$

If we further define $G_0(x, y) = A_0(x, y) = L(x, y)$, it is easy to verify that G_t and A_t are continuous in t. They are also positive and even in t.

THEOREM 2. *If x and y are positive and t is real, then*

(6) $$G_t(x, y) < L(x, y) < A_t(x, y), \qquad t(x - y) \neq 0.$$

The first and third members are respectively decreasing and increasing functions of $|t|$, and the sharpness of the inequalities is measured by

(7) $$A_t^2(x, y) - G_t^2(x, y) = \tfrac{1}{4} t^2 (x - y)^2.$$

Proof. In (2) replace x by x^t and y by y^t and multiply by the positive quantity $t(x - y)/(x^t - y^t)$ to get (6). By straightforward calculation,

$$t \frac{dG_t}{dt} = G_t \left(1 - \frac{A_t}{L}\right), \quad t \frac{dA_t}{dt} = A_t - \frac{G_t^2}{L},$$

from which it follows by (6) that A_t increases with $|t|$ while G_t decreases. Incidentally, a second differentiation shows that A_t is convex and $1/G_t$ is log convex in t.

COROLLARY 1. *If x and y are positive and unequal and n is a nonnegative integer, then*

$$(8) \qquad (xy)^{2^{-n-1}} \prod_{m=1}^{n} \alpha_m(x, y) < L(x, y) < \alpha_n(x, y) \prod_{m=1}^{n} \alpha_m(x, y),$$

where

$$\alpha_m(x, y) = \frac{x^{2^{-m}} + y^{2^{-m}}}{2}.$$

The products are taken to be unity if $n = 0$. The first and third members of (8) are respectively increasing and decreasing functions of n, and the difference of their squares is $2^{-2n-2}(x - y)^2$.

Proof. Choose $t = 2^{-n}$ in Theorem 2 and note that

$$x - y = (x^{2^{-n}} - y^{2^{-n}}) \prod_{m=1}^{n} (x^{2^{-m}} + y^{2^{-m}}).$$

The inequalities (8) reduce to (2) if $n = 0$ and to the inner inequalities of (4) if $n = 1$. As $n \to \infty$ we obtain the following infinite product.

COROLLARY 2. *If x and y are positive numbers, then*

$$(9) \qquad L(x, y) = \prod_{m=1}^{\infty} \alpha_m(x, y).$$

An equality or inequality for $L(x, y)$ of course implies a corresponding result for $\log x$ obtained by putting $y = 1$. For example, (6) gives

$$(10) \qquad \frac{2}{t} \frac{x^t - 1}{x^t + 1} < \log x < \frac{x^t - 1}{tx^{t/2}}, \quad t \neq 0, \ x > 1,$$

with reversed inequalities if $0 < x < 1$. The inequalities become sharper as $|t|$ decreases and as $|x - 1|$ decreases. Likewise (9) implies, for $x > 0$,

$$(11) \qquad \log x = (x - 1) \prod_{m=1}^{\infty} \frac{2}{1 + x^{2^{-m}}}.$$

Finally we give an algorithm for computing $L(x, y)$ or $\log x$ by recurrence relations. As $t \to 0$ we find by developing (5) in powers of t that A_t and G_t differ from L by terms of order t^2, but

$$(12) \qquad L(x, y) = \tfrac{1}{3}\{A_t(x, y) + 2G_t(x, y)\} \{1 + \delta_t(x, y)\},$$

where δ_t is of order t^4. Since

$$(13) \qquad A_{t/2} = \tfrac{1}{2}(A_t + G_t), \quad G_{t/2} = (A_{t/2}G_t)^{\frac{1}{2}},$$

extraction of one square root cuts t in half and ultimately reduces the fractional error δ_t by a factor of 16. For small t it is difficult to calculate A_t and G_t directly from (5) owing to cancellation in $x^t - y^t$, but use of (13) avoids this problem. We define $a_n = A_t$ and $g_n = G_t$, where $t = 2^{1-n}$, and proceed as follows.

ALGORITHM. *If x and y are positive numbers, let*

(14)
$$a_1 = \tfrac{1}{2}(x + y), \qquad g_1 = (xy)^{\frac{1}{2}},$$

$$a_{n+1} = \tfrac{1}{2}(a_n + g_n), \qquad g_{n+1} = (a_{n+1}g_n)^{\frac{1}{2}}, \quad n = 1, 2, 3, \cdots.$$

Then the common limit of a_n and g_n as $n \to \infty$ is the logarithmic mean $L = L(x, y)$ defined by (1). Moreover,

(15)
$$L = \tfrac{1}{3}(a_n + 2g_n)(1 + \varepsilon_n)^{-1},$$

where

(16)
$$0 \leqq \varepsilon_n \leqq \frac{2^{-4n}}{180} \left(\frac{x - y}{g_n}\right)^4 \leqq \frac{2^{-4n-2}(x - y)^4}{45 x^2 y^2}.$$

The recurrence relations (14) are those of Borchardt's algorithm [4]. We omit the proof of the error bounds (16) by expansion in power series, because a method of further speeding the convergence will be discussed elsewhere [5]. An algorithm with slower convergence is given in [4, Eq. (2.4)].

Note added in proof. Corollary 1 provides a solution of the second of two problems proposed by D. S. Mitrinović, Problem 5626, this MONTHLY, 75 (1968) 911–912. See also [2, pp. 383–384].

This work was performed in the Ames Laboratory of the U. S. Atomic Energy Commission,

References

1. B. Ostle and H. L. Terwilliger, A comparison of two means, Proc. Montana Acad. Sci., 17 (1957) 69–70.
2. D. S. Mitrinović, Analytic Inequalities, Springer-Verlag, Berlin, 1970, p. 273.
3. B. C. Carlson, Some inequalities for hypergeometric functions, Proc. Amer. Math. Soc., 17 (1966) 32–39.
4. ———, Algorithms involving arithmetic and geometric means, this MONTHLY, 78 (1971) 496–505.
5. ———, An algorithm for computing logarithms and arctangents, Math. Comp., April 1972.

Is Ln the Other Shoe?

Byron L. McAllister
J. Eldon Whitesitt

Byron L. McAllister has taught mathematics at several universities and colleges since 1949. He is now a Professor at Montana State University. His field is point set topology, and he is especially interested in historical studies in that area.

J. Eldon Whitesitt is Professor of Mathematics at Montana State University. He is the author of three textbooks in mathematics, written primarily for prospective teachers. His particular interests are algebra and mathematics education.

When the person in the apartment directly above drops a shoe on the floor, one expects that sooner or later another shoe will drop. Waiting for this is notoriously nerve-wracking.

Dropping the formula $\int x^n \, dx = x^{n+1}/(n+1) + C$ on a class of beginners is a little like dropping a single shoe, since we must exclude the case $n = -1$. Students see that the formula couldn't apply, but they expect x^{-1} to turn out eventually to have some integral. Later, another shoe drops in the shape of the formula $\int x^{-1} \, dx = \ln x + C$. (We consider only $x > 0$, to avoid absolute values, etc.)

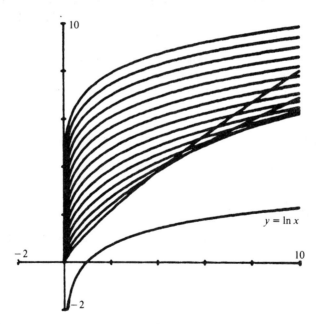

$y = \ln x$

Figure 1. Graphs of several of the curves $y = x^{n+1}/(n + 1)$, showing that as the curves recede toward infinity, the shapes increasingly resemble that of $y = \ln x$.

But not everyone's nerves are soothed thus. This second shoe doesn't seem to match the first. The former had an algebraic ring to it, while the second sounds definitely transcendental. There may be more going on upstairs than we first realized.

Of course the first thing to check is whether it may happen that the limit as $n \to -1$ of $x^{n+1}/(n+1)$ is $\ln x$. But it doesn't happen. In fact, this limit doesn't exist. The graph of $f(x,n) = x^{n+1}/(n+1)$ "goes to infinity" at every point of the line $n = -1$.

Well, we neglected the constant. It should be $x^{n+1}/(n+1) + C$ that we are testing. At first glance this appears not to help either, since $x^{n+1}/(n+1) + C$ also "goes to infinity" on the line $n = -1$. But all is not lost. The constant C arises because $dC/dx = 0$, so that $(d/dx)[x^{n+1}/(n+1) + C] = x^n$. But this can be regarded as a different formula for each value of n, and C need not be the same for each n. In other words, we may, if we wish, regard C as a function $C(n)$ of n.

The effect on the graph of $g(x,n) = x^{n+1}/(n+1) + C(n)$ of letting C change with n is, for each n_0, to raise the cross section along the line $n = n_0$ by the distance $C(n_0)$. If we make sure to choose $C(n)$ negative, so that we raise by a negative amount—i.e., lower—the graph along these lines, we may be able to complete the choice of $C(n)$ in such a way as to obtain

$$\lim_{n \to -1}\left[\frac{x^{n+1}}{n+1} + C(n) \right] = \lim_{n \to -1}\left[\frac{x^{n+1} + (n+1)C(n)}{n+1} \right] = \ln x.$$

Clearly this is only possible if the numerator approaches zero (since the denominator certainly does), so that we must have

$$\lim_{n \to -1} (n+1)C(n) = -1.$$

The quick-and-easy choice of $C(n)$ is thus $(-1)/(n+1)$, and an application of L'Hospital's rule shows that this choice works.

Could we obtain the same result with a different choice of $C(n)$? Well, if we can make $C(n)$ asymptotic to $(-1)/(n+1)$ and, at the same time, $C'(n)$ asymptotic to $(n+1)^{-2}$, we have

$$\lim_{n \to -1}\left[\frac{x^{n+1} + (n+1)C(n)}{n+1} \right] = \lim_{n \to -1}\left[\frac{x^{n+1}\ln x + C(n) + (n+1)C'(n)}{1} \right] = \ln x.$$

So the answer appears to be yes.

A second question of the same type is whether we might, by choosing $C(n)$ still differently, obtain a limit $h(x)$ not differing by a constant from $\ln x$. If $C(n)$ is completely unrestricted so that L'Hospital's rule can't be used, this question appears to be difficult. But if we stick to $C(n)$ "nice enough" that L'Hospital's rule applies, the answer is no. For, besides the condition that $\lim_{n \to -1} C(n) = (-1)/(n+1)$ we

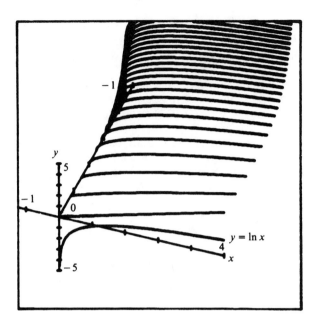

Figure 2. Sections of the surface $y = x^{n+1}/(n+1)$ for n between -1 and 0. The curve $y = \ln x$ is also drawn in the xy-plane.

have to have

$$\lim_{n \to -1} \left[x^{n+1}\ln x + C(n) + (n+1)C'(n) \right] = h(x),$$

from which

$$\lim_{n \to -1} \left[C(n) + (n+1)C'(n) \right] = h(x) - \ln x.$$

Since the left side is independent of x, so must the right be, and $h(x)$ differs from $\ln x$ by a constant.

The third natural question is, "Isn't there an easier way to put $\ln x$ and $x^{n+1}/(n+1)$ together smoothly?" The answer is yes, but by a method that loses us a little of the fun. Still, it's a method that gets over the matter quickly and easily, namely to work with indefinite integrals rather than antiderivatives. For

$$\lim_{n \to -1} \int_a^x t^n \, dt = \lim_{n \to -1} \frac{x^{n+1} - a^{n+1}}{n+1} = \lim_{n \to -1} \frac{x^{n+1}\ln x - a^{n+1}\ln a}{1}$$

$$= \ln x - \ln a = \int_a^x t^{-1} \, dt,$$

showing that $\int_a^x t^n \, dt$ is a continuous function of n. We need only take $a = 1$ to put every "shoe" in place.

The Place of ln x Among the Powers of x

HENRY C. FINLAYSON,
*Department of Mathematics and Astronomy, University of Manitoba, Winnipeg, Manitoba,
Canada R3T 2N2*

Many calculus texts introduce ln x by means of the definition

$$\int_1^x \frac{1}{t}\, dt = \ln x \tag{1}$$

which, they observe, fills the gap (k cannot be zero) in the set of formulas

$$\int t^{k-1}\, dt = \frac{t^k}{k} + C. \tag{2}$$

It is perhaps worth making explicit the observation that ln x is not quite so isolated from the power functions x^k/k as might at first sight seem to be the case. For the selection of a specific set of antiderivatives in (2) yields

$$\int_1^x t^{k-1}\, dt = \frac{x^k - 1}{k}. \tag{3}$$

One would guess from (1) and (3), and verify by l'Hopital's rule that

$$\lim_{k \to 0} \frac{x^k - 1}{k} = \ln x. \tag{4}$$

Sketches of a few graphs of the functions $f_k(x) = (x^k - 1)/k$ along with that of ln x show ln x fitting in nicely among these power functions.

BIBLIOGRAPHIC ENTRIES: LOGARITHMIC FUNCTIONS

1. *Monthly* Vol. 81, No. 8, pp. 879–883. Tung-Po Lin, The power mean and the logarithmic mean.

Sequel to Carlson, Monthly Vol. 79, pp. 615–618, reproduced above, p. 72.

2. TYCMJ Vol. 5, No. 3, p. 58. Norman Schaumberger, Some comments on the exceptional case in a basic integral formula.

The integral of x^α as $\alpha \to -1$.

3. TYCMJ Vol. 9, No. 3, pp. 136–140. Bruce S. Babcock and John W. Dawson, Jr., A neglected approach to the logarithm.

This approach may not be well known, but it is used in the calculus texts by E. Landau (Chelsea, 1951) and by W. Maak (Holt, Rinehart and Winston, 1963).

(d)

EXPONENTIAL AND HYPERBOLIC FUNCTIONS

AN ELEMENTARY DISCUSSION OF THE TRANSCENDENTAL NATURE OF THE ELEMENTARY TRANSCENDENTAL FUNCTIONS

R. W. Hamming, Bell Telephone Laboratories

When the elementary transcendental functions are introduced in the calculus course, it is usually stated that they are not algebraic functions, but little indication is given either as to what this means or how it can be proved. The purpose of this note is to fill this gap partially.

Probably the most convenient approach to this matter is the increasingly common path of introducing the $\ln x$ as

$$\ln x = \int_1^x dt/t,$$

which is equivalent to $d/dx(\ln x) = 1/x$; $\ln 1 = 0$.

We first prove that $\ln x$ is not a rational function, that is

$$\ln x \neq \frac{N(x)}{D(x)},$$

where $N(x)$ and $D(x)$ are polynomials with no common factor. If it were a rational function, then upon differentiating both sides of the equality we would have $1/x = (DN' - ND')/D^2$, or $D^2 = x(DN' - ND')$. We see that $D(x)$ has a factor x. Let

$$D(x) = x^k D_1(x), \quad D_1(0) \neq 0, \quad k \geq 1.$$

Substituting and dividing out x^k, we get

$$x^k D_1^2 = x D_1 N' - k N D_1 - x N D_1',$$

from which we see that $N(x)$ is divisible by x. The common factor of x in both $N(x)$ and $D(x)$ leads to a contradiction.

We are now ready to prove that

$$y = \ln x$$

is not algebraic, that is, there is no polynomial in x and y with real or complex coefficients such that $f(x, y) = 0$, or what is the same thing,

$$\sum_{k=0}^{N} P_k(x)y^k = 0, \qquad P_N(x) \not\equiv 0,$$

where the $P_k(x)$ are polynomials in x. There is an essentially unique equation of minimum degree N because if there were two equations of the same degree differing by more than a multiplicative factor then by eliminating the highest power of ln x between them we would have a lower degree equation. Assuming we have chosen the function $f(x, y)$ for which N is the smallest possible we can write the equation as

$$(\ln x)^N + \frac{P_{N-1}}{P_N}(\ln x)^{N-1} + \cdots + \frac{P_0}{P_N} = 0, \qquad N \geq 2$$

and differentiate to get

$$N(\ln x)^{N-1} + x\left(\frac{P_{N-1}}{P_N}\right)'(\ln x)^{N-1} + \cdots = 0.$$

If all the terms in $(\ln x)^k$, $k = 0, 1, \cdots, N-1$, do not vanish identically, then we have a lower degree polynomial (in ln x), a contradiction. If all the terms do vanish, then in particular

$$\frac{N}{x} + \left(\frac{P_{N-1}}{P_N}\right)' = 0.$$

Integrating this, we find that ln x is a rational function, which we just proved is impossible. Hence, ln x is not an algebraic function.

There is a second idea of an algebraic function that the student needs to consider, namely that any finite combination of additions, subtractions, multiplications, divisions, and radicals with rational exponents of algebraic functions is still algebraic. In particular the student asks, "Is it possible that

$$\ln x = \frac{\sqrt{x^2 + 1} - \pi\sqrt[3]{x^2 - 1} + 2\sqrt[5]{x^2 - 2x + 3} + x^2}{x^{1/3}\sqrt[7]{x + 1} + 9\sqrt[11]{x^2 + 3} - (\sqrt[13]{x^2 + \pi^2})(\sqrt[7]{x^2 + 1})}$$

or something like it?" Note we are excluding x^π etc.

We now indicate the proof that this second definition is included in the first one. Consider the sum of two expressions, each of which is a root of a polynomial. Let this sum be

$$\alpha_1(x) + \beta_1(x),$$

where $\alpha_1(x)$ is a solution of $f_1(x, y) = 0$ with the complete set of solutions

$$\alpha_1(x), \alpha_2(x), \cdots, \alpha_r(x).$$

Let $\beta_1(x)$ be a solution of $f_2(x, y) = 0$ with the complete set of solutions

$$\beta_1(x), \beta_2(x), \cdots, \beta_s(x).$$

Now consider the set of rs functions $\alpha_i(x) + \beta_j(x)$ and the corresponding polynomial in y,

$$\prod_{i,j} (y - \alpha_i - \beta_j) = 0$$

having these rs factors as solution. This is a symmetric function in both the α_i and the β_j. We now use the theorem that every rational symmetric function is expressible rationally in terms of the elementary symmetric functions:

$$
\begin{aligned}
p_1 &= \alpha_1 + \alpha_2 + \cdots + \alpha_s \\
p_2 &= \alpha_1\alpha_2 + \alpha_1\alpha_3 + \cdots + \alpha_{s-1}\alpha_s \\
&\vdots \\
p_s &= \alpha_1\alpha_2\alpha_3 \cdots \alpha_s.
\end{aligned}
$$

But these are in turn rational expressions in the coefficients of $f_1(x, y) = 0$. Similarly for the $\beta_j(x)$. Thus we have

$$\prod_{i,j} (y - \alpha_i - \beta_j)$$

as a rational expression in the coefficients of $f_1(x, y) = 0$, $f_2(x, y) = 0$ and assorted integers that arose in the algebraic manipulations are indicated.

Similar arguments show that differences, products, quotients, and radicals of algebraic expressions are again algebraic, and we have therefore shown that the second definition of an algebraic function is included in the first.

We now turn to the inverse function of $\ln x$, namely e^x. Since $\ln x$ does not satisfy any polynomial

$$f(x, y) = 0$$

we have merely to set $x = e^t$ to get $f(e^t, t) = 0$ as an equivalent impossibility.

For $\sin x$, $\cos x$, $\tan x$, etc., if one of them, say $\sin x$, satisfied

$$f(x, \sin x) = 0,$$

then the polynomial $f(x, 0) = 0$ would have an infinite number of zeros, namely $x = 0$, $\pm\pi$, $\pm 2\pi$, $\cdots$. Thus since the trigonometric functions have an infinite number of zeros they cannot be algebraic functions.

From the argument we used for the exponential function, we see that the corresponding inverse functions arcsin x, arccos x, arctan x, etc. also cannot be algebraic functions.

This presentation, *except* possibly that of the inclusion of the second definition of algebraic functions in the first, is readily presented in a calculus course to the better prepared students; the less prepared usually don't care, being willing to believe that the functions are transcendental (not algebraic).

Using similar methods, and slight extensions of them, integrals like

$$\int_0^z e^{-t^2}dt, \qquad \int_z^\infty (e^t/t)dt$$

can be shown to be transcendental. However, at this point in the development the real question is: Can these new functions be expressed as finite combinations of the algebraic and elementary transcendental functions we now have on hand? Unfortunately the above simple methods seem to be inadequate for this purpose. (See J. F. Ritt, Integration in Finite Terms, Columbia University Press, 1948, for a more sophisticated and more powerful treatment.)

Thanks for help in discussing this topic are due to A. J. Goldstein and Jessie MacWilliams.

A MATTER OF DEFINITION

M. C. MITCHELMORE

The investigation to be described below arose out of the following problem*.

PROBLEM. *Find all positive values of x for which*

$$x^{x^{x^{\cdot^{\cdot^{\cdot}}}}} = 2,$$

where the x's continue to infinity.

To save the printer any further headaches, we shall write t for such an "infinite tower" of x's.

Several students solved this problem as follows:

$$t = 2$$
$$\Rightarrow x^t = 2$$
$$\Rightarrow x^2 = 2$$
$$\Rightarrow x = \sqrt{2}.$$

However, none of them questioned whether an infinite tower of $\sqrt{2}$'s really was equal to 2. In the subsequent discussion, it soon became clear where the real problem lay: We had no definition of the value of an infinite tower of x's, only a vague intuition. We tried two ways to eliminate this shortcoming.

First approach. The only property used in the solution of $t = 2$ above is

(1) $$x^t = t.$$

This property derives from the standard convention that finite towers of exponents are evaluated "from the top down," a convention which we decided to maintain for infinite towers. (The reader is invited to define infinite towers for other evaluation conventions.) Equation (1) is equivalent to

(2) $$x = t^{1/t}.$$

A graph of this relation is shown in Fig. 1. The derivative at the point (t, x) is $t^{-2+1/t}(1 - \ln t)$. The graph is therefore horizontal at the origin, rises monotonically to a maximum at $(e, e^{1/e})$ [$\approx (2.718, 1.445)$], and descends asymptotically to $x = 1$ as $t \to \infty$. To sketch this graph required little more than a knowledge of the limits of $(\ln y)/y$ as $y \to 0+$ and as $y \to \infty$.

We saw immediately that t cannot be defined when $x > e^{1/e}$; that t is uniquely defined when $0 \leq x \leq 1$ and when $x = e^{1/e}$; and that there are two

---

* See Exercise 12-7 on p. 383 of Apostol's *Mathematical Analysis*, Addison-Wesley, 1957. [ED]

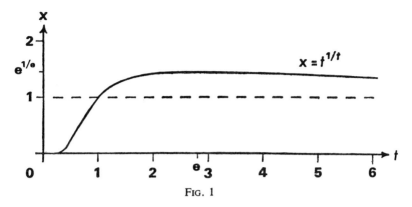

FIG. 1

values of t when $1 < x < e^{1/e}$. For example, $1 < \sqrt{2} < e^{1/e}$ and $\sqrt{2} = t^{1/t}$ has the two solutions $t = 2$ and $t = 4$. Which of these is the value of an infinite tower of $\sqrt{2}$'s? We seemed to be no closer to finding a definition than when we started.

Second approach. Our next idea was to try regarding t as the limit of the sequence

$$x, \; x^x, \; x^{x^x}, \; x^{x^{x^x}}, \cdots .$$

Writing t_n for the nth term of this sequence, we have $t_1 = x$ and

$$t_{n+1} = x^{t_n}.$$

If t_n tends to a limit as $n \to \infty$, then this limit is a solution for t in equation (1). Perhaps we could obtain a unique value for t by defining it by this limit.

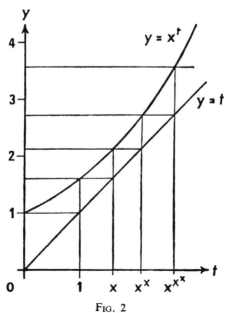

FIG. 2

To find when (t_n) converges, we drew graphs of $y = x^t$ and $y = t$ on the same axes. Fig. 2 and Fig. 3 show two possibilities. In both cases, the zigzag starting at (x, x) gives the successive values of t_n on the axes. In Fig. 2, t_n increases without limit, whereas in Fig. 3, t_n converges to the t-coordinate of the "lower" point of intersection. The convergence of the sequence depends on whether the two curves intersect, which they do if and only if $x^t = t$ has a real solution for t. The t-coordinates of the two points of intersection are therefore given by the two real solutions for t in equation (1).

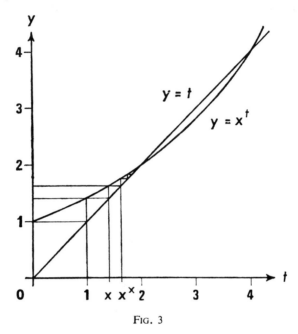

FIG. 3

This showed clearly which of the two possibilities should be chosen when $1 < x < e^{1/e}$. It can be checked rigorously that the zigzag always converges to the point with the smaller coordinates. Thus the value of an infinite tower of $\sqrt{2}$'s is 2, not 4.

Crisis and resolution. We were just about to adopt the definition of t as the limit of t_n, where it exists, when the question was raised: "What happens when $0 < x < 1$?" This did not seem troublesome, because equation (1) has a unique solution in this range; but it was as well to check. When $0 < x < 1$, the graphs of $y = x^t$ and $y = t$ intersect at this solution, as shown in Fig. 4. The sequence (t_n) is now given by a spiral instead of a zigzag.

Put this way, it ceased to be obvious whether (t_n) converges. We found that the gradient of the graph of $y = x^t$ at the point where it intersects $y = t$ is $\ln t$, and this is less than -1 if and only if $x < e^{-e}$ (≈ 0.066). The sequence (t_n) therefore

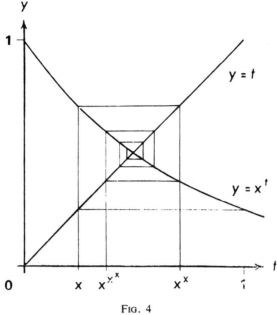

FIG. 4

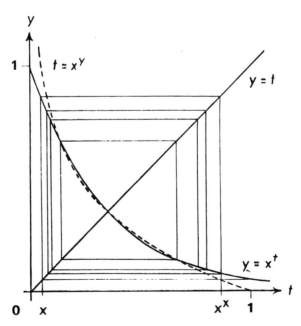

FIG. 5

does not converge when $x < e^{-e}$, even though equation (2) has a unique solution in this interval. (In fact, the sequence now has two limit points, given by the intersections of $y = x^t$ and $t = x^y$ which do not lie on $y = t$. The solution to equation (2) is given by the intersection on $y = t$. This is shown in Fig. 5.) So our second approach also did not work for all values of x.

We were left with the following resolution.

DEFINITION. For all real x such that $0 \leqq x \leqq e^{1/e}$, $x^{x^{x^{\cdot^{\cdot}}}}$ is the real solution for t of the equation $x^t = t$, and in case this equation has two solutions, then the lesser one.

Conclusion. Here was an investigation with apparently innocuous beginnings but which ranged widely over several concepts and techniques of elementary analysis. It differs from most problems in having a definition as the end-point instead of the starting point. Perhaps problems like this could be used more liberally in college courses as an antidote to the type of mental paralysis which sometimes results from over-exposure to the more dogmatic aspects of the standard expository method.

Acknowledgements. I am grateful to my former colleague, Dr. Petr Liebl, now at Charles University, Prague, for the part he played in the discussion, and for his several constructive criticisms of the first draft of this note; also to my referee for suggesting the inclusion of Fig. 5 and for making me draw the graphs accurately.

DEPARTMENT OF MATHEMATICS, MICO COLLEGE, KINGSTON 5, JAMAICA.

The Relationship Between Hyperbolic and Exponential Functions

Roger B. Nelsen, Lewis and Clark College, Portland, OR

In most calculus texts, therefore presumably in most calculus courses, hyperbolic functions are defined in terms of exponential functions: $\cosh\theta = (e^\theta + e^{-\theta})/2$ and $\sinh\theta = (e^\theta - e^{-\theta})/2$. Then certain identities are verified, and the source of the name "hyperbolic" is revealed: the points $(\cosh\theta, \sinh\theta)$ lie on the right-hand branch of the unit hyperbola $x^2 - y^2 = 1$. What seems to be unjustified or lacking here is a rationale for choosing these particular combinations of exponential functions for defining $\cosh\theta$ and $\sinh\theta$.

To answer this, first recall that the circular functions are generally *defined* as coordinates of points on the unit circle. If θ represents the radian measure of the signed angle from the positive x-axis to the radius drawn to a point P on the unit circle, then the coordinates of P are defined to be $(\cos\theta, \sin\theta)$. This is equivalent to denoting by $\theta/2$ the signed area of the circular sector swept out by the radius OP (Figure 1).

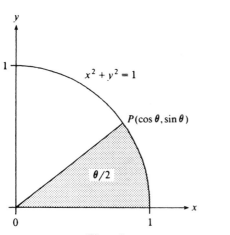

Figure 1.

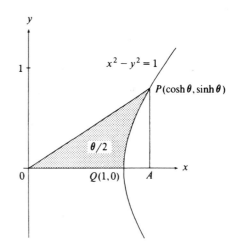

Figure 2.

It is well known that there is an analogous relationship between hyperbolic functions and areas of hyperbolic sectors (see, for example, B. M. Saler's "Inverse Hyperbolic Functions as Areas" [CMJ 16 (March 1985) 128–131]). If $\theta/2$ denotes the signed area of the region swept out by the "radius" OP to the right-hand branch of the hyperbola $x^2 - y^2 = 1$ (the area is taken to be positive when P is in the first quadrant and negative when P is in the fourth quadrant), then the coordinates of P *define* the hyperbolic functions, as in Figure 2.

Now, let us pursue this further. Since we already have $\cosh^2\theta - \sinh^2\theta = 1$, we seek a second equation involving $\cosh\theta$ and $\sinh\theta$. Figure 2 suggests finding an alternate expression for the area $\theta/2$ of the shaded hyperbolic sector, or for the area

$$\frac{1}{2}\cosh\theta \sinh\theta - \frac{1}{2}\theta$$

89

of the unshaded portion of $\triangle OAP$. As we shall see, a 45° counterclockwise rotation of the shaded region in Figure 2 (or, equivalently, a 45° clockwise rotation of the axes) yields a region whose area can be evaluated easily by use of the natural logarithm function. Since the area of a region is unchanged by such a rotation, this will provide us with a second expression for $\theta/2$ as a natural logarithm function of $\cosh\theta$ and $\sinh\theta$.

With standard results on rotation of axes, the new $\bar{x} - \bar{y}$ coordinates after the 45° rotation are related to the original $x - y$ coordinates by

$$\bar{x} = \frac{\sqrt{2}}{2}(x - y) \qquad \bar{y} = \frac{\sqrt{2}}{2}(x + y).$$

If necessary, this transformation can be easily derived. But perhaps the geometric "proof without words" illustrated in Figure 3 is sufficient at this time, leaving the general case until later in the course.

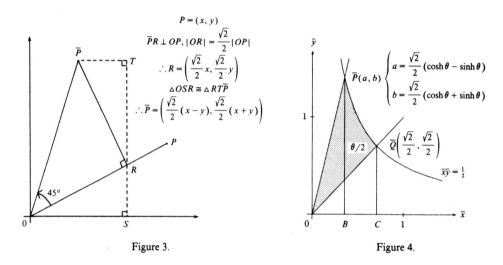

Figure 3.   Figure 4.

After the rotation by 45°, we have the situation illustrated in Figure 4.
Now the area $\theta/2$ of the shaded sector can be expressed as (replacing $\bar{x}$ by x in the integral)

$$\frac{\theta}{2} = \int_a^{\sqrt{2}/2} \frac{1}{2x}\,dx + \text{area}\,(\triangle O\bar{P}B) - \text{area}\,(\triangle O\bar{Q}C).$$

Since the area of each triangle is $1/4$, the area of the shaded sector is precisely the area under the hyperbola from $\bar{P}$ to $\bar{Q}$, and, hence,

$$\theta = \int_a^{\sqrt{2}/2} \frac{1}{x}\,dx = \ln(\sqrt{2}/2) - \ln(a)$$

$$= -\ln(\cosh\theta - \sinh\theta).$$

Recalling that $(\cosh\theta, \sinh\theta)$ lies on $x^2 - y^2 = 1$, we thus have the system

$$\cosh^2\theta - \sinh^2\theta = 1$$
$$\cosh\theta - \sinh\theta = e^{-\theta}.$$

This has as solution the familiar expressions, $\cosh\theta = (e^\theta + e^{-\theta})/2$ and $\sinh\theta = (e^\theta - e^{-\theta})/2$, for the hyperbolic cosine and sine.

BIBLIOGRAPHIC ENTRIES: EXPONENTIAL AND HYPERBOLIC FUNCTIONS

1. *Monthly* Vol. 77, No. 9, pp. 995–998. J. van Yzeren, A rehabilitation of $\left(1 + \dfrac{z}{n}\right)^n$.

An approach to $\exp(z)$ for both real and complex z.

2. *Monthly* Vol. 82, No. 8, pp. 842–844. Z. A. Melzak, On the exponential function.

An approach via the arithmetic and geometric mean inequality.

4

LIMITS AND CONTINUITY

Some Thoughts about Limits

RAY REDHEFFER
UCLA
Los Angeles, CA 90024

When I was a student at MIT one of my professors—a distinguished foreign scholar —expressed consternation at having received the following response on a written examination:

> Question: What is the meaning of $\lim_{x \to a} f(x) = L$?
> Answer: For every ε there is a δ.

This brings us to the purpose of this note, which is twofold. A first objective is to simplify the formulation and proof of theorems on limits. A second is to give precise definitions to, and so legitimize, certain modes of expression that are avoided by algebraists but embraced by analysts.

There is a sense in which theorems on limits are much easier than many of the standard homework problems in elementary calculus. For example, I would defend the view that it's easier to prove the theorem about the limit of a sum than to integrate $(\sec x)^5$. Yet many people have the opposite perception, as suggested by the anecdote above.

Why is this? Although various answers can be given, certainly a good deal of the trouble is caused by a piling up of logical terms: *For any ε there exists a δ such that for all x* of a certain kind something happens.... Quantifiers are the quicksand in which the path ends. Here we give a formulation in which quantifiers are kept separate, so that not too many of them pile up in one place. The approach has been classroom tested at UCLA, and the improvement in student performance is wholly out of proportion to the modest mathematical content of the innovation. The latter consists solely in the formulation and use of Theorem 1 (which follows).

Turning to the second objective, there seems to be a consensus that certain modes of expression common in analysis should be kept below stairs, in the scullery, and not admitted to polite company. As an extreme case, I recall reading somewhere that we musn't say a function is "increasing" because, after all, it's one and the same function all the time, so can't increase (or decrease or change in any other way). Still less, then, can one use phrases like "for x near a" or "for large x."

The view taken here is that such phrases have an honorable tradition in analysis, and express concepts of great importance, and should be used. If you want to say that $1/x$ is bounded near $x = 1$, or is less than 0.001 for large x, go ahead and say it. All that's missing is a precise definition. And not even that will be missing if you read far enough.

The main definitions. Although the methods suggested here apply to situations of great generality, it is assumed for simplicity that all functions and variables are real valued. We begin with the following:

DEFINITION 1. *A condition C holds for x near a if there exists a positive constant δ such that C holds when $x \neq a$ and $|x - a| < \delta$.*

For example, a function $f(x)$ is *bounded for x near a* if there exists a constant M such that $|f(x)| \leqslant M$ for x near a. Here the condition C is expressed by the inequality $|f(x)| \leqslant M$. Another example is given next:

DEFINITION 2. *The statement $\lim_{x \to a} f(x) = L$ means the following: If $\varepsilon > 0$ then $|f(x) - L| < \varepsilon$ holds for x near a.*

Using lim as an abbreviation for $\lim_{x \to a}$ we shall develop the theory of limits by a series of examples. The first two of these depend only on the definitions.

Example 1 (boundedness). If $\lim f(x) = L$, then $f(x)$ is bounded for x near a. To see this, let $\varepsilon > 0$ be given. The condition $|f(x) - L| < \varepsilon$ holds for x near a and gives

$$|f(x)| = |f(x) - L + L| \leqslant |f(x) - L| + |L| < \varepsilon + |L|.$$

This shows that the bound M can be any number larger than $|L|$.

Example 2 (boundedness of reciprocal). If $L \neq 0$ in the above discussion then not only $f(x)$ but also $1/f(x)$ is bounded for x near a. This follows from

$$|f(x)| = |L - (L - f(x))| \geqslant |L| - |f(x) - L| > |L| - \varepsilon$$

with $0 < \varepsilon < |L|$. Taking reciprocals, we see that the bound for $|1/f(x)|$ could be any number larger than $|1/L|$.

Since Examples 1 and 2 both assert that a certain condition holds "for x near a" they involve Definition 1. The same applies to several of the examples given below. The fact that Definition 1 allows such a simple formulation is already worth the price of admission. However, the full scope of the definition is seen only in conjunction with the following theorem.

The main theorem. If C_1 and C_2 are two conditions, then $C_1 \wedge C_2$ holds when both C_1 and C_2 hold simultaneously. For example, if C_1 is the condition $f(x) > 2$, and C_2 is the condition $f(x) \leqslant 5$, then $C_1 \wedge C_2$ is the condition $2 < f(x) \leqslant 5$.

When two conditions C_1 and C_2 *each* hold for x near a, one would expect that C_1 and C_2 *both* hold for x near a. The following theorem shows that Definition 1 is consistent with this use of everyday language:

THEOREM 1. *If C_1 and C_2 each hold for x near a, then $C_1 \wedge C_2$ holds for x near a.*

Proof. Let C_i hold for $x \neq a$ and $|x - a| < \delta_i$, where $i = 1$ or 2 and where each $\delta_i > 0$. Let $\delta = \min(\delta_1, \delta_2)$. Then $\delta > 0$, and $|x - a| < \delta$ implies $|x - a| < \delta_i$ for $i = 1, 2$. Hence $x \neq a$ and $|x - a| < \delta$ imply that C_1 and C_2 both hold. This completes the proof.

A corresponding result holds for any finite number of conditions $C_1, C_2, \ldots, C_n$. All we have to do is let i range over the indices $1, 2, \ldots, n$ instead of $1, 2$. Alternatively, since $C_1 \wedge C_2 \wedge C_3 = (C_1 \wedge C_2) \wedge C_3$, the general result can be obtained by repeated use of Theorem 1 as it stands. In most applications, two or three conditions C_i suffice.

Further examples. Here we give some examples that require the main theorem. The first expresses the fact that $\lim f(x)$ depends only on the local behavior of f, that is, on the behavior near a. This is perhaps the most important of all the properties of limits:

Example 3 (localization). If $f(x) = g(x)$ for x near a, and if $\lim f(x) = L$, then $\lim g(x) = L$. For proof, let $\varepsilon > 0$ be given. Then $|f(x) - L| < \varepsilon$ holds for x near a by Definition 2 and $f(x) = g(x)$ holds for x near a by hypothesis. Hence Theorem 1 tells us that *both* conditions hold for x near a. When both hold we have

$$|g(x) - L| = |f(x) - L| < \varepsilon.$$

Hence $|g(x) - L| < \varepsilon$ for x near a, and $\lim g(x) = L$ by Definition 2.

The localization theorem underlies familiar calculations like

$$\lim_{x \to 1} \frac{x^2 - 1}{x - 1} = \lim_{x \to 1} (x + 1) = 2.$$

Here $f(x) = x + 1$ and $g(x)$ is the fraction on the left. It is not true that $f = g$ in the sense of functional equality, but we do have $f(x) = g(x)$ for x near 1 in the sense of Definition 1.

As another example, if $f(x)$ is the constant function $f(x) = c$ then it is a trivial consequence of the definition that $\lim f(x) = c$. But the localization theorem gives the same conclusion under the much weaker hypothesis that $f(x) = c$ for x near a.

The next result is an aid in proving the theorem about the limit of a product and it also gives results when the latter does not apply:

Example 4 (preservation of zero limit). If $f(x)$ is bounded near a and if $\lim g(x) = 0$, then $\lim f(x)g(x) = 0$. For proof choose $M > 0$ so that $|f(x)| \leqslant M$ for x near a and let $\varepsilon > 0$ be given. Then $\varepsilon/M > 0$ and hence $|g(x)| < \varepsilon/M$ for x near a. Thus, the conditions $|g(x)| < \varepsilon/M$ and $|f(x)| \leqslant M$ *each* hold for x near a. By Theorem 1 the conditions *both* hold for x near a. When both hold we have

$$|f(x)g(x)| < M\frac{\varepsilon}{M} = \varepsilon.$$

This gives $|f(x)g(x)| < \varepsilon$ for x near a and the result follows.

The next example requires the extension of Theorem 1 to three conditions C_i, as discussed above:

Example 5 (squeeze theorem). If $f(x) \leqslant g(x) \leqslant h(x)$ holds for x near a, and if $\lim f(x) = \lim h(x) = L$, then $\lim g(x) = L$. For proof, let $\varepsilon > 0$ be given. Then the conditions $|f(x) - L| < \varepsilon$ and $|h(x) - L| < \varepsilon$ each hold for x near a by Definition 2, and $f(x) \leqslant g(x) \leqslant h(x)$ holds for x near a by hypothesis. Hence *all three* conditions hold for x near a. When all three conditions hold we have

$$-\varepsilon < f(x) - L \leqslant g(x) - L \leqslant h(x) - L < \varepsilon.$$

This gives $|g(x) - L| < \varepsilon$ for x near a and completes the proof.

Example 6 (limit of a sum). If $\lim f_i(x) = L_i$ for $i = 1, 2$, then

$$\lim[f_1(x) + f_2(x)] = L_1 + L_2.$$

Proof. Let $\varepsilon > 0$ be given and note that also $\varepsilon/2 > 0$. The conditions $|f_i(x) - L_i| < \varepsilon/2$ *each* hold for x near a and $i = 1, 2$. Hence, by Theorem 1, *both* conditions hold for x near a. When both hold we have

$$|f_1(x) + f_2(x) - (L_1 + L_2)| = |f_1(x) - L_1 + f_2(x) - L_2|$$

$$\leqslant |f_1(x) - L_1| + |f_2(x) - L_2| < \frac{\varepsilon}{2} + \frac{\varepsilon}{2}.$$

Hence the left side is $< \varepsilon$ for x near a, and this gives the conclusion.

Example 7 (limit of a product). If $\lim f_i(x) = L_i$ for $i = 1, 2$, then $\lim f_1(x)f_2(x) = L_1L_2$. For proof, it follows from Definition 2 that $f_1(x)$ and $f_2(x)$ are *each* defined for x near a and hence by Theorem 1 they are *both* defined for x near a. When both are defined we have the identity

$$f_1(x)f_2(x) - L_1L_2 = f_1(x)[f_2(x) - L_2] + L_2[f_1(x) - L_1].$$

By Examples 1 and 4 each of the two terms on the right has the limit 0. Hence the sum also has limit 0 and the result follows. Here and in Example 8 below, we use the fact that $\lim f(x) = L$ is equivalent to $\lim[f(x) - L] = 0$. This is an immediate consequence of the definition.

Example 8 (limit of reciprocal). If $\lim f(x) = L \neq 0$ then $\lim(1/f(x)) = 1/L$. By Example 2, the function $1/f(x)$ is defined for x near a and we can write

$$\frac{1}{f(x)} - \frac{1}{L} = \frac{1}{Lf(x)}(L - f(x)).$$

The conclusion now follows from Examples 2 and 4. Writing $g/f = (1/f)g$ we see that a theorem regarding the limit of a quotient follows from Examples 7 and 8.

Example 9 (uniqueness). If $\lim f(x) = L_i$ for $i = 1$ and 2, then $L_1 = L_2$. For proof let $\varepsilon > 0$ and note that the two conditions $|f(x) - L_i| < \varepsilon$ for $i = 1$ and 2 *each* hold for x near a, hence *both* hold for x near a. If x is a value for which both hold we have

$$|L_1 - L_2| = |L_1 - f(x) + f(x) - L_2| \leqslant |L_1 - f(x)| + |f(x) - L_2| < 2\varepsilon.$$

Since ε is arbitrary, and the left side is independent of ε, the left side must be 0. This gives $L_1 = L_2$.

On logical grounds this theorem should probably have been presented first. However, its postponement is justified on the psychological ground that, of all theorems on limits, it is perhaps the least exciting. Two questions: Are the results of the previous examples meaningful before we know the uniqueness property asserted in Example 9? And can we get uniqueness by applying Example 3 (the localization theorem) to the two functions f and g, taking $f = g$? The answers are given at the end of this note.

In the following example the abbreviation lim for $\lim_{x \to a}$ is not used, because the limits involve two points, a and b.

Example 10 (composite function). Let $\lim_{t \to b} f(t) = L$, $\lim_{x \to a} g(x) = b$, and suppose further that $g(x) \neq b$ for x near a. Then $\lim_{x \to a} f[g(x)] = L$. For proof let $\varepsilon > 0$ be given and choose $\eta > 0$ so that the two conditions $|t - b| < \eta$ and $t \neq b$ together imply $|f(t) - L| < \varepsilon$. (Here we are using Definitions 1 and 2, with the roles of x, a, δ taken respectively by t, b, η). The two conditions $g(x) \neq b$ and $|g(x) - b| < \eta$ each hold for x near a, hence *both* do. When both hold the variable $t = g(x)$ satisfies the conditions required for t above and we get

$$|f[g(x)] - L| = |f(t) - L| < \varepsilon.$$

This shows that the left side is $< \varepsilon$ for x near a and completes the proof.

A remark on continuity. By definition, the function f is *continuous* at $x = a$ if $\lim f(x) = f(a)$. Hence, with one exception, theorems on continuous functions follow immediately from theorems on limits, as established above. The exception concerns the continuity of composite functions, the analog of Example 10. Namely, if $f(t)$ is continuous at $t = g(a)$, and if $g(x)$ is continuous at $x = a$, then $f[g(x)]$ is continuous at $x = a$. The trouble is that Example 10 requires the extraneous condition $g(x) \neq g(a)$ for x near a, which is not needed here.

To see what is going on let us interpret the equation $\lim f(x) = f(a)$ by Definition 2. Namely, if $\varepsilon > 0$ the inequality $|f(x) - f(a)| < \varepsilon$ holds for x near a. In the definition of limit the point $x = a$ has to be excluded, but since $|f(a) - f(a)| = 0$, the defining inequality for continuity holds at $x = a$ automatically and the exclusion is no longer necessary. In the case at hand, suppose $f(t)$ is continuous at $t = b$. Then we can allow $g(x) = b$ without harm and a repetition of the proof of Example 10 gives the correct theorem on continuity of composite functions.

One-sided conditions. If the inequality $x \neq a$ in the side condition in Definition 1 is replaced by $x < a$ or $x > a$, it is said that C holds for x near $a -$ or $a +$, respectively. Similar notation applies to concepts that depend on Definition 1. For example, $f(x)$ is bounded near $a +$ if there exists a constant M such that $|f(x)| \leqslant M$ for x near $a +$. If a is replaced by $a -$ or $a +$ in Definition 2 the result is a definition of the *one-sided limits*

$$\lim_{x \to a-} f(x) \quad \text{or} \quad \lim_{x \to a+} f(x),$$

respectively. Definitions and theorems involving $a -$ or $a +$ are termed "one-sided" to distinguish them from the two-sided conditions discussed above. There is no need to repeat the proofs, since one-sided results are obtained by the purely mechanical process of writing $a -$ or $a +$ for a wherever it occurs.

It is obvious that, if a condition C holds for x near a, then C must hold both for x near $a -$ and for x near $a +$. The following theorem states a converse:

Theorem 2. *If a condition C holds for x near $a -$ and for x near $a +$, then C holds for x near a.*

Proof. Choose $\delta_1 > 0$ so that C holds for $x < a$ and $|x - a| < \delta_1$. Also choose $\delta_2 > 0$ so C holds for $x > a$ and $|x - a| < \delta_2$. Let $\delta = \min(\delta_1, \delta_2)$. Then $\delta > 0$, and the condition $|x - a| < \delta$ implies $|x - a| < \delta_i$ for $i = 1$ and 2. Hence C holds if $|x - a| < \delta$ and $x < a$, and also if $|x - a| < \delta$ and $x > a$. Thus C holds if $|x - a| < \delta$ and $x \neq a$, and this completes the proof.

We illustrate the theorem by two examples.

Example 11 (boundedness). If $f(x)$ is bounded for x near $a -$ and also for x near $a +$, then $f(x)$ is bounded for x near a. To see this, choose M^- so that $|f(x)| < M^-$ holds for x near $a -$, and choose M^+ so $|f(x)| < M^+$ holds for x near $a +$. Let $M = \max(M^-, M^+)$. Then $|f(x)| < M$ holds for x near $a -$ and for x near $a +$. By Theorem 2 the same inequality holds for x near a and the result follows.

Example 12 (limits). If $\lim f(x) = L$ as $x \to a -$ and also as $x \to a +$, then $\lim f(x) = L$ as $x \to a$. For proof let $\varepsilon > 0$ be given. Then the condition $|f(x) - L| < \varepsilon$ holds for x near $a -$, and also for x near $a +$. By Theorem 2 we have $|f(x) - L| < \varepsilon$ for x near a and this completes the proof.

Since the converse of Example 12 is trivial, we can say that $\lim f(x)$ exists if, and only if, the left- and right-hand limits exist and have the same value. This gives the best procedure to establish nonexistence of limits in typical textbook problems. For example, $x/|x|$ has no limit as $x \to 0$ because the right- and left-hand limits are not equal. A direct proof of the nonexistence by going back to the (ε, δ) definition is harder.

The left- and right-hand limits as $x \to a$ are often denoted by $f(a-)$ and $f(a+)$, respectively. By definition, $f(x)$ is *continuous from the left* or *right* at $x = a$ if $f(a) = f(a-)$ or $f(a) = f(a+)$, respectively. It is a consequence of Example 12 that $f(x)$ is continuous at $x = a$ if, and only if, it is continuous both from the left and from the right. This gives a simple test for continuity, analogous to the test for existence of a limit noted above.

Associated with the left- and right-hand limits are the corresponding left- and right-hand derivatives,

$$D^- f(a) = \lim_{x \to a-} \frac{f(x) - f(a)}{x - a}, \qquad D^+ f(a) = \lim_{x \to a+} \frac{f(x) - f(a)}{x - a}.$$

By Example 12 the derivative $Df(a) = f'(a)$ exists if, and only if, the left- and right-hand derivatives exist and have the same value. For example, this shows *by inspection* that $|x|$ is not differentiable at $x = 0$. As in the cases above, an (ε, δ) proof by reference to the definition of derivative is harder.

Large x. Following the pattern of Definition 1 we say that *a condition C holds for large x* if there exists a constant N such that C holds for $x > N$. The analog of Definition 2 is that $\lim_{x \to \infty} f(x) = L$ means the following: If $\varepsilon > 0$, then $|f(x) - L| < \varepsilon$ holds for large x. The analog of Theorem 1 is that if two conditions C_1 and C_2 each hold for large x, then $C_1 \wedge C_2$ also holds for large x. Development of the theory parallels Examples 1–12 and is in some respects simpler, because we no longer have to worry about the side condition $x \neq a$. In the present case a corresponds to ∞ and we have $x \neq a$ automatically since x is a real number.

If x is confined to integer values n one generally writes $f(x) = f_n$ and the role of x is now taken by n. Thus, $\lim_{n \to \infty} f_n = L$ means the following: If $\varepsilon > 0$, the inequality $|f_n - L| < \varepsilon$ holds for large n. The fact that n is an integer is understood and need not be emphasized, just as in the former cases it was understood that x is real. The resulting development parallels that for limits as $x \to a$ and yields the theory of limits of sequences.

Uniformity. Although we have not emphasized it by the notation, the condition C in Definition 1 is understood to be a condition depending on x; thus, $C = C(x)$. Furthermore, the statement that C holds for $x \neq a$, $|x - a| < \delta$ means that C holds *for all* x satisfying these conditions. Hence C holds on any smaller set of the form $x \neq a$, $|x - a| < \delta$, and that is why we did not have to insist that δ be small. For example, a function bounded on a set is bounded on any subset, and the inequality $|f(x)| \leqslant M$ expressing boundedness is a condition of form $C(x)$. By contrast, the statement that f is *unbounded* is not of the form $C(x)$, is not preserved by passage to subsets, and would not be a suitable condition C in Definition 1.

Sometimes $C(x)$ involves another parameter t so that $C = C(x, t)$. The parameter t is assumed to range over a specified set E. The statement that $C(x, t)$ holds *for x near a* is defined just as before. Namely, it means that there exists $\delta > 0$ such that $C(x, t)$ holds for $x \neq a$, $|x - a| < \delta$. In general δ depends on t, so that $\delta = \delta(t)$. If δ can be chosen independently of t, for all $t \in E$, it is said that $C(x, t)$ holds *uniformly for x near a*. A similar definition is used for large x. That is, $C(x, t)$ holds *uniformly for large x* if $C(x, t)$ holds for $x > N$, where N can be chosen independently of t. When x is restricted to integral values we have $C(x, t) = C(n, t)$ and the condition for uniformity is that $C(n, t)$ holds for $n > N$ where N is independent of t. When Theorem 1 and its analogs are extended to this situation it is advisable to require that the various conditions $C_i(x, t)$ be associated with one and the same set E. Aside from this, there is no significant change.

As an illustration, if $C(x, t) = C(n, t)$ is a condition of the form

$$|f_n(t) - L(t)| < \epsilon, \qquad t \in E,$$

it will be found that the methods used here yield a substantial part of the theory of uniform convergence of sequences. We do not give details, since the details so closely parallel what has already been done.

The domain-dependent definition of limit. There is a certain awkwardness in defining continuity of a function $f(x)$ on a closed interval $a \leqslant x \leqslant b$ because the two-sided limits at a and b need not exist. Thus, we have to require $\lim_{t \to x} f(t) = f(x)$ for $a < x < b$ together with the one-sided conditions $f(a) = f(a+)$, $f(b) = f(b-)$. This problem becomes more acute in more complicated situations, for example, if we want to define continuity of a function $f(x, y)$ in a closed region of the (x, y) plane.

One way of dealing with problems of this kind is to introduce what we shall call the *domain-dependent definition* of limit. In this definition the inequality $|f(x) - L| < \varepsilon$ is required for x satisfying the three conditions $x \neq a$, $|x - a| < \delta$, $x \in D(f)$ where $D(f)$ is the domain of f. For example, if $D(f)$ is the closed interval $a \leqslant x \leqslant b$ then the limits at $x = a$ and $x = b$ are on the same footing as the limits at any other point of this interval and the condition for continuity is $\lim_{t \to x} f(t) = f(x)$, $a \leqslant x \leqslant b$. With the domain-dependent definition, the distinction between one-sided and two-sided limits at the endpoints of the domain disappears.

To study the domain-dependent definition by the methods of this note one can introduce a set F for x in Definition 1 analogous to the set E for t above, and require the side condition $x \in F$. It is said then that "C holds for x in F and near a." When applying this form of Definition 1 to the theory of limits via Definition 2, one takes $F = D(f)$.

If we prefer not to complicate Definition 1 as described above, another method is the following: Define a function f^* by $f^*(x) = f(x)$ when $x \in D(f)$, otherwise

$f^*(x) = L$. Then $\lim f(x) = L$ holds with the domain-dependent definition if, and only if, $\lim f^*(x) = L$ holds in the sense of Definition 2. The foregoing results apply to the latter problem and hence to the former.

Further discussion of domain dependence. It is of some interest to contrast the domain-dependent definition with Definition 2, and this is done now. For brevity we introduce the abbreviations

DD = the domain-dependent definition of limit

SD = the simple definition of limit.

More specifically, SD stands for the definition given in Definition 2 or for its one-sided extensions involving $a-$ or $a+$. Use of the word "simple" in this connection suggests an implied judgment about DD vs. SD that I hope to justify in the following discussion.

Let us begin with the question of uniqueness. If we want to falsify the statement "$\lim f(x) = L$" using DD we must pick a suitable $\varepsilon > 0$ and then, for each $\delta > 0$, we must exhibit an $x \in D(f)$ satisfying $x \neq a$, $|x - a| < \delta$ for which the defining inequality $|f(x) - L| < \varepsilon$ fails. But if no point $x \neq a$ of $D(f)$ is in the interval $|x - a| < \delta$ then we can find no x of this kind, the statement $\lim f(x) = L$ cannot be falsified, and we must admit that L is a value of $\lim f(x)$. This happens whenever some set of the form $|x - a| < \delta$ contains no points of $D(f)$ except possibly a itself; in more technical language, it happens whenever a is not a limit point of $D(f)$. In that case $\lim f(x)$ is not a number but a set; in fact, it is the entire real axis.

Although in principle there is nothing wrong with set-valued expressions, their introduction in the present context can lead to strange results. For instance let $f(x) = 0$ for x rational and let $f(x)$ be undefined otherwise. Let $g(x) = 0$ for x irrational and let $g(x)$ be undefined otherwise. Then using DD we have $\lim f(x) = \lim g(x) = 0$ at every value a, without exception. But the sum function $f + g$ is defined for no x and hence, at every value a, the expression $\lim[f(x) + g(x)]$ is the entire real axis! This is a spectacular failure of the expected theorem regarding limit of a sum.

Since most analysts would find situations such as the above intolerable, it is customary to agree that $\lim f(x)$ is undefined at any point a which is not a limit point of $D(f)$. Such an agreement solves the problem, but it is a complication.

Another complication arises when we try to give a simple statement of the localization theorem, Example 3 above. At first glance one might think that all would be well if we just require $D(f) = D(g)$ except for the point a, which might or might not belong to either set. But this is not a good idea, because $D(f)$ is a *global* concept and the whole purpose of the exercise is to show that $\lim f(x)$ is a *local* concept. Suppose, for example, that f and g are defined by the formulas

$$f(x) = x, \qquad g(x) = x\frac{x - 1000}{x - 1000}$$

with their natural domains. If we have ascertained that $\lim f(x) = 0$ as $x \to 0$ we can't use the alleged localization theorem to conclude anything about $\lim g(x)$ as $x \to 0$ because, no matter whether we include the point 0 or not, the functions f and g do not have the same domain. Since $g(x) = f(x)$ for $|x| < 1000$ this example describes a spectacular failure to come to grips with the problem of localization.

To deal effectively with such examples in the vocabulary of this paper, one can introduce the two conditions

$$C_1\text{: } g(x) = f(x) \text{ when } x \text{ is in } D(f)$$

$$C_2\text{: } g(x) \text{ is undefined when } x \text{ is not in } D(f)$$

and declare that "$f(x) = g(x)$ for x near a" means "$C_1 \wedge C_2$ holds for x near a." The latter phrase is defined unambiguously in Definition 1. For those familiar with the concepts of *deleted neighborhood* and *restriction* f_A of a function to a set A, an equivalent formulation is "$f_A = g_A$ in some deleted neighborhood A of a." The above methods lead to an adequate localization theorem for DD, but there is no denying that they involve extra complication.

Similar problems arise when we want to develop theorems about the limit of sums, products, or quotients. One method (which is actually used in textbooks) is to require $D(f) = D(g)$. To see why this is inappropriate let f and g be the functions defined by the formulas

$$f(x) = x, \qquad g(x) = \frac{x}{x - 1000}$$

with their natural domains. If we have ascertained that $\lim f(x) = \lim g(x) = 0$ as $x \to 0$, we cannot use the alleged limit-of-sum theorem to deduce anything about the limit of $f(x) + g(x)$, because f and g have different domains. Of course the fact that $g(x)$ is undefined at $x = 1000$ has nothing to do with the limit as $x \to 0$, and the correct hypothesis for theorems of this kind is that a is a limit point of $D(f) \cap D(g)$. But that too is a complication.

Let us see next what happens when DD is used to define the derivative, which is, after all, a principal motive for discussing limits in the first place. One of the things we lose is the familiar fact that, if $f(x)$ has a maximum or minimum at a point c where $f'(c)$ exists, then $f'(c) = 0$. For example, let $f(x) = x$ for $0 \leqslant x \leqslant 1$, the latter interval being $D(f)$. If DD is used, the defining limit for $f'(x)$ exists for every x on $0 \leqslant x \leqslant 1$, including the end points, and has the value 1. But $x = 0$ gives an absolute minimum of $f(x)$ and $x = 1$ gives an absolute maximum.

A more striking example is the following. Let $f(x)$ be undefined for x rational and let $f(x) = 1$ for x irrational. Then (if DD is used to define the derivative) $f'(x)$ exists at every point in $D(f)$ and in fact $f'(x) = 0$, $x \in D(f)$. But if we had said that $f(x) = 0$ for rational x, instead of being undefined, the resulting function would be discontinuous at every value in its domain and would not have a derivative anywhere. Yet the only "advantage" the former function has over the latter, as regards smoothness, is that the former is undefined at some points where the latter is defined.

It is perhaps worthwhile to reflect why it is that none of these problems are encountered with SD. The reason is that existence of $\lim f(x)$ with SD automatically implies that a is a limit point of $D(f)$, and it also implies that $f(x)$ is defined for x near a, or $a-$, or $a+$ as the case may be. When two or more functions are involved, Theorem 1 and its analogs ensure that all of them are defined at the relevant values of x and questions about the domain do not arise.

The above discussion lends color to the view that DD should not be used in introductory courses on calculus. If you begin by learning SD you will have no trouble in understanding DD when the need for it arises at a later time. But if you start with DD, chances are that you will never understand either DD or SD.

Answers to questions. If we do not know the uniqueness theorem, we must allow the possibility that $\lim f(x)$ is set-valued and interpret statements accordingly. Aside from this it can be said that the results of Examples 1–8 retain their validity. For a specific illustration let us consider Example 2, the theorem concerning boundedness of $1/f(x)$. In our presumed state of ignorance the theorem would say that if L is a value of $\lim f(x)$ and $L \neq 0$, then $1/f(x)$ is bounded for x near a, and the bound can be any number larger than $1/|L|$. This is still true, and the proof is identical to the proof given above.

The answer to the second question, whether uniqueness can be deduced from localization, is more subtle. At first glance it would seem that one could reason as follows: Suppose $\lim f(x)$ has two values L_1 and L_2. Then we can say $\lim f(x) = L_1$, and with $g(x) = f(x)$ we can also say $\lim g(x) = L_2$. By the localization theorem $\lim f(x) = \lim g(x)$, hence $L_1 = L_2$.

Despite its plausibility, this line of thought is incorrect. So long as uniqueness is in doubt the localization theorem gives $\lim f(x) = \lim g(x)$ only in the sense of set equality; if a value L is in one of these sets, it is also in the other. (An examination of the argument will show that this is what is actually proved.) To clinch the matter, suppose we had defined "$\lim f(x) = L$" by use of the inequality $|f(x) - L| < 1$ rather than $|f(x) - L| < \varepsilon$. Then localization would still hold in the sense of set equality. But uniqueness fails, as shown by the simplest examples.

BIBLIOGRAPHIC ENTRIES: LIMITS AND CONTINUITY

1. *Monthly* Vol. 77, No. 3, pp. 303–306. F. M. Pavlick, A comparson of two approaches to teaching limit theory.

2. *Monthly* Vol. 81. No. 7, pp. 739–743, P. Cameron, J. G. Hocking, and S. A. Naimpally, Nearness—a better approach to continuity and limits

3. *Monthly* Vol. 82, No. 1, pp. 63–64. P. Ramankutty and M. K. Vamanamurthy, Limit of the composite of two functions.

4. *Math. Mag.* Vol. 48, No. 2, p. 101. R. B. Darst and E. R. Deal, Remarks on limits of functions.

Functions of several variables.

5. TYCMJ Vol. 5, No. 3, pp. 12–13. Peter A. Lindstrom, A note on epsilons and deltas.

Uses definition of limit to show that $x^n \to a^n$ as $x \to a$.

6. TYCMJ Vol. 7, No. 3, p. 18. Larry F. Bennett. Another note on epsilons and deltas.

Sequel to the previous entry.

7. TYCMJ Vol. 11, No. 4, pp. 263–266. A. L. Yandl, Delta, epsilon, and polynomials.

8. TYCMJ Vol. 14, No. 1, pp. 42–47. Larry King, The ϵ-δ connection.

5

DIFFERENTIATION
(a)
THEORY

AN INTRODUCTION TO DIFFERENTIAL CALCULUS

D. G. Herr, Duke University

How do you begin a first course in calculus on a note of excitement and anticipation? How do you convey the essence of what the course is about in two or three introductory lectures? These are some of the questions which faced two of us this spring in planning a revamped calculus sequence. We decided to give three introductory lectures: 1. An Introduction to Differential Calculus, 2. An Introduction to Integral Calculus, and 3. An Introduction to the Real Numbers. The order of the first two is rather arbitrary, but we felt that since a lecture on the real numbers was potentially the driest of the three we would build up some interest in the properties of the real numbers (especially inequalities) by having the introduction to calculus precede the lecture on real numbers. So far our ideas paralleled those in several texts.

The next problem we had to face was the content of the three lectures. Taking the last first, we decided on an axiomatic treatment of the real numbers with some historical notes such as the discovery of the irrationality of π in 1761 by Lambert [3] and the solution of the problem of "squaring the circle" as a result of C. L. F. Lindemann's proving in 1882 that π is not algebraic [3]. The lecture on the integral was not difficult to construct in view of T. M. Apostol's excellent introduction to integration in his calculus text [1]. However, the lecture on differential calculus was a problem.

The usual introduction to differential calculus makes use of physical concepts such as velocity and thus immediately the focus of the discussion is on the limit of the difference quotient. This seems to us unfortunate for two reasons. First this attention to the derivative does not generalize satisfactorily to functions of several variables in that the existence of partial derivatives is not a generalization of the existence of a derivative. In addition this immediate obsession with the difference quotient obscures the actual problem being attacked and as a consequence such concepts as differentials often remain a mystery.

We take the position that the principal concern of differential calculus is the linearizing of functions, or more completely the identification of linearizable (differentiable) functions and their subsequent linearization. To be more explicit suppose f is a real valued function defined on the real line, R. In geometric terms linearizing f will mean finding a straight line which best approximates f. A student can readily visualize such a linearization and furthermore simple examples quickly demonstrate the futility of trying to obtain one straight line to approximate f over all of R in any reasonable way. In this way the point that we are concerned with local properties of functions in differential calculus is clearly and forcefully made. The problem now is to move from the geometric argument to an analytic one which will be explicit enough to answer two questions: "Which functions are linearizable at a point?" and "If a function is linearizable, what is the 'best' straight line approximation?" To effect this transition we ask the student to help formulate a definition of "best straight line approximation."

Since a nonvertical straight line in the plane has the general equation $y = ax + b$ and functions of the form $A = \{(x, y); y = ax + b\}$ are called affine functions, we want the "best affine approximation to f at a point."

It is not difficult to get the students to agree that one condition that a best affine approximation to f at x_0 should satisfy is

(I) $$A(x_0) = f(x_0),$$

where A is the approximating affine function. This establishes that $b = f(x_0) - ax_0$ so that we need only specify a, the slope of the line, to have A completely determined. A second reasonable condition to impose on A is that

(II) $$|A(x) - f(x)| \to 0 \quad \text{as} \quad |x - x_0| \to 0,$$

i.e., the error made by approximating f by A at x becomes arbitrarily small as x tends to x_0. Examples easily demonstrate that (II) may pose no restriction whatever on A and is thus not a strong enough condition for our purposes. However, although the error $|A(x) - f(x)|$ gets small for any A satisfying (I) (f continuous at x_0) as $x \to x_0$, examples are again enough to demonstrate that the rate at which the error gets small is not the same for all A. If the relative error $|A(x) - f(x)| / |x - x_0|$ is used as a measure of the rate at which the error tends to zero, it is natural enough to require the best approximation to have the best rate, i.e.,

(III) $$|A(x) - f(x)| / |x - x_0| \to 0 \quad \text{as} \quad |x - x_0| \to 0.$$

Note here that (III) implies (II). By noting that if A satisfies (I), $|A(x) - f(x)| = |f(x) - f(x_0) - a(x - x_0)|$, the student can readily see that (III) is equivalent to

(IV) $$\left| \frac{f(x) - f(x_0)}{x - x_0} - a \right| \to 0 \quad \text{as } x \to x_0.$$

It can then be pointed out that if the "best affine approximation to f at x_0" is defined to be any A which satisfies (I) and (III), then the slope of the best fitting straight line, a, satisfies (IV) which means that a and thus A are uniquely determined. Furthermore (IV) then provides an answer to both original questions, i.e., the functions f which are linearizable at x_0 are precisely those for which the limit as $x \to x_0$ of $[f(x) - f(x_0)]/(x - x_0)$ exists as a real number, say $f'(x_0)$, and the best affine approximation is $A = \{(x, y): y = f'(x_0)(x - x_0) + f(x_0)\}$. Thus the best affine approximation to f at x_0 is defined to be any A which satisfies (I) and (III).

With this foundation the student can see limits of difference quotients, derivatives, and differentiability in perspective. Also the differential of f at x_0 as the linear part of the best affine approximation to f at x_0 is a straightforward concept without any of the mystery usually associated with a differential. Finally these ideas generalize directly to vector-valued functions of several variables with the understanding that linearizability and differentiability are synonymous [2].

References

1. T. M. Apostol, Calculus, Volume 1, Blaisdell, Waltham, Mass., 1961.

2. R. H. Crowell and R. E. Williamson, Calculus of Vector Functions, Prentice-Hall, Englewood Cliffs, N. J., 1962, 196–207.

3. G. H. Hardy and E. M. Wright, An Introduction to the Theory of Numbers, 4th ed., Oxford, London, 1960; 47, 173.

AN ELEMENTARY PROOF OF A THEOREM IN CALCULUS

DONALD E. RICHMOND

500 Fulton St., #201, Palo Alto, CA 94301

A fundamental theorem states that if $f'(x) = 0$ at every point of an interval $[a, b]$, then $f(x)$ is constant on $[a, b]$. Most proofs use the Mean Value Theorem, but several proofs have been given that are independent of this theorem (Bers [1], Halperin [3], Powderly [4]). The present note proves our fundamental theorem in an elementary way.

Let $f'(x) \equiv 0$ on $[a, b]$. If $f(x)$ is not constant on $[a, b]$ as the theorem states, then for some $u < v$ on $[a, b]$, $f(v) \neq f(u)$. Hence the chord joining $(u, f(u))$ and $(v, f(v))$ has a slope different from zero. That is

$$(1) \qquad f(v) - f(u) = C(v - u), \quad \text{or} \quad \Delta f = C\Delta x, \quad \text{with } C \neq 0.$$

Assume that $C > 0$. Bisect $[u, v]$ at w. If

$$f(v) - f(w) < C(v - w)$$

and

$$f(w) - f(u) < C(w - u),$$

then

$$f(v) - f(u) < C(v - u),$$

which contradicts (1) with $C > 0$. Hence on at least one of the intervals $[u, w]$ and $[w, v]$, $\Delta f / \Delta x \geq C$. Call this interval $[u_1, v_1]$. By repeated bisection one obtains a nested sequence of intervals $[u_n, v_n]$ over each of which $\Delta f / \Delta x \geq C$. $[u_n, v_n]$ converges to some x in $[u, v]$. Hence if $C > 0$, $\Delta f / \Delta x$ cannot approach $f'(x) = 0$ as required.

The possibility $C < 0$ is excluded by reversing the inequalities. Hence $C = 0$, contrary to the assumption that $f(v) \neq f(u)$.

The same argument shows that if $f'(x) \geq 0$ on $[a, b]$, then $f(v) \geq f(u)$, and if $f'(x) \leq 0$ on $[a, b]$, then $f(v) \leq f(u)$.

The argument may be generalized to show that if $m \leq f'(x) \leq M$ on $[a, b]$, then

$$(2) \qquad m(v - u) \leq f(v) - f(u) \leq M(v - u).$$

It is sufficient to replace $C > 0$ by $C > M$ and $C < 0$ by $C < m$. (2) has been called the "weak" form of the Mean Value Theorem [2].

References

1. L. Bers, On avoiding the mean value theorem, this MONTHLY, 74 (1967) 583.
2. R. P. Boas, Who needs those mean-value theorems, anyway?, Two-Year College Mathematics Journal, 12 (1981) 178–181.
3. I. Halperin, A fundamental theorem of the calculus, this MONTHLY, 61 (1954) 122–123.
4. M. Powderly, A simple proof of a basic theorem of the calculus, this MONTHLY, 70 (1963) 544.

Inverse Functions and their Derivatives

ERNST SNAPPER

Department of Mathematics and Computer Science, Dartmouth College, Hanover, NH 03755

If the concept of inverse function is introduced correctly, the usual rule for its derivative is visually so obvious, it barely needs a proof. The reason why the standard, somewhat tedious proofs are given is that the inverse of a function $f(x)$ is usually graphed by flipping the graph of $f(x)$ over the 45° line. And although this procedure is formally correct, it hides the visual obviousness of the formula for the derivative of the inverse. In short, graphing the inverse by flipping over the 45° line is an error in pedagogy.

The concept of the inverse of a one-to-one and onto function should first be explained on the set-theoretic level. This means that domains (of definition) and ranges should not be restricted to subsets of the real numbers, but arbitrary sets, and also subsets of the plane and 3-space should be used. It should be stressed that the inverse is obtained by simply interchanging domain and range. If this is done properly, the students should feel irritated that their time is being wasted on such trivialities.

What does this mean for the graph of the inverse of a real variable function $f(x)$ in one variable? Suppose the graph of $f(x)$ is given:

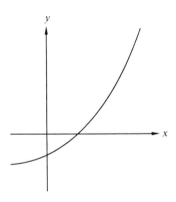

FIG. 1. The graph of $y = f(x)$.

How does one find the graph of the inverse $g(x)$ of $f(x)$? Since the students know that this is a matter of interchanging domain and range, they will quickly and correctly say: Interchange the x-axis and y-axis. Hence the graph is:

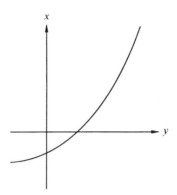

FIG. 2. The graph of the inverse $y = g(x)$.

The instructor should stress that the fact that the two axes are now in an unnatural position is entirely immaterial. This is the graph of the inverse which is helpful and effective. For example, the standard theorem that $f(x)$ is strictly increasing (decreasing) iff its inverse has the same property is now visually obvious. The same holds for all other theorems. Let us work it out for the derivative.

The students have learned that the derivative of $f(x)$ at a point P on the curve $y = f(x)$ is $\tan \alpha$, where α is the angle which the positive x-axis makes with the upward direction of the tangent of $y = f(x)$ at P:

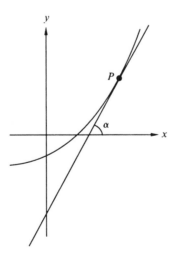

FIG. 3. $f'(P) = \tan \alpha$.

Interchanging the *x*-axis and *y*-axis gives, by the same rule, that the derivative of the inverse $g(x)$ of $f(x)$ at P is $\tan \beta$:

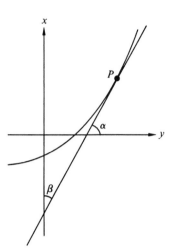

FIG. 4. $g'(P) = \tan \beta$.

Since $\alpha + \beta = 90°$, $\tan \beta = 1/\tan \alpha$ and the formula for the derivative of the inverse has been obtained.

This method of interchanging the *x*-axis and *y*-axis is also very efficient for viewing the graph of the inverse function with the axes in their *natural* position. Simply graph the function $f(x)$ on a piece of paper, putting a little pressure on the pencil, interchange the *x* and *y* and now look at the graph through the back of the paper with the axes in their natural position. What you see is the customary graph of the inverse of $f(x)$. It is very effective to do this in class, choosing for $f(x)$ some trigonometric function or a polynomial function such as $y = x^2$ or $y = x^3$. You can easily look through the back of the paper by holding it up to a window on a sunny day or up to a light. No visual aids are necessary or even helpful.

Of course, once the above has been done, there is nothing against making the students observe that the graph of the inverse of $f(x)$, viewed through the back of the paper, is the same as the graph of $f(x)$ flipped over the 45° line. This flipping over the 45° line can again be done easily with pencil and paper but, at least with some programs, computers can do it too. The fact remains, however, that for the purpose of understanding the laws which govern the inverse, nothing beats interchanging the *x*-axis and *y*-axis in the graph of $f(x)$ and doing nothing else.

An Elementary Result on Derivatives

David A. Birnbaum
Northrup Fowler III

David A. Birnbaum has taught at Amherst College and is now an Assistant Professor of Mathematics at Hamilton College. He received his Ph.D. in mathematics from the University of Illinois. His current mathematical interests are functional analysis and mathematical statistics.

Northrup Fowler III received his Ph.D. in mathematics from Rutgers University in 1973. After one year as an Instructor at Rutgers, he joined the faculty of Hamilton College as an Assistant Professor of Mathematics. His current research interest is recursive function theory.

A standard problem in an elementary calculus course is to compute the derivative at $x = 0$ of continuous functions such as

$$g(x) = \begin{cases} x^2 + 3x + 2, & \text{if } x \geq 0, \\ 2x^3 + 3x + 2, & \text{if } x < 0. \end{cases}$$

One usually applies the definition of the derivative and one-sided limits to determine the solution. Some of our students, however, often reason as follows: for x larger than 0, $g'(x)$ equals $2x + 3$ which approaches 3 as x decreases to 0. Similarly, for x less than zero, $g'(x)$ equals $6x^2 + 3$ which also approaches 3 as x increases to 0. Conclusion: $g'(0)$ equals 3. Interestingly enough, this argument is essentially correct because of the following result:

Theorem. *Suppose* $y = f(x)$ *is continuous over the open interval* I *containing the point* a. *Suppose further* $f'(x)$ *exists over the set* $I \setminus \{a\}$. *If* $\lim_{x \to a} f'(x) = L$, *then* $f'(a)$ *exists and* $f'(a) = L$.

Proof: We show that $\lim_{h \to 0^+} (f(a + h) - f(a))/h = L$. Since f is continuous on I and differentiable on $I \setminus \{a\}$, f is continuous on $[a, a + h] \subseteq I$ and differentiable on $(a, a + h) \subseteq I$. By the Mean Value Theorem applied to $f(x)$ on the interval $[a, a + h] \subseteq I$, we note that $(f(a + h) - f(a))/h = f'(c)$ for some c in $(a, a + h)$. Taking the limit as h decreases to zero of both sides of the last equality, we see that

$$f'_+(a) = \lim_{h \to 0^+} \frac{f(a + h) - f(a)}{h}$$

$$= \lim_{h \to 0^+} f'(c)$$

$$= \lim_{c \to a^+} f'(c) = L.$$

Similarly, one shows that $f'_-(a) = L$. The conclusion now follows.

The reader may wish to show that the function

$$h(x) = \begin{cases} x^2 \sin(1/x), & x \neq 0, \\ 0, & x = 0, \end{cases}$$

provides an example where $f'(a)$ exists but where $\lim_{x \to a} f'(x)$ does-not, and hence the partial converse fails.

We have found this result to be a nice classroom application of the Mean Value Theorem which is non-trivial yet readily accessible to most students. It is easily motivated, supports their intuitive view of derivatives and enhances their understanding of the Mean Value Theorem.

Mapping Diagrams, Continuous Functions and Derivatives

Thomas J. Brieske

Thomas Brieske is Associate Professor of Mathematics at Georgia State University, Atlanta, where he has been teaching since 1969. He received his Ph.D. from the University of South Carolina in 1969. His special areas of interest are learning in mathematics, mathematics curriculum, and teacher preparation.

Introduction. It is difficult to imagine teaching calculus without using graphs of functions. Important concepts such as functions increasing or decreasing on an interval, continuous functions, local extrema of functions, concavity, inflection points, the derivative, the mean value theorem, and the definite integral are effectively visualized by using graphs. However, other concepts, such as composition of functions, continuity of composite functions, derivatives of composite functions, the change of variables formula, and the generalization of some of these concepts to vector valued functions of a vector variable are not readily visualized by using graphs. The visual representations of these concepts are best rendered by mapping diagrams.

In an article published in 1963 [4], D. E. Richmond related mapping diagrams to derivatives. More recently some textbooks (see the bibliography) have used mapping diagrams in a limited way. The purpose of this article is to argue for a more enthusiastic use of mapping diagrams by demonstrating how effective they are in teaching some concepts in calculus.

In Figure 1 the mapping diagrams for the functions illustrate some important properties of these functions. For example, from the diagrams it is clear that: g is one-to-one and onto while f is neither; g has an inverse (reverse the arrows) and f does not; the image of an interval under g is always an interval twice as long as the pre-image—that is, g has the constant *scale factor* of 2; f also maps intervals onto intervals but has no constant scale factor.

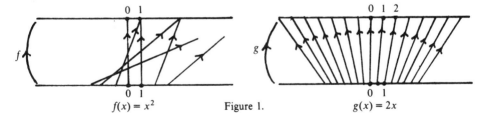

$$f(x) = x^2 \qquad \text{Figure 1.} \qquad g(x) = 2x$$

Continuity. A function $s: R \to R$ is said to be continuous at a if for every open interval B centered at $s(a)$ there is an open interval A centered at a such that the image of A is contained in B. If r and s are continuous functions, the proof that the

113

composite function, $r \circ s$, is continuous is almost transparent when mapping diagrams are used. As Figure 2 illustrates, for every interval C centered at $r(s(a))$ there is an open interval B centered at $s(a)$ such that $r(B) \subset C$ since r is continuous; also there is an open interval A centered at a such that $s(A) \subset B$ since s is continuous. Consequently $(r \circ s)(A) \subset C$ and $r \circ s$ is continuous at a. In my experience many students familiar with mapping diagrams have produced this proof on their own.

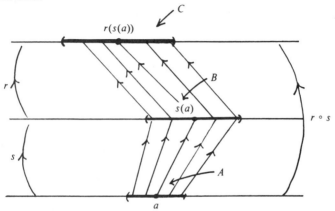

Figure 2.

In the case of linear functions of the form

$$l : R \to R,$$
$$l(x) = mx, \quad m \neq 0,$$

mapping diagrams are especially vivid. Since $l(b) - l(a) = m(b - a)$, the length of the range interval is $|m|$ times the length of the mapped interval $[a, b]$.

The mapping diagram in Figure 3 gives the argument for continuity a geometric flavor: l is continuous at c since if P is the open interval of length p centered at $l(c)$, the open interval of length $p/|m|$ centered at c is mapped onto P.

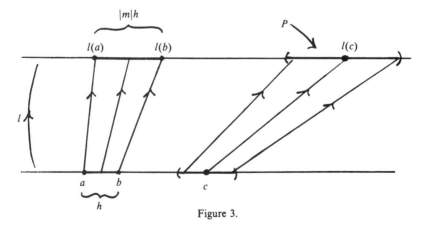

Figure 3.

An important reason for using mapping diagrams to study R to R functions is to prepare students for their use in studying R^2 to R^2, R^2 to R^3, R^3 to R^2, and R^3 to R^3 functions. In particular, mapping diagrams facilitate the generalization of the concepts of continuity and derivative. For example the mapping diagram for the function

$$F : R^2 \to R^2,$$
$$F(x, y) = (2x, 2y),$$

shows that F is an enlargement of the plane with scale factor 2, which makes it easy to show that F is continuous at the point (a, b) (see Figure 4). Let K be the open disc of radius r centered at $(2a, 2b)$; then L, the disc of radius $r/2$ centered at (a, b), is mapped onto K by F. (The similarity to $f : R \to R$, $f(x) = 2x$ is obvious.)

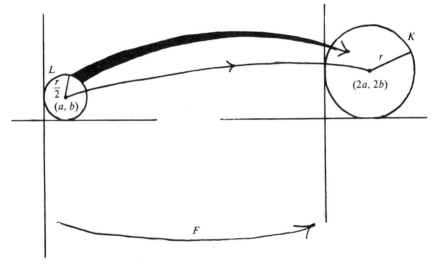

Figure 4.

The proof that the composition of two continuous R^2 to R^2 functions is continuous is clear from a similar two step mapping diagram where R^2 regions replace the linear intervals of Figure 2.

Derivatives.   Since derivatives are closely related to linear functions and linear functions have particularly vivid representations as mapping diagrams, it is natural to use mapping diagrams to represent derivatives. The usual definition of the derivative of the function $f : R \to R$ at a,

$$\lim_{h \to 0} \frac{f(a + h) - f(a)}{h} = f'(a),$$

is equivalent to

$$\lim_{h \to 0} \frac{f(a + h) - f(a) - f'(a)h}{h} = 0.$$

Since a is fixed the numerator is a function of h only: call it $o(h)$; then $f(a + h) - f(a) - f'(a)h = o(h)$ implies $f(a + h) - f(a) = f'(a)h + o(h)$, where $\lim_{h \to 0} o(h)/h = 0$. The result means simply that for sufficiently small $|h|$, $|o(h)|$ is much smaller than $|h|$, so the approximation $f(a + h) - f(a) \approx f'(a)h$ is valid. (For example, if $f(x) = x^2$, $f(a + h) - f(a) = (a + h)^2 - a^2$ implies that $f(a + h) - f(a) = 2ah + h^2$ which may be approximated as $f(a + h) - f(a) \approx 2ah$, where $f'(a) = 2a$ and $o(h) = h^2$.)

The correspondence $h \to f'(a)h$ defines a linear function of h with the scale factor $f'(a)$. The mapping diagram in Figure 5 illustrates its meaning; if $|h|$ is small, the difference $f(a + h) - f(a)$ is well approximated by the product of the scale factor $f'(a)$ and h. So, very near a, f has the effect of either enlarging or reducing the lengths of intervals by the absolute value of the scale factor, $f'(a)$.

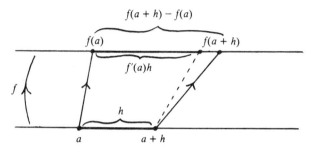

Figure 5.

The chain rule can be made plausible through mapping diagrams (see Figure 6). Suppose s and r are differentiable functions. To find the derivative $(r \circ s)'(a)$ one must find the scale factor for $r \circ s$ at a. It is clear that the scale factor is the product of the scale factors of the individual mappings; that is, $(r \circ s)'(a) = r'(s(a)) \cdot s'(a)$.

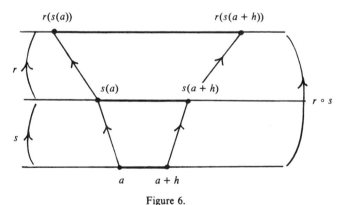

Figure 6.

In my opinion heuristic arguments of this type are very important in convincing students that a theorem is plausible. Even if students do not discover this argument for the chain rule, they are left with the feeling, after seeing it, that they might have produced it on their own. Contrast this reaction with the all too frequent dumfounded response to the arguments of the "consider a function of the form . . ." genre. Also as we shall see, the above heuristic argument can be generalized to R^2 to R^2 functions.

If $F : R^2 \rightarrow R^2$ is differentiable at $A = (a_1, a_2)$, and $H = (h_1, h_2)$, then $F(A + H) - F(A) = F'(A)H + o(H)$, where $F'(A)$ is a linear function from R^2 to R^2 and

$$\lim_{H \to 0} \frac{\|o(H)\|}{\|H\|} = 0.$$

The linear function $F'(A)$ can be represented by a 2×2 matrix of partial derivatives denoted by $[F'(A)]$. (When using this matrix in computation, the ordered pair (a, b) is written $\binom{a}{b}$.) For example, if

$$F : R^2 \rightarrow R^2,$$
$$F(x_1, x_2) = \left(x_1^2, x_2^2\right),$$

F is comprised of two R^2 to R coordinate functions, $F_1(x_1, x_2) = x_1^2$ and $F_2(x_1, x_2) = x_2^2$; that is, $F(x_1, x_2) = (F_1(x_1, x_2), F_2(x_1, x_2))$. This function is differentiable at $A = (a_1, a_2)$ and

$$[F'(A)] = \begin{bmatrix} \dfrac{\partial F_1}{\partial x_1}(A) & \dfrac{\partial F_1}{\partial x_2}(A) \\ \dfrac{\partial F_2}{\partial x_1}(A) & \dfrac{\partial F_2}{\partial x_2}(A) \end{bmatrix} = \begin{bmatrix} 2a_1 & 0 \\ 0 & 2a_2 \end{bmatrix}.$$

Therefore, for H close to 0 the difference $F(A + H) - F(A)$ is closely approximated by $F'(A)H$, where the linear function $F'(A)$ consists of a stretch along the horizontal axis by a factor of $2a_1$ and a stretch along the vertical axis by a factor of $2a_2$. This approximation is represented in the mapping diagram of Figure 7.

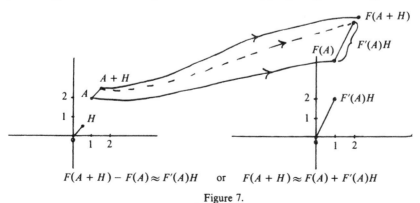

$$F(A + H) - F(A) \approx F'(A)H \quad \text{or} \quad F(A + H) \approx F(A) + F'(A)H$$

Figure 7.

The chain rule for R^2 to R^2 functions is a direct generalization of the R to R case. If $S : R^2 \rightarrow R^2$ and $T : R^2 \rightarrow R^2$ are differentiable functions, then using two

R^2 to R^2 mapping diagrams and the $R \rightarrow R$ chain rule as guides, it is reasonable to conjecture that

$$T(S(A + H)) - T(S(A)) = T(S(A) + S(A + H) - S(A)) - T(S(A))$$

$$\approx T'(S(A))[S(A + H) - S(A)]$$

$$\approx T'(S(A))S'(A)H,$$

which suggests that $(T \circ S)'(A) = T'(S(A)) \circ S'(A)$, where $(T \circ S)'$, T', and S' are linear functions from R^2 to R^2. This is indeed the case. As in the R to R case the derivative of the composite function is the composition of the derivative functions, or expressed in terms of matrices

$$[(T \circ S)'(A)] = [T'(S(A))] \cdot [S'(A)];$$

the matrix of the derivative of the product (composition) of two differentiable functions is the product (matrix product) of the matrices of the derivatives.

Conclusion. I believe that mapping diagrams belong in the mathematics curriculum from the beginning—that is, from elementary school onwards. That this is not now the case is unfortunate but should not prevent their use in the calculus. The extra time required to include them in the calculus sequence pays several dividends. The student's concept of an R to R function is broadened in a way that is especially useful when studying $R^2 \rightarrow R^2$, $R \rightarrow R^2$, $R^2 \rightarrow R$, $R^2 \rightarrow R^3$, $R^3 \rightarrow R^2$, and $R^3 \rightarrow R^3$ functions. More immediately, if students use mapping diagrams to study R to R functions they see the concepts of one to one functions, onto functions, inverses of functions, continuous functions, and derivatives of functions in an alternative mode of visualization in addition to the traditional approach through graphs. This is beneficial for two reasons: it enables students who have failed to understand the traditional approach to these concepts and related theorems to have a second chance at their mastery—the R to R chain rule for example; and it ultimately forces students to accommodate their understanding of all these concepts to include both of these representations thus strengthening their understanding. With this stronger understanding the generalization of concepts to the various R^N to R^Q functions previously mentioned is a much more active process for students, especially if they are first given a short review of the corresponding R to R mapping diagram visualization of the relevant concepts.

REFERENCES

1. T. J. Brieske, Functions, mappings, and mapping diagrams, The Mathematics Teacher, 66, May 1973.
2. A. G. Fadell, Calculus with Analytic Geometry, Van Nostrand, Princeton, N.J., 1964.
3. F. J. Flanigan and J. L. Kazdan, Calculus Two: Linear and Non-Linear Functions, Prentice-Hall, Englewood Cliffs, N.J., 1971.
4. D. E. Richmond, Calculus: a new look, Amer. Math. Monthly, 7, 1963.
5. A. W. Roberts, Introductory Calculus, Academic Press, N.Y., 1972.
6. S. L. Salas and Einar Hille, Calculus: One and Several Variables, Xerox, Lexington, Mass., 1974.
7. E. W. Swokowski, Calculus with Analytic Geometry, Prindle, Weber, and Schmidt, Boston, Mass., 1975.

A Self-contained Derivation of the Formula $\frac{d}{dx}(x^r) = rx^{r-1}$ for Rational r

Peter A. Lindstrom, North Lake College, Irving, TX

To establish that $\frac{d}{dx}(x^r) = rx^{r-1}$ for rational r, calculus texts invoke earlier rules of differentiation (quotient rule, chain rule with implicit differentiation, etc.) to first prove this for special cases of r. The purpose of this note is to show that it is possible to establish this result directly, without having to resort to earlier theorems on differentiation. This proof can serve as an instructive exercise for capable students.

Let $r = m/n$, where m and n are positive integers. Then

$$f'(x) = \lim_{h \to 0} \frac{(x+h)^{m/n} - (x)^{m/n}}{h} = \lim_{h \to 0} \frac{\left[\left\{(x+h)^{1/n}\right\}^m - \left\{x^{1/n}\right\}^m\right]}{\left[\left\{(x+h)^{1/n}\right\}^n - \left\{x^{1/n}\right\}^n\right]}. \tag{1}$$

By letting $a = (x+h)^{1/n}$ and $b = x^{1/n}$ in the difference formula

$$a^N - b^N = (a-b)(a^{N-1} + a^{N-2}b + \cdots + ab^{N-2} + b^{N-1}),$$

and separately considering $N = m$ and $N = n$, we see that (1) becomes

$$f'(x) = \lim_{h \to 0} \frac{\left\{(x+h)^{1/n} - x^{1/n}\right\} \sum_{i=1}^{m} \left\{(x+h)^{1/n}\right\}^{m-i}\left\{x^{1/n}\right\}^{i-1}}{\left\{(x+h)^{1/n} - x^{1/n}\right\} \sum_{i=1}^{n} \left\{(x+h)^{1/n}\right\}^{n-i}\left\{x^{1/n}\right\}^{i-1}}$$

$$= \frac{\sum_{i=1}^{m} x^{(m-i)/n} \cdot x^{(i-1)/n}}{\sum_{i=1}^{n} x^{(n-i)/n} \cdot x^{(i-1)/n}}.$$

In particular,

$$f'(x) = \frac{\sum_{i=1}^{m} x^{(m-1)/n}}{\sum_{i=1}^{n} x^{(n-1)/n}} = \frac{mx^{(m-1)/n}}{nx^{(n-1)/n}} = (m/n)x^{(m/n)-1}. \tag{2}$$

Now consider the case of $r = -m/n$, where m and n are positive integers. Here

$$f'(x) = \lim_{h \to 0} \frac{(x+h)^{-m/n} - x^{-m/n}}{h} = \lim_{h \to 0} \frac{\left\{(1/(x+h))^{1/n}\right\}^m - \left\{(1/x)^{1/n}\right\}^m}{h},$$

119

which can be recast as (3):

$$f'(x) = \lim_{h \to 0} \frac{x^{m/n} - (x+h)^{m/n}}{h(x+h)^{m/n}x^{m/n}} = \lim_{h \to 0} \frac{-1}{(x+h)^{m/n}x^{m/n}} \cdot \lim_{h \to 0} \frac{(x+h)^{m/n} - x^{m/n}}{h}.$$

By definition, the last limiting quotient in (3) is the right-hand expression in (2). Hence,

$$f'(x) = \lim_{h \to 0} \frac{-1}{(x+h)^{m/n} \cdot x^{m/n}} \cdot (m/n)x^{(m/n)-1} = -(m/n)x^{-(m/n)-1}.$$

$(x^n)' = nx^{n-1}$: Six Proofs

Russell Jay Hendel, Dowling College, Oakdale, NY 11769

A perusal of calculus textbooks published during the past twenty years reveals six distinct approaches to the proof that the derivative of x^n is $(x^n)' = nx^{n-1}$ for a positive integer n. These proofs use important techniques that should be in the repertoire of every calculus student.

I. Proof with induction and the product rule. First, $(x)' = 1$ by definition. Assuming that $(x^{n-1})' = (n-1)x^{n-2}$ and using the product rule we have

$$(x^n)' = (x^{n-1}x)' = (x^{n-1})'x + x^{n-1}(x)' = nx^{n-1}.$$

[L. Bers with F. Karal, *Calculus*, Holt, Rinehart, and Winston, 1976, p. 87; N. Friedman, *Basic Calculus*, Scott, Foresman, 1968, p. 112; S. Salas and E. Hille, *Calculus: One and Several Variables with Analytic Geometry*, Wiley, 1974, p. 84]

II. Proof using properties of logarithms. If $y = x^n$, $\ln y = n \ln x$ and logarithmic differentiation yields $y'/y = n/x$ which immediately implies $y' = nx^{n-1}$. [L. Hoffmann, *Calculus for Business, Economics, and the Social and Life Sciences*, McGraw-Hill, 1986, p. 255]

III. Proof by estimation. We use the big O notation common in analytic proofs: We say $f(h) = O(g(h))$ if there is some constant C such that, if h is sufficiently close to 0, then $f(h) < C|g(h)|$. This is slightly nonstandard (h goes to zero instead of infinity) but it clarifies the proof.

Lemma. $(x + h)^n = x^n + nx^{n-1}h + O(h^2)$.

The lemma has a direct inductive proof independent of the binomial expansion. The lemma is used to simplify the numerator of the difference quotient $[(x+h)^n - x^n]/h$. A similar simplification occurs in proofs IV and V. [S. Stein, *Calculus and Analytic Geometry*, McGraw-Hill, 1987, p. 85]—Stein, however, does not use the big O notation.

IV. Proof using factoring. We use the factor formula for the difference of nth powers, $(x+h)^n - x^n = (x+h-x)\Sigma(x+h)^{n-1-i}x^i$, to simplify the difference quotient. [S. Grossman, *Calculus (International Edition)*, Academic Press, 1981, p. 135; L. Loomis, *Calculus*, Addison-Wesley, 1974, p. 100; A. Spitzbart, *Calculus with Analytic Geometry*, Scott, Foresman, 1975, pp. 85–86]

V. Proof using the binomial theorem. We use the binomial expansion, $(x+h)^n = \Sigma \binom{n}{i} x^{n-i} h^i$, to simplify the difference quotient. This seems to be the most popular method. [H. Anton, *Calculus with Analytic Geometry*, Wiley, 1988, pp. 159–160; J. Fraleigh, *Calculus with Analytic Geometry*, Addison-Wesley, 1980, p. 45; L. Leithold, *The Calculus with Analytic Geometry*, Harper and Row, 1986, p. 189; G. Simmons, *Calculus with Analytic Geometry*, McGraw-Hill, 1985, pp. 63–64; E. Swokowski, *Calculus with Analytic Geometry*, Prindle, Weber and Schmidt, 1983, pp. 99–100; A. Willcox, R. Buck, H. Jacob, and D. Bailey, *Introduction to Calculus 1 and 2*, Houghton Mifflin, 1971, pp. 63–64]

VI. "Proof" by example.   Finally, not every formula is totally proven in math courses. In such cases, computation of several examples or proof of several subcases of the main theorem is a welcome procedure [L. Goldstein, D. Lay, and D. Schneider, *Calculus and Its Applications*, Prentice-Hall, 1980, pp. 57–61]. Verification by examples or subcases has intrinsic value even when a general proof is given. For example, some calculus students can prove $(x^n)' = nx^{n-1}$ by the binomial theorem for general n, but still fumble when asked to prove it for the case $n = 2$ or $n = 3$.

Occasionally, a book does present several of the above proofs [H. Flanders and J. Price, *Calculus with Analytic Geometry*, Academic Press, 1978, p. 65; L. Bers, *Calculus*, Holt, Rinehart and Winston, 1969, p. 169]. Some proofs (II, III) are mentioned rarely. Even proof I, despite the importance of induction, occurs in only about 20% of the books. I suggest that freshman calculus courses be enriched by presenting all the above proofs.

A Note on Differentiation

Russell Euler, Northwest Missouri State University, Maryville, MO

The following technique illustrates an alternate method for deriving the product rule for differentiation.

If $f^2(x)$ is a differentiable function of x, then

$$\left[f^2(x)\right]' = \lim_{h \to 0} \frac{f^2(x+h) - f^2(x)}{h}$$

$$= \lim_{h \to 0} \frac{\left[f(x+h) + f(x)\right]\left[f(x+h) - f(x)\right]}{h}$$

$$= 2f(x)f'(x). \tag{$*$}$$

Now, for differentiable functions f and g, the identity

$$f(x)g(x) = \tfrac{1}{2}\left(\left[f(x) + g(x)\right]^2 - f^2(x) - g^2(x)\right)$$

and ($*$) give

$$\left[f(x)g(x)\right]' = \tfrac{1}{2}\left(2\left[f(x) + g(x)\right]\left[f'(x) + g'(x)\right] - 2f(x)f'(x) - 2g(x)g'(x)\right),$$

which simplifies to

$$\left[f(x)g(x)\right]' = f(x)g'(x) + g(x)f'(x).$$

Editor's Note: Once students know that the quotient of differentiable functions is a differentiable function, they may appreciate Marie Agnesi's 1748 proof of the quotient rule: If $h = f/g$, then $hg = f$ and (by the product rule) $hg' + h'g = f$; it remains only to substitute f/g for h and solve for h'.

Differentials and Elementary Calculus

D. F. Bailey, Trinity University, San Antonio, TX

The purpose of this capsule is to provide an example illustrating the use of the differential in basic calculus. The commonly used examples seem badly out of date when our students either own or have ready access to computers and sophisticated calculators. Examples of the type given below will make the topic more relevant and meaningful as well as giving new meaning and importance to the topic of implicit differentiation. The method is well known and is studied in most elementary courses in differential equations, namely the Euler "tangent line" method.

In the typical calculus text one finds $df = f'(x)\,dx$ and $\Delta f \doteq df$, for well-behaved functions, and thus

$$f(x+h) \doteq f(x) + f'(x)h. \tag{1}$$

To illustrate the utility of equation (1) let us consider the equation

$$\sin x \cos y = y. \tag{2}$$

This equation defines y implicitly as a function of x. Using implicit differentiation we have

$$y' = \frac{\cos x \cos y}{1 + \sin x \sin y}. \tag{3}$$

Most students are uncomfortable at this point because they feel the underlying function $y(x)$ is somehow less real than functions that can be exhibited explicitly. They think it is less real because the defining rule is too complex; they would prefer of course to be able to solve for y! To combat this idea I usually encourage my students to attempt to find at least one point on the curve defined by a given equation. Sometimes this is very difficult but in the present case students can determine after a while that $(0,0)$ is a point on the curve defined by equation (2). It is then possible to lead them to approximate $y(1)$, $y(2.3)$, or what have you as we show below.

Combining (1) and (3) we obtain

$$y(x+h) \doteq y(x) + h\left(\frac{\cos x \cos y(x)}{1 + \sin x \sin y(x)}\right). \tag{4}$$

Thus taking $h = 0.05$ in (4) and recalling that $y(0) = 0$ we find

$$y(0.05) \doteq y(0) + 0.05\left(\frac{\cos 0 \cos 0}{1 + \sin 0 \sin 0}\right) = 0.05.$$

Of course at this stage of the game students still question how valuable our ability to approximate values of y really is. This questioning disappears only when they see that we can continue our procedure to approximate $y(x)$ for x far away from 0. For instance, since we have deduced above that $(0.05, 0.05)$ is very nearly on the curve we seek, we have

$$y(0.1) \doteq y(0.05) + 0.05\left(\frac{\cos 0.05 \cos 0.05}{1 + \sin 0.05 \sin 0.05}\right)$$

124

or $y(0.1) \doteq 0.0997508$, and we can continue this process as long as we like. The approximate values of $y(x)$ for $x = 0, 0.05, 0.1, 0.15, \ldots, 0.95, 1$ are shown in Table 1.

Table 1

x	$y(x)$
0.0	0.0
0.05	0.05
0.10	0.0997508
0.15	0.1487664
0.20	0.1965994
0.25	0.2428634
0.30	0.2872467
0.35	0.3295173
0.40	0.3695203
0.45	0.4071697
0.50	0.4424365
0.55	0.4753372
0.60	0.5059219
0.65	0.5342638
0.70	0.5604510
0.75	0.5845799
0.80	0.6067498
0.85	0.6270597
0.90	0.6456055
0.95	0.6624783
1.0	0.6777631

To obtain a rough feeling for the accuracy of an approximation one should not, of course, go into a rigorous error analysis. Students will readily believe that accuracy depends on the size of h and one can reinforce this belief as follows. Table 1 shows $y(1) \doteq 0.6777631$ and equation (2) asserts that, were our values of y exact, we would have

$$\sin 1 \cos 0.6777631 = 0.6777631.$$

In fact

$$\sin 1 \cos 0.6777631 = 0.6554868$$

which means our approximation is not very accurate. If however one uses $h = 0.01$ in equation (4) the approximation given for $y(1)$ is 0.6660564 and

$$\sin 1 \cos 0.6660564 = 0.6616187$$

indicating that this approximation can be very accurate indeed.

This example illustrates one way to proceed when teaching differentials. There are many benefits to be derived from introducing Euler's method at the time one introduces the differential. Perhaps most important is the simple fact that students are convinced differentials have some utility. In addition students learn an important technique which will reappear in later courses in a more sophisticated form. Finally, the computer may be worked into the elementary calculus in a significant way.

THE DIFFERENTIABILITY OF a^x

J. A. EIDSWICK

A "from scratch" proof of the differentiability of a^x, $a > 0$, is avoided by essentially all modern-day authors. A slick and popular way of handling the problem is to define a^x as $e^{x \log a}$, its differentiability and other properties following from that of the functions e^x and $\log x$. Unfortunately, the usual definitions of e^x and $\log x$ involve relatively sophisticated ideas (e.g., integration or power series). Furthermore, the student, having heard of e, the natural logarithm base, at an early stage of his development, is hardly enlightened when he is told that e is e^1. He would have a much better feeling for the "naturalness" of e if it were defined as that number a for which $(a^x)' = a^x$.

The purpose of this note is to provide a direct and relatively simple way of getting at the differentiability of a^x. We define $a^x = \lim a^r$ as $r \to x$ through rational values of r from which continuity and other basic properties follow (see e.g., [1, p. 63]). The differentiability question obviously reduces to showing that the function $F(x) = (a^x - 1)/x$ has a limit at 0. Since $F(-x) = a^{-x}F(x)$, it suffices to show only that the right-hand limit exists. By a similar observation, we may assume that $a > 1$. As a final reduction, we note that, for $a > 1$, F is bounded below on $(0, \infty)$ and, hence, it is sufficient to show that F is increasing on $(0, \infty)$.

Define $S(x, n) = 1 + x + \cdots + x^{n-1}$ so that $S(x, n)(x - 1) = x^n - 1$. Since

$$n(a^{1/n} - a^{1/(n+1)}) = na^{1/(n+1)}(a^{1/n(n+1)} - 1)$$

$$> S(a^{1/n(n+1)}, n)(a^{1/n(n+1)} - 1)$$

$$= a^{1/(n+1)} - 1,$$

the sequence $\{F(1/n)\}$ is decreasing. Therefore, for positive rational numbers $m/n < p/q$, we have

$$F(m/n) = F(1/pn)S(a^{1/pn}, pm)/pm$$

$$< F(1/qm)S(a^{1/qm}, pm)/pm$$

$$= F(p/q).$$

In other words, F is increasing on the positive rationals. By continuity, F is increasing on $(0, \infty)$.

We conclude by noting that $(a^x)'$ is proportional to a^x and that the constant of proportionality can be taken to be 1 if a is chosen suitably, leading to an appealing definition of e (cf. [**2**, p. 41–44]).

References

1. Casper Goffman, Introduction to Real Analysis, Harper & Row, New York, 1966.
2. Edmund Landau, Differential and Integral Calculus, 2nd ed., Chelsea, New York, 1960.

DEPARTMENT OF MATHEMATICS AND STATISTICS, UNIVERSITY OF NEBRASKA, LINCOLN, NE 68508.

BIBLIOGRAPHIC ENTRIES: THEORY

1. *Monthly* Vol. 76, No. 7, pp. 816–817. W. R. Jones and M. D. Landau, The relation between the derivatives of f and f^{-1}.

2. *Monthly* Vol. 94, No. 4, pp. 354–356. J. B. Wilker, The nth derivative as a limit.

3. *Monthly* Vol. 98, No. 1, pp. 40–44. Stephen Kuhn, The derivative á la Caratheodory.

> A treatment similar to that in Apostol's *Mathematical Analysis*, 2nd ed., pp. 105–107 (Addison-Wesley, 1974).

4. *Math. Mag.* Vol. 44, No. 4, pp. 214–216, Simeon Reich, Schwarz differentiability and differentiability.

5. *Math. Mag.* Vol. 59, No. 5, pp. 275–282. L. B. Rall, The arithmetic of differentiation.

6. TYCMJ Vol. 5, No. 2, pp. 68–70. Roland E. Larson, Continuous deformation of a polynomial into its derivative.

7. TYCMJ Vol. 7, No. 1, pp. 38–39. Lewis G. Maharam and Edward P. Shaughnessy, When does $(fg)' = f'g'$?

8. TYCMJ Vol. 10, No. 1, p. 37. Dan Kalman, Differentiation and synthetic division.

9. TYCMJ Vol. 11, No. 2, pp. 102–106. John W. Dawson, Jr., Wavefronts, box diagrams, and the product rule: a discovery approach.

> Variations on an idea, going back to Leibniz, of interpreting a product as the area of a rectangle.

10. TYCMJ Vol. 13, No. 5, pp. 328–329. Michael W. Ecker, The sum of zeros of polynomial derivatives.

11. CMJ Vol. 17, No. 2, pp. 133–143. Irl C. Bivens, What a tangent line is when it isn't a limit.

> (This paper won a Polya award).

(b)

APPLICATIONS TO GEOMETRY

The Width of a Rose Petal

S. C. Althoen and M. F. Wyneken
Department of Mathematics, University of Michigan, Flint, MI 48502

"Don't hurry, don't worry. You're only here for a short visit. So be sure to stop and smell the roses" (cf. [1]). Roses are a beautiful class of polar graphs too often passed by too quickly for lack of an interesting remark. (However, see [2] for a thorough discussion of roses and computer graphics.) We investigate three aspects of the famous cosine roses

$$r = \cos n\theta, \qquad n = 2, 3, 4, \ldots, \tag{1}$$

and two generalizations: the wonderful yet less well-known "double" roses

$$r = a + b \cos n\theta, \qquad n = 2, 3, 4, \ldots, \qquad 0 < a < b, \tag{2}$$

which generalize the limaçon, $r = a + b \cos \theta$; and the family

$$r^\varepsilon = \cos n\theta, \qquad n = 2, 3, 4, \ldots, \qquad \varepsilon > 0, \tag{3}$$

which generalizes the lemniscate, $r^2 = \cos n\theta$.

Graphs. First let's look at a few graphs:

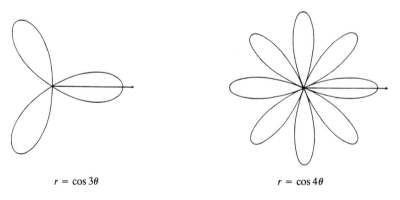

$r = \cos 3\theta$ $\qquad\qquad\qquad\qquad$ $r = \cos 4\theta$

Fig. 1

On the one hand, the little petals in Figure 2 grow longer as $a \to 0$. Thus $r = \cos 2k\theta$ has $4k$ petals, while $r = \cos(2k + 1)\theta$ has only $2k + 1$ petals. On the

128

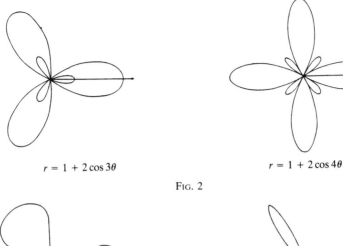

$r = 1 + 2\cos 3\theta$ $r = 1 + 2\cos 4\theta$

FIG. 2

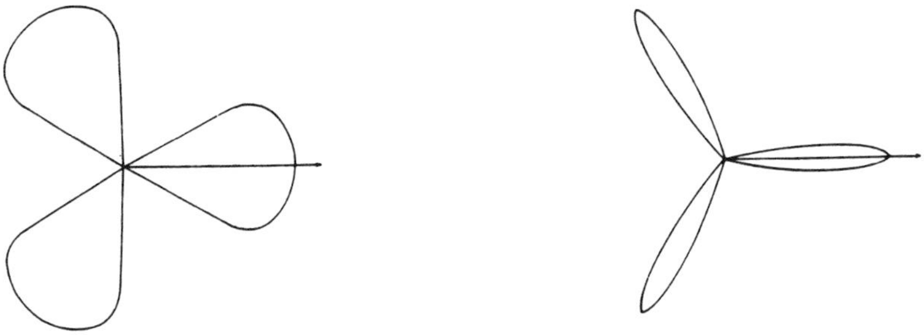

$r^8 = \cos 3\theta$ $r^{1/8} = \cos 3\theta$

FIG. 3

other hand, these little petals shrivel up as $a \to b$. Thus $r = 1 + \cos n\theta$ is an n-petalled rose for $n = 2, 3, \ldots$.

Petal widths. The graphs of equations (1) and (3) are symmetric about the line containing the polar axis and look generally like the graph in FIGURE 4 for $-\pi/2n \leqslant \theta \leqslant \pi/2n$.

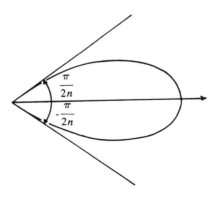

FIG. 4

A natural problem is to find the maximum width $2y_n$ of the petals of $r = \cos n\theta$. Although one can approach this problem by first converting to rectangular coordinates, it is most satisfying to attack the problem intrinsically. To this end consider FIGURE 5.

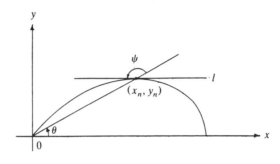

FIG. 5

The angle ψ is the angle from the radius vector to the tangent and, as is shown in calculus, $\tan \psi = r/(dr/d\theta)$. For (x_n, y_n) to be a maximum we must have the tangent line l parallel to the polar axis. Thus, we require $\psi = \pi - \theta$, so $\tan \psi = \tan(\pi - \theta) = -\tan \theta$ and so

$$\frac{r}{dr/d\theta} = -\tan \theta$$

or

$$r + \tan \theta \frac{dr}{d\theta} = 0. \tag{4}$$

In the case of the four-leaf rose $r = \cos 2\theta$, (4) yields

$$2 \tan \theta = \cot 2\theta = \frac{1 - \tan^2 \theta}{2 \tan \theta},$$

so $\tan \theta = \sqrt{1/5}$. Then the width is $2y_n = 2r \sin \theta = 2 \cos 2\theta \sin \theta = 2\sqrt{6}/9$.

In general, if $r = \cos n\theta$, (4) yields $n \tan \theta = \cot n\theta$ which is difficult to solve because the formula that expresses $\cot n\theta$ as a rational function of $\tan \theta$ is complicated. Another approach to finding the width uses the fact that $\cos n\theta$ is a polynomial in $\cos\theta$. Because $\cos n\theta$ is the real part of $e^{in\theta}$:

$$\cos n\theta = \mathrm{Re}(\cos\theta + i\sin\theta)^n$$

$$= \mathrm{Re}\sum_{0\leqslant k\leqslant n}\binom{n}{k}i^k\sin^k\theta\cos^{n-k}\theta$$

$$= \sum_{0\leqslant m\leqslant n/2}\binom{n}{2m}(-1)^m\sin^{2m}\theta\cos^{n-2m}\theta$$

$$= \sum_{0\leqslant m\leqslant n/2}\binom{n}{2m}(-1)^m(1-\cos^2\theta)^m\cos^{n-2m}\theta.$$

We shall denote this polynomial by P_n, so that $P_n(\cos\theta) = \cos n\theta$. Observe that when n is even each term of P_n has even degree, and when n is odd each term of P_n has odd degree. Then, at least in principle, one can find the petal width of $r = \cos n\theta$ by doing the following. First differentiate the equation $y = r\sin\theta = \cos n\theta\sin\theta$ with respect to θ and set $dy/d\theta$ equal to zero. Denoting the smallest positive critical value by θ_n, we have

$$\cos\theta_n\cos n\theta_n = n\sin\theta_n\sin n\theta_n. \tag{5}$$

If we then add $n\cos\theta_n\cos n\theta_n$ to both sides of this equation and apply the difference formula for cosine, we obtain

$$\cos\theta_n P_n(\cos\theta_n) = \frac{n}{n+1}P_{n-1}(\cos\theta_n),$$

which is a polynomial equation in $\cos\theta_n$ of degree $n+1$ and with the degrees of terms all of the same parity. In the case when n is odd we can view this as a polynomial equation in $\cos^2\theta_n$ of degree $(n+1)/2$, and in the case when n is even we can divide both sides by $\cos\theta_n$ and then view this as a polynomial equation in $\cos^2\theta_n$ of degree $n/2$. One could, in theory, solve these equations exactly in terms of radicals for $n = 2, 3, \ldots, 8$. For arbitrary n, Newton's method could be applied to find the approximate petal width.

The reader can also do the width calculation with the equation $r = a + b\cos n\theta$, $0 < a < b$. For example, for $n = 2$ the maximum petal width of the large petal in FIGURE 6 is $(2\sqrt{6}/9)(b+a)^{3/2}b^{-1/2}$. The width of the little petal is $(2\sqrt{6}/9)(b-a)^{3/2}b^{-1/2}$. We leave the case $r^2 = a + b\cos 2\theta$, $0 < a < b$, as an exercise.

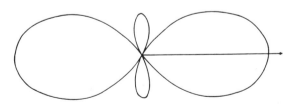

FIG. 6

Asymptotic results. In the search for asymptotic results, standard calculus limits have interesting geometric applications. First observe that $0 < n\theta_n < \pi/2$, so $\theta_n \to 0$ as $n \to \infty$. Then from (5) we obtain

$$(n\theta_n)^{-1} \cos \theta_n \cos n\theta_n = (n\theta_n)^{-1} n \sin \theta_n \sin n\theta_n$$

$$= \frac{\sin \theta_n}{\theta_n} \sin n\theta_n;$$

thus $\alpha^{-1} \cos \alpha = \sin \alpha$ or

$$\alpha^{-1} = \tan \alpha, \tag{6}$$

where α is a limit point of the sequence $\{n\theta_n\}_{n \geq 1}$. From (6) we observe that α cannot equal 0 or $\pi/2$ so that α must lie in the open interval $(0, \pi/2)$, and must in fact be the limit of the entire sequence since the graphs of α^{-1} and $\tan \alpha$ intersect in only one place on this interval. The corresponding x-coordinates $x_n = \cos n\theta_n \cos \theta_n$ will converge to $\cos \alpha$, and a calculator reveals $\alpha = .860333589\ldots$ and $\cos \alpha = .652184623\ldots$.

Now consider the equation $r^\varepsilon = \cos n\theta$, $\varepsilon > 0$. In solving for the smallest positive critical value $\theta_{n,\varepsilon}$ equations (5) and (6) become

$$\cos \theta_{n,\varepsilon} \cos n\theta_{n,\varepsilon} = \frac{n}{\varepsilon} \sin \theta_{n,\varepsilon} \sin n\theta_{n,\varepsilon} \tag{5_ε}$$

and

$$\varepsilon \alpha_\varepsilon^{-1} = \tan \alpha_\varepsilon \tag{6_ε}$$

where $n\theta_{n,\varepsilon} \to \alpha_\varepsilon$ as $n \to \infty$. Let $x_{n,\varepsilon}$ be the x-coordinate of the critical value $\theta_{n,\varepsilon}$, so $x_{n,\varepsilon} = (\cos n\theta_{n,\varepsilon})^{1/\varepsilon} \cos \theta_{n,\varepsilon}$, and let x_ε be the limit of $x_{n,\varepsilon}$ as $n \to \infty$, so $x_\varepsilon = (\cos \alpha_\varepsilon)^{1/\varepsilon}$. It is interesting to compute the limit of x_ε as $\varepsilon \to 0$. From (6_ε) one can see not only that $\alpha_\varepsilon \to 0$ as $\varepsilon \to 0$, but one can also obtain implicitly the derivative of α_ε with respect to ε. Hence, we may apply l'Hôpital's rule as follows:

$$\lim_{\varepsilon \to 0} \ln x_\varepsilon = \lim_{\varepsilon \to 0} \frac{\ln \cos \alpha_\varepsilon}{\varepsilon}$$

$$= \lim_{\varepsilon \to 0} - \tan \alpha_\varepsilon \frac{d\alpha_\varepsilon}{d\varepsilon}$$

$$= \lim_{\varepsilon \to 0} \frac{-\tan \alpha_\varepsilon}{\alpha_\varepsilon \sec^2 \alpha_\varepsilon + \tan \alpha_\varepsilon}$$

$$= \lim_{\varepsilon \to 0} \frac{-\sin \alpha_\varepsilon / \alpha_\varepsilon}{\sec \alpha_\varepsilon + (\sin \alpha_\varepsilon / \alpha_\varepsilon)}$$

$$= -\tfrac{1}{2},$$

and so $x_\varepsilon \to 1/\sqrt{e}$ as $\varepsilon \to 0$.

We note that these calculations can also be applied to the other intersection points of α^{-1} with $\tan \alpha$. These intersection points relate to the x-coordinates of the horizontal tangents of the other petals in the limit as $n \to \infty$.

The reader can easily generalize to the equations $r^\varepsilon = a + b \cos n\theta$, $0 < a < b$, $a + b = 1$, $\varepsilon > 0$, to again obtain $\lim_{\varepsilon \to 0} x_\varepsilon = 1/\sqrt{e}$. If $a + b > 1$, the graph blows up as $\varepsilon \to 0$, and if $a + b < 1$, the graph shrinks to the pole as $\varepsilon \to 0$.

REFERENCES

1. Apologies to the late golfer Walter C. Hagen (1892–1969), *New York Times* May 22, 1977, sec. v, p. 4.
2. P. M. Maurer, A rose is a rose ..., this MONTHLY, 94 (1987) 631–645.

Differentiating Area and Volume

Jay I. Miller

Jay I. Miller has taught at the Milwaukee and Parkside campuses of the University of Wisconsin. Currently he is an Assistant Professor of Mathematics at Marquette University . He earned his Ph.D. in Mathematics at the University of Illinois, Urbana, in 1976.

In a first semester calculus course, once the students acquire some facility at differentiation, there is usually someone who recognizes that $D_r(\pi r^2) = 2\pi r$ and $D_r(4\pi r^3/3) = 4\pi r^2$. I.e., when expressed as functions of their respective radii, the derivative of the area of a circle equals its circumference, and the derivative of the volume of a sphere equals its surface area. These results are noted as interesting, but all too often, that is where the matter ends.

This note shows, however, that the circle and sphere are not anomalies. In fact, using the correct analogs of the radius, the desired property holds for all regular polygons in the plane and for the five regular (Platonic) solids in 3-space. (A polygon is *regular* if all its sides have the same length and all its interior angles are equal. A *regular* solid is one in which all the faces are congruent regular polygons and all the polyhedral angles are equal.) The variables with respect to which we differentiate are the apothem (the perpendicular distance from the center of a regular polygon to a side) and its analog in 3-space, the perpendicular distance from the center of a regular solid to a face. Note that these are the radii of the incircle and insphere, respectively. With this in mind, we obtain the following theorems:

Theorem 1.   *The derivative of the area of a regular polygon equals its perimeter.*

Theorem 2.   *The derivative of the volume of a regular solid equals its surface area.*

Proof of Theorem 1.   Assume the polygon has n sides. Denote the apothem by a and half the length of one side by b. (See Fig. 1.)

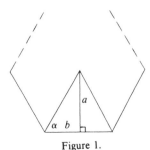

Figure 1.

From the figure, we have $b = a/\tan \alpha$, where α is half of an interior angle of the polygon. The area of a central triangle is ab, and thus the total area of the polygon is $nab = na^2/\tan \alpha$. Since n and α are constants for a given regular polygon, the derivative of this with respect to a is $2na/\tan \alpha = 2nb$, the perimeter of the polygon. Q.E.D.

Proof of Theorem 2. Assume the solid has m faces, each of which is a regular polygon of n sides. To compute the volume of the solid, we first compute the volume of a pyramid whose vertices are those of one face of the solid plus the point at the center of the solid.

Let angle α and lengths a and b be as in Figure 1, and let h be the perpendicular distance from the center of the solid to a face. (See Fig. 2.)

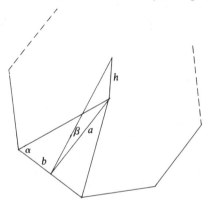

Figure 2.

The volume of the pyramid is $hB/3$, where B is the area of the base, and $B = na^2/\tan \alpha$ from Theorem 1. From Figure 2, $a = h/\tan \beta$, where β is a polyhedral angle determined by the base and one side of the pyramid. Thus, $B = nh^2/(\tan \alpha)(\tan^2 \beta)$, and the total volume becomes $mhB/3 = mnh^3/3(\tan \alpha)(\tan^2 \beta)$. With m, n, α, and β being constants, differentiating the volume with respect to h yields $mnh^2/(\tan \alpha)(\tan^2 \beta) = mB$, the total surface area of the solid. Q.E.D.

Convexity in Elementary Calculus: Some Geometric Equivalences

Victor A. Belfi

Victor A. Belfi received the B.A. and Ph.D. degrees in mathematics from Rice University with research in differential topology. His current research interest is in Banach algebras. He has been on the faculty of the Department of Mathematics at Texas Christian University since 1969.

What is convexity? For the beginning student of calculus, it is a concept usually introduced to explain the geometric significance of the second derivative. It is easy enough to illustrate graphically or describe verbally ("holds water"), but what sort of *definition* is appropriate in this setting? There are several equivalent ways to formulate the definition and relate it to the sign of the second derivative, but many widely used textbooks do not reveal the variety of alternatives or, in the case of some more advanced ones, give proofs of equivalence which are unnecessarily technical or nonintuitive. For texts which contain a more thorough discussion of convexity, see [1], [2]. Here we will look at six well-known formulations of convexity which have easily visualized geometric interpretations and prove their equivalence in an elementary way. We hope that the presentation of some of this (admittedly optional) material in the classroom may hint at the fundamental character of convexity and help to dispel the possible suspicion that the second derivative test for convexity is little more than a disguised definition.

Though infrequently seen in calculus texts, perhaps the most common definition of convexity for an arbitrary function is as follows:

Condition I. *A function f is convex (also called concave upward)* on an interval J if the graph of f lies below the chord on any subinterval (a, b) of J (Figure 1).*

Since the chord joining $(a, f(a))$ and $(b, f(b))$ has slope $m = \dfrac{f(b) - f(a)}{b - a}$, we can express Condition I analytically (via the point-slope form of a line) as either

$$f(x) < f(a) + \frac{f(b) - f(a)}{b - a}(x - a) \tag{1a}$$

or

$$f(x) < f(b) + \frac{f(b) - f(a)}{b - a}(x - b) \tag{1b}$$

for all $x \in (a, b)$. That (1a) and (1b) are equivalent (both are equivalent to Condition I) also follows by observing that $f(x) < f(a) + m(x - a)$ if and only if

---

*The definition of *concave* (or *convex downward*) would replace "below the chord" by "above the chord." It will be clear that our results for concave upward have analogs for concave downward.

136

$$f(x) < f(a) + \left[f(b) - f(a) \right] + m(x - a) = f(b) + m(x - b).$$

For $x \in (a, b)$, inequalities (1a) and (1b) can be easily combined as the continued inequality

$$\frac{f(x) - f(a)}{x - a} < \frac{f(b) - f(a)}{b - a} \tag{2a}$$

$$< \frac{f(b) - f(x)}{b - x} . \tag{2b}$$

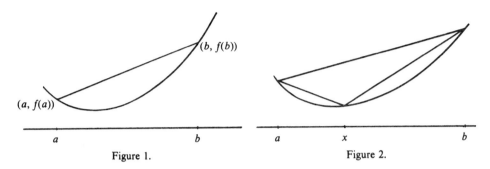

Figure 1.   Figure 2.

The equivalent inequalities (2a) and (2b) themselves make the following geometric statement (Figure 2) about the graph of f:

Condition II.   *For any three points a, x, b of the interval J with $a < x < b$, the slope of the chord between a and b is* (i) *larger than the slope of the chord between a and x, and* (ii) *smaller than the slope of the chord between a and b.* (Of course, either of the conditions (i) or (ii) individually is equivalent to Condition I.)

An apparently weaker condition states that of any two "consecutive" chords the latter has the larger slope. More precisely (Figure 3):

Condition III.   *For any three points a, x, b of the interval J with $a < x < b$, the slope of the chord between a and x is less than the slope of the chord between x and b.*

Analytically, this can be written

$$\frac{f(x) - f(a)}{x - a} < \frac{f(b) - f(x)}{b - x} \tag{3}$$

for all $x \in (a, b)$. Clearing fractions from (3) and rearranging terms, we have $(b - a)f(x) - bf(a) < (x - a)f(b) - xf(a)$. This, after adding $af(a)$ to both sides, can be written

$$\frac{f(x) - f(a)}{x - a} < \frac{f(b) - f(a)}{b - a} .$$

Therefore, (3) implies (2a). Thus (1a), (1b), (2a), (2b), and (3) are all equivalent.

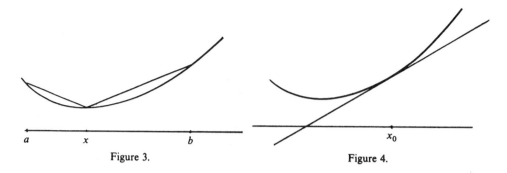

Figure 3.   Figure 4.

Now we will look at a class of formulations which involve derivatives and tangent lines, beginning with (Figure 4):

Condition IV. *A differentiable function f is convex on an interval J if for any point x_0 in J, the tangent line to the graph at $(x_0, f(x_0))$ lies below the graph throughout J except at x_0.*

Analytically,

$$f(x_0) + f'(x_0)(x - x_0) < f(x) \tag{4a}$$

for all $x \neq x_0$ in J. This becomes

$$f'(x_0) < \frac{f(x) - f(x_0)}{x - x_0} \qquad \text{for} \quad x > x_0 \tag{4b}$$

and

$$\frac{f(x_0) - f(x)}{x_0 - x} < f'(x_0) \qquad \text{for} \quad x < x_0. \tag{4c}$$

If $a, b \in J$ and $a < b$, we can take $b = x > x_0 = a$ in (4b) and let $a = x < x_0 = b$ in (4c) in order to obtain

$$f'(a) < \frac{f(b) - f(a)}{b - a} < f'(b). \tag{5}$$

The geometric interpretation of (5) is the following (Figure 5):

Condition V. *For any points $a, b \in J$ with $a < b$, the slope of the chord between a and b is (i) greater than the slope of the tangent line at $(a, f(a))$, and (ii) less than the slope of the tangent line at $(b, f(b))$.*

To see that Condition I implies Condition V when f is differentiable, choose a point $c \in (a, b)$ and apply (2b) twice to obtain

$$\frac{f(b) - f(a)}{b - a} < \frac{f(b) - f(c)}{b - c} < \frac{f(b) - f(x)}{b - x} \qquad \text{for} \quad x \in (c, b).$$

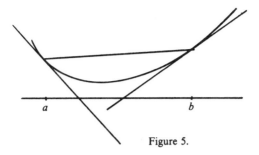

Figure 5.

Since

$$\frac{f(b) - f(c)}{b - c} < \frac{f(b) - f(x)}{b - x}$$

for all x increasing toward b,

$$\frac{f(b) - f(c)}{b - c} \leqslant \lim_{x \to b} \frac{f(b) - f(x)}{b - x} = f'(b),$$

and we obtain the second inequality in (5). A symmetric argument, using (2a), yields the first inequality in (5). The converse requires the Mean Value Theorem and the observation that (5) implies f' is increasing. Assuming $a < x < b$, we choose $c \in (a, b)$ such that $\dfrac{f(b) - f(a)}{b - a} = f'(c)$. If $c \leqslant x$, then

$$\frac{f(b) - f(a)}{b - a} = f'(c) \leqslant f'(x) < \frac{f(b) - f(x)}{b - x},$$

and so (2b) holds. On the other hand, if $c > x$, then

$$\frac{f(b) - f(a)}{b - a} = f'(c) > \frac{f(x) - f(a)}{b - a},$$

and we have the equivalent (2a).

The convexity condition most directly related to the sign of the second derivative for a twice differentiable function is the following (Figure 6):

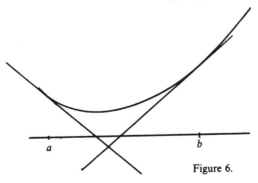

Figure 6.

Condition VI.   *The differentiable function f is convex on J if f' is (strictly) increasing on J. (Thus, f" > 0 when f' is differentiable.)*

As noted above, Condition VI is an immediate consequence of (5). But the converse also follows quickly by applying the Mean Value Theorem to f on $[a, b]$.

Other variations on these conditions are possible, such as requiring the second of any two non-overlapping chords to have the larger slope. All six conditions discussed above have now been shown equivalent, differentiability being assumed only when the statement of a condition requires it. Since conditions IV and VI are most often cited as definitions, it would be illuminating simply to point out to students the equivalence of at least one of the first three conditions, for there is a certain satisfaction in using properties of the derivative to verify a condition which can be formulated naturally without explicit mention of derivatives.

REFERENCES

1. Lynn H. Loomis, Calculus, 2nd ed., Addison-Wesley, 1977.
2. Michael Spivak, Calculus, rev. ed., Publish or Perish, Berkeley, 1980.

Does "holds water" Hold Water?

R. P. Boas, Northwestern University, Evanston, IL

A number of calculus books give the mnemonic that a curve which is concave upward "holds water," whereas one which is concave downward "spills water." Recently, a student asked one of my colleagues why the graph of $|x|^{1/2}$ is not concave upward in an interval containing 0, because it would evidently hold water. Indeed, there are actual glasses that have a similar shape (with the addition of a stem) and do hold wine (if not water). The problem arises because the mnemonic, like most mnemonics, is flawed: "holding water" is neither necessary nor sufficient for a given curve to be concave upward. The student's example can be augmented by $y = -|x|^{1/2}$, which clearly "spills water."

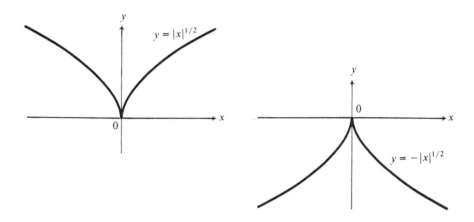

The late Professor W. R. Ransom used to use "bowl shaped" and "dome shaped." I have often followed his lead without realizing that this mnemonic is equally flawed. Wide bowls with sides that curve downward are not uncommon, and the domes on Eastern Orthodox churches are not concave downward at the extreme top. I have also seen the mnemonic that a smile is concave up and a frown (or scowl) is concave down; this connects nicely with the colloquial use of "up" and "down." Here smiles are to be thought of as in cartoons and graffiti; actual smiles can be misleading, especially when they are crooked.

I am indebted to A. M. Trimbinska for telling me about the puzzled student.

Transitions

Jeanne L. Agnew
James R. Choike

A Canadian by birth, Jeanne Agnew did her undergraduate work at Queen's University, Kingston, Ontario, and obtained her Ph.D. from Harvard, where her major professor was Dr. G. D. Birkhoff. During World War II she worked for the National Research Council in Canada. Her great love is teaching. Two years ago she achieved the status of Professor Emeritus after thirty years with the Department of Mathematics at Oklahoma State University. These thirty years were survived by her husband, five children, and many readers of this journal. For the past ten years, she has been involved with developing curricular materials based on real problems from industry. Her transition to retirement has been smooth and continuous, since she is currently writing coordinator for the AIM project. The learning modules developed in this project are for use at the high school level and also feature Applications in Mathematics, obtained from Industry.

James R. Choike is Professor of Mathematics at Oklahoma State University, where he has taught since 1970. He received a B.S. from the University of Detroit, an M.S. from Purdue University, and a Ph.D. at Wayne State University in 1970. His major scholarly interests are research in complex analysis and research and curricular development in applied mathematics. Professor Choike was the Writing Coordinator on the MAWIS and TEAM projects of The Mathematical Association of America. He currently serves as President of the Oklahoma Council of Teachers of Mathematics.

While a great deal of attention at present centers on the role of discrete processes in the mathematics curriculum, many, if not most, natural processes remain continuous. (The apparent position of the sun changes gradually as one moves along a parallel of latitude, whether or not its location is recorded in artificially discrete time zones. The aging process proceeds continuously from birth to death despite the fact that the record of aging is kept in terms of discrete numbers of years or months.) This article is concerned with making mathematical observations about some continuous curves, called transitions, encountered in experiences well known to us all. A *transition* or *transition curve* in a plane is a curve which joins smoothly two line segments in the plane having different slopes.

The Transition Parabola. Consider first the transitions involved in highway design [2]. As a highway lying in a vertical plane proceeds from a downhill segment to an uphill segment, it is clearly impossible to avoid some kind of transition curve from one straight line portion to the other. For this transition, engineers use a parabola, primarily because the rate of change of its slope is constant. A typical transition problem is the following (Figure 1):

> A straight segment of highway with a grade of -5% is to be joined smoothly to a segment with a grade of 3%. The parabola to accomplish this transition is to begin at a point with elevation 1250 ft. The horizontal distance from the first point of the parabolic curve to its last point is to be 1000 ft. Using a suitable coordinate system, determine: the equation of the parabolic curve; the elevation at which the parabolic curve meets the 3% grade line; and the best location for a drain to remove accumulating water.

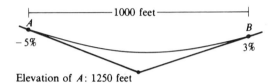

Figure 1.   Vertical highway transition.

This problem has been given to first-year calculus students. Once they decide where to place the origin of their xy-coordinate system, they usually proceed to solve the problem in a straightforward manner.

Although almost every text and teacher presents the reflection property of the parabola, other very simple and useful properties remain unnoticed. One such property, rediscovered by a freshman calculus student in the process of answering the above question, states (Figure 2):

> *tangents drawn from any two points on a vertical-axis parabola*
> *intersect on a vertical line midway between the points of tangency.*

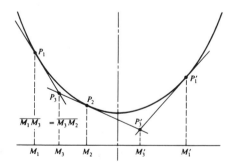

Figure 2.   Tangents to a parabola.

Engineers use this property (in preference to a simple calculus technique) to determine the location at which the parabola meets the 3% grade line. Students should be encouraged to prove this tangent-intersection property, and then use both methods to determine the transition parabola.

Our next example concerns the distance a driver can see in negotiating an uphill (or downhill) transition. It illustrates an interesting property of the differential.

The "vertical offset" between points (x, y_1) and (x, y_2) on a common vertical line is defined to be $|y_2 - y_1|$. Now consider the rule of offsets for a parabola (Figure 3):

> *Suppose a tangent line is drawn through a point* $P(x_0, y_0)$ *on the parabola* $y = ax^2 + bx + c$, *and the vertical line* $x = x_0 + d$ *intersects the parabola at* P_1 *and the tangent line at* P_2. *Then the vertical offset between* P_1 *and* P_2 *is* $|a|d^2$.

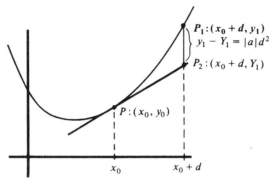

Figure 3.   Vertical offset.

This offset property for parabolas, which says that $|y - dy| = |a|(dx)^2$, is needed to solve problems involving the distance a driver is able to see in negotiating uphill or downhill transitions.

A driver's line of sight from a point A on a crest parabola is a line through the point 5 feet above point A (the height of the driver's eyes) and tangent to the parabola. The sight distance S from a point A is the horizontal distance of the driver's line of sight as it extends from the point 5 feet above A to a point 5 feet above a curve beyond the point of tangency of the driver's line of sight (Figure 4).

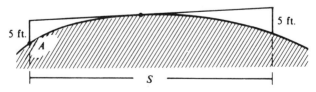

Figure 4.   Sight distance.

As a follow-up to the transition problem presented above, students should be asked to apply the rule of offsets to find the sight distance S from point $A = (0, 1200)$ on the parabolic crest given by the equation $y = -(1/21780)x^2 + (2/100)x + 1200$, where x and y are measured in feet.

The Transition Spiral. There are many curves that can be used to join smoothly straight roads of different directions. In the horizontal plane, what type of transition is required for a road which changes direction? The obvious initial choice for such a smooth transition is a circle. Unfortunately, the use of a circle causes the transition to change radius of curvature abruptly when an object moves from a straight path (which has infinite radius of curvature) to a circular path (which has finite radius of curvature). If the change in the turning radius for an automobile traveling a stretch of highway is too abrupt, the automobile has a tendency to veer from its traffic lane. This is dangerous.

The question, just what is the most desirable transition curve between two straight roads of different directions in the horizontal plane?, generated a great deal of interest at the turn of the century, when the problem arose in the building of railroads. In *The Railway Transition Spiral*, published in 1901, Arthur Newell Talbot (Illinois professor of mechanical engineering) wrote: "A transition curve is a curve of varying radius used to connect circular curves with tangents for the purpose of avoiding the shock and disagreeable lurch of trains due to an instant change in the relative position of cars, trucks, and draw-bars." So, according to Talbot, the most desirable transition curve is one which provides a continuous change in the radius of curvature throughout the transition.

The shape of a curve (its flatness or sharpness) at any point depends on the rate of change of direction $d\phi/ds$, where ϕ is the angle the tangent to the curve makes with the x-axis, and s is the arc length of the curve measured from some point. As a point P moves a distance Δs to the point Q, the angle ϕ changes by an amount $\Delta\phi$. The average rate of change is $\Delta\phi/\Delta s$. The *curvature at P*, defined as $d\phi/ds = \lim_{\Delta s \to 0} (\Delta\phi)/(\Delta s)$, is measured in radians per unit arc length. The reciprocal of the curvature, $\rho = 1/(d\phi/ds)$, is called the *radius of curvature*. Talbot suggests that for a smooth transition, the curvature at each point of the transition curve should be proportional to the arc length of the curve measured from the point where the transition curve begins. Thus, the intrinsic equation of the transition curve is $\rho s = \text{constant}$. Since such curves form spirals in the plane, they are often called transition spirals.

Let $\rho s = 1/a$, where a is a constant. Thus, $d\phi/ds = as$. Suppose we draw a coordinate system such that the negative x-axis coincides with a straight path, the x-axis is tangent to the transition curve at the origin, and the origin is the point of departure of the transition curve from the negative x-axis. Then $\phi = 0$ when $s = 0$, and integration gives us the angle ϕ in terms of s:

$$\phi = as^2/2.$$

Using arc length s as the parameter to describe the transition curve $x = x(s)$ and $y = y(s)$, our initial conditions yield $(x, y) = (0, 0)$ for $s = 0$. Now recall that the unit tangent vector to a curve is $(dx/ds, dy/ds)$. Since ϕ is the angle the tangent makes with the x-axis, we have $dx/ds = \cos\phi$ and $dy/ds = \sin\phi$. Then integration yields

$$x = \int_0^s \cos(at^2/2)\, dt, \qquad y = \int_0^s \sin(at^2/2)\, dt. \qquad (1)$$

These integrals certainly look forbidding. It is clear that the integration cannot be performed in closed form. The theoretical part of Talbot's book is devoted to writing the sine and cosine in series form and integrating term by term. Since the theoretical solution, in series form, cannot be applied to practical situations, an even greater part of the book is devoted to making a discrete model and calculating the position of the transition curve at specified intervals.

Suppose for small values of s, we use only the first term of the series to approximate $\cos(at^2/2)$ and $\sin(at^2/2)$. Then integration yields $x = s$ and $y = as^3/6$. Thus, the transition curve which provides a continuous change in the radius of curvature from the negative x-axis is approximately a cubical parabola, a fact predicted by Granville, Smith, and Longley [5, pp. 152–153], and used in practical situations by engineers. Modern computers, of course, are undaunted by forbidding-looking integrals. Figure 5 (with $a = \pi$) compares the actual transition curve $\rho s = 1/\pi$ to the cubical parabola $y = \pi x^3/6$. The drawing was done by a computer on a plotter.

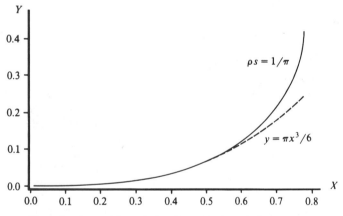

Figure 5.   A transition spiral and its cubical parabola.

For curves in the horizontal plane, centripetal force is an additional complication. On a railway curve, this may cause flange pressure against the rails. On a highway pavement, this force must be provided exclusively by friction. To reduce the reliance upon flange pressure or friction and to resist the tendency to overturn or skid, it is customary to raise the outer rail or incline the highway surface, so that a banking effect is produced.

Figure 6 depicts an object of weight W traveling with speed v over a circular track of radius ρ, banked at an angle θ. The object has a force of magnitude N normal to the surface of the inclined plane. This normal force N, when resolved into horizontal and vertical components, yields $N \sin \theta$ for the horizontal component and $N \cos \theta = W$ for the vertical component. The object's force of friction is a force parallel to the inclined plane with a magnitude μN, where μ is the coefficient of friction.

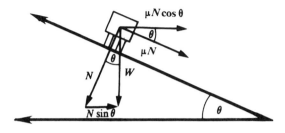

Figure 6. Banked track.

The sum of the horizontal forces $\mu N \cos \theta$ and $N \sin \theta$, which are directed toward the center of the circular track, constitutes the object's centripetal force $(Wv^2)/(g\rho)$. Thus,

$$\frac{Wv^2}{g\rho} = \mu N \cos \theta + N \sin \theta$$

$$= \mu W + W \tan \theta,$$

or

$$\tan \theta = \frac{v^2}{g\rho} - \mu. \tag{2}$$

Hence, the required angle of bank depends on the curvature, the velocity, and the coefficient of friction.

By using transition spirals $\rho s = 1/a$, road builders are able to execute the road's angle of bank in a gradual, hence, safe manner. For example, in the case of constant speed v and constant coefficient of friction μ, the correct angle of bank θ of an inclined plane depends solely on the radius of curvature ρ. According to equation (2), as ρ changes continuously along a transition spiral, the required bank angle θ for the road which follows the track of the transition spiral will also change in a continuous fashion. The transition spiral $\rho s = 1/a$ gives the engineer length in which to gradually introduce the angle of bank. Since $\rho s = 1/a$, and $\phi = as^2/2$, we see that $\phi = s/(2\rho)$, or $s = 2\rho\phi$. Thus, the length of the transition curve is twice the length of a circular arc of radius ρ and central angle ϕ.

Integrals such as (1), which describe a transition spiral, had been studied at least as early as 1743 by the celebrated Leonhard Euler [1]. Euler was interested not just in a small part of the curve but in the curve as a whole. In particular, he asked: "Does the integral $\int_0^\infty \sin s^2 \, ds$ converge?" It took him about 38 years to prove that it does.

As a result of advances made by Euler, and others like Cauchy (who laid the foundations of the theory of complex variables), the present day student would solve

Euler's problem by applying the theory of residues of complex analysis [4] to show that

$$\int_0^\infty \sin s^2 \, ds = \int_0^\infty \cos s^2 \, ds = \frac{1}{2}\sqrt{\pi/2}\,.$$

As often happens in mathematics, the property that makes this transition curve appropriate for railways and roads also arises in a completely different setting. For example, the spiral occurs in the calculation of vibration curves in optics.

The Sidestep Maneuver. The banking of a road to minimize the effects of centripetal force as a car or train negotiates a turn is different from, yet similar to, the banking of the wings of an airplane. When an airplane is flying at constant speed v in a straight line in level flight, the forces acting on the airplane are:

  L, a lifting force upward perpendicular to the wings
  $W = mg$, weight acting downward
  T, a forward thrusting force provided by the engines
  D, a drag opposite to the forward thrust (D is attributable to the resistance of the aircraft and its airfoils as it moves through the air).

In level flight at constant speed, the forces are in balance: $L = W$ and $T = D$. If we assume that the airplane is flying at constant speed in level flight and that its wings maintain a constant angle of bank, the thrust and drag are still equal and opposite forces, but the lift is tilted from the vertical position. (See Figure 7.)

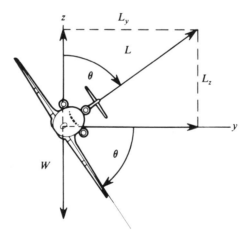

Figure 7. Forces in a bank.

The vertical component L_z of the lift must equal the weight W. Thus, the horizontal component of lift is $L_y = L_z \tan \theta = W \tan \theta$. If a constant bank angle θ is maintained, this horizontal force is the centripetal force that pulls the airplane into a circular arc. Since the magnitude of the centripetal force is $Wv^2/(g\rho)$ (where v^2/ρ is the magnitude of the acceleration vector directed toward the center of curvature), we have $Wv^2/(g\rho) = W\tan\theta$, or

$$\rho = \frac{v^2}{(g\tan\theta)}.$$

Constant θ and v imply constant ρ, and hence circular motion.

But an airplane cannot go directly from "wings level" to "wings banked." Thus, one is led to ask: What is the path of the airplane as it moves gradually from level wings to wings banked? The Federal Aviation Administration first asked this question in order to ensure safe landing conditions at an airport which had two parallel runways that were constructed too close together to allow the simultaneous use of Instrument Landing Systems (ILS) aligned on each of the runways. Efficient use of these runways could be made if one ILS (Figure 8) was aligned along the center line of runway A, whereas the other ILS for runway B was aligned on a parallel line not at the center of runway B.

The placement of the nonaligned ILS for runway B must insure minimum separation of 4300 ft. between two aircraft which are making a simultaneous landing approach on instruments, one approaching runway A and the other approaching runway B. At some point in a landing approach, the pilot must have visual contact with the runway for touchdown or he must abort the landing. The pilot approaching on ILS to runway B must have time when he emerges from the clouds to align the aircraft with the runway. This will involve a banking roll to the left then back to the right into a straight line path. This type of S-curve is called a "Sidestep Maneuver."

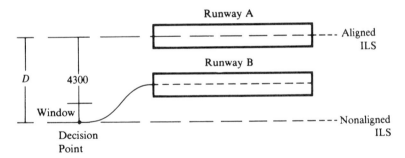

Figure 8.   Aligned and nonaligned ILS.

One important need for the exact form of this transition curve is to locate the decision point for landing with a nonaligned ILS (Figure 8).

If we have the equations for the x- and y-coordinates for the sidestep maneuver, then the decision point can be located relative to the airplane's touchdown on the center of the runway. The decision point is one endpoint of a segment, called the decision window, that locates the zone from which safe and comfortable landings are possible. An airplane must pass through the decision window with the runway in view in order to continue and complete the landing.

Suppose that the velocity of the airplane remains constant and its bank angle θ changes at a constant rate. Let us look at the first portion of the maneuver, in which the plane is turning to the left. Suppose we draw a coordinate system in which the x-axis is tangent to the straight-line portion of the path, and the origin is at the beginning of the maneuver. Then the length of the path is $s = vt$. As the bank angle $\theta = kt$ of the airplane increases, the angle ϕ that the tangent to the curve makes with the positive x axis is also increasing. Since the curvature $1/\rho = (g/v^2)\tan\theta$ and since $\theta = ks/v$, we have

$$\frac{d\phi}{ds} = \left(\frac{g}{v^2}\right)\tan\left(\frac{ks}{v}\right).$$

For the given initial conditions $s = 0$ and $\phi = 0$, integration yields

$$\phi = \left(\frac{g}{vk}\right)\ln\left|\sec\left(\frac{ks}{v}\right)\right|.$$

As in the case of the transition spiral given by equations (1), the coordinates x and y are also given by integrals:

$$x = \int_0^s \cos\left[\left(\frac{g}{vk}\right)\ln\left|\sec\left(\frac{kt}{v}\right)\right|\right]dt$$

$$y = \int_0^s \sin\left[\left(\frac{g}{vk}\right)\ln\left|\sec\left(\frac{kt}{v}\right)\right|\right]dt$$

$$(3)$$

Again, we can use numerical integration to obtain the position of the plane at any time during the maneuver [3].

Figure 9 shows a graph of a sidestep maneuver for an airplane with $v = 120$ knots, $k = 10$ degrees/second, and a maximum bank angle of 15 degrees. The

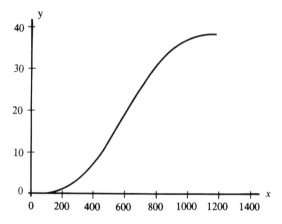

Figure 9.   Sidestep maneuver ($t = 6$, $v = 120$, $k = 10$, Max bank = 15). Distances x, y measured in feet.

aircraft rolls left at a uniform rate of 10 degrees/second, up to a bank angle of 15 degrees. This portion of the maneuver takes 1.5 seconds. Upon reaching a bank angle of 15 degrees to the left, the airplane then rolls in the opposite direction up to a bank angle of 15 degrees. This portion of the maneuver takes 3 seconds. Finally, the airplane rolls back to wings level. The total time for the entire maneuver is 6 seconds.

The integrals in (3) seem to be unrelated to those obtained in (1) for the case of railway transitions. However, consider the series expansion of $\ln|\sec\theta|$. This begins with the term $\theta^2/2$. Using the first term as an approximation of $\ln|\sec(ks/v)|$, and letting $a = kg/v^3$, the integrals take the form

$$x = \int_0^s \cos\left(\frac{at^2}{2}\right) dt, \qquad y = \int_0^s \sin\left(\frac{at^2}{2}\right) dt.$$

As s approaches $v\pi/2k$, we see that $\cos(ks/v)$ approaches 0, and hence $\ln|\sec(ks/v)|$ becomes infinite. However, the integrals in (3) exist for $s = v\pi/2k$, since the integrands are continuous and bounded on the interval $[0, v\pi/2k)$. Although the integrals in (3) are related to the integrals in (1), they are much worse than those in (1). Numerical methods can be applied to evaluate

$$\int_0^{v\pi/2k} \cos\left(\frac{g}{vk} \ln\left|\sec\left(\frac{kt}{v}\right)\right|\right) dt. \tag{4}$$

However, the authors find themselves in a position similar to that of Euler in 1743. We know the integral in (4) exists and we can approximate it numerically, but we are unable to evaluate it exactly.

Concluding Remarks. These examples illustrate mathematics as a means for modeling familiar real world phenomena. As we have seen, sometimes a transition curve can be modeled with elementary mathematics; at other times the mathematics required is advanced and, in some cases, yet to be developed.

REFERENCES

1. R. C. Archibald, "Euler's Integrals and Euler's Spiral—Sometimes Called Fresnel Integrals and the Clothoide or Cornu's Spiral," American Mathematical Monthly 25 (1918) 276–282.
2. James R. Choike, *A Path to Applied Mathematics: Highway Slope Design*, Student Book and Teacher Book bound together, Teaching Experiential Applied Mathematics (TEAM) Project, The Mathematical Association of America, Washington, D.C., 1984.
3. James R. Choike, *A Path to Applied Mathematics: Aircraft Sidestep Maneuver*, Student Book and Teacher Book bound together, Teaching Experiential Applied Mathematics (TEAM) Project, The Mathematical Association of America, Washington, D.C., 1984.
4. R. V. Churchill and J. W. Brown, *Complex Variables and Applications*, McGraw-Hill, New York, 1984.
5. W. A. Granville, P. F. Smith, and W. R. Longley, *Elements of the Differential and Integral Calculus*, Ginn and Company, Boston, 1929.
6. A. N. Talbot, *The Railway Transition Spiral*, Gazette Press, Champaign, Ill., 1901.

A Note on Parallel Curves

Allan J. Kroopnick, Social Security Administration, Baltimore, MD

Recently, F. M. Stein [TYCMJ, 11 (1980) 239–246] discussed the notion of parallel curves. Specifically, two smooth curves C_u and C_g are parallel if the normal at each point $P \in C_u$ meets C_g at a point $\bar{P}$ such that $|P\bar{P}|$ is constant and the tangent lines to the curves at $P \in C_u$ and $\bar{P} \in C_g$ are parallel. (Note, therefore, that every smooth curve is parallel to itself.)

Our objective is to present a simple, functional characterization of parallelism for curves. Using this, we readily obtain Stein's earlier results.

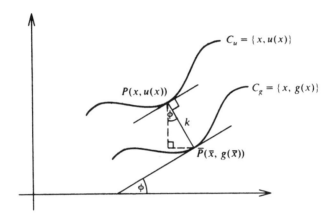

The curves C_u, C_g are parallel if and only if $|P\bar{P}|$ has constant length k and $u'(x) = g'(\bar{x})$ for each pair of abscissas $x \in P$ and $\bar{x} \in \bar{P}$. If ϕ denotes the angle of inclination of the tangent line to C_u at P, then $\bar{x} = x + k \sin \phi$ and the preceding conditions for parallelism can be written as an identity:

$$u'(x) = g'\left(x + k \cdot u'(x)/\sqrt{1 + \{u'(x)\}^2}\right) \qquad (*)$$

for all x where u' is defined.

By way of illustration, suppose $g(x) = ax + b$ and $u(x) = Ax + B$. Then the lines C_g, C_u are parallel if and only if $u'(x) = A$ equals $g'(x + k \cdot A/\sqrt{1 + A^2}) = a$. Thus, we see anew that straight lines are parallel if and only if they have the same slope.

Example 1.   Stein showed that a curve parallel to a parabola is not a parabola. To see this using $(*)$, let $g(x) = x^2$ and suppose that the curve C_u determined by $u(x) = ax^2 + bx + c$ is parallel to C_g. Since the normal at $(0,0) \in C_g$ intersects $(0,c) \in C_u$, we must have $u'(0)$ equal to $g'(0)$. Thus, $b = 0$. Now, invoking $(*)$, we obtain

$$2ax = 2\left(x + k \cdot \frac{2ax}{\sqrt{1 + 4a^2x^2}} \right)$$

which reduces to

$$a = 1 + 2k \cdot \frac{a}{\sqrt{1 + 4a^2x^2}}$$

for all x. This identity in x, however, is true if and only if $k = 0$ and $a = 1$ (that is, if and only if $u(x) = x^2 = g(x)$).

Example 2. Parallel sine curves must be identical. Let $g(x) = \sin(x)$ and suppose that C_u is a sine curve which is parallel to C_g. Since C_u must have the same periodicity as C_g, we may assume that C_u is determined by $u(x) = a \cdot \sin(x + b) + c$. From (∗), we have

$$a \cdot \cos(x + b) = \cos\left[x + k \cdot \frac{a \cdot \cos(x + b)}{\sqrt{1 + a^2 \cdot \cos^2(x + b)}} \right]$$

for all x. For $x = (\pi/2) - b$, this reduces to $\cos\{(\pi/2) - b\} = 0$. Thus, $b = n\pi$ for some integer n. Therefore, $u(x) = (-1)^n a \cdot \sin(x) + c$. Since C_g and C_u are assumed to be parallel, the *normal* distance between the curves when $x = \pi/2$ must be the same as the *normal* distance between the two curves when $x = 3\pi/2$. Accordingly, $u(\pi/2) - \sin(\pi/2) = c + (-1)^n \cdot a - 1$ equals $u(3\pi/2) - \sin(3\pi/2) = c - (-1)^n \cdot a + 1$. Hence $(-1)^n \cdot a = 1$ and $u(x) = \sin(x) + c$. This requires that $c = 0$. (Otherwise, the normal distance between $(0, 0) \in C_g$ and $(-1, 1) \in C_u$ is strictly less than the normal distance $u(\pi/2) - \sin\pi/2 = c$.)

BIBLIOGRAPHIC ENTRIES: APPLICATIONS TO GEOMETRY

   1. *Math. Mag.* Vol. 43, No. 4, pp. 211–212. Peter Hagis, Jr., Axis rotation via partial derivatives.
   2. *Math. Mag.* Vol. 44, No. 4, pp. 212–214. Clifford A. Long, Two definitions of tangent plane.
   3. TYCMJ Vol. 11, No. 4, pp. 239–246. F. Max Stein, The curve parallel to a parabola is not a parabola.
   4. TYCMJ Vol. 12, No. 5, pp. 332–333. Michael W. Ecker, Must a "dud" necessarily by an inflection point?
   5. TYCMJ Vol. 13, No. 1, pp. 52–55. Jane T. Grossman and Michael P. Grossman, Dimple or no dimple.
   6. TYCMJ Vol. 13, No. 3, pp. 186–190. Herb Holden, Chords of the parabola.
   7. CMJ Vol. 21, No. 3, pp. 208–215. Wilbur J. Hildebrand, Connecting the dots parametrically: An alternative to cubic splines.

(c)

APPLICATIONS TO MECHANICS

Velocity Averages

GERALD T. CARGO

Syracuse University
Syracuse, NY 13210

A particle travels on a straight line from time $t = a$ to $t = b$. What is its average velocity? If $f(t)$ denotes the position of the particle at time t, the conventional answer is $(f(b) - f(a))/(b - a)$. If f is differentiable, one can also consider the average of the velocities, $(f'(a) + f'(b))/2$, or the velocity at the average time, $f'((a + b)/2)$. Beginning students of calculus often confuse the latter two averages with the first. Their problems are exacerbated by the fact that for a uniformly accelerated particle the results are identical. (This follows easily from the general form $f(t) = At^2 + Bt + C$ for motion under uniform acceleration.) The purpose of this note is to show that these three averages are equal *only* in the case of uniform acceleration.

We will actually prove a bit more: *If any two of the three expressions — average velocity, average of the velocities, and velocity at the average time — agree on each time interval $[a, b]$, then f'' exists and is constant.*

First, suppose that the average velocity equals the average of the velocities on each interval $[c, d]$. Then

(1) $$(f(t) - f(a))/(t - a) = (f'(a) + f'(t))/2 \quad (t \neq a).$$

Differentiation of (1) shows that $f''(t)$ exists for every $t \neq a$. Since a is arbitrary, $f''(t)$ exists for every t. Similar reasoning shows that f has derivatives of all orders.

From (1) we conclude that

(2) $$2(f(t) - f(a)) = (t - a)(f'(a) + f'(t))$$

for all real numbers a and t. By differentiating (2) twice, we discover that $(t - a)f'''(t) = 0$. Hence, $f'''(t) = 0$ for every $t \neq a$. Since a is arbitrary, $f'''(t) = 0$ for every t; and f'' is constant.

Here is an alternative proof based on the chord trapezoidal rule of approximate integration. This rule says that, whenever g is twice differentiable on an interval $[a, b]$, there exists a point s such that $a < s < b$ and

(3) $$\int_a^b g(t)dt = (1/2)(b - a)[g(a) + g(b)] - (1/12)(b - a)^3 g''(s).$$

Setting $g = f'$ and using the fundamental theorem of calculus to evaluate the left side of (3), we obtain

$$f(b) - f(a) = (1/2)(b - a)[f'(a) + f'(b)] - (1/12)(b - a)^3 f'''(s).$$

It follows from (1) (with $t = b$) that in this case the chord trapezoidal approximation is exact, that is,

154

the "error" term, $-(1/12)(b-a)^3 f'''(s)$, must vanish. Thus, in each interval $[a, b]$ there exists a point s such that $f'''(s) = 0$. Since f''' is continuous, f''' is identically zero, as desired.

A variation on this argument can be used to verify the second case in our three-part proposition. The tangent trapezoidal rule states that, if g is twice differentiable on the interval $[a, b]$, then there exists a point s such that $a < s < b$ and

(4)
$$\int_a^b g(t)dt = (b-a)g((a+b)/2) + (1/24)(b-a)^3 g''(s).$$

Suppose now that the average velocity equals the velocity at the average time on each compact interval. Setting $g = f'$, we infer from (4) that $f'''(s) = 0$ and conclude the proof as above.

Here is a more elementary verification of the second case. By hypothesis,

(5)
$$(t-a)f'((a+t)/2) - f(t) + f(a) = 0.$$

If we differentiate both sides of (5) three times with respect to t, we obtain

(6)
$$(1/8)(t-a)f^{(4)}((a+t)/2) + (3/4)f^{(3)}((a+t)/2) - f^{(3)}(t) = 0.$$

Setting $t = a$ in (6), we conclude that $f^{(3)}(a) = 0$. Hence, f'' is constant.

Our final case is more difficult. If we assume only that the velocity at the average equals the average of the velocities, then we have

(7)
$$F((a+b)/2) = (1/2)(F(a) + F(b))$$

where $F = f'$. The only continuous solutions to this equation, known as Jensen's equation, are linear functions; such solutions have constant derivatives, as desired. However, the existence of nonlinear solutions of (7) can be established with the aid of Zermelo's axiom of choice. So to complete our proof we must verify that, even though f' need not be *a priori* continuous, it must nevertheless be linear. This follows from two theorems of real analysis: f' is Lebesgue measurable (since it is a derivative [1, p. 288]); and, because it is convex, it must be continuous [2, p. 96]. Hence, as above, it must be linear.

References
[1]   R. R. Goldberg, Methods of Real Analysis, Blaisdell, New York, 1964.
[2]   G. H. Hardy, J. E. Littlewood, and G. Pólya, Inequalities, 2nd ed., Cambridge, London, 1952.

Travelers' Surprises

R. P. Boas

R. P. Boas was born in 1912 in Walla Walla, Washington, received a Ph.D. from Harvard in 1937, and was a National Research Fellow during 1937–38, at Princeton and at Cambridge (England). He has taught at Duke; the U. S. Navy Pre-Flight School in Chapel Hill, North Carolina; Harvard; MIT; and since 1950 at Northwestern. During 1945–50 he was Executive Editor of Mathematical Reviews. *He has published an unnecessarily large number of articles in professional journals, three advanced books and the Carus Monograph "A primer of real functions." He has held various offices in both the AMS and MAA, and is currently the Editor of the* American Mathematical Monthly.

1. First question. Suppose that you travel (in a differentiable way) for a considerable distance at an average speed of 50 mph. Is there some instant during your journey at which your instantaneous speed is precisely 50 mph? Many people are mildly surprised, the first time they encounter this question, to learn that the answer is "yes." The problem is often given as an application of the mean-value theorem for derivatives. Let us introduce some notation so that from time 0 to time t you cover a distance $s(t)$ miles. If "average speed" has its everyday meaning of distance divided by time, saying that your average speed from $t = a$ to $t = b$ is 50 means that

$$\frac{s(b) - s(a)}{b - a} = 50.$$

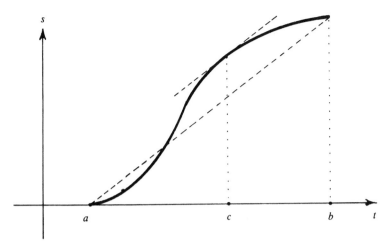

The mean-value theorem says that if the instantaneous speed $s'(t)$ exists for all t between a and b (which is what I meant by "travelling in a differentiable way") then there is at least one time c between a and b at which $s'(c) = 50$.

If $s(t)$ were not differentiable, but merely continuous, the original question would make no sense. If you could travel so that your distance $s(t)$ is given by a nowhere differentiable function, you would never have an instantaneous speed. Travel of this kind is not physically realizable, but we can think about the mathematical problem that it suggests. If we do this, it becomes interesting to ask another question.

2. Second question. Is there at least a very short interval during which the average speed is 50? It will be helpful to think about this geometrically. The mean-value theorem said that if there is a tangent at every point of the graph of $s(t)$, then there is surely a point at which the tangent is parallel to the chord that joins the initial and final points. (A chord is a line segment with both ends on the

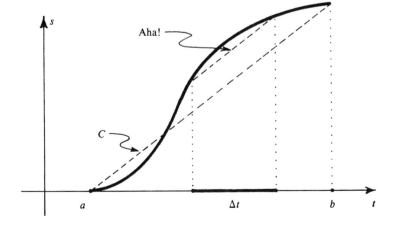

graph; it is irrelevant whether or not it meets the graph at other points.) Call the chord C. Our first idea might be to look for a chord of very short length parallel to C, but this is not altogether a good idea. In fact the length of a chord has no natural interpretation for a graph of distance against time: what would $\sqrt{(s^2 + t^2)}$ mean, when s is distance and t is time? What we should be looking for is a chord, parallel to C, over a very short time span, where "span" means the time interval between the endpoints of the chord (Δt, if you like). It seems pretty obvious geometrically that there always are such short chords, but it is not quite easy to give a formal proof. One is outlined in §5, below, in case you want to see it. As soon as we are convinced of the existence of arbitrarily short chords parallel to C, we know that the answer to Question 2 is also "yes."

3. Third question.   Apparently it was only quite recently that anyone thought of asking what turns out to be a more subtle question: (3a) if you travel for time h, more than one hour, and average 50 mph for the trip, is there necessarily some one continuous hour during which you covered exactly 50 miles?

An alternative question (3b) can be asked about reciprocal speeds, which are sometimes used to describe relatively slow activities: suppose you run a considerable number m of miles and average 8 minutes per mile; is there necessarily some one continuous mile (like a "measured mile" on a highway) that you covered in exactly 8 minutes? [3]

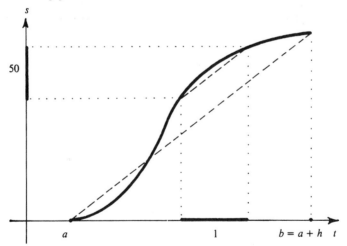

The mean-value theorem no longer provides answers to these questions, and the answers turn out to be rather unexpected. In each case, the answer is "yes" if h or m is an integer; but "not necessarily" otherwise—provided that we make reasonable assumptions about how you travel. For (3a) it is altogether reasonable to assume that s is a continuous function of t: "continuous" since discontinuous motion is not observed in macroscopic situations; "function," since you cannot be in two places at the same time. (There is no objection to your stopping for a while, or even backing up.) Question (3b) is different because, on the one hand, time has to increase; on the other hand, if you stood still for a while, t would change discontinuously in terms of s; furthermore, you can easily be in the same place at different times, so t is not necessarily a (single-valued) function of s. I shall assume that in fact $t = t(s)$ is an increasing continuous function; this precludes stopping or backing up. Actually the result is true without this assumption, but it is considerably harder to prove, and I shall not try to prove it here.

Instead of obtaining the answers to (3a) and (3b) directly, I am going to derive both of them from the so-called universal chord theorem. A horizontal chord of a continuous function f means a line segment of slope 0 with both ends on the graph of f (remember that a chord might have other points on the graph too). The theorem says that if there is a horizontal chord of length L, then there are always horizontal chords of lengths L/n for integral n, but not necessarily for nonintegral n. In formulas, if $f(b) = f(a)$ and $b - a = L$, then given $n > 1$ there is at least one x between a and b for which $f(x + L/n) = f(x)$ provided n is an integer, but there is not necessarily such an x if n is not an integer. Any given continous function with a horizontal chord of length L of course has *some* horizontal chords that are not of length L/n, n = integer; what the negative half of the theorem says is that, given n which is not an integer, we can find *some* function f for which $f(x + L/n) \neq f(x)$ for any x.

The universal chord theorem was proved by P. Lévy in 1934, but it did not get into the textbooks and consequently is rediscovered every few years. There are a number of variants and related results; for references see [1], p. 163, note 16. For the convenience of the reader I give a proof in §6; it is not difficult, but not completely obvious either.

We now find the answers to questions (3a) and (3b). First, let your distance from your starting point be $s(t)$, where $0 \leqslant t \leqslant h$, h is an integer, and $s(0) = 0$. To say

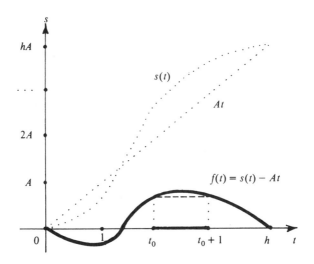

that your average speed is A means that $s(h) = Ah$. Consider the function $s(t) - At$. This takes the same value (0, in fact) at $t = 0$ and at $t = h$; in other words it has a horizontal chord of length h. The universal chord theorem says that this function has horizontal chords of all lengths h/n for integral n, and in particular one of length $h/h = 1$. That is, there is some t_0 such that

$$s(t_0 + 1) - A(t_0 + 1) = s(t_0) - At_0,$$

that is

$$s(t_0 + 1) - s(t_0) = A.$$

This says that in the hour between t_0 and $t_0 + 1$ you went exactly A miles.

More generally, if two travelers start together when $t = 0$ and arrive together when $t = h$, and h is an integer, there is some one hour during which each of them goes exactly A miles (not usually the same A miles however). To see this, consider $s_1(t) - s_2(t)$ instead of $s(t) - At$. The original case $(s(t) - At)$ corresponds to considering a second traveler who travels at constant speed A.

Before looking for an example to show that the result fails when h is not an integer, let us look at question (3b), where we describe how fast we travel in terms of reciprocal speed, that is, in terms of hours/mile instead of miles/hour. With the assumption that time is an increasing continuous function of distance, to say that you average A hours per mile means that if $t(0) = 0$ and you go m miles, $t(m) = mA$. The same argument as before shows that some one mile is covered in exactly A hours.

4. Counterexamples. We now show that when h or m is not an integer, both questions (3a) and (3b) have to be answered "not necessarily." This is slightly harder for t since we have to find an example in which t is an increasing function.

We are then looking for an increasing function $t = t(s)$ such that $t(0) = 0$, $t(m) = mA$, m is not an integer, and $t(s + 1) - t(s)$ is never equal to A for any s. We shall do somewhat more: we shall find t so that $t(s + 1) - t(s)$ is never less than or equal to A; then if you travel so that your time is given by this function, not only will you fail to cover any one whole mile in your average time, you will even fail to cover any one whole mile in less than your average time.

We start from a counterexample for the universal chord theorem (§6), adjusted to an interval $[0, m]$:

$$f(t) = \frac{t}{m} \sin^2 m\pi - \sin^2 \pi t.$$

For this function, $f(0) = f(m) = 0$, but $f(t + 1) - f(t) = (1/m)\sin^2 m\pi \neq 0$. Let $s(t) = kf(t) + At$, where k is a (small) positive number. If k is small enough, $s(t)$ increases because its derivative is positive:

$$s'(t) = kf'(t) + A = k\left\{(1/m)\sin^2 m\pi - \pi \sin 2\pi t\right\} + A,$$

and this is positive if k is small enough. Moreover, if $s(t + 1) - s(t) < A$ we would have $kf(t + 1) - kf(t) + A < A$, that is, $k\{f(t + 1) - f(t)\} < 0$; but $f(t + 1) - f(t)$ $= (1/m)\sin^2 m\pi > 0$.

In terms of the original question (3b), this means that if you run a nonintegral number of miles, and you average 8 minutes per mile, not only may there not be any one (connected) mile that you cover in 8 minutes, there may be no one mile that you cover in *less* than 8 minutes. This seems paradoxical; but of course there must be nonintegral distances that you cover at better than your usual speed.

The same example, with only a change in notation, also shows that only integral times work in question (3a).

5. Short parallel chords.

Here I shall outline a formal proof that, given a chord C of a continuous function, there are chords parallel to C and of arbitrarily short span. Your first reaction may well be that for every span shorter than that of C there is a parallel chord of that span. If you make this guess, you are not thinking of a sufficiently complicated distance function.

Suppose, for example, that $s(t) = \sin t$ for $0 \leqslant t \leqslant 2\pi$. Then there is a chord (along the t-axis) of span 2π, but no parallel chord of span L (with both ends between 0 and 2π) for $\pi < L < 2\pi$. This is obvious from a sketch, and only a little less obvious from formulas.

Now let's consider how we could prove the existence of chords of arbitrarily short span parallel to a given chord C. Let C have the (linear) equation $s = c(t)$ (we don't need to find $c(t)$ explicitly), and consider $g(t) = s(t) - c(t)$. We have $g(a)$ $= 0$, $g(b) = 0$, since the chord meets the graph at a and b. Hence g has a maximum (or else a minimum) between a and b. A horizontal line starting on the graph near a maximum of g must meet the graph again on the other side of the maximum; so it determines a horizontal chord of g. If the maximum is a proper maximum we can make the horizontal chord as short as we please; if the maximum is improper, i.e. if the graph has a flat top, we can take a short line-segment coinciding with part of the tangent line through the maximum. A chord determined in this way corresponds to a chord of s parallel to C. The reader may want to fill in the details of this argument.

6. Proof of the universal chord theorem.

For simplicity of notation I take $a = 0$, $b = 1$, so $L = 1$. Suppose first that n is an integer greater than 1, that $f(0) = f(1)$, and that f has no horizontal chord of length $1/n$. Consider the continuous function $g(x) = f(x + 1/n) - f(x)$, $0 \leqslant x \leqslant 1 - 1/n$. Since we supposed that $f(x + 1/n)$ is never equal to $f(x)$, it follows that $g(x)$ is never 0. Since g is continuous, this means that g cannot change sign (here we appeal to the property that a continuous function cannot get from one value to another without taking on all the values in between). Suppose that $g(x) > 0$. (If not, consider $-g(x)$ instead.)

In particular,

$$g(1 - 1/n) > 0,$$
$$g(1 - 2/n) > 0,$$
$$\cdots \cdots$$
$$g(1 - n/n) = g(0) > 0.$$

Writing these inequalities in terms of f, we have

$$f(1) - f(1 - 1/n) > 0,$$
$$f(1 - 1/n) - f(1 - 2/n) > 0.$$
$$\cdots \cdots$$
$$f(1/n) - f(0) > 0.$$

If we add these inequalities we get $f(1) - f(0) > 0$, whereas we assumed to begin with that $f(1) - f(0) = 0$. Consequently $g(x)$ must in fact be 0 for some x, and for that x we have $f(x + 1/n) - f(x) = 0$. In other words, f has a horizontal chord of length $1/n$ starting at x.

Notice that the proof doesn't work if n is not an integer. This does not prove that the theorem is false when n is not an integer, but it encourages us to look for a counterexample. It is not easy to find one geometrically, but Lévy provided us with a simple formula. Let p *not* be an integer. Then $f(t) = t \sin^2 p\pi - \sin^2 p\pi t$ is 0 at 0 and 0 at 1, so it has a horizontal chord of length 1. But

$$f(t + 1/p) - f(t) = (t + 1/p)\sin^2 p\pi - \sin^2 p\pi(t + 1/p) - t \sin^2 p\pi + \sin^2 p\pi t$$

$$= (1/p)\sin^2 p\pi,$$

which is independent of t and not zero. Hence for each p that is *not* an integer we can find a continuous function with a horizontal chord of length 1 but none of length $1/p$, for that particular p.

To answer question (3b) without restrictions on how t depends on s, we would need the more general result that any continuous curve that has a chord of length 1 has a parallel chord of length $1/n$ if n is an integer [2]. I do not know of any elementary proof of this.

REFERENCES

1. R. P. Boas, A Primer of Real Functions, Carus Mathematical Monographs, No. 13, Mathematical Association of America, Washington, D.C., 1972.
2. H. Hopf, Uber die Sehnen ebener Kontinuen und die Schliefen geschlossener Wege, Comment. Math. Helv., 9 (1937) 303–319.
3. J. D. Memory, Kinematics problem for joggers, Amer. J. Physics, 41 (1973) 1205–1206.

Related Rates and the Speed of Light

S. C. Althoen
J. F. Weidner

Steven C. Althoen is Associate Professor of Mathematics at the University of Michigan–Flint, where he has been since 1975. Prior to that he taught for two years at Hofstra University. He received his Ph.D. in 1973 from the City University of New York and his B.A. in 1969 from Kenyon College. His primary research interest is in the classification of real division algebras.

John F. Weidner is presently a Systems Engineer at MITRE Corporation, in McLean, Virginia, while on leave from Hofstra University. He earned his Ph.D. from Purdue University in 1973. His mathematical interests include statistics and artificial intelligence. He has been interested in problems involving the speed of light ever since reading M. A. Rothman's article "Things that go faster than light," Scientific American 203 (1960), pp. 142–152.

Standard calculus texts often include a related rates problem involving light cast onto a straight line by a revolving light source (for example, a lighthouse's beam hitting a straight shoreline) or a shadow cast onto a straight wall by an object moving in a circle (for example, a running horse's shadow on a fence tangent to a circular track which has a light source at the center). These problems always ignore the fact that light has a finite speed. In a reasonably scaled, real-life problem, assuming that light has an infinite speed has no measurable effect on any calculations. Although ignoring the finiteness of the speed of light has no significant effect, when one considers the true speed of light, there are some interesting mathematical aspects to these problems, both in the solution and in the method by which that solution is obtained.

In a general formulation of these problems (Figure 1), a light source moves with constant angular velocity $d\theta/dt = k$ around a circle (with center 0 and radius r) casting light onto a line L which is q units away from 0. The beam is always cast directly away from the center. We are to determine the rate of motion of the point where the light strikes the line L.

If we assume that light travels instantaneously (i.e., that the speed of light is infinite), we solve this problem by differentiating $s = q \tan \theta$ with respect to t. Much more work is required, however, when we assume that the light has a finite speed c, because the light traveling toward L_1 when the light source is at A_1 has not yet reached L_1. That is, at that instant, the points 0, A_1, and the point where the light is then striking the line are not collinear.

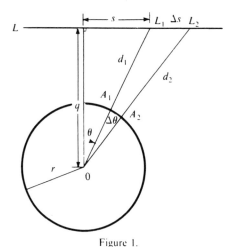

Figure 1.

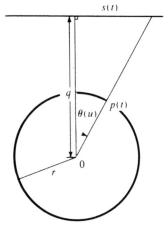

Figure 2.

One approach to determining ds/dt is to calculate Δs and Δt directly in terms of θ and $\Delta\theta$. The change in s is $\Delta s = q\tan(\theta + \Delta\theta) - q\tan\theta$. The light reaches L_1 exactly d_1/c seconds after the source is at A_1. It takes $\Delta\theta/k$ seconds for the light source to reach A_2 and then d_2/c seconds for the light to reach L_2. Thus, $\Delta t = \Delta\theta/k + (d_2/c) - (d_1/c)$. (Note that Δt is now the difference of the times from when the light strikes L_1 to when the light strikes L_2. It is *not* the time it takes for the light source to go from A_1 to A_2, as it was when the infinite speed of light was assumed.)

Since $d_2 - d_1 = [q\sec(\theta + \Delta\theta) - r] - [q\sec\theta - r]$, we have

$$\frac{\Delta s}{\Delta t} = \frac{ckq\big[\tan(\theta + \Delta\theta) - \tan\theta\big]}{c(\Delta\theta) + kq\big[\sec(\theta + \Delta\theta) - \sec\theta\big]} . \qquad (*)$$

Now divide the numerator and denominator by $\Delta\theta$, and use the definition of derivative and the chain rule to evaluate the limit as $\Delta\theta$ approaches zero. This yields the finite speed of light result

$$\frac{ds}{dt} = \frac{ckq\sec^2\theta}{c + kq\sec\theta\tan\theta} . \qquad \text{(FSL)}$$

This result can also be obtained by implicit differentiation (Figure 2). Let $s(t)$ denote the position of the light on the line at time t, and let $p(t)$ denote the distance from that point to 0. Also, let $\theta(u)$ denote the measure of θ at time u. Then $s(t) = q\tan\theta(u)$ and $u + (p/c) = t$. Differentiating s with respect to t yields

$$s'(t) = q\big[\sec^2\theta(u)\big]\theta'(u)u'(t).$$

Now $\theta'(u) = k$ and $u'(t) = 1 - (1/c)p'(t)$. Since $p^2 = s^2 + q^2$, we have $p'(t) = (s/p)s'(t)$. Thus,

$$s'(t) = q\left[\sec^2\theta(u)\right]k(1 - (s/cp)s'(t)).$$

Solving for $s'(t)$ and recalling that $s/p = \sin\theta(u)$, we obtain (FSL).

We retrieve the answer to the traditional textbook exercises (where it is assumed that light has infinite speed) by allowing c to approach infinity in (FSL), thereby yielding the infinite speed of light result

$$\frac{ds}{dt} = kq\sec^2\theta = \frac{d\theta}{dt}q\sec^2\theta. \tag{ISL}$$

Of course, this is the answer one obtains by differentiating the equation $s = q\tan\theta$, as mentioned above.

Assuming light to have finite speed, formula (ISL) shows that the velocity of the light's intersection with L is symmetric about $\theta = 0$ and increases without bound as $\theta \to \pi/2$. By contrast, assuming light to have finite speed, formula (FSL) reveals that the actual situation is not symmetric. It is easier to investigate the finite speed of light case by multiplying numerator and denominator of (FSL) by $\cos^2\theta$. This yields

$$s'(t) = \frac{ckq}{c\cos^2\theta + kq\sin\theta}. \tag{FSL$'$}$$

Here, the limit of $s'(t)$ as $\theta \to \pi/2$ is easily seen to be c. This agrees with intuition, since as $\theta \to \pi/2$, the beam of light striking the shore is more and more nearly parallel to the shore and q becomes negligible in comparison to $s(t)$. Therefore, in this case, the light may be thought of as traveling along the line.

If $\theta \to -\pi/2$, the limit of $s'(t)$ is $-c$. Thus, for θ slightly larger than $-\pi/2$ (but keeping the direction of rotation clockwise), $s'(t)$ is negative. This means that the point where the light strikes the line appears to be moving backward, counter to the direction of the rotation of the light source. Figure 3 helps show why this is true. The distance to L_1 is so much greater than the distance to L_2 that the light will actually strike L_2 first, even though it "left" first toward L_1.

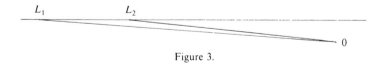

Figure 3.

Let's analyze the velocity given by formula (FSL)$'$. First of all, the velocity is not defined where $c\cos^2\theta + kq\sin\theta = 0$. Such a place always exists, since $\cos^2\theta/\sin\theta$ takes on all negative values, including $-kq/c$, as θ goes from $-\pi/2$ to 0. If $kq \geqslant 2c$, then $d(s'(t))/d\theta < 0$ and so $s'(t)$ is always decreasing for $\theta \in [-\pi/2, \pi/2]$. When $kq < 2c$ (the more reasonably scaled case), $s'(t)$ drops to a minimum where $\sin\theta = kq/2c$ and then continues to increase past that point. Figure 4 shows a graph of $s'(t)$ as a function of θ for the two essentially different cases: $kq < 2c$ and $kq \geqslant 2c$. (Figure 4 actually illustrates the case when $kq < c$. The case $c \leqslant kq < 2c$ has essentially the same graph; the y-intercept is at kq and there is still a minimum below c for some $\theta > 0$.)

This problem can be generalized to the case of a light moving along a curve given parametrically by $x = f(\theta)$, $y = g(\theta)$. Here too, the light is cast directly away from the origin onto a given line. The calculations are similar, but involve differentiating an arc length integral to find the distance from A_1 to A_2 along the curve. Briefly, let $d(\theta) = \sqrt{f^2(\theta) + g^2(\theta)}$ be the distance from the origin to the point A_1.

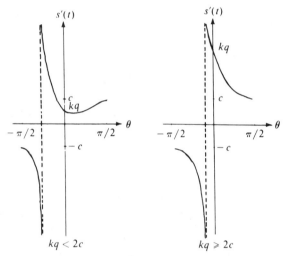

Figure 4.

Then, in formula ($*$), replace $c(\Delta\theta)$ by $\int_{\theta}^{\theta+\Delta\theta} \sqrt{(f'(t))^2 + (g'(t))^2}\, dt$ and add the term $- k(\sqrt{f^2(\theta + \Delta\theta) + g^2(\theta + \Delta\theta)} - \sqrt{f^2(\theta) + g^2(\theta)})$ to the denominator. This yields, in the same way we obtained (FSL),

$$\frac{ds}{dt} = \frac{ckqd(\theta)\sec^2\theta}{kqd(\theta)\sec\theta\tan\theta - k[f(\theta)f'(\theta) + g(\theta)g'(\theta)] + cd(\theta)\sqrt{(f'(\theta))^2 + (g'(\theta))^2}}.$$

It is even possible to allow the light to fall on an arbitrary curve and to allow the rate of revolution or the speed of light to vary. The form of this answer is also similar to (FSL).

The authors would like to thank Larry Kugler and numerous referees for their help in the preparation of this article.

Intuition Out to Sea

William A. Leonard, California State University, Fullerton, CA

Using only intuition and *without* writing anything, try to answer the following question:

> A rope, attached to boat B and passing over pulley C, is drawn to the right one foot at A. Does the boat move more than one foot, less than one foot, or exactly one foot to the right?

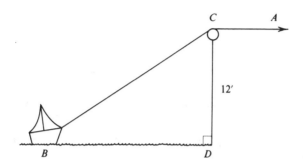

It is interesting to note that, when using intuition alone, the majority of calculus teachers queried answered incorrectly. Would you be confident of your answer under these conditions?

Perhaps we should begin by looking at some special cases. Let x denote the distance from B to the base of the pier CD, and let $h = \sqrt{x^2 + 12^2}$ denote the initial length of the rope from B to C. If B' represents the final position of the boat after A has moved one foot to the right, then $h - 1$ is the length of the shortened rope from B' to C.

For any initial $x \leqslant 5$, we have $h - 1 = \sqrt{144 + x^2} - 1 \leqslant 12$. Therefore, the boat (being lifted out of the water if and only if $x < 5$) moves exactly x feet to the right. In particular, the boat can move less than, equal to, or more than one foot to the right.

Analytic Solution. Since $x^2 + 12^2 = h^2$, implicit differentiation yields

$$\frac{dx}{dh} = \frac{h}{x} > 1.$$

Therefore, for $x > 1$, the boat moves more than one foot to the right.

Even after confirming the above conclusion, it may still not "feel right." The diagram below may help if we reason as follows: Assuming that B is pulled right up the rope, out of the water to point P, the effect of gravity will be to force it to "swing" down into the water again. Thus, more than A's horizontal component A_x is at work on B.

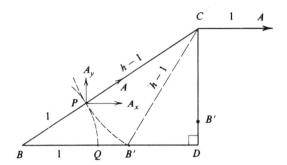

Visual Solution. For $x > 1$, the circular arcs $\overset{\frown}{PQ}$ and $\overset{\frown}{PB'}$ meet only at P. Therefore, $|\overline{BB'}| > |\overline{BQ}| = 1$ for $h - 1 > 12$, and $|\overline{BD}| > |\overline{BQ}| = 1$ for $h - 1 \leqslant 12$.

The author would like to thank Leonard Gillman for his insight into the visual solution.

BIBLIOGRAPHIC ENTRIES: APPLICATIONS TO MECHANICS

1. *Monthly* Vol. 76, No. 1, pp. 16–23. M. S. Klamkin and D. J. Newman, Flying in a wind field, I.

> Straight line course gives not only the shortest path but also the path of minimum flight time.

2. *Monthly* Vol. 76, No. 9, pp. 1013–19. M. S. Klamkin and D. J. Newman, Flying in a wind field, II.

3. *Monthly* Vol. 78, No. 10, pp. 1122–1126. L. B. Williams, Fly around a circle in a wind.

> Determines magnitude of wind drift.

4. *Monthly* Vol. 84, No. 1, pp. 40–43. Rochelle W. Meyer, Theory vs. mechanics in an application of calculus to biology.

5. *Monthly* Vol. 91, No. 1, pp. 3–17. Donald G. Saari and John B. Urenko, Newton's method, circle maps, and chaotic motion.

6. *Math. Mag.* Vol. 45, No. 5, pp. 246–253. Michael A. B. Deakin, Walking in the rain.

> How fast should you walk in the rain to stay as dry as possible?

(d)

DIFFERENTIAL EQUATIONS

A FACT ABOUT FALLING BODIES

WILLIAM C. WATERHOUSE, Cornell University

Suppose a body of mass m falls in a constant gravitational field g and encounters air resistance proportional to the nth power of its velocity v. Clearly v is governed by the equation

$$m \frac{dv}{dt} = mg - kv^n.$$

Calculus books often solve the equation for $n = 1$ or $n = 2$ and then point out that the velocity approaches a limit proportional to m or to $\sqrt{m}$.

It is actually easy to deduce this more generally. We have

$$\frac{dt}{dv} = \frac{1}{g[1 - (k/mg)v^n]},$$

so

$$t = \frac{1}{g} \int_0^v \frac{dv}{1 - (k/mg)v^n} + t_0.$$

Since $1 - (k/mg)v^n$ is a bounded factor times $1 - (k/mg)^{1/n}v$, the integral diverges; that is, t gets arbitrarily large as v approaches $(mg/k)^{1/n}$. Hence that is the limiting velocity.

This uses nothing beyond the grasp of a calculus class, and is a good example of deriving information about a problem without knowing an explicit solution. It also furnishes what is all too often lacking, a nontrivial application of improper integrals.

The Homicide Problem Revisited

David A. Smith

David A. Smith is Associate Professor of Mathematics at Duke University, where he has been a member of the faculty since 1962. He is the author of a number of articles on algebra, combinatorial theory, mathematics education, applied mathematics, and one book, "Interface: Calculus and the Computer" (Houghton Mifflin, 1976). The present article was written while he was a Visiting Associate Professor at Case Western Reserve University during a sabbatical leave from Duke.

What if Columbo knew a little calculus? Hurley [2] has presented a possible answer to this question with his application of Newton's law of cooling to a situation in which the "cooling body" is *corpus delicti*. Having used this example in the classroom, I can endorse Hurley's thesis that students find a great deal more interest in determining the time of death of a homicide victim than they do in plunging a hot object into a cooling solution. The purpose of this note is to point out that a very modest extension of Hurley's problem can be used to motivate two other important points for a class in calculus, differential equations, or applied mathematics: (a) selection of the best possible coordinate system for the problem at hand, and (b) use of elementary numerical methods when analytic or algebraic methods fail.

Recall that Newton's law of cooling may be formulated as

$$\frac{dy}{dt} = -k(y - a), \tag{1}$$

where y = temperature of body at time t, a = ambient temperature, and k is a positive proportionality constant. It is assumed that $y = 37°C$ (or $98.6°F$) at the time of death, and as many measurements of y as necessary may be made at or after the time of discovery of the body. The problem is to determine the interval between time of death and time of discovery.

The separation-of-variables solution of (1) is

$$\ln \frac{y - a}{y_0 - a} = -kt, \tag{2}$$

where y_0 is the temperature at time $t = 0$. One's first inclination may be to set $t = 0$ at the time of death (as Hurley does), and to solve for the "unknown" time of discovery $\bar{t}$, using measurements of y at $\bar{t}$ and at one other time, say $\bar{t} + 1$. Following Hurley's example, if $y = 31$ at time of discovery and $y = 29$ one hour later, with $a = 21$ (all in degrees Celsius), then (2) leads to the following simultaneous, nonlinear equations:

$$k\bar{t} = \ln 1.6, \quad k(\bar{t} + 1) = \ln 2, \tag{3}$$

which are easy to solve. However, if we set $t = 0$ at the *known* time of discovery, the problems of determining k and the time of death are no longer intertwined. Given the same data, equation (2) leads immediately to:

$$k = \ln 1.25 = 0.22314, \tag{4}$$

from which we may solve for t when $y = 37$:

$$t = -(\ln 1.6)/k = -2.10630, \tag{5}$$

or 2 hours and 6 minutes before time of discovery.

This computation is only slightly simpler than that given by Hurley, but separation of the two problems to be solved (one being a necessary step before pursuing the other) is a conceptual simplification that may ease the way for the student. When the problem becomes slightly more complicated, this step may be essential for success.

For example, suppose that the body is discovered outdoors, in which case the assumption of constant ambient temperature is likely to be incorrect. In this case, equation (1) may be rewritten as

$$\frac{dy}{dt} + ky = ka(t), \tag{6}$$

where $a(t)$ is the ambient temperature at time t, presumably known from a nearby weather station recording device. We no longer have a variable separable equation, but (6) is a first-order, linear, nonhomogeneous equation with constant coefficients, and hence solvable by a variety of methods, such as "integrating factors" or "homogeneous solution plus particular solution."

For sake of illustration, suppose the body is found at midnight, and its temperature is 30°C. Suppose the air temperature has been dropping 1° per hour since sundown, and it has just reached 0° at midnight. Suppose further that the body temperature is 25° at one A.M. and the air temperature is $-1°$. If we set $t = 0$ at time of death (with $y_0 = 37$) and $t = \bar{t}$ at midnight, we have $a(t) = \bar{t} - t$. Multiplication of (6) by the integrating factor e^{kt} leads to

$$\frac{d}{dt}\left(ye^{kt}\right) = k(\bar{t} - t)e^{kt}. \tag{7}$$

The right-hand side may be integrated by parts and the initial condition $y_0 = 37$ substituted to obtain the solution

$$y = \left(37 - t - \frac{1}{k}\right)e^{-kt} + \bar{t} - t + \frac{1}{k}. \tag{8}$$

Substitution of $y = 30$ at $t = \bar{t}$ and $y = 25$ at $t = \bar{t} + 1$ yields the following simultaneous, nonlinear equations:

$$\left(37 - \bar{t} - \frac{1}{k}\right)e^{-k\bar{t}} + \frac{1}{k} = 30, \tag{9}$$

$$\left(37 - \bar{t} - \frac{1}{k}\right)e^{-k(\bar{t}+1)} + \frac{1}{k} = 26. \tag{10}$$

If one solves for the term involving $\bar{t}$ in equation (9), the corresponding expression in equation (10) may be replaced by $30 - 1/k$ to obtain an equation in k alone:

$$\left(30 - \frac{1}{k}\right)e^{-k} - 26 + \frac{1}{k} = 0. \tag{11}$$

However, this step may not be transparent to the student, and in any case, equations (9) and (10) are unnecessarily intimidating.

On the other hand, if we set $t = 0$ at midnight, we have $a(t) = -t$, and we can concentrate first on finding y as an explicit function of t from

$$\frac{dy}{dt} + ky = -kt, \, y_0 = 30. \tag{12}$$

The solution of (12) is found in exactly the same manner as that of (7), with $\bar{t}$ replaced by zero. That solution is

$$y = \left(30 - \frac{1}{k}\right)e^{-kt} - t + \frac{1}{k}. \tag{13}$$

Substitution of $y = 25$ at $t = 1$ leads directly to equation (11). That completes our discussion of point (a) and brings us to point (b). Obviously equation (11) cannot be solved algebraically for k, but it poses no real difficulty for the student who has been exposed to Newton's method (or any numerical root-finding method).

We note that the computation can be simplified by clearing fractions first:

$$30k - 1 + (1 - 26k)e^k = 0. \tag{14}$$

If we denote the left-hand side of (14) by $f(k)$ then

$$f'(k) = 30 - (25 + 26k)e^k. \tag{15}$$

We have $f(0) = 0$ and $f(1) < 0$ from (14) and $f'(0) > 0$ from (15), so f rises from the origin and falls to a negative value at $k = 1$. It follows that there is a root in $(0, 1)$. (In fact, the root is unique, which may be shown by computing f'' and verifying that the graph is concave downward throughout the interval.)

The iteration for Newton's method [5, pp. 463–467] is:

$$k_{n+1} = k_n - f(k_n)/f'(k_n). \tag{16}$$

Starting with $k_0 = 0.5$, we get the following results:

n	k_n	$f(k_n)$	$f'(k_n)$
0	0.5	-5.785	-32.65
1	0.322836	-1.526	-16.12
2	0.228162	-0.3514	-8.860
3	0.188497	-0.0552	-6.103
4	0.179454	-0.00275	-5.497
5	0.178954	-8.3×10^{-6}	-5.464
6	0.178952	-6.0×10^{-8}	-5.464

All the numbers shown are rounded to the indicated number of digits from intermediate calculations retaining the least seven significant decimal digits, and k_6 is the value of k to six decimal places. These results were obtained by computer, but the student using a hand calculator with an exponential function and a word or two of memory could find this answer faster than cards could be punched for a computer.

The final step is to substitute the computed value of k and $y = 37$ in equation (13) and solve for t, the time of death. Of course, this results in another numerical problem, similar to (14), namely

$$(37k - 1 + kt)e^{kt} - 30k + 1 = 0. \qquad (17)$$

Denoting the left-hand side of (17) by $g(t)$, and using Newton's method with $t_0 = -1$, we find:

n	t_n	$g(t_n)$	$g'(t_n)$
0	-1.0	0.1820	0.9640
1	-1.188778	0.0035	0.9271
2	-1.192560	1.4×10^{-6}	0.9263
3	-1.192562	0.0	0.9263

Hence the time of death was about one hour and 12 minutes before midnight.

It is perhaps worth noting that a problem equivalent to the one just discussed appears as an exercise in at least one sophomore calculus text [1, p. 548, ex. 7]. The context is that of plunging a piece of hot metal into a cooler oil bath whose temperature is allowed to rise at a uniform rate. Interestingly, a numerical answer is given in the back of the book, but there is no hint to the student that a numerical root-finding method is required to arrive at the answer.

Postscript.   One might reasonably ask whether the "application" of Newton's law of cooling to determination of time of death has any connection with the way this problem is solved in the "real" world. A standard reference of a generation ago [3] described a more pragmatic method that would not have required the use of a pocket calculator (as yet unavailable). Taking advantage of the fact that the body does not cool uniformly after death, Taber described a way to divide the leg of the deceased into 10 parts so that, if part n is colder than part $n + 1$, the body could be assumed to have been dead for n hours. "Experiments conducted in temperatures between 40° and 80°F proved fairly accurate in over 100 examinations." Whatever that means, the method is now apparently discredited, since the latest edition of the same work [4] gives the following prescription:

> "Take the rectal temperature. In general the body loses one degree of Fahrenheit temperature each hour following death. Of course the rate of heat loss varies with the temperature of the surrounding air, water, or snow."

The last sentence suggests some connection with Newton's law of cooling, but of course the constant rate of heat loss is inconsistent with it. (Incidentally, rectal temperature is generally considered more consistent and more accurately measurable than oral temperature, dead or alive. However, "normal" rectal temperature averages 1°F higher than oral, so one would assume 37.56°C at time of death.) Could it be that the necessary research has not been done yet to determine accurately the rate of heat loss after death?

A call to a local coroner's office produced the following additional information. While the coroner knew about and believed in Newton's law of cooling (under conditions of constant ambient temperature), he was unwilling to make judgments of time of death more precise than to within a few hours. Besides the obvious difficulties regarding ambient temperature and body temperature at time of death, there is the less obvious one that metabolic processes may continue for an unknown period after death, and cooling begin only when these processes cease. While conceding the possibility of taking several rectal temperatures and extrapolating back to time of death, his own judgments are based on temperature determined by feel, observation of whether *rigor mortis* is present, whether eyes are glazed, etc. The "degree per hour" formula quoted above he considered a gross oversimplification.

REFERENCES

1. H. Flanders, R. R. Korfhage, and J. J. Price, A Second Course in Calculus, Academic Press, New York, 1974.
2. J. F. Hurley, An application of Newton's Law of Cooling, Mathematics Teacher, 67 (1974) 141–142.
3. C. W. Taber, Taber's Cyclopedic Medical Dictionary, 7th ed., F. A. David Co., Philadelphia, 1956.
4. C. L. Thomas (ed.), Taber's Cyclopedic Medical Dictionary, 12th ed., F. A. David Co., Philadelphia, 1973.
5. G. B. Thomas, Jr., Calculus and Analytic Geometry, Alternate ed., Addison-Wesley, Reading, Mass., 1972.

A Linear Diet Model

Arthur C. Segal, University of Alabama at Birmingham, Birmingham, AL

Calculus courses usually include a unit on simple differential equation models of population growth, radioactive decay, Newtonian cooling, and so forth. To supplement this standard fare, consider the following linear diet model.

A person's weight depends both on the daily rate of energy intake, say, C calories per day, on the daily rate of energy consumption, which is typically between 15 and 20 calories per pound per day depending on age, sex, metabolic rate, etc. Using an average value of 17.5 calories per pound per day, a person weighing w pounds expends $17.5w$ calories per day. If $C = 17.5w$, then weight remains constant, and weight gain or loss occurs according to whether C is greater or less than $17.5w$.

How fast will weight gain or loss occur? The most plausible physiological assumption is that dw/dt is proportional to the net excess (or deficit) $C - 17.5w$ in the number of calories per day. In the equation

$$\frac{dw}{dt} = K(C - 17.5w),$$

the left side has units of pounds/day, and $C - 17.5w$ has units calories/day. Hence, the units of K are pounds/calorie. Therefore, we need to know how many pounds each excess or deficit calorie puts on or takes off. The commonly used dietetic conversion factor is that 3500 calories is equivalent to one pound. Thus, $K = (1/3500)$ pound/calorie.

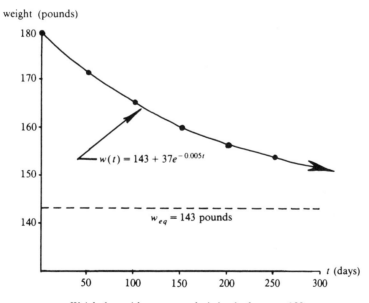

weight (pounds)

$w(t) = 143 + 37e^{-0.005t}$

$w_{eq} = 143$ pounds

t (days)

Weight loss with constant caloric intake for $w_0 = 180$.

175

Now, the differential equation modeling weight gain or loss is

$$3500\left(\frac{dw}{dt}\right) = C - 17.5w.$$

If C is constant, the equation is separable and its solution is

$$w(t) = \frac{C}{17.5} + \left[w_0 - \frac{C}{17.5}\right]e^{-0.005t},$$

where w_0 is the initial weight and t is in days. Note that the equilibrium weight is $w_{eq} = C/17.5$.

Example. If someone initially weighing 180 pounds adopts a diet of 2500 calories per day, then $w_{eq} \cong 143$ pounds and the weight function is $w(t) = 143 + 37e^{-0.005t}$. Notice how long it takes to even get close to the asymptotic weight of 143 pounds. The half-life for the process is $(\ln 2)/0.005 \cong 138.6$ days, about twenty weeks. (It would take about 583 days, or 83 weeks, to get to 145 pounds.) This may be why so many dieters give up in frustration.

BIBLIOGRAPHIC ENTRIES: DIFFERENTIAL EQUATIONS

1. *Monthly* Vol. 82, No. 2, pp. 159–162. Tom M. Apostol, Explicit formulas for solutions of the second-order matrix differential equation $Y'' = AY$.

2. *Monthly* Vol. 92, No. 6, pp. 422–423. David Scott, When is an ordinary differential equation separable?

Necessary and sufficient conditions for a first order differential equation to be separable.

3. *Monthly* Vol. 95, No. 4, p. 344. J. L. Brenner, An elementary approach to $y'' = -y$.

4. *Math. Mag.* Vol. 45, No. 5, pp. 241–246, M. N. Brearley, The long jump miracle of Mexico City.

Analysis of dramatic increase in a world record.

5. *Math. Mag.* Vol. 50, No. 4, pp. 186–197. David A. Smith, Human population growth. Stability or explosion?

6. *Math. Mag.* Vol. 58, No. 2, pp. 78–83. R. B. Borelli, C. S. Coleman, and D. D. Hobson, Poe's pendulum.

Mathematical model describing different aspects of Poe's pendulum motion.

7. TYCMJ Vol. 4, No. 1, pp. 72–75, H. L. Kung, On particular solutions of $P_n(D)Y = 0$.

8. TYCMJ Vol. 4, No. 2, pp. 1–15. Colin Clark, Some socially relevant applications of elementary calculus.

9. TYCMJ Vol. 10, No. 3, pp. 200–201. R. S. Luthar, Another approach to a standard differential equation.

Second order with constant coefficients.

(e)

PARTIAL DERIVATIVES

USING THE MULTIVARIABLE CHAIN RULE

FRED HALPERN

10602 Stone Canyon Rd., #260, Dallas, TX 75230

Finding the derivative of an expression with many occurrences of x is a major application of the multivariable chain rule. The usual computation, using extraneous variables, introduces various side calculations and bookkeeping chores. This paper restates the multivariable chain rule to provide a simple straightforward method of computation. Informally, the computation states that

The derivative of an expression with many occurrences of x is the sum of the derivatives of the expression with respect to each separate occurrence of x.

Thomas and Finney [1] (pp. 600–605) contains a standard statement of the multivariable chain rule and its application to computing derivatives.

We consider functions $f(x_1, \ldots, x_n)$ and are interested in $Df(x, \ldots, x)$ where D denotes differentiation with respect to x. Since $\partial x_i / \partial x = 1$ ($i = 1, \ldots, n$),

$$Df(x, \ldots, x) = \sum \left. \frac{\partial f}{\partial x_i} \right|_{(x, \ldots, x)} .$$

When computing each $\partial f / \partial x_i$ the other variables are kept constant. This is signified by replacing the other variables by c, x being resubstituted for c after the differentiation.

With this agreement about the relation between x and c, we obtain the following application of the multivariable chain rule to functions of the form $f(x, \ldots, x)$.

THEOREM. *If each of*

$$Df(x, c, \ldots, c), Df(c, x, \ldots, c), \ldots, Df(c, c, \ldots, x)$$

exists and is continuous in an interval containing c, then

$$Df(x, x, \ldots, x) = Df(x, c, \ldots, c) + Df(c, x, \ldots, c) + \cdots + Df(c, c, \ldots, x).$$

The usual product and quotient rules for derivatives easily follow from the Theorem. In addition, an "exponential rule" for $Df(x)^{g(x)}$ can be obtained. Here are some applications of the Theorem.

EXAMPLES:

(1)
$$Dx\, e^x / \sin x = Dx\, e^c / \sin c + Dc\, e^x / \sin c + Dc\, e^c / \sin x$$

$$= e^c / \sin c + ce^x / \sin c + ce^c (-1/\sin^2 x) \cos x$$

$$= e^x / \sin x + xe^x / \sin x - xe^x \cos x / \sin^2 x.$$

(2)
$$Dx^{x^x} = Dx^{c^c} + Dc^{x^c} + Dc^{c^x}$$

$$= c^c x^{(c^c-1)} + (\ln c)\, c^{x^c} cx^{c-1} + (\ln c)\, c^{c^x}(\ln c\, c^x)$$

$$= x^x x^{(x^x-1)} + x^x x^{x^x} \ln x + x^x x^{x^x} \ln^2 x.$$

(3)
$$D\int_a^{f(x)} g(x,t)\, dt = D\int_a^{f(x)} g(c,t)\, dt + D\int_a^{f(c)} g(x,t)\, dt$$

$$= g(c,f(x))f'(x) + \int_a^{f(c)} \frac{\partial g}{\partial x}(x,t)\, dt$$

$$= g(x,f(x))f'(x) + \int_a^{f(x)} \frac{\partial g}{\partial x}(x,t)\, dt.$$

(4)
$$D\int_0^x \int_0^x f(u,v)\, du\, dv = D\int_0^c dv \int_0^x f(u,v)\, du + D\int_0^x dv \int_0^c f(u,v)\, du$$

$$= \int_0^c f(x,v)\, dv + \int_0^c f(u,x)\, du$$

$$= \int_0^x f(x,v)\, dv + \int_0^x f(u,x)\, du.$$

The Theorem also applies to expressions involving determinants, inner products, and cross products.

Reference

1. G. B. Thomas and R. L. Finney, Calculus and Analytic Geometry, 5th ed., Addison-Wesley, Reading, MA, 1980.

BIBLIOGRAPHIC ENTRIES: PARTIAL DERIVATIVES

1. *Monthly* Vol. 76, No. 1, pp. 76–77. Donald H. Trahan, The mixed partial derivatives and the double derivative.
2. *Monthly* Vol. 80, No. 10, pp. 1134–1135. A. G. Fadell, A proof of the chain rule for derivatives in *n*-space.

> Editor's note in *Monthly* Vol. 81, p. 1098 points out that the same proof appears in Lang's *A Second Course in Calculus* (2nd ed.), pp. 527–528.

3. *Monthly* Vol. 92, No. 9, pp. 663–665. Michael W. Botsko and Richard A. Gosser, On the differentiability of functions of several variables.
4. CMJ Vol. 21, No. 5, pp. 370–378. M. R. Cullen, Moire fringes and the conic sections.

6

MEAN VALUE THEOREM FOR DERIVATIVES, INDETERMINATE FORMS

(a)

MEAN VALUE THEOREM

A VERSATILE VECTOR MEAN VALUE THEOREM

D. E. SANDERSON, Iowa State University

If a particle moves smoothly in n-space and at two points in time its velocity is orthogonal to a given direction, then so must its acceleration be at some intermediate time. The following easily proved extension of Rolle's theorem embodies this principle for arbitrary dimension and orders of differentiation (the one-dimensional case reduces to Rolle's theorem if orthogonality is interpreted as meaning the (inner) product of the vectors is zero). The two-dimensional version affords a simple way to present the elementary applications or forms of the usual mean value theorems.

THEOREM 1. *Suppose $v: [a, b] \to R^n$ is a k times differentiable n-dimensional vector-valued function and $v(a)$, $v(b)$ and the first $k-1$ derivatives of v at a are orthogonal to a non-zero vector v_0. Then for some c between a and b, $v^{(k)}(c)$ is orthogonal to v_0.*

Proof. Let $F(t) = v(t) \cdot v_0$ denote the inner (dot) product of the vectors $v(t)$ and v_0. Then, since the vanishing of $F^{(m)}(t) = v^{(m)}(t) \cdot v_0$ is equivalent to orthogonality of $v^{(m)}(t)$ and v_0, we have $F(b) = F(a) = F'(a) = \cdots = F^{(k-1)}(a) = 0$. Successive applications of Rolle's theorem give points $c_0 = b, c_1, \cdots, c_k = c$ such that $F^{(m)}(c_m) = 0$ and $a < c_m < c_{m-1}$ for $m = 1, \cdots, k$. Thus $v^{(k)}(c)$ is orthogonal to v_0 and the proof is complete.

To illustrate the ease with which standard mean value results can be obtained from this theorem (with $n = 2$) let us simplify the form by translating coordinates in the domain and range of v so that a is replaced by 0, b by $h = b - a$, and $v(0)$ by the origin of R^2. If we write $v(t) = (f(t), g(t))$ where $f(0) = g(0) = 0$, and assume $v(h)$ is non-zero, then we may use $(g(h), -f(h))$ for v_0 so that $F(t) = f(t)g(h) - g(t)f(h)$ and the orthogonality condition in the conclusion becomes $f(h)g^{(k)}(c) = g(h)f^{(k)}(c)$. This remains true, trivially, but of little use if $v(h)$ is the zero vector.

Applications. (1) The ordinary mean value theorem for a function f, differentiable on $[0, h]$ (where $f(0) = 0$) is obtained by setting $k = 1$, $g(t) = t$: $f(h) = hf'(c)$.

(2) The Cauchy or generalized mean value theorem results from setting $k = 1$: $f(h)g'(c) = g(h)f'(c)$ (where $f(0) = g(0) = 0$).

(3) From (2) and appropriate conditions on f and g, one can of course write $f(h)/g(h) = f'(c)/g'(c)$ and derive L'Hospital's Rule.

For applications involving values of k greater than one (and $n = 2$, still) it should be observed that the condition on v and its first $k-1$ derivatives at a requires them to all be parallel. In particular, the theorem is applicable whenever the values of f and its first $k-1$ derivatives at a are equal to the respective values of g and its first $k-1$ derivatives at a. We state this as the next application, continuing to use the notationally simpler case, $a = 0$.

(4) If $f^{(m)}(0) = g^{(m)}(0)$ for $m = 0, 1, \cdots, k-1$ ($f^{(0)} = f$, etc.) and $f^{(k)}(t)$, $g^{(k)}(t)$ exist for $t \in [0, h]$, then $f(h)g^{(k)}(c) = g(h)f^{(k)}(c)$ for some c between 0 and h.

(5) Taylor's Formula for a k times differentiable function ϕ follows from (4) if we set $f(t) = \phi(t) - \sum_{s=0}^{k-1} \phi^{(s)}(0)t^s/s!$ and $g(t) = t^k$.

Proof. Since $f^{(m)}(t) = \phi^{(m)}(t) - \sum_{s=m}^{k-1} \phi^{(s)}(0)t^{s-m}/(s-m)!$, we have $f^{(m)}(0) = 0 = g^{(m)}(0)$ for $m = 0, 1, \cdots, k-1$ and (4) applies, giving $f(h)k! = h^k f^{(k)}(c) = h^k \phi^{(k)}(c)$, hence

$$\phi(h) = \sum_{s=0}^{k-1} \phi^{(s)}(0)h^s/s! + \phi^{(k)}(c)h^k/k!.$$

(6) The standard formula for the error in Simpson's Rule for approximating the integral of a four times differentiable function ϕ on the interval $[-h, h]$ follows from Corollary 4 by setting

$$f(t) = (t/3)[\phi(-t) + 4\phi(0) + \phi(t)] - \int_{-t}^{t} \phi \quad \text{and} \quad g(t) = t^5.$$

Proof. Differentiating, one finds that f and its first three derivatives vanish at 0. In particular, $f'''(t) = [\phi'''(t) - \phi'''(-t)]t/3$. Applying (4) with $k = 3$ and using the mean value theorem (i.e., (1) modified to apply to the interval $[-c, c]$) gives

$$f(h) \cdot 60c^2 = [\phi'''(c) - \phi'''(-c)]h^5c/3 = 2c\phi^{(4)}(\bar{c})h^5c/3,$$

or

$$f(h) = h^5 \phi^{(4)}(\bar{c})/90,$$

where $\bar{c} \in (-c, c) \subset (-h, h)$. This is the standard formula for the error $f(h)$ in Simpson's Rule.

Note that in the proof of (6) we could just as well apply the theorem with $k = 4$, and it would be more natural to do so. However, this leads to the more complicated form

$$f(h) = [2\phi^{(4)}(\bar{c}) + \phi^{(4)}(c) + \phi^{(4)}(-c)]h^5/360$$

and the same estimate $|f(h)| < M h^5/90$, where M is the maximum of $|\phi^{(4)}(t)|$ for $-h < t < h$.

(7) The standard formula for the error in the Trapezoidal Rule for approximating the integral of a twice differentiable function ϕ on the interval $[-h, h]$ follows from (4) by setting

$$f(t) = [\phi(-t) + \phi(t)]/2 - \int_{-t}^{t} \phi \quad \text{and} \quad g(t) = t^5 \quad (\text{and} \quad k = 1).$$

The proof of (7) parallels that of (6), the error formula being $\frac{2}{3} h^3 \phi''(\bar{c})$ for some $\bar{c} \in (-h, h)$. The corresponding formulas for an arbitrary interval divided into several (equal) subintervals are easily obtained if $\phi^{(4)}$ (respectively, ϕ'') is continuous on the interval (see problem 9 section 8.22 of [1]). The fact that the hypothesis of the theorem is satisfied for a higher value of k than is used in the proofs of (6) and (7) suggests that a sharper error estimate may be possible but the note preceding (7) does not bear this out.

Using Theorem 1 with $k = 1$ in much the same way that Rolle's theorem was used in proving Theorem 1, the following variation can be proved:

THEOREM 2. *Suppose* $v: [a, b] \to R^n$ *is a k times differentiable n-dimensional vector-valued function which is orthogonal to a non-zero vector* v_0 *at* $k + 1$ *distinct points of* $[a, b]$. *Then for some c between a and b,* $v^{(k)}$ *(c) is orthogonal to* v_0.

Theorem 2 can be used to obtain the error formula for polynomial interpolation given in Theorem 8–3 of [1].

Reference

1. T. M. Apostol, Calculus, vol. 2, Blaisdell, New York, 1962.

Who Needs Those Mean-Value Theorems, Anyway?

Ralph P. Boas

R. P. Boas was born in 1912 in Walla Walla, Washington, received a Ph.D. from Harvard in 1937, and was a National Research Fellow during 1937–38, at Princeton and at Cambridge (England). He has taught at Duke; the U. S. Navy Pre-Flight School in Chapel Hill, North Carolina; Harvard; MIT; and since 1950 at Northwestern. During 1945–50 he was Executive Editor of "Mathematical Reviews." He has published articles in professional journals, three advanced books and the Carus Monograph "A primer of real functions." He has held various offices in both the AMS and MAA, and is currently the Editor of the "American Mathematical Monthly."

If we are to believe the textbooks, every student of calculus is supposed to learn the mean-value theorem ("the law of the mean," as it was called when I was a student). Let me remind you that this theorem says that if f is differentiable then

$$f(b) - f(a) = (b - a)f'(c), \tag{1}$$

where c is some value between a and b, and we usually don't know any more about where c actually is.

Geometrically this says that every chord has a parallel tangent: (Figure 1 once appeared on the Graduate Record Examination with 5 suggested answers for the question "Which theorem does the picture remind you of?")

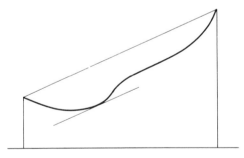

Figure 1.

I claim, in common with Dieudonné ([5, pp. 142, 154]), that (1) enjoys too high a status and that we would be better off with the mean-value inequality

$$(b - a)\min f'(x) \leqslant f(b) - f(a) \leqslant (b - a)\max f'(x) \tag{2}$$

(where the max and min refer to the interval (a, b)).

182

We can alternatively write (2) as

$$(b - a)\min f'(x) \leqslant \int_a^b f'(t)\,dt \leqslant (b - a)\max f'(x) \tag{3}$$

if we suppose, as is appropriate in a first course in calculus, that f is the integral of its derivative (Figure 2).

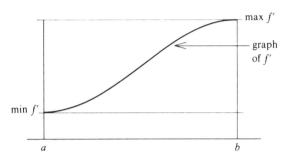

Figure 2.

The first advantage of (2) over (1) is that it avoids the perennial problem that we can't say where the point c is on (a, b). Many students are bothered by the indetermination. (They think that we *could* tell them where c is, if we only would. This belief is only reinforced by exercises that ask them to find c in special cases. Such exercises may be good for something else, but they don't help the understanding of the mean-value theorem.)

Second, (2) is more intuitive than (1) if we think of x as time and $f(x)$ as the distance you have traveled up to time x. Then (1) says that at some instant you are moving at exactly your average speed (cf. [3]). This seems not to be very intuitive. But (2) says that the average speed is between the minimum speed and the maximum speed, or that the distance traveled is no greater than the maximum speed times the time, and no less than the minimum speed times the time; and what could be more intuitive?

In the third place, most of the applications of (1) are in proofs of theorems. For example, to prove that a function with a positive derivative increases, we argue that

$$f(b) - f(a) = (b - a)f'(c).$$

Since f' is positive everywhere, it is positive at c, wherever c may be. Hence $f(b) > f(a)$.

But we can just as well appeal to (2):

$$f(b) - f(a) \geqslant (b - a) \min_{a \leqslant x \leqslant b} f'(x) > 0,$$

and there is no reason (except for a century or two of tradition) for dragging in the nebulous point c.

In any case, proving theorems ought not to be a principal aim (probably not even a proper aim) of a first course in calculus. (Cf. [2] for further discussion of this point.)

Fourth, some textbooks (especially older ones) make much of (1) for computational purposes, preferring it to the tangent approximation, namely,

$$f(b) - f(a) \approx (b - a)f'(a), \qquad (4)$$

often written as

$$f(x + \Delta x) - f(x) \approx f'(x)\Delta x,$$

or even as

$$dy = f'(x)\,dx.$$

The trouble with the tangent approximation is presumably that (4) provides no error bounds, whereas (1) does provide them—but only via (2), and again it seems simpler to use (2) directly instead of going around through (1).

However, this application no longer has much point for simple problems like $\sqrt{26}$ or $\sin 61°$, since such numbers can now be read from any decent pocket calculator with more accuracy than most practical problems require.

One might expect that (2) would be useful for one of the many functions that aren't (yet) available on calculators. Let us look at this possibility geometrically. If we are working with an unfamiliar function, there is no reason why we might not encounter a situation like Figure 3.

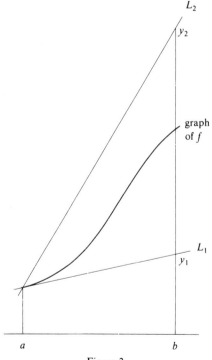

Figure 3.

In Figure 3, L_1 and L_2 are lines whose slopes are the minimum and maximum slopes of the curve, drawn through the initial point $(a, f(a))$. The mean-value theorem (in either form) tells us that the graph hits the vertical line $x = b$ somewhere between y_1 and y_2. If f' happens to have a maximum or minimum hiding somewhere between a and b, the distance between y_1 and y_2 might be enormous; then we wouldn't get much information about $f(b)$, and might not even realize that we are not getting it. This situation (in a less extreme form than indicated in Figure 3) occurs, for example, for the Bessel function J_0, if we know J_0 and J_0' at $x = 1.5$ and 2.0 and want to estimate $J_0(1.9)$.

Finally, (1) is no longer true for vector-valued functions, whereas an appropriate generalization of (2) is; see, for example, [5, p. 154].

If the intermediate point c causes trouble in the ordinary mean-value theorem, it causes even more trouble in the generalized mean-value theorem,

$$\frac{f(b) - f(a)}{g(b) - g(a)} = \frac{f'(c)}{g'(c)} . \tag{5}$$

The applications of (5) demand that, although we don't know where c is, it must be the same upstairs and downstairs. As far as I know, the *only* application of (5) in elementary calculus is to Lhospital's rule (I give the Marquis his own spelling, even if he did plagiarize the rule from John Bernoulli: see [8] or [9]).

Now, as somebody said, each generation has to make its own discoveries. One such discovery ([6], 1923; [7], 1936; [1], 1969) is that the generalized mean-value theorem (5) is utterly unnecessary for deriving Lhospital's rule at the elementary calculus level. All you need is the idea of a limit and an integral inequality something like (3), namely,

$$\min h(x) \int_a^b g'(x)\, dx \leqslant \int_a^b h(x) g'(x)\, dx \leqslant \max h(x) \int_a^b g'(x)\, dx, \qquad g'(x) \geqslant 0 \quad (6)$$

(this follows directly from the definition of the integral as the limit of a sum). In (6), $g'(x)$ plays the role of dx in (3), and $h(x)$ takes the place of $f'(x)$. All the effort that previously went into proving and comprehending (5) can now be saved for understanding more fundamental principles. For details, see [1]. (I am not arguing against using (5) to prove a more general form of the rule in a more advanced course.)

REFERENCES

1. R. P. Boas, Lhospital's rule without mean value theorems, Amer. Math. Monthly, 76 (1969) 1051–1053.
2. ——, Calculus as an experimental science, Amer. Math. Monthly, 78 (1971) 664–667 = this Journal, 2 (1971) 36–39.
3. ——, Travelers' surprises, this Journal, 10 (1979), 82–88.
4. T. J. I'A. Bromwich, An introduction to the theory of infinite series, Macmillan, London, 1st ed., 1908, 2nd ed., 1926.
5. J. Dieudonné, Foundations of Modern Analysis, Academic Press, New York and London, 1960.
6. E. V. Huntington, Simplified proof of l'Hospital's theorem on indeterminate forms (abstract), Bull. Amer. Math. Soc., 29 (1923) 207.
7. F. Lettenmeyer, Über die sogenannte Hospitalsche Regel, J. Reine Angew. Math., 174 (1936) 246–247.
8. O. Spiess, ed., Der Briefwechsel von Johann Bernoulli, vol. 1, Birkhäuser, Basel, 1955.
9. C. Truesdell, The new Bernoulli edition, Isis, 49 (1958) 54–62.

BIBLIOGRAPHIC ENTRIES: MEAN VALUE THEOREM FOR DERIVATIVES

1. *Monthly* Vol. 86, No. 6, pp. 484–485. Alexander Abian, An ultimate proof of Rolle's theorem.

2. *Monthly* Vol. 86, No. 6, p. 486. Hans Samelson, On Rolle's theorem.

3. *Math. Mag.* Vol. 52, No. 3, pp. 157–158. W. S. Hall and M. L. Newell, The mean value theorem for vector valued functions: A simple proof.

4. TYCMJ Vol. 4, No. 3, pp. 11–17. Rodney G. Gentry, Generalizing Rolle's theorem in elementary calculus.

5. TYCMJ Vol. 8, No. 1, pp. 51–53. Roy E. Myers, Some elementary results related to the mean value theorem.

6. TYCMJ Vol. 10, No. 2, pp. 114–115. Norman Schaumberger, Another application of the mean value theorem.

7. TYCMJ Vol. 11, No. 5, pp. 329–330. M. Powderly, A geometric proof of Cauchy's law of the mean.

8. CMJ Vol. 16, No. 5, pp. 397–398. Norman Schaumberger, More applications of the mean value theorem.

> Another proof of the arithmetic mean geometric mean inequality.

9. CMJ Vol. 17, No. 5, pp. 403–406. Robert S. Smith, Rolle over Lagrange—another shot at the mean value theorem.

10. CMJ Vol. 20, No. 4, p. 323. Herb Silverman, A simple auxiliary function for the mean value theorem.

(b)

INDETERMINATE FORMS

LHOSPITAL'S RULE WITHOUT MEAN-VALUE THEOREMS

R. P. Boas, Jr., Northwestern University

The "rule" that goes by the name of the Marquis De Lhospital (to give him, for once, the spelling that he himself used [4]), but which was actually discovered by John Bernoulli (see [4] or [5]), is usually proved by using the generalized mean-value theorem. I shall show that, with slightly stronger hypotheses that suffice for all applications, it can be proved quite simply without any mean-value theorems at all; this proof seems to have some pedagogical advantages, as well as suggesting some results that are not covered by the usual formulation.

We are to prove that *if f and g are real functions with continuous derivatives, if f(x) and g(x) both approach 0 or both approach ∞ as x→a, if g'(x)≠0, and if f'(x)/g'(x)→L as x→a then f(x)/g(x)→L*; all limits are taken on one side of a. We shall take $a = \infty$ and L finite, but only formal modifications are required for other cases.

We shall need only (i) the definition of a limit; (ii) that a continuous function that is never 0 has a fixed sign; and (iii) that the integral over an interval of a positive continuous function is positive (or alternatively that a function with a positive derivative is increasing; for proofs of the latter fact without mean-value theorems see [1], [3]).

Given $\epsilon > 0$, we have

$$-\epsilon < \{f'(t)/g'(t)\} - L < \epsilon$$

if t is sufficiently large. Since g' is continuous and never zero, it has a fixed sign; suppose for definiteness that $g'(t) > 0$. Then

(1) $$-\epsilon g'(t) < f'(t) - Lg'(t) < \epsilon g'(t).$$

Consider the right-hand half of this inequality; it says that

(2) $$f'(t) - (\epsilon + L)g'(t) < 0$$

for sufficiently large t. Since a negative function has a negative integral, we have for sufficiently large x and y, with $x > y$,

(3) $$f(x) - f(y) - (\epsilon + L)\{g(x) - g(y)\} < 0.$$

Suppose first that $f(x)$ and $g(x) \to 0$; fixing y and letting $x \to \infty$, we obtain

(4) $$-f(y) + (\epsilon + L)g(y) \leqq 0.$$

187

Since $g'(y) > 0$ and $g(y) \to 0$, we have $g(y) < 0$ for large y. Hence (4) says

$$-\frac{f(y)}{g(y)} + L \geqq -\epsilon$$

for sufficiently large y. Similarly the left-hand half of (1) yields

$$-\frac{f(y)}{g(y)} + L \leqq \epsilon,$$

and the last two inequalities together say that $f(y)/g(y) \to L$ as $y \to \infty$.

If $f(x)$ and $g(x) \to \infty$, we have $g(x) > 0$ for large x, since we assumed $g'(x) > 0$. We rewrite (3) as

$$\frac{f(x)}{g(x)} - (\epsilon + L) < \frac{f(y) - (\epsilon + L)g(y)}{g(x)}.$$

Fix y; for sufficiently large x the right-hand side is less than ϵ, and so

$$\frac{f(x)}{g(x)} - L < 2\epsilon.$$

Similarly $-2\epsilon < [f(x)/g(x)] - L$, and again $f(x)/g(x) \to L$.

The hypothesis that g' is continuous is actually redundant, although it is always satisfied in practice and makes the proof more comprehensible. All that we really use in (ii) is that a derivative that is different from 0 on an interval has a fixed sign there.

Since our proof did not use any mean-value theorems, it opens up the possibility of extending the rule to cases where mean-value theorems are not available. As an illustration, I state the sequence analogue of Lhospital's rule (see [2], 1st ed., pp. 377 ff.; 2nd ed., pp. 413 ff.).

Let $\{a_n\}$ and $\{b_n\}$ be two real sequences that both approach zero or both approach ∞; let $\Delta g_n = g_n - g_{n+1}$ have a fixed sign, and let $\Delta a_n / \Delta b_n \to L$; then $a_n / b_n \to L$.

The proof is the same, substituting differences for derivatives; in going from (2) to (3) we use

$$\sum_{k=n}^{\infty} (c_k - c_{k+1}) = c_n,$$

which is true when $c_k \to 0$. For example, suppose that $\sum x_n$ and $\sum y_n$ are two convergent series of positive terms; put $a_n = \sum_n^\infty x_k$, $b_n = \sum_n^\infty y_k$, the remainders. Then $x_n/y_n \to L$ implies $a_n/b_n \to L$, a result that is sometimes useful in dealing with infinite series.

It is possible to formulate a theorem that includes both Lhospital's rule and this discrete analogue; the reader is invited to find such a theorem for himself.

References

1. L. Bers, On avoiding the mean value theorem, this MONTHLY, 74 (1967) 583.
2. T. J. I'a. Bromwich, An introduction to the theory of infinite series, Macmillan, London, 1st ed., 1908; 2nd ed., 1926.
3. L. W. Cohen, On being mean to the mean value theorem, this MONTHLY, 74 (1967) 581–582.
4. O. Spiess (editor), Der Briefwechsel von Johann Bernoulli, Band I, Birkhäuser, Basel, 1955.
5. C. Truesdell, The new Bernoulli edition, Isis, 49 (1958) 54–62.

INDETERMINATE FORMS OF EXPONENTIAL TYPE

JOHN V. BAXLEY AND ELMER K. HAYASHI

When $\lim f(x)^{g(x)}$ yields an indeterminate of the form 0^0, ∞^0, or 1^∞, the usual procedure is to consider $g(x)\log f(x)$ and apply L'Hospital's Rule (see, for instance [1], [2], [4]). However, this method requires that f and g be differentiable and even so it may fail as in the examples

$$\lim_{x\to 0+} \left(2\sin\sqrt{x} + \sqrt{x}\, \sin(1/x)\right)^x$$

and

$$\lim_{x\to 0} \left(1 + x\sin(1/x^4)e^{-1/x^2}\right)e^{1/x^2}.$$

See [3]. Using the results in this paper, it is easy to see at a glance that both of the above limits are 1; in fact it will be clear why 0^0 and ∞^0 almost always have limit 1. These results are simple and make no direct use of L'Hospital's rule; in fact no smoothness conditions are required of the functions f and g. Only their orders of magnitude are significant. All results will be stated for x approaching 0. The corresponding results for x approaching a or $\pm\infty$ are obtained by using the appropriate change of variables.

THEOREM 1. *If* $\lim_{x\to 0+} g(x) = 0$, *if there exists a real number* α *such that* $b(x) = f(x)/x^\alpha$ *is positive, bounded, and bounded away from* 0 *as* $x\to 0+$, *and if* $\gamma = \alpha \lim_{x\to 0+} g(x)\log x$ *exists or is* $\pm\infty$, *then*

$$\lim_{x\to 0+} f(x)^{g(x)} = e^\gamma,$$

where $e^\infty = \infty$ *and* $e^{-\infty} = 0$.

Proof. Using the hypotheses on f, it follows that $g(x)\log f(x) = g(x)\log(x^\alpha b(x)) = \alpha g(x)\log x + g(x)\log b(x)$. But $\log b(x)$ is bounded and $g(x)\to 0$ as $x\to 0+$. Thus

$$\lim_{x\to 0+} g(x)\log f(x) = \lim_{x\to 0+} \alpha g(x)\log x = \gamma.$$

The theorem follows by exponentiation.

Note that when $\alpha > 0$ the indeterminate form is 0^0, and when $\alpha < 0$ the indeterminate form is ∞^0. Further, if $f(x)^{g(x)}\to k\neq 0, \infty$, then $g(x)\log f(x)\to\log k$ so the rate of growth of $\log f(x)$ is closely tied to the rate at which $g(x)\to 0$; in fact, the reader may easily develop a converse to the theorem.

If the order of magnitude of f is the same as x^α, then according to Theorem 1 $\lim_{x\to 0+} f(x)^{g(x)} = 1$ whenever $\lim_{x\to 0} g(x)\log x = 0$. More generally, the battle between 0 and 1 as the rightful value of 0^0 and between 1 and ∞ as the value of ∞^0 is determined by the rate at which $g(x)\to 0$. If (i) $g(x)\to 0$ faster than $1/\log x$, then $f(x)^{g(x)}\to 1$; if (ii) $g(x)\to 0$ slower than $1/\log x$, then $f(x)^{g(x)}\to 0$ if $\alpha > 0$, and $f(x)^{g(x)}\to\infty$ if $\alpha < 0$; if (iii) $g(x)\to 0$ at the same rate as $1/\log x$, then $\lim_{x\to 0} f(x)^{g(x)}$ is ambiguous. Since case (i) almost always obtains, the indeterminate form 0^0 or ∞^0 is almost always 1. The following corollary is a particular case.

COROLLARY 1. *If f is as in Theorem* 1, *and if* $g(x) = x^\beta c(x)$ *where* $c(x)$ *is bounded and* $\beta > 0$, *then*

$$\lim_{x\to 0+} f(x)^{g(x)} = 1.$$

In using Theorem 1 or Corollary 1 to evaluate an indeterminate form, one has only to identify the orders of magnitude of f and g. This technique is easily used in the first example, where $\alpha = 1/2$, $\beta = 1$, $b(x) = 2x^{-\frac{1}{2}}\sin\sqrt{x} + \sin(1/x)$, and $c(x) = 1$. Since $b(x)$ and $c(x)$ are bounded and $b(x)$ is bounded away from 0, the limit in this case is 1 by Corollary 1.

The technique is further illustrated in the following simplification of the work of previous authors.

COROLLARY 2. *If $f(0) = g(0) = 0$, if f is analytic at 0 and positive as $x \to 0+$, and if g is differentiable at 0, then*

$$\lim_{x \to 0+} f(x)^{g(x)} = 1.$$

Proof. Since $g'(0) = \lim_{x \to 0} g(x)/x$, $g(x) = xc(x)$ where $c(x)$ is bounded as $x \to 0$. Since f is analytic at 0 and $f(0) = 0$, there is a positive integer α such that $f(x) = x^{\alpha} b(x)$ where b is analytic at 0 and $b(0) \neq 0$. Thus b is bounded and bounded away from 0 as $x \to 0$, and by Corollary 1, $\lim_{x \to 0+} f(x)^{g(x)} = 1$.

The next theorem concerns the case 1^{∞}.

THEOREM 2. *If $\lim_{x \to 0} f(x) = 1$, if $\lim_{x \to 0} g(x) = \infty$, and if $\lim_{x \to 0} g(x)(f(x) - 1) = \gamma$, then $\lim_{x \to 0} f(x)^{g(x)} = e^{\gamma}$.*

Proof. The technique we use here is motivated by a common proof of the chain rule. Let $y = f(x)^{g(x)}$. Then $\log y = g(x) \log f(x) = g(x) \log(f(x) - 1 + 1)$. Now recall that

$$\lim_{x \to 0} \frac{\log(1+x)}{x} = \log'(1) = 1.$$

So define

$$H(x) = \begin{cases} (1/x)\log(1+x), & \text{if } x \neq 0, \\ 1, & \text{if } x = 0. \end{cases}$$

Then H is continuous and $\log(1 + x) = xH(x)$ for all $x > -1$. Thus

$$\lim_{x \to 0} g(x) \log f(x) = \lim_{x \to 0} g(x)(f(x) - 1) \lim_{x \to 0} H(f(x) - 1) = \gamma \cdot 1 = \gamma,$$

and by exponentiation the proof is complete.

In the second example given at the beginning of this paper, $g(x)(f(x) - 1) = x \sin(1/x^4) \to 0$ as $x \to 0$. Hence $\lim_{x \to 0} f(x)^{g(x)} = e^0 = 1$. If instead,

$$f(x) = 1 + e^{-1/x^2} \text{Arctan}(1/x^2) + x \sin(1/x^4) e^{-1/x^2},$$

then $g(x)(f(x) - 1) \to \pi/2$ as $x \to 0$ and $\lim_{x \to 0} f(x)^{g(x)} = e^{\pi/2}$.

Note that if $f(x) \to 1$ and $f(x)^{g(x)} \to k \neq 1, \infty$, then $g(x) \log f(x) \to \log k$. Hence $g(x) \log(f(x) - 1 + 1) \to \log k$; therefore $g(x)(f(x) - 1) \to \log k$. Thus a converse of Theorem 2 holds and the rate of growth of $g(x)$ is closely tied to the rate at which $f(x) - 1 \to 0$.

A general case when 1^{∞} has the value 1 is expressed in the following corollary, whose proof is left to the reader.

COROLLARY 3. *If there exist $\alpha > \beta > 0$ such that $f(x) = 1 + x^{\alpha} b(x)$ and $g(x) = x^{-\beta} c(x)$ where $b(x)$ and $c(x)$ are bounded as $x \to 0+$, then*

$$\lim_{x \to 0+} f(x)^{g(x)} = 1.$$

References

1. H. Korn and L. M. Rotando, The indeterminate form 0^0, Math. Mag., 50 (1977) 41–42.
2. L. J. Paige, A note on indeterminate forms, this MONTHLY, 61 (1954) 189–190.
3. N. W. Rickert, A calculus counterexample, this MONTHLY, 75 (1968) 166.
4. G. C. Watson, A note on indeterminate forms, this MONTHLY, 68 (1961) 490–492.

DEPARTMENT OF MATHEMATICS, WAKE FOREST UNIVERSITY, WINSTON-SALEM, NC 27109.

COUNTEREXAMPLES TO L'HÔPITAL'S RULE

R. P. Boas

Department of Mathematics, Northwestern University, Evanston, IL 60201

1. Introduction. I am not, of course, claiming that L'Hôpital's rule is wrong, merely that unless it is both stated and used very carefully it is capable of yielding spurious results. This is not a new observation, but it is often overlooked.

For definiteness, let us consider the version of the rule that says that if f and g are differentiable in an interval (a, b), if

$$\lim_{x \to b-} f(x) = \lim_{x \to b-} g(x) = \infty,$$

and if $g'(x) \neq 0$ in some interval (c, b), then

$$\lim_{x \to b-} f'(x)/g'(x) = L$$

implies that

$$\lim_{x \to b-} f(x)/g(x) = L.$$

If $\lim f'(x)/g'(x)$ does not exist, we are not entitled to draw any conclusion about $\lim f(x)/g(x)$. Strictly speaking, if g' has zeros in every left-hand neighborhood of b, then f'/g' is not defined on (a, b), and we ought to say firmly that $\lim f'/g'$ does not exist. There is, however, the insidious possibility that f' and g' contain a common factor: $f'(x) = s(x)\psi(x)$, $g'(x) = s(x)\omega(x)$, where s does not approach a limit and $\lim \psi(x)/\omega(x)$ exists. It is then quite natural to cancel the factor $s(x)$. This is just what we must not do in the present situation: it is quite possible that $\lim \psi(x)/\omega(x)$ exists but $\lim f(x)/g(x)$ does not.

This claim calls for an example. A number of textbooks give one, but it is (as far as I know) always the same example. The aim of this note is both to emphasize the necessity of the condition $g'(x) \neq 0$ and to provide a systematic method of constructing counterexamples when this condition is violated. I consider the case when $b = +\infty$, since the formulas are simpler than when b is finite.

2. A construction. Take a periodic function λ (not a constant) with a bounded derivative, for example $\lambda(x) = \sin x$. Let

$$f(x) = \int_0^x \{\lambda'(t)\}^2 \, dt.$$

It is clear that $f(x) \to +\infty$ as $x \to +\infty$. Now choose a function φ such that $\varphi(\lambda(x))$ is bounded and both $\varphi(\lambda(x))$ and $\varphi'(\lambda(x))$ are bounded away from 0. There are many such functions φ; for example,

$$\varphi(x) = e^x \quad \text{or} \quad (x + c)^2 \quad \text{or} \quad 1/(c + x),$$

provided $|\lambda(x)| < c$ and $|\lambda'(x)| < c$. Take $g(x)$ to be $f(x)\varphi(\lambda(x))$. Since $\inf \varphi(\lambda(x)) > 0$, we have $g(x) \to \infty$ as $x \to \infty$.

Now try to apply L'Hôpital's rule to $f(x)/g(x)$. We have to consider $f'(x)/g'(x)$, where

$$f'(x) = \{\lambda'(x)\}^2,$$

$$g'(x) = \{\lambda'(x)\}^2 \varphi(\lambda(x)) + f(x)\varphi'(\lambda(x))\lambda'(x).$$

Here $g'(x) = 0$ whenever $\lambda'(x) = 0$, i.e., g' has zeros in every neighborhood of ∞, and consequently we are not entitled to apply L'Hôpital's rule at all. However, this conclusion seems rather pedantic; let us go ahead anyway. If we cancel the factor $\lambda'(x)$, we obtain

$$\frac{f'(x)}{g'(x)} = \frac{\lambda'(x)}{\lambda'(x)\varphi(\lambda(x)) + f(x)\varphi'(\lambda(x))} .$$

Now $\lambda'(x)$ is bounded (by hypothesis), $\lambda'(x)\varphi(\lambda(x))$ is bounded, $\varphi'(\lambda(x))$ is bounded away from 0, but $f(x) \to \infty$, so $f'(x)/g'(x) \to 0$. Yet $f(x)/g(x) = 1/\varphi(\lambda(x))$ does not approach zero, since $\varphi(\lambda(x))$ is bounded!

3. Discussion. What went wrong? If you will study any proof of L'Hôpital's rule, you will find a place where it used (or should have used) the assumption that $g'(x)$ did not change sign infinitely often in a neighborhood of ∞. Our example shows that, at least sometimes, L'Hôpital's rule actually fails when this hypothesis is not satisfied.

The phenomenon just described was discovered more than a century ago by O. Stolz [1], [2]. His example was $\lambda(x) = \sin x$, $\varphi(x) = e^x$; it has been repeated in all the modern discussions that I have seen. It was wondering whether there *are* any other examples that led to this note.

One can verify that it is the changes of sign of $\lambda'(x)$ that cause the trouble, not the mere presence of zeros of λ'. In other words, if $\lambda'(x) \geq 0$, the cancellation process still leads to a correct result, as Stolz pointed out. However, it seems wildly improbable that an example of either kind will occur in practice, especially for limits at a finite point. Differentiable functions with infinitely many changes of sign in a finite interval are rarely encountered outside notes like this one; all the less, functions with infinitely many double zeros.

4. History. Guillaume François Antoine de Lhospital, Marquis de Sainte-Mesme (1651–1704) published (anonymously) in 1691 the world's first textbook on calculus, based on John Bernoulli's lecture notes. He seems to have written his name as above, but it is more familiar as L'Hospital (old French spelling) or L'Hôpital (modern French); I prefer the latter, since it stops students from pronouncing the s (which Larousse's dictionary says is not to be pronounced).

References

1. O. Stolz, Ueber die Grenzwerthe der Quotienten, Math. Ann., 15 (1879) 556–559.
2. _____ , Grundzüge der Differential- und Integralrechnung, vol. 1, Teubner, Leipzig, 1893, pp. 72–84.

L'Hôpital's Rule Via Integration

Donald Hartig

Mathematics Department, California Polytechnic State University, San Luis Obispo, CA 93407

In elementary calculus texts L'Hôpital's rule is usually proven only for the case $0/0$, $x \to x_0$ (finite), by applying the Cauchy mean value theorem. Extension to $x \to \infty$ is then accomplished by replacing x with $1/x$. Verification of the rule for the ∞/∞ indeterminate form is regarded as too difficult and may be discussed in an exercise, an appendix, or not at all. In this note we give a proof for the ∞/∞ case that does not make use of the Cauchy mean value theorem. Instead, we require that the functions have continuous derivatives and take advantage of the order properties of the definite integral. The argument adapts nicely to the case $0/0$ as well.

L'HÔPITAL'S RULE. ∞ / ∞. *Let f and g have continuous derivatives with $g'(x) \neq 0$. If* $\lim_{x \to \infty} f(x) = \infty$, $\lim_{x \to \infty} g(x) = \infty$, *and* $\lim_{x \to \infty} f'(x)/g'(x) = L$, *then* $\lim_{x \to \infty} f(x)/g(x) = L$ *also.*

Proof. We assume that L is finite; the other case can be handled in a similar fashion. The limit hypothesis on g allows us to assume that it is a positive function. Moreover, since g' is continuous and nonvanishing it too must be positive.

Let ε be some positive number. Choose M so that

$$\left| \frac{f'(x)}{g'(x)} - L \right| < \varepsilon$$

whenever $x > M$. Since $g'(x)$ is positive we have

$$|f'(x) - Lg'(x)| < \varepsilon g'(x),$$

so that

$$\left| \int_a^b [f'(x) - Lg'(x)] \, dx \right| \leqslant \int_a^b |f'(x) - Lg'(x)| \, dx < \int_a^b \varepsilon g'(x) \, dx$$

whenever $M < a < b$. Therefore, for such a and b,

$$|f(b) - f(a) - L[g(b) - g(a)]| < \varepsilon [g(b) - g(a)]. \qquad (*)$$

Dividing through by the positive number $g(b)$ we obtain

$$\left| \frac{f(b)}{g(b)} - \frac{f(a)}{g(b)} - L\left[1 - \frac{g(a)}{g(b)}\right] \right| < \varepsilon \left[1 - \frac{g(a)}{g(b)}\right] < \varepsilon.$$

It follows easily that

$$\left| \frac{f(b)}{g(b)} - L \right| < \varepsilon + \frac{|f(a)|}{g(b)} + |L| \frac{g(a)}{g(b)}.$$

194

As b increases, $g(b)$ grows larger and larger without bound. Consequently, both $|f(a)|/g(b)$ and $|L|g(a)/g(b)$ will eventually become (and remain) smaller than ε, implying that

$$\left| \frac{f(b)}{g(b)} - L \right| < 3\varepsilon$$

for all b sufficiently large. This shows that $\lim_{x \to \infty} f(x)/g(x) = L$. $\square$

Since our proof of this version of L'Hôpital's rule makes no use of the assumption $\lim_{x \to \infty} f(x) = \infty$, that condition can be dropped from the hypotheses. A straightforward variation of the preceding proof works when $x \to x_0$; alternatively, that case can be derived from the $x \to \infty$ case by considering $F(x) = f(x_0 + 1/x)$ and $G(x) = g(x_0 + 1/x)$.

This type of proof also applies to the indeterminate form $0/0$. For example, if we assume that L is finite and g' is a positive function (as was the case above), then allowing b to increase without bound in inequality ($*$) will force $f(b)$ and $g(b)$ towards 0, implying that

$$|-f(a) - L[-g(a)]| \leqslant \varepsilon[-g(a)].$$

Dividing by the (positive) number $-g(a)$ reveals that

$$\left| \frac{f(a)}{g(a)} - L \right| \leqslant \varepsilon$$

whenever $a > M$.

As you can see, the algebra for $0/0$ is a bit simpler making this proof even more suitable for popular consumption.

Acknowledgment. The author wishes to thank the referee for helpful comments on the arrangement of the proofs.

Indeterminate Forms Revisited

R. P. BOAS
3540 NE 147th Street
Seattle, WA 98155

1. Introduction You must all have seen at least one calculus textbook. It may surprise some of you that three centuries ago no such book existed: the very first book that was in any sense a calculus text was published, anonymously, in 1696, under the rather forbidding title *Analysis of the Infinitely Small* [5]. It was well known in European mathematical circles that the author was a French marquis, Guillaume de L'Hôpital. (I give him the modern French spelling, which at least keeps students from pronouncing the silent s in L'Hospital.) The book was hardly easy reading. It began with propositions like this: "One can substitute, one for the other, two quantities which differ only by an infinitely small quantity; or (what amounts to the same thing) a quantity that is increased or decreased only by another quantity infinitely less than it, can be considered as remaining the same." This sort of presentation gave calculus a reputation, which has survived to modern times, of being unintelligible.

Sylvester, writing in about 1880 [10, vol. 2, pp. 716-17], said that when he was young (around 1830) "a boy of sixteen or seventeen who knew his infinitesimal calculus would have been almost pointed out in the streets as a prodigy like Dante, who had seen hell." (Here and now, we would very likely find students of the same age feeling much the same; but Sylvester, in 1870, was teaching students who dealt casually with topics that we would now describe as advanced calculus.) When I was in high school (somewhat later), calculus was thought of, by otherwise well-educated people, as being as deep and mysterious as (say) general relativity is thought of today. My parents knew somebody who was reputed to know calculus, but they had no idea what that was (and they were college teachers—of English). Nowadays there are perhaps too many calculus books, but some of the answers that students give to examination questions make me wonder whether the subject has even now become sufficiently intelligible.

In his own time, and for long afterwards, L'Hôpital had an impressive reputation. Today he is remembered only for "L'Hôpital's rule," which evaluates limits like

$$\lim_{x \to 1} \frac{\left(2x - x^4\right)^{1/2} - x^{1/3}}{1 - x^{3/4}}$$

(L'Hôpital's own example) by replacing the numerator and denominator by their derivatives and hoping for the best.

L'Hôpital's rule seems to have fallen somewhat out of favor; I have heard it claimed that all it is useful for is as an exercise in differentiation.

It has been known for some time that many of L'Hôpital's results, including the rule, were purchased (quite literally) from John (= Jean = Johann) Bernoulli. Immediately after L'Hôpital's death in 1704, Bernoulli published an article claiming that he had communicated the rule for $0/0$ to L'Hôpital, along with other material, before L'Hôpital had published it. This claim was disbelieved for some two hundred years; sceptics wondered why Bernoulli had not advanced his claim earlier. The reason for the delay eventually became clear when Bernoulli's correspondence with L'Hôpital came to light in the early 1900s. Bernoulli gave the rule to L'Hôpital only after L'Hôpital had promised to pay for it, had repeatedly asked for it, and had finally come across with the first installment. We now also know that there are records of Bernoulli's having lectured on the rule before L'Hôpital's book was published.

In the preface to his book, L'Hôpital says, "I acknowledge my debt to the insights of MM Bernoulli, above all to those of the younger [John], now Professor at Groningen. I have unceremoniously made use of their discoveries and of those of M Leibnis [sic]. Consequently I invite them to claim whatever they wish, and will be satisfied with whatever they may leave me." Considering what we now know, this seems somewhat disingenuous, especially since L'Hôpital was clearly unable to discover for himself how to prove the rule of which, as Plancherel once said of his own theorem, he had "the honor of bearing the name."

You can find the whole story in the 1955 volume of Bernoulli's correspondence [7], or in Truesdell's review of the volume [11].

I used to wonder, from time to time, what kind of proof L'Hôpital had used, but never when I was where I could find a copy of his book. Recently I happened to mention this question to Professor Alexanderson—who promptly produced his own copy. Professor Underwood Dudley, who is more resourceful than I am, also found a copy, and has translated it into modern terminology [4], but retaining its geometric character (L'Hôpital thought of functions as curves). L'Hôpital actually considered only $\lim_{x \to a} f(x)/g(x)$, where a is finite, $f(a) = g(a) = 0$, and both $f'(a)$ and $g'(a)$ exist, are finite, and not zero. In analytical language, what L'Hôpital did amounts to writing

$$\lim_{x \to a} \frac{f(x)}{g(x)} = \lim_{x \to a} \frac{f(x) - f(a)}{g(x) - g(a)} = \lim_{x \to a} \frac{f'(a) + \varepsilon(x)}{g'(a) + \delta(x)} (\varepsilon \to 0, \delta \to 0) = \frac{f'(a)}{g'(a)} .$$

It is not trivial to extend such a proof to the cases when $f'(a)$ and $g'(a)$ do not exist (but have limits as $x \to a$), or are both zero, or $f(a) = g(a) = \infty$, or $a = \infty$. I do not know when or by whom these refinements were added, but the complete theory was in place by 1880 [8, 9].

2. A common modern proof If you saw a proof of L'Hôpital's rule in a modern calculus class, the probability is about 90% that it is Cauchy's proof. This proof appeals to mathematicians because it is elegant, but often fails to appeal to students because it is subtle. It depends on knowing Cauchy's refinement of the mean-value theorem, namely that (with appropriate hypotheses)

$$\frac{f(x)-f(a)}{g(x)-g(a)} = \frac{f'(c)}{g'(c)}, \quad c \text{ between } a \text{ and } b. \tag{1}$$

Given this, L'Hôpital's rule becomes obvious.

In spite of its elegance, Cauchy's proof seems to me to be inappropriate for an elementary class. Any proof that begins with a lemma like Cauchy's mean value theorem, that says "Let us consider...," repels most students. Students are also uncomfortable with the nebulous point c. They want to know where it is, and feel that the instructor is deliberately keeping them in the dark. Of course, the exact location of c is completely irrelevant (although numerous papers have been written about it).

3. A caution   Cauchy's proof tacitly assumes that there is a (one-sided) neighborhood of the point a in which $g'(x) \neq 0$. Strictly speaking, if there is no such neighborhood, the limit in (1) is not defined, and we would have no business talking about it. However, if f' and g' are given by explicit formulas, they may happen to share a common factor that is zero at a, and the temptation to cancel this factor is irresistible. One can obtain a spurious result in this way [8, 9; 3, p. 124, ex. 24; 1].

Let me give you a specific example, just to emphasize that there is a reason for the requirement that $g'(x) \neq 0$. Let $f(x) = 2x + \sin 2x$, $g(x) = x \sin x + \cos x; a = +\infty$. Then $f'(x) = 4\cos^2 x$, $g'(x) = x \cos x$, and $f'(x)/g'(x) \to 0$, whereas $f(x)/g(x)$ does not approach a limit. The trouble comes from cancelling a factor that changes sign in every neighborhood of the point a; it would have been legitimate to cancel a quadratic factor.

Some writers think that the difficulty arises only in artificial cases that would never occur in practice. But then, what happens to our claim to be giving correct proofs?

You might not guess from Cauchy's proof that there is a discrete analog of L'Hôpital's rule; see, for example, [6]. This was known to Stolz in the 1890's, and has often been rediscovered.

4. A more satisfactory proof   I want now to show you a proof of L'Hôpital's rule that avoids the difficulties of Cauchy's and establishes a good deal more. It may seem more complicated, but not if you include a proof of Cauchy's mean value theorem as part of Cauchy's proof. This proof is also quite old; Stolz knew it, but preferred Cauchy's proof, perhaps because of Cauchy's reputation. It has been published several times by people (including me) who failed to search the literature.

Let us suppose that $f(x)$ and $g(x)$ approach 0 as $x \to a$ from the left, where a might be $+\infty$; it does no harm to define (if necessary) $f(a) = g(a) = 0$. We may suppose that $g'(x) > 0$ (otherwise consider $-g(x)$). Now let $f'(x)/g'(x) \to L$, where

$0 < L < \infty$. Then, given $\varepsilon > 0$, we have, if x is sufficiently near a, and a is finite,

$$L - \varepsilon \leqslant \frac{f'(x)}{g'(x)} \leqslant L + \varepsilon,$$

$$(L - \varepsilon)g'(x) \leqslant f'(x) \leqslant (L + \varepsilon)g'(x) \quad (\text{since } g'(x) > 0).$$

Integrate on (x, a) to get

$$-(L - \varepsilon)g(x) \leqslant -f(x) \leqslant -(L + \varepsilon)g(x)$$

(notice that since g increases to 0, we have $g(x) < 0$). Since $-g$ and $-f$ are positive near a,

$$L - \varepsilon \leqslant \frac{f(x)}{g(x)} \leqslant L + \varepsilon,$$

$$\lim_{x \to a} \frac{f(x)}{g(x)} = L.$$

Only formal changes are needed if $a = +\infty$ or if $L = 0$ or ∞. For the ∞/∞ case, we get, in the same way, with $\delta > 0$,

$$L - \varepsilon < \frac{f(a - \delta) - f(x)}{g(a - \delta) - g(x)} < L + \varepsilon,$$

$$L - \varepsilon < \frac{\dfrac{f(a - \delta)}{f(x)} - 1}{\dfrac{g(a - \delta)}{g(x)} - 1} \cdot \frac{f(x)}{g(x)} < L + \varepsilon.$$

Letting $x \to a$, we obtain

$$L - \varepsilon \leqslant \liminf_{x \to a} \frac{f(x)}{g(x)} \leqslant \limsup_{x \to a} \frac{f(x)}{g(x)} \leqslant L + \varepsilon.$$

Letting $\varepsilon \to 0$, we obtain $f(x)/g(x) \to L$.

If it happens that $f'(a) = g'(a) = 0$, one repeats the procedure with f'/g', and so on. If $f^{(n)}(a) = g^{(n)}(a) = 0$ for every n (which can happen with f and g not identically zero), the procedure fails. Otherwise, the limit can be handled more simply in a single step, as we shall see below.

5. Generalizations If f and g are defined only on the positive integers, we can reason in a similar way with differences instead of derivatives to conclude that if the differences of g are positive, and $f(n)$ and $g(n)$ approach zero as $n \to \infty$, then if

$$\frac{f(n) - f(n-1)}{g(n) - g(n-1)} \to L \qquad \text{as } n \to \infty,$$

it follows that $f(n)/g(n) \to L$. This is sometimes called Cesàro's rule. For more detail, and illustrations, see [6]. A possibly more familiar version is as follows: If $a_n \to 0$ and $b_n \to 0$, and $a_n/b_n \to L$, then

$$\frac{\displaystyle\sum_{k=1}^{n} a_k}{\displaystyle\sum_{k=1}^{n} b_k} \to L.$$

The key point in the proof of L'Hôpital's rule is the principle that the integral of a nonnegative function ($\not\equiv 0$) is positive. More precisely, if $f(x) \geqslant 0$ on $p \leqslant x \leqslant q$ then

$$\int_p^q f(t) \, dt > 0 \qquad \text{if } p < x < q \text{ and } f(t) \not\equiv 0.$$

Repeated integration with the same lower limit has the same property, as we see by rewriting the n-fold iterated integral as a single integral:

$$\frac{1}{(n-1)}! \int_p^x (t-p)^{n-1} f(t) \, dt.$$

This suggests the appropriate treatment of the case of L'Hôpital's rule when $f'(a) = g'(a) = 0$, or more generally when $f^{(k)}(a) = g^{(k)}(a) = 0, k = 1, 2, \ldots, n-1$, but not both of $f^{(n)}(a)$ and $g^{(n)}(a)$ are 0. The positivity of iterated integration then yields the conclusion of L'Hôpital's rule in a single step.

An operator that carries positive functions to positive functions is conventionally called a positive operator. If F is an invertible operator whose inverse is positive, we can conclude that if $F[f(x)]/F[g(x)] \to L$ and $F[g] > 0$, then $f(x)/g(x) \to L$.

As an example of the use of operators, consider $D + P(x)I$, where $D = d/dx$ and I is the identity operator. This is the operator that occurs in the theory of the linear first-order differential equation $y' + P(x)y = Q(x)$. The solution of this differential equation, with $y(a) = 0$, is

$$y = \exp\left\{\int_a^x P(t) \, dt\right\} \int_a^x Q(t) \exp\left\{\int_a^t P(u) \, du\right\} dt. \tag{2}$$

In other words, (2) provides the inverse Λ of $D + P(x)I$.

The explicit formula shows that if $Q(x) \geqslant 0$ we have $\Lambda[Q] > 0$, so that Λ is a positive operator. Hence we may conclude that if

$$\frac{[D + PI]f}{[D + PI]g} \to L$$

and $[D + PI]g > 0$ then

$$(L - \varepsilon)[D + PI]g < [D + PI]f < (L + \varepsilon)[D + PI]g.$$

A positive linear operator evidently preserves inequalities. Consequently, if we apply Λ to both sides, we obtain

$$(L - \varepsilon)g(x) < f(x) < (L + \varepsilon)g(x)$$

and hence

$$f(x)/g(x) \to L.$$

Thus $D + P(x)I$ can play the same role as D in L'Hôpital's rule. It is at least possible that $D + P(x)I$ might be simpler than D.

Since some forms of fractional integrals and derivatives are defined by positive operators, one could also formulate a fractional L'Hôpital's rule.

This article is the text of an invited address to a joint session of the American Mathematical Society and the Mathematical Association of America, January 14, 1989.

REFERENCES

1. R. P. Boas, Counterexamples to L'Hôpital's rule, *Amer. Math. Monthly* 94 (1986), 644–645.
2. _____, *Indeterminate Forms Revisited*, videotape, Amer. Math. Soc., Providence, RI, 1989.
3. R. C. Buck and E. F. Buck, *Advanced calculus*, 3rd ed., McGraw-Hill, New York, etc., 1978.
4. U. Dudley, Review of Calculus with Analytic Geometry, *Amer. Math. Monthly* 95 (1988), 888–892.
5. [G. de L'Hôpital], *Analyse des infiniment petits, pour l'intelligence des lignes courbes*, Paris, 1696; later editions under the name of le Marquis de L'Hôpital.
6. Xun-Cheng Hwang, A discrete L'Hôpital's rule, *College Math. J.* 19 (1988), 321–329.
7. [O. Spiess, ed.], *Die Briefwechsel von Johann Bernoulli*, vol. 1, Birkhäuser, Basel, 1955.
8. O. Stolz, Über die Grenzwerthe der Quotienten, *Math. Ann.* 15 (1889), 556–559.
9. _____, *Grundzüge der Differential- und Integral-rechnung*, vol. 1, Teubner, Leipzig, 1893.
10. J. J. Sylvester, *The Collected Mathematical Papers*, Cambridge University Press, 1908.
11. C. Truesdell, The new Bernoulli edition, *Isis* 49 (1958), 54–62.

L'Hôpital's Rule and the Continuity of the Derivative

J. P. King, Lehigh University, Bethlehem, PA

This note can be used to make students aware of some of the subtleties involved in the proper use of L'Hôpital's rule:

Let f and g be differentiable in some neighborhood of $x = a$. If $\lim_{x \to a} f(x) = 0 = \lim_{x \to a} g(x)$ and $\lim_{x \to a} f'(x)/g'(x) = L$ exists, then $\lim_{x \to a} f'(x)/g'(x) = L$.

It is particularly difficult to have students understand that one must first show that $\lim_{x \to a} f'(x)/g'(x)$ exists before it can be concluded that $\lim_{x \to a} f(x)/g(x)$ exists and the two limits are equal.

One can, however, exploit L'Hôpital's rule to motivate a circle of ideas which lead to an important result rarely taught in elementary calculus, namely the intermediate value theorem for derivatives. The procedure is as follows:

Let f be differentiable on an interval containing $x = a$. Then

$$f'(a) = \lim_{x \to a} \frac{f(x) - f(a)}{x - a},$$

and (since f is continuous at $x = a$) both the numerator and the denominator of the difference quotient approach zero. L'Hôpital's rule then gives

$$f'(a) = \lim_{x \to a} f'(x), \tag{*}$$

a result which apparently shows that f' is continuous at $x = a$.

At this stage it is natural to present $f(x) = x^2 \sin(1/x)$ with $f(0) = 0$ in order to show that not every differentiable function has a continuous derivative. One may then challenge students to find the flaw in the argument or to explain why, in view of the example, that equation (*) is valid. A properly conducted discussion should lead students to understand that, because of the hypothesis of L'Hôpital's rule, equation (*) is true only in the sense that "if $\lim_{x \to a} f'(x)$ exists, then $\lim_{x \to a} f'(x) = f'(a)$."

The existence of $\lim_{x \to a} f'(x)$ for each a in an interval I means that f' has the intermediate value property (i.e., f' assumes all values between any two values it takes on I). Actually, f' cannot fail to have this property, since the intermediate value theorem for derivatives asserts that, if f' exists everywhere on I and takes on any two values on I, then f' takes every possible value between them. [See, for example, W. Rudin, *Principles of Mathematical Analysis*, McGraw-Hill (1964), p. 93.]

Some Subtleties In L'Hôpital's Rule

Robert J. Bumcrot, Hofstra University, Hempstead, NY

The use of L'Hôpital's Rule (actually a theorem of Johann Bernoulli) to calculate $\lim_{x \to 0} \frac{x - \sin x}{x^3}$, for example, is almost always presented as a simple string of equalities:

$$\lim_{x \to 0} \frac{x - \sin x}{x^3} = \lim_{x \to 0} \frac{1 - \cos x}{3x^2} \tag{1}$$

$$= \lim_{x \to 0} \frac{\sin x}{6x} \tag{2}$$

$$= \lim_{x \to 0} \frac{\cos x}{6} \tag{3}$$

$$= \frac{1}{6}.$$

Of course, the first three equalities are not justified until the last equality has been reached. A more complete presentation would add, "if this limit exists" after the second, third, and fourth limits (a limit of $\pm\infty$ is considered here to "exist"), and the conclusion would be: "Since $\lim_{x \to 0} \frac{\cos x}{6} = \frac{1}{6}$, Equation (3) holds; therefore Equation (2) holds; therefore Equation (1) holds; hence $\lim_{x \to 0} \frac{x - \sin x}{x^3} = \frac{1}{6}$."

We do not advocate such a complete presentation for every example of L'Hôpital's Rule, but we do suggest that students be made aware of the importance of the existence hypothesis. Although the existence of $\lim_{x \to a} \frac{f'(x)}{g'(x)}$ is a sufficient condition for the existence of $\lim_{x \to a} \frac{f(x)}{g(x)}$, very little is usually said about whether or not this is a necessary condition for the existence of $\lim_{x \to a} \frac{f(x)}{g(x)}$. Thus, it may be enticing for students to believe that when a chain of "L'Hôpital equalities" leads to a limit that does *not* exist, then the original limit also does not exist. The following examples show that such a conclusion may or may not hold.

Example 1 ($\frac{0}{0}$ indeterminate forms). If $f(x) = x^2 \sin(x^{-1})$ and $g(x) = \sin x$, then $\lim_{x \to 0} \frac{f'(x)}{g'(x)}$ does not exist, whereas $\lim_{x \to 0} \frac{f(x)}{g(x)} = 0$. On the other hand, if $f(x) = x \sin(x^{-1})$ and $g(x) = \sin x$, then neither $\lim_{x \to 0} \frac{f'(x)}{g'(x)}$ nor $\lim_{x \to 0} \frac{f(x)}{g(x)}$ exists.

Example 2 ($\frac{\infty}{\infty}$ indeterminate forms).   If $f(x) = x(2 + \sin x)$ and $g(x) = x^2 + 1$,

then $\lim_{x \to \infty} \dfrac{f'(x)}{g'(x)}$ does not exist, whereas $\lim_{x \to \infty} \dfrac{f(x)}{g(x)} = 0$. On the other

hand, if $f(x) = x(2 + \sin x)$ and $g(x) = x + 1$, then neither $\lim_{x \to \infty} \dfrac{f'(x)}{g'(x)}$ nor

$\lim_{x \to \infty} \dfrac{f(x)}{g(x)}$ exists.

Editor's Note: For a variation on this theme, see J. P. King's Classroom Capsule "L'Hôpital's Rule and the Continuity of the Derivative," TYCMJ 10 (June 1979), 197–198.

BIBLIOGRAPHIC ENTRIES: INDETERMINATE FORMS

1. *Monthly* Vol. 83, No. 4, pp. 239–242. A. M. Ostrowski, Note on the Bernoulli-L'Hospital rule.

2. *Monthly* Vol. 95, No. 3, pp. 253–254. William P. Cooke, L'Hôpital's rule in a Poisson distribution.

3. *Monthly* Vol. 97, No. 6, pp. 518–527. Gianluca Gorni, A geometric approach to L'Hôpital's rule.

4. *Math. Mag.* Vol. 44, No. 4, pp. 217–218. Men-Chang Hu and Ju-Kwei Wang, On the L'Hôpital rule for indeterminate forms ∞/∞.

5. *Math. Mag.* Vol. 50, No. 1, pp. 41–42. Louis M. Rotando and Henry Korn, The indeterminate form 0^0.

Addresses the question: When is $0^0 \neq 1$?

6. CMJ Vol. 21, No. 3, pp. 222–224. Stephan C. Carlson and Jerry M. Metzger, A recursively computed limit.

Finds limit as $x \to 0 +$ of $(1/\sin^t x - 1/x^t)$ for each real t.

7

TAYLOR POLYNOMIALS, BERNOULLI POLYNOMIALS AND SUMS OF POWERS OF INTEGERS

(a)

TAYLOR POLYNOMIALS

FROM CENTER OF GRAVITY TO BERNSTEIN'S THEOREM

RAY REDHEFFER

Department of Mathematics, University of California, Los Angeles CA 90024

Let $w(x)$ be continuous for $x \geqslant 0$ and positive for $x > 0$. If $w(x)$ is thought to be the density of a rod at point x, then

$$\frac{\int_0^x tw(t)\, dt}{\int_0^x w(t)\, dt}$$

is the center of gravity of the part of the rod on $[0, x]$ and, on physical grounds, this expression must be an increasing function of x. The mathematical reason is that the factor t in the numerator is increasing. In fact if ϕ is increasing for $x > 0$ and ϕw is continuous for $x \geqslant 0$, then

$$(1) \qquad \frac{\int_0^x \phi(t) w(t)\, dt}{\int_0^x w(t)\, dt} \quad \text{is increasing for } x > 0.$$

This is seen when we differentiate by the quotient rule; the numerator of the resulting expression is

$$w(x)\left(\phi(x)\int_0^x w(t)\, dt - \int_0^x w(t)\phi(t)\, dt\right) \geqslant 0.$$

More generally, for any continuous function h let

$$J_n(h, x) = \int_0^x \int_0^{t_n} \cdots \int_0^{t_2} h(t_1)\, dt_1 \cdots dt_{n-1} dt_n, \quad n \geqslant 2$$

$$J_1(h, x) = \int_0^x h(t)\, dt.$$

Then, with w and ϕ as above,

(2) $$\frac{J_n(\phi w, x)}{J_n(w, x)} \quad \text{is increasing for } x \geqslant 0.$$

This is true for $n = 1$ by (1). If it holds for $n - 1$ then $J_{n-1}(\phi w, x) = \Phi(x)J_{n-1}(w, x)$ where Φ is increasing. With $W(x) = J_{n-1}(w, x)$ the numerator and denominator in (2) are respectively

$$\int_0^x \Phi(t)W(t)\, dt, \quad \int_0^x W(t)\, dt.$$

Hence, (2) follows from (1) applied to (Φ, W) instead of (ϕ, w).

Suppose now that f is a real-valued function satisfying $f^{(n)}(x) \geqslant 0$ for $0 \leqslant x < \rho$, where $\rho > 0$ and $n = 0, 1, 2, \ldots$. Bernstein's theorem states that the Taylor series for f converges to $f(x)$ on $[0, \rho)$ (and hence f is the restriction of an analytic function $f(z)$, $|z| < \rho$).

To prove Bernstein's theorem, let us integrate $f^{(n)}(x)$ from 0 to x repeatedly as in [3]. The result is

(3) $$f(x) = f(0) + f'(0)x + f''(0)\frac{x^2}{2!} + \cdots + f^{(n-1)}(0)\frac{x^{n-1}}{(n-1)!} + R_n(x)$$

where for $n \geqslant 2$

$$R_n(x) = \int_0^x \int_0^{t_n} \cdots \int_0^{t_2} f^{(n)}(t_1)\, dt_1 \cdots dt_{n-1} dt_n = J_n(f^{(n)}, x).$$

Since $f^{(n+1)}(x) \geqslant 0$ the function $\phi = f^{(n)}$ is increasing and (2) with $w = 1$ shows that $R_n(x)/x^n$ is increasing also. Hence

(4) $$\frac{R_n(x)}{x^n} \leqslant \frac{R_n(r)}{r^n}, \quad 0 < x \leqslant r < \rho.$$

The hypothesis $f^{(n)}(x) \geqslant 0$ shows that $R_n(x) \geqslant 0$ and Equation (3) with $f^{(j)}(0) \geqslant 0$, $j = 0, 1, 2, \ldots, n - 1$ gives $R_n(r) \leqslant f(r)$. Thus,

(5) $$0 \leqslant R_n(r) \leqslant f(r).$$

Combining (4) and (5) gives $0 \leqslant R_n(x) \leqslant (x/r)^n f(r)$, hence $R_n(x) \to 0$ for $0 \leqslant x < r$, and this is Bernstein's theorem.

In comparing with other procedures [1], [2] it should be stated that, once monotonicity of $R_n(x)/x^n$ is established, completion of the argument by (4) and (5) is common to all. The novelty here consists in the proof of monotonicity.

References

1. Tom Apostol, Mathematical Analysis, 2nd ed., Addison-Wesley, 1975, pp. 242–244.

2. Emil Artin, Calculus and Analytic Geometry, CUPM, 1957, pp. 82–90.

3. I. S. Sokolnikoff and R. M. Redheffer, Mathematics of Physics and Modern Engineering, 2nd ed., McGraw-Hill 1966, p. 36.

A Simple Derivation of the Maclaurin Series for Sine and Cosine

DENG BO

Coal Exploration No. 174, Team of Gui Zhou, Zhi Jin, Gui Zhou Province, People's Republic of China

The standard derivations of the Maclaurin series for sine and cosine via differentiation are well-known. In this note I show how integration can be used to obtain the two series and prove that they converge to the sine and cosine functions.

Clearly, it suffices to do this for $x \geqslant 0$. Because $\cos x \leqslant 1$, $\int_0^x \cos y \, dy \leqslant \int_0^x 1 \, dy$. That is, $\sin x \leqslant x$. From this it follows that $\int_0^x \sin y \, dy \leqslant \int_0^x y \, dy$. That is, $1 - \cos x \leqslant x^2/2!$. Continuing in this fashion, we obtain the inequalities

$$1 - \frac{x^2}{2!} \leqslant \cos x \leqslant 1$$

$$x - \frac{x^3}{3!} \leqslant \sin x \leqslant x$$

$$1 - \frac{x^2}{2!} + \frac{x^4}{4!} - \frac{x^6}{6!} \leqslant \cos x \leqslant 1 - \frac{x^2}{2!} + \frac{x^4}{4!}$$

$$x - \frac{x^3}{3!} + \frac{x^5}{5!} - \frac{x^7}{7!} \leqslant \sin x \leqslant x - \frac{x^3}{3!} + \frac{x^5}{5!}.$$

Now, still assuming $x \geqslant 0$, induction can be used to prove that $S(2r) \leqslant \cos x \leqslant S(2r - 1)$ and $T(2r) \leqslant \sin x \leqslant T(2r - 1)$ for $r = 1, 2, \ldots,$ where

$$S(k) = 1 + \sum_{i=1}^{k-1} (-1)^i \frac{x^{2i}}{(2i)!} \quad \text{and} \quad T(k) = \sum_{i=0}^{k-1} (-1)^i \frac{x^{2i+1}}{(2i+1)!}.$$

The upper and lower bounds differ by $x^m/m!$, which, for fixed x, can easily be shown to approach 0 as $m \to \infty$.

An editor's note in the *Monthly* Vol. 98, p. 364, points out that the method is not new and appears in several calculus texts. An earlier reference is J. P. Ballatine, Monthly Vol. 44 (1937), pp. 470–472, reproduced in Selected Papers, Vol. 1, pp. 214–217. See also the paper by Leonard and Duemmel reproduced on pp. 396–397 of this volume.

Trigonometric Power Series

JOHN STAIB
Drexel University

Most modern calculus texts have little to say about the power series for $\tan x$ (or $\sec x$). A few coefficients are given, the difficulty of obtaining more through the formula $a_n = f^{(n)}(0)/n!$ is made clear, and perhaps a reference is provided. If the student pursues the reference, he will very likely be confronted with a somewhat involved connection between these coefficients and the Bernoulli (or Euler) numbers. The following single path to both the tangent and secant coefficients is more direct and it serves as a good exercise in the method of generating functions.

We begin by letting $f(x) = \sec x + \tan x$, and then observe that $f(x) = \cos x \cdot f'(x)$. Introducing power series for each of these latter functions, we have

$$\sum_0^\infty a_n x^n = \sum_0^\infty b_n x^n \cdot \sum_0^\infty c_n x^n.$$

Identifying coefficients of like terms, and noting that $b_n = 0$ when n is odd, we obtain

$$a_n = \sum_{j=0}^{[n/2]} b_{2j} c_{n-2j}.$$

Replacing b_{2j} by its known value, and noting that $c_n = (n+1)a_{n+1}$, we arrive at

$$a_n = \sum_{j=0}^{[n/2]} \frac{(-1)^j}{(2j)!} (n+1-2j) a_{n+1-2j}.$$

Although this equation can be used to generate the a_k, a better recursion formula is possible for the related numbers A_k, where $A_k = (k!)a_k$. These A_k turn out to be positive integers. To see this we make the appropriate substitutions and then multiply through by $n!$ to obtain

$$A_n = \sum_{j=0}^{[n/2]} (-1)^j \binom{n}{2j} A_{n+1-2j}.$$

Finally, we solve for A_{n+1}. For $n = 0$ and 1, we obtain $A_1 = A_0$ and $A_2 = A_1$. For $n > 1$, we have

$$A_{n+1} = A_n + \binom{n}{2} A_{n-1} - \binom{n}{4} A_{n-3} + \cdots,$$

where the sum ends either with A_1 or A_2. Then, beginning with $A_0 = (0!)a_0 = f(0) = 1$, we may generate the A_k:

$$1, 1, 1, 2, 5, 16, 61, 272, 1385, \ldots.$$

It follows that

$$\sec x + \tan x = 1 + x + \frac{1}{2!} x^2 + \frac{2}{3!} x^3 + \frac{5}{4!} x^4 + \frac{16}{5!} x^4 + \cdots.$$

But $\sec x$ is even and $\tan x$ is odd. Therefore,

$$\sec x = 1 + \frac{1}{2!}\, x^2 + \frac{5}{4!}\, x^4 + \frac{61}{6!}\, x^6 + \frac{1385}{8!}\, x^8 + \cdots \quad \text{and}$$

$$\tan x = x + \frac{2}{3!}\, x^3 + \frac{16}{5!}\, x^5 + \frac{272}{7!}\, x^7 + \cdots .$$

The technique used above can be exploited to produce a variety of similar exercises for the student in advanced calculus. One has only to seize on another of the many identities involving $\tan x$ or $\sec x$. For example, the identity $\tan x \cos x = \sin x$ leads to

$$\sum_{k=0}^{n} (-1)^k \binom{2n+1}{2k+1} A_{2k+1} = 1,$$

while $\sec x \cos x = 1$ leads to

$$\sum_{k=0}^{n} (-1)^k \binom{2n}{2k} A_{2k} = 0 \quad \text{for} \quad n > 0.$$

However, it should be noted that these formulas lead separately to the coefficients of $\tan x$ and $\sec x$. The special beauty of the particular identity used here is that it forcefully demonstrates that the somewhat obscure and hardly related coefficients of $\tan x$ and $\sec x$ are in fact the odd and even subsequences of the same sequence!

For futher insight into this matter, it is instructive to look at $\sin x$ and $\cos x$. Here some connection between their coefficients appears from the very beginning. This connection is ultimately established by the identity $e^{ix} = \cos x + i \sin x$. But what is more pertinent here is that the coefficients of $\sin x$ and $\cos x$ can be found in a manner paralleling that used above for $\tan x$ and $\sec x$.

Taking $f(x) = \cos x + i \sin x$ we note first that $f'(x) = if(x)$. Introduction of $\sum_{n=0}^{\infty} a_n x^n$ for $f(x)$ leads to the recursion formula

$$a_{n+1} = \frac{i}{n+1}\, a_n,$$

and from $a_0 = f(0) = 1$ we get our start. Thus we obtain

$$\cos x + i \sin x = 1 + ix - \frac{1}{2!}\, x^2 - \frac{i}{3!}\, x^3 + \cdots,$$

from which the usual series are obtained by matching real and imaginary parts. Note that the relatedness between $\sin x$ and $\cos x$, despite appearances, is not so direct as that between $\tan x$ and $\sec x$. That is, we cannot say that the coefficients of $\sin x$ and $\cos x$ form the odd and even subsequences of the same sequence but only that this is true for $i \sin x$ and $\cos x$.

Finally, the "togetherness" of $\tan x$ and $\sec x$ suggests that the function $(\tan x + \sec x)$ should have an existence of its own independent of its relationship to $\tan x$ and $\sec x$. This turns out to be the case: the numbers A_n give the solution to the so-called zigzag problem. A **zigzag** is a permutation of 1 through n in which the listed numbers successively rise and fall. For example, for $n = 7$, one zigzag is $(7, 1, 4, 2, 6, 3, 5)$. This zigzag starts with a "fall." By replacing each number by its complement relative to $n + 1 = 8$, we obtain $(1, 7, 4, 6, 2, 5, 3)$, which is a zigzag beginning with a "rise." In this way, all zigzags (for a given n) may be paired. Thus, there are just as many zigzags beginning with a rise as with a fall. And, curiously, the count of either class is A_n. The mathematical link between the zigzag count and the coefficients of $(\tan x + \sec x)$ is elegantly established in [1, pp. 64–69]; see also [2, p. 258].

However, this proof seems not to be based on any intuitive considerations. After all, why should that number which gives half the count of all the zigzags arising from $(1, 2, \ldots, n)$ be the same number as that which gives the nth derivative evaluated at 0 for the function $(\tan x + \sec x)$? Perhaps some reader could suggest why this "ought" to be.

References

[1]   Heinrich Dorrie, 100 Great Problems of Elementary Mathematics, Dover, New York, 1965.
[2]   L. Comtet, Advanced Combinatorics, D. Reidel, Dordrecht, 1974.

Rediscovering Taylor's Theorem

Dan Kalman

Dan Kalman studied mathematics at Harvey Mudd College, earning a B.S. in 1974, and at the University of Wisconsin-Madison, earning a Ph.D. in 1980. He taught at Lawrence University in Appleton, Wisconsin, and is an assistant professor at the University of Wisconsin-Green Bay. His interests span the undergraduate mathematics curriculum and extend to preparation of elementary school teachers, history and philosophy of mathematics, and computer science. Currently, he is visiting Augustana College while on leave from the University of Wisconsin-Green Bay.

"The proof of Taylor's formula is not too hard to follow, but it is a whopper to think up. We have put an outline of it in the problem set." [2, p. 475]

The Interpolation Problem and Taylor's Theorem

Taylor's theorem is something of an anomaly in a first calculus course. For many of the important theorems such as the chain rule, the mean value theorem, and the fundamental theorem, there are compelling geometric interpretations that belie the technical difficulties. Even if the student is not ready for a rigorous proof, it is possible to show "what is going on." In contrast, the most common approach to Taylor's theorem derives Lagrange's formula for the error using a contrived application of Rolle's Theorem which is difficult to motivate. One can certainly check that the steps are correct, but discovering the proof seems to depend on knowing in advance exactly how it all works out. In this note, we discuss an approach to polynomial approximation that leads one to stumble on Taylor's theorem and its proof. The proof that emerges is essentially that of Fisher and Ziebur [1, p. 113]. As in other developments of Taylor's theorem, we assume the mean value theorem is known.

Imagine that the usual introduction to Taylor polynomials has been followed this far: the geometric interpretation of equality of the values of two functions and their first n derivatives has been presented, and the polynomial formula for approximating $f(x)$ near $x = a$,

$$p(x) = \sum_{k=0}^{n} \frac{1}{k!} f^{(k)}(a)(x - a)^k,$$

has been derived. We observe that adding additional terms seems to improve the approximation in the vicinity of a. But what can be done to get a better approximation throughout some interval (a, b)? One idea is to force agreement between $p(x)$ and $f(x)$ at $x = b$ in the hope that this will push $p(x)$ closer to $f(x)$ throughout (a, b). Given this motivation, we consider the following interpolation problem:

Find a polynomial $h(x)$ that agrees with $f(x)$ and its first n derivatives at $x = a$, and also agrees with $f(x)$ at $x = b$.

211

By our previous study of Taylor polynomials we know that $h(x)$ must agree with $p(x)$ at least through the nth power of $(x - a)$. Our expectation is that the additional condition $h(b) = f(b)$ may be met by appending one additional term to $p(x) = f(a) + f'(a)(x - a) + \cdots + \dfrac{f^{(n)}(a)}{n!}(x - a)^n$. Therefore, let $h(x) = p(x) + A(x - a)^{n+1}$, where the constant A is to be determined. Setting $h(b)$ equal to $f(b)$ leads to $A = \dfrac{f(b) - p(b)}{(b - a)^{n+1}}$. Thus,

$$h(x) = p(x) + \frac{f(b) - p(b)}{(b - a)^{n+1}}(x - a)^{n+1}. \tag{1}$$

The geometric interpretation of Taylor polynomials convinces us that $p(x)$ is a good approximation to $f(x)$ near $x = a$. In passing to $h(x)$, we invested an additional power of $(x - a)$ to purchase some accuracy near $x = b$. Does this additional accuracy extend over the interval (a, b) as well? Beginning with a simple case, consider the situation when $n = 0$. Here, $p(x) = f(a)$ is a constant and (1) becomes $h(x) = f(a) + \dfrac{f(b) - f(a)}{b - a}(x - a)$. Geometrically, p is a horizontal line striking the graph of f at $x = a$, while h is the line intersecting the graph of f at both $x = a$ and $x = b$. Depending on how f behaves between these points, $h(x)$ may or may not be a better approximation to $f(x)$ over (a, b) than $p(x)$ is. For example, in the interval $(0, 1)$, taking $f(x) = x^k$ leads to $h(x) = x$ and $p(x) = 0$ independent of k. As Figure 1 illustrates, $h(x)$ appears to be a better approximation than $p(x)$ for $f(x) = x^2$, but the situation is reversed for $f(x) = x^{1000}$.

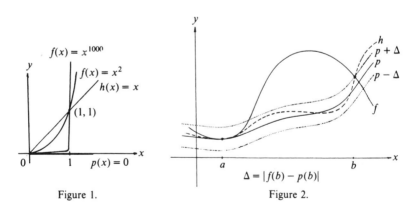

Figure 1.   Figure 2.

More generally, since $|h(x) - p(x)| = |f(b) - p(b)|(\frac{x-a}{b-a})^{n+1}$ is no greater than $|f(b) - p(b)|$ throughout (a, b), we may easily imagine the situation depicted in Figure 2. The graph of h can't wander too far from the graph of p, but the graph of f may stray far away.

As an attempt to discover how far $f(x)$ may stray from $h(x)$, let us try to find $\max_{a<x<b}|f(x) - h(x)|$. This maximum must be an extremum of $\Delta(x) = f(x) - h(x)$. Since $\Delta(a) = \Delta(b) = 0$, the extremum necessarily occurs at a value $c \in (a,b)$ where $\Delta'(c) = 0$. In other words, we are after a point c where $f'(c) = h'(c)$.

To make this situation more concrete, let's look once again at the special case $n = 0$. Here, $h'(x) = \dfrac{f(b) - f(a)}{b - a}$ is constant and, at the point c which we seek,

$$f'(c) = \frac{f(b) - f(a)}{b - a}. \tag{2}$$

This is where we stumble. Equation (2) does not give any useful information about the location of c or the magnitude of $\Delta(c)$. However, we regain our footing a bit by recognizing that (2) is the conclusion of the Mean Value theorem. Indeed, since we already know that the existence of c in (a,b) is assured, we have evidently rederived the Mean Value theorem.

And now, feet firmly beneath us once again, we are tempted to wander off in a new direction. If taking $n = 0$ leads to a proof of the Mean Value theorem, what happens if we take $n = 1$? In this case, $p(x)$ is the linear Taylor polynomial and $h(x)$ is a quadratic. Since $f'(a) = h'(a)$ by construction of h, the existence of $c > a$ satisfying $f'(c) = h'(c)$ implies (by Rolle's theorem) that $f''(c_1) = h''(c_1)$ at some point $c_1 \in (a,c)$. But h'' is constant (as was h' in the preceding case) and so we have $f''(c_1) = \dfrac{2[f(b) - p(b)]}{(b - a)^2}$ accordingly. Thus,

$$f(b) - p(b) = \frac{f''(c_1)}{2}(b - a)^2.$$

This is new! Although we still haven't learned anything about $|f(x) - h(x)|$, we have discovered new information about $f(b) - p(b)$.

The path to generalization is now almost obvious. Let $p(x)$ be the nth order Taylor polynomial for f at $x = a$, and let $h(x) = p(x) + \dfrac{f(b) - p(b)}{(b - a)^{n+1}}(x - a)^{n+1}$. Since $f(a) = h(a)$ and $f(b) = h(b)$, we have $f'(c_1) = h'(c_1)$ at some $c_1 \in (a,b)$. Similarly, $f'(a) = h'(a)$ and $f'(c_1) = h'(c_1)$ yield a point $c_2 \in (a,b)$ where $f''(c_2) = h''(c_2)$. This process continues through the nth derivative because $f^{(n)}(a) = h^{(n)}(a)$. Thus, we derive a point $c_{n+1} \in (a,b)$ where $f^{(n+1)}(c_{n+1}) = h^{(n+1)}(c_{n+1})$. Since $p(x)$ is a polynomial of degree n, its $(n + 1)$th derivative is zero. Therefore,

$$h^{(n+1)}(c_{n+1}) = (n + 1)! \frac{f(b) - p(b)}{(b - a)^{n+1}}$$

equals $f^{(n+1)}(c_{n+1})$. In particular,

$$f(b) - p(b) = \frac{f^{(n+1)}(c_{n+1})}{(n + 1)!}(b - a)^{n+1},$$

which is the Lagrange remainder formula. On replacing b by x and c_{n+1} by c, we have the familiar form of Taylor's theorem:

$$f(x) = p(x) + \frac{f^{(n+1)}(c)}{(n+1)!}(x-a)^{n+1} \qquad \text{for some} \quad c \in (a, x).$$

If the foregoing presentation is to be credible, it is not enough to stop here, admit defeat in the analysis of $|h(x) - f(x)|$, and console ourselves with the "accidental" discovery of Taylor's theorem. Indeed, reviewing this discovery suggests a new approach. If by studying $h(x)$ we discoverd an error formula for $p(x)$, perhaps by studying a polynomial of still higher degree, one that agrees with both $f(x)$ and $f'(x)$ at $x = b$, we can find an error formula for $h(x)$. This idea leads to a more general polynomial interpolation problem, and a corresponding generalization of Taylor's theorem. Whereas the preceding discussion has emphasized the process of discovery, in the interest of brevity, results and proofs are presented in the sequel without further embellishment.

A Generalization of Taylor's Theorem

We shall be concerned here with the following interpolation problem:

Construct a polynomial h whose first n derivatives agree with those of f at $x = a$ and whose first m derivatives agree with those of f at $x = b$.

This is a special case of Hermite or osculatory interpolation (see [3] p. 192, or [4]). The case previously considered corresponds to $m = 0$, where $h(x) = p(x) + (x - a)^{n+1}A$. For $m > 0$, the constant A is replaced by a polynomial in $(x - b)$.

In the discussion to follow, we let $f \overset{na}{=} g$ denote the the fact that $f^{(k)}(a) = g^{(k)}(a)$ for $0 \leq k \leq n$. Given a sufficiently differentiable function f, our task is to produce a polynomial h for which $h \overset{na}{=} f$ and $h \overset{mb}{=} f$. In constructing h, the following easily established results are useful.

$$\text{If } f_1 \overset{na}{=} g_1 \text{ and } f_2 \overset{na}{=} g_2, \text{ then } f_1 + f_2 \overset{na}{=} g_1 + g_2. \qquad (*)$$

$$\text{If } f \overset{na}{=} g \text{ and } h \text{ is } n \text{ times differentiable at } a, \text{ then } hf \overset{na}{=} hg. \qquad (**)$$

Now we can prove the following generalization of Taylor's theorem.

Theorem. *Let $f(x)$ be differentiable n times at $x = a$ and differentiable m times at $x = b$, and let $g(x) = \dfrac{f(x) - p(x)}{(x-a)^{n+1}}$, where $p(x)$ is the nth order Taylor polynomial for $f(x)$ at $x = a$. Then*

$$h(x) = p(x) + \left[\sum_{k=0}^{m} \frac{1}{k!} g^{(k)}(b)(x-b)^k \right](x-a)^{n+1} \qquad (3)$$

satisfies $h \overset{na}{=} f$ and $h \overset{mb}{=} f$. Moreover, if $x > a$ and $f^{(m+1)}$ exists in the interval between *x and b, then*

$$f(x) - h(x) = \frac{1}{(m+1)!} g^{(m+1)}(c)(x-b)^{m+1}(x-a)^{n+1}$$

for some c between x and b.

Proof. Note that $h(x) = p(x) + q(x)(x-a)^{n+1}$, where $q(x)$ is the *m*th order Taylor polynomial for $g(x)$ at $x = b$. In particular, $q \overset{mb}{=} g$. Using (*) and (**), we have $q(x-a)^{n+1} \overset{mb}{=} g(x-a)^{n+1}$ and $p + q(x-a)^{n+1} \overset{mb}{=} p + q(x-a)^{n+1}$. But this reduces to $h \overset{mb}{=} f$. For the situation at a, note $(x-a)^{n+1} \overset{na}{=} 0$ and so $q(x-a)^{n+1} \overset{na}{=} 0$. This latter result, added to $p \overset{na}{=} f$, yields $p + q(x-a)^{n+1} \overset{na}{=} f + 0$, or $h \overset{na}{=} f$.

To establish the remainder formula, apply Taylor's theorem to the function g. We have $g(x) - q(x) = \frac{1}{(m+1)!} g^{(m+1)}(c)(x-b)^{m+1}$ for some c between x and b. Thus, since $f(x) = p(x) + g(x)(x-a)^{n+1}$, it follows that

$$f(x) - h(x) = [g(x) - q(x)](x-a)^{n+1}$$

$$= \frac{1}{(m+1)!} g^{(m+1)}(c)(x-b)^{m+1}(x-a)^{n+1}$$

as required.

Example. Consider $f(x) = \sqrt{x}$ on the interval $[1,4]$, and let $n = 2$ and $m = 1$. Then (3) yields

$$h(x) = 1 + \frac{(x-1)}{2} - \frac{(x-1)^2}{8} + \frac{5(x-1)^3}{216} - \frac{(x-1)^3(x-4)}{216}.$$

Note, for example, that $h(3) = 1.7\overline{2}$. In contrast, the fourth degree Taylor polynomial at $x = 1$,

$$p_4(x) = 1 + \frac{(x-1)}{2} - \frac{(x-1)^2}{8} + \frac{(x-1)^3}{16} - \frac{5(x-1)^4}{128},$$

gives $p_4(3) = 1.375$. In fact, the average difference between $f(x)$ and $h(x)$ over the interval $[1,4]$, calculated as $\frac{1}{3}\int_1^4 [f(x) - h(x)]\,dx$, is .004, whereas the average difference between $f(x)$ and $p_4(x)$ over $[1,4]$ is about 1.1. These results suggest that $h(x)$ is a better fourth degree approximation to $f(x)$ than $p_4(x)$ is throughout the interval $[1,4]$. In fact, a numerical investigation indicates that p_4 is much better for x near 1, and has a slight advantage for x between 1.3 and 1.8, but in the interval as a whole, h is far superior to p_4. The greatest error in approximating $\sqrt{x}$ by $h(x)$ seems to occur around $x = 2.5$ and is on the order of .01.

The procedure above may be iterated to provide for agreement between f and a polynomial h at each of k points with specified agreement of derivatives. The interested reader may wish to compare the resulting formulation with that of any standard numerical analysis text (e.g., [3] p. 256).

REFERENCES

1. Robert Fisher and Allen Ziebur, Calculus and Analytic Geometry (3rd ed.) Prentice-Hall, Engle-wood Cliffs, NJ, 1975.
2. Philip Gillett, Calculus and Analytic Geometry, D. C. Heath, Lexington, MA, 1981.
3. Eugene Isaacson and Herbert Bishop Keller, Analysis of Numerical Methods, John Wiley, New York, 1966.
4. Dan Kalman, The generalized VanderMonde matrix, Mathematics Magazine, 57 (January 1984).

BIBLIOGRAPHIC ENTRIES: TAYLOR POLYNOMIALS

1. *Monthly* Vol. 81, No. 6, pp. 592–601. Tyre A. Newton, Using a differential equation to generate polynomials.

> A class of differential equations that can be used to illustrate Taylor polynomial approxima-tions to functions with the help of electronic analog computers.

2. *Monthly* Vol. 86, No. 8, pp. 681–684, Richard J. Bagby, Taylor polynomials and difference quotients.

3. *Monthly* Vol. 89, No. 5, p. 311–312. Alfonso G. Azpeitia, On the Lagrange remainder of the Taylor formula.

4. *Monthly* Vol. 94, No. 5, pp. 453–455. Fred Bauer, A simplification of Taylor's theorem.

> Treatment of the remainder term is similar to that given on pp. 278–280 of Apostol's *Calculus*, Vol. 1, 2nd. ed., John Wiley & Sons (1967). Apostol gives both upper and lower bounds for the error, while this paper gives only upper bounds for the absolute value of the error.

5. *Monthly* Vol. 97, No. 3, pp. 233–235. G. B. Folland, Remainder estimates in Taylor's theorem.

> Shows that it is not necessary to use derivative estimates to obtain remainder estimates.

6. CMJ Vol. 20, No. 5, pp. 435–436. David P. Kraines, Vivian Y. Kraines, and David A. Smith, Taylor polynomials.

> Uses software to illustrate graphically that the Taylor polynomials approximate a given function.

(b) BERNOULLI POLYNOMIALS AND SUMS OF POWERS OF INTEGERS

(No papers reproduced in this section)

BIBLIOGRAPHIC ENTRIES: BERNOULLI POLYNOMIALS AND SUMS OF POWERS OF INTEGERS

1. *Monthly* Vol. 77, No. 5, pp. 840–847. L. S. Levy, Summation of the series $1^n + 2^n + \cdots + x^n$ using elementary calculus.

2. *Monthly* Vol. 78, No. 9, p. 987. L. S. Levy. Addendum.

3. *Monthly* Vol. 79, No. 1, pp. 44–51. H. W. Gould, Explicit formulas for Bernoulli numbers.

4. *Monthly* Vol. 91, No. 7, pp. 394–403. B. L. Burrows and R. F. Talbot, Sums of powers of integers.

5. *Monthly* Vol. 95, No. 10, pp. 905–911, D. H. Lehmer, A new approach to Bernoulli polynomials.

6. *Math. Mag.* Vol. 53, No. 2, pp. 92–96. Barbara Turner, Sums of powers of integers via the binomial theorem.

7. *Math. Mag.* Vol. 61, No. 3, pp. 189–190. Dumitri Acu, Some algorithms for the sums of integer powers.

8. CMJ Vol. 18, No. 5, pp. 406–409. Michael Carchidi, Two simple recursive formulas for summing $1^k + 2^k + 3^k + \cdots + n^k$.

8

MAXIMA AND MINIMA

A STRONG SECOND DERIVATIVE TEST

J. H. C. Creighton

Department of Mathematics, Pahlavi University, Shiraz, Iran.

Definition. A *dense zero* of a function $g(x)$ is a point x_0 such that any interval containing x_0 contains a zero of $g(x)$ other than x_0. That is, x_0 is a limit point of the zeros of $g(x)$.

Except for the trivial case that $g(x)$ is zero on an entire interval, this phenomenon is a pathology for one variable calculus and accounts for much that is unsatisfactory in the usual presentation. For example, in the absence of dense zeros of the first derivative an easy proof of the chain rule is possible (see David Gans [1] p. 150; note that $y' \neq 0$ implies $\Delta y \neq 0$) and in the absence of dense zeros of the second derivative, a simple and natural definition of inflection point is possible (see below). Furthermore, the possibility of a dense zero of the second derivative precludes a satisfactory classification of critical points. By ruling out this pathology, we obtain in Theorem A of this note a complete classification in terms of the second derivative.

Let $f(x)$ be a real valued function of a real variable. We assume throughout that $f''(x)$ is continuous. Consequently, if x_0 is not a dense zero of $f''(x)$, the sign of $f''(x)$ is constant on sufficiently small intervals of the form $(x_0, x_0 + \varepsilon)$ or $(x_0 - \varepsilon, x_0)$. Then we define x_0 to be an *inflection point* of $f(x)$ if this sign changes; that is, if the concavity of $f(x)$ changes at x_0. This definition is equivalent, *in the absence of dense zeros of the second derivative*, to the various inequivalent standard definitions. Thus the disagreement among the standard definitions revolves about the pathology of dense zeros. Of course, one might not adopt our viewpoint that dense zeros are pathological in which case he must choose among the standard definitions of inflection point (see the articles of G. M. Ewing [1] p. 155 and A. W. Walker [1] p. 161). By a *strict extremum* of $f(x)$ we mean the defining inequality to be strict.

Theorem A. *If $f'(x_0) = 0$ and x_0 is not a dense zero of the (continuous) second derivative, then x_0 is a strict relative extremum or an inflection point. Furthermore, the second derivative distinguishes which:*

(a) x_0 *is a strict relative minimum* $\Leftrightarrow f''(x) \geqq 0$ *on a neighborhood of* x_0.

(b) x_0 *is a strict relative maximum* $\Leftrightarrow f''(x) \leqq 0$ *on a neighborhood of* x_0.

(c) x_0 *is an inflection point* $\Leftrightarrow f''(x)$ *changes sign at* x_0.

THEOREM B. *If x_0 is a dense zero of $f(x)$ then x_0 is a dense zero of $f'(x)$.*

Thus in Theorem A, x_0 is not a dense zero of $f'(x)$ nor of $f(x)$ nor is $f(x)$ constant at x_0 (i.e., constant on a neighborhood of x_0).

THEOREM C. *If x_0 is a relative extremum of $f(x)$ and not a dense zero of the first derivative, then x_0 is a strict relative extremum.*

Theorems B and C are elementary applications of Rolle's Theorem and the Mean Value Theorem, respectively.

Proof of Theorem A. The hypotheses guarantee that one of the conditions on $f''(x)$ given in (a), (b) or (c) holds, thus we need only prove these equivalences. We prove only part (a), assuming $f''(x_0) = 0$. The other cases are immediate from standard theorems or follow by analogy to this case.

Suppose $f''(x) \geq 0$ on (a, b). Then since x_0 is an isolated zero of $f''(x)$, we may assume equality holds only at x_0. If $x_1 \in (x_0, b)$, by the Mean Value Theorem there is a $\xi \in (x_0, x_1)$ such that

$$f'(\xi) = \frac{f(x_1) - f(x_0)}{x_1 - x_0}.$$

Since $f''(x) \geq 0$, $f'(x)$ is increasing on (a, b) thus $f'(\xi) \geq 0$. But $f'(\xi) \neq 0$, for otherwise by Rolle's Theorem $f''(x)$ has a root between x_0 and ξ. Hence $f(x_1) > f(x_0)$. Similarly, if $x_2 \in (a, x_0)$, then $f(x_2) > f(x_0)$. Hence x_0 is a strict relative minimum.

Conversely, suppose that in every neighborhood of x_0 $f''(x) < 0$ for some x. Then by continuity, $f''(x) \leq 0$ in a neighborhood of x_0 or $f''(x)$ changes sign at x_0 and we are reduced to part (b) or (c). This completes the proof of part (a).

We now show that the existence of a dense zero of the *second* derivative is indeed the pathology to be avoided. That is, we provide a counter example to Theorem A, where it is assumed only that x_0 is not a dense zero of $f'(x)$. Define $g(x)$ on $[0, 1]$ to be the saw-tooth function that is zero at $\{1/n\}$ and whose value at the midpoint of $[1/(2n + 1), 1/2n]$ is $1/n$ and at the midpoint of $[1/2n, 1/(2n - 1)]$ is $-1/2n$. Extend to $[-1, 1]$ by $g(x) = x$ for $x \leq 0$. Then $g(x)$ is continuous and has the following two properties:

(1) $\displaystyle \int_0^t g(x)dx > 0, \qquad t \neq 0,$

(2) *on any interval of the form $(0, \varepsilon)$, $g(x)$ takes on both positive and negative values.*

Now let $f(u) = \int_0^u \int_0^t g(x)dx\, dt$. Then by (1) zero is an *isolated* zero of $f'(u)$ but is not a relative extremum of $f(u)$ since in fact $f(u)$ is strictly increasing. Further, by (2), zero is a dense zero of $f''(u)$ and therefore is not an inflection point of $f(u)$.

Reference

1. T. M. Apostol et al., Selected Papers on Calculus, MAA, 1968.

L. H. LANGE

To my teacher, George Pólya, for his 89th birthday

A. A long (geometric minimum) story short. An ancient geometer might simply present the following figure using only the word "Behold!"

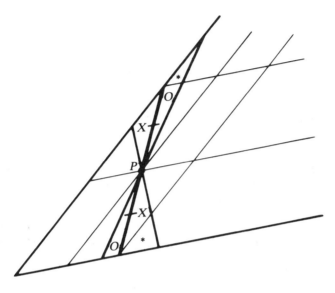

Here is an explanatory note. If we wish to prove some theorems about polygons of minimum area circumscribed about a given convex set, as in [2], for example, we need a helping theorem which tells us *how to cut off a triangle of minimum area with a line passing through a given point P inside an angle.* What the figure tells us (*sans* calculus and *sans* any linear transformation theory) is that of all segments through P determining corner triangles, it is that segment for which P is the *midpoint* which will yield the corner triangle of minimum area, for all other eligible triangles are seen to be bigger than that particular one. (The angle does not have to be acute.)

B. Note from the prelude. Years ago, after using "$f'(x) = 0$" to solve the related old calculus book problem where P is a point (a, b) in the first quadrant, a right angle, and then lingering over the geometric interpretation of the fact that the area of the minimum triangle in that first quadrant corner turns out to be $2ab$, I had noticed that I could solve that special old problem with a rather quick, direct thrust, *without* the tools of calculus, as indicated in Figure 2.

Later, when I needed to deal with the non-right angle problem and brought to mind the familiar properties of affine transformations of the plane—e.g., "midpoints go into midpoints" and "ratios of areas are preserved" so that the property of being a minimum triangle in a collection is preserved—I

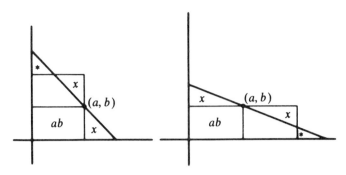

knew I had the solution. It was still later that it occurred to me to make the delightfully *direct construction* of the proof as shown in Figure 1. (For a convenient listing of some of the planar affine properties involved here, see [2, p. 61]. The easy three-dimensional analogues are needed when we necessarily exploit affinity in our next section.)

C. Some notes from the postlude. It is natural to look for related problems and make guesses of possible theorems. (One of Professor Pólya's "synthetic proverbs" from [8] leaps to mind: "Look around when you have got your first mushroom or made your first discovery: they grow in clusters." See also [10].) The main one I settle on here is this particular three-dimensional generalization: *Through a fixed point P in a given trihedral angle, identify the plane which will cut off a corner tetrahedron of minimum volume.* Solution? We yield to an irresistible analogy and pick that plane through P which is such that P is the *centroid* of the triangle which is determined when this plane cuts across that given trihedral angle.

For the proof, we would ignore affinity only with pain, for it is an unhappy circumstance that an easy analogue to the direct construction in Figure 1 is difficult. Relying on affinity, then we restrict our attention to the case where the given point P finds itself in the right trihedral angle represented by the first octant of an XYZ coordinate system. We shall use the fact that an equation for the tangent plane to a surface $F(X, Y, Z) = 0$ at the point $P = (x, y, z)$ is given (usually) by

$$\frac{\partial F}{\partial x}(X - x) + \frac{\partial F}{\partial y}(Y - y) + \frac{\partial F}{\partial z}(Z - z) = 0.$$

(Here, for example, $\partial F/\partial x$ is the value of $\partial F/\partial X$ at P, of course.)

Here first is yet another preparatory proof of the (right angle) case considered in Figure 2, using the two-dimensional analogues for the equations just mentioned. For sound heuristic reasons, we now look at the family of hyperbolas $XY = a^2$. Each choice of a ($a > 0$, for convenience) determines a specific hyperbola whose first quadrant branch we'll call $H(a)$. Then, in an XY system, the tangent line at a point (x, y) on $H(a)$ is given by $y(X - x) + x(Y - y) = 0$, or $yX + xY = 2xy = 2a^2$. This then yields $(X/2x) + (Y/2y) = 1$ as a convenient equation for this tangent line at (x, y). The line intersects the axes at $(2x, 0)$ and $(0, 2y)$, and this tells us, in turn, that (x, y) is the midpoint of the corner segment and that the resulting corner triangle has area

$$A(a) = \frac{(2x)(2y)}{2} = 2a^2.$$

We've thus encountered an heuristic prelude note from analytic geometry days: the hyperbola

$xy = a^2$ is precisely the locus of the midpoints of all those corner segments which determine corner triangles of constant area $2a^2$.

Now, to finish the proof concerning the minimum triangle in the corner, let σ be the tangential segment to the hyperbola $H(a)$ at the given point $P = (x, y)$. P is the centroid of σ and determines $H(a)$, we know. Furthermore, $H(a)$ cuts the plane into three natural separate pieces $D(a)$, $H(a)$, and $D'(a)$—where $D(a)$ is the maximal domain containing the origin and lying below $H(a)$, while $D'(a)$ is the maximal domain "out beyond" $H(a)$. It is useful to note that a first quadrant point (X, Y) belongs to the planar portion $D(a)$, $H(a)$, or $D'(a)$, respectively, if and only if $XY < a^2$, $XY = a^2$, or $XY > a^2$.

Now consider any *other* line through P, and the resulting first quadrant line segment σ' which this line determines. (That is, σ' is simply the intersection of this other line through P with the first quadrant.) There then exists a point P', other than P, which is the centroid of σ' and determines a hyperbola $H(a')$. Finally, after working a bit (we do this in two ways below) to observe that P' belongs to $D'(a)$, we have $a' > a$, yielding the desired result $A(a') > A(a)$, and settling the two-dimensional case. It is σ that gives us the minimum triangle in the corner.

Here is a way of assuring ourselves that the point P', the centroid of σ' in the preceding paragraph, belongs to $D'(a)$. There exists a segment σ'' which is parallel to σ' and tangent to $H(a)$ at a point P'', which, by our earlier work, is the centroid of σ''. The ray from the origin through P'' will go out into $D'(a)$ and there intersect σ' in *its* centroid, namely P'. (Happily, this reasoning readily applies to the higher dimensional cases.)

We record a few more observations concerning the present two-dimensional situation.

For example, we can prove with a pleasant simplicity and without our earlier appeal to certain calculus formulae, that at $P = (x, y)$, the segment σ, whose points (X, Y) satisfy $(X/2x) + (Y/2y) = 1$, is *tangent* to our hyperbolic curve $H(a)$ given by $XY = a^2$. For, suppose that some point $P' = (x', y')$ is, like P, in the intersection of σ and $H(a)$. Then we have $(x'/2x) + (y'/2y) = 1$ and also $x'y' = a^2 = xy$. This tells us that

$$1 = \left(\frac{x'}{x}\right)\left(\frac{y'}{y}\right) = \left[\frac{(x'/x) + (y'/y)}{2}\right]^2 = 1,$$

and so, from the geometric-arithmetic mean inequality theorem (in its simplest, two-dimensional form) we conclude that it must be that $x'/x = y'/y = 1$. Thus P' and P are one and the same point; our tangency proof is in hand. (See [7]; also [4].)

Now suppose $P' = (x', y')$, $P' \neq P$, is the centroid of the segment σ' discussed in our minimum area proof above. We needed to know that $P' \in D'(a)$, a fact which we shall confirm once more if we manage to show, for example, that $x'y' > a^2$. Well, since P is on σ', we have $(x/2x') + (y/2y') = 1$; and since $P' \neq P$, we have

$$\frac{a^2}{x'y'} = \frac{xy}{x'y'} < \left[\frac{x}{2x'} + \frac{y}{2y'}\right]^2 = 1.$$

This yields $a^2 < x'y'$, and we are finished with our preparations for higher dimensional analogues.

For the proof of the three-dimensional case, finally, we may now merely mimic what we have just done. Here we consider the first-octant branches of the family of surfaces $XYZ = a^3$, $a > 0$. In the XYZ system, the tangent plane to the surface $H(a)$ at a point $P = (x, y, z)$ in $H(a)$ is given by

$$yz(X - x) + zx(Y - y) + xy(Z - z) = 0;$$

$$yzX + zxY + xyZ = 3xyz;$$

$$\frac{X}{3x} + \frac{Y}{3y} + \frac{Z}{3z} = 1.$$

(Of course, the tangency of this plane can, here too, be established via the geometric-arithmetic mean theorem.) The corner tetrahedron thus is seen to have vertices at $(0, 0, 0)$, $(3x, 0, 0)$, $(0, 3y, 0)$, and $(0, 0, 3z)$; it has volume

$$V(a) = \frac{(3x)(3y)(3z)}{6} = \frac{9a^3}{2};$$

and $P = (x, y, z)$ is indeed the *centroid* of its outer face. Where we formerly had the tangential line segment σ, determined by the intersection with the first quadrant of a tangent line through P, we now have a triangle τ, determined by the intersection with the first octant of a tangent plane through P. The hyperbolas $H(a)$ are here replaced by surfaces $H(a)$, the maximal domains $D(a)$ and $D'(a)$ take on three-dimensional meanings, areas $A(a)$ are replaced by volumes $V(a)$ of the convenient sort where $a' > a$ implies $V(a') > V(a)$, and "it all goes through, word for word." It is τ that gives us the minimum tetrahedron in the corner; we have solved completely the (general) trihedral angle problem with which we began this section. *The minimum does exist and we know how to identify it.*

On the other hand, there exists an inductive proof (due to the reviewer) which does not require the three axes to be orthogonal, but requires one assume existence of a minimizing section. To show how the three dimensional case follows from the two dimensional case, consider the three rays *OX*, *OY*, *OZ* with origin *O* forming the edges of the solid angle, a point *P* interior to the angle, and a plane section through *P* intersecting *OX*, *OY*, *OZ* in points *A*, *B*, *C*, respectively. Assume this section cuts minimum volume from the solid angle. The ray *CP* intersects the side *AB* of triangle *ABC* in a point *Q*. If *Q* is not the midpoint of *AB*, then the two dimensional case implies that there are points *A'* on *OX* and *B'* on *OY* with *Q* the midpoint of *A'B'* and triangle *OA'B'* has smaller area than triangle *OAB*. Then the volume of tetrahedron *OA'B'C* is smaller than the volume of *OABC*, a contradiction since the plane *A'B'C* also contains *P*. Thus the ray *CP* must be along a median of triangle *ABC*. Similarly *AP* and *BP* are along medians. Hence *P* is the centroid of triangle *ABC*. (Parts of this section are reminiscent of Pólya's treatment of "tangent level lines." See [9]; also [4], [11], [12].)

D. A selection of guesses, problems, and parting comments. How does one cut off a triangle of minimum *perimeter* with a line passing through a point *P* inside an angle?

We can see quickly that this problem is probably more difficult than the minimum area problem, because affine transformations do not preserve lengths, only ratios of parallel lengths. It develops that this problem has been treated by Pólya in [11], and with his usual elegance. Chamberlain and Moore [3] have suggested and solved the following related problem: find the slope of the line through the first quadrant point (a, b) which forms the corner triangle of minimum perimeter. And here is a problem which I leave to the reader: given a point *P* in an *n*-hedral angle in three-space, find a plane through *P* which will determine a corner solid with minimum *surface area*.

In [13, p. 47] Jakob Steiner observes that the planes which cut a constant volume from a given solid angle envelope a certain curvilinear surface *S*, and each of the resulting planar sections is tangent to *S* at its centroid. A footnote then calls attention to the (rectangular) special case we have treated in the discussion of our minimum volume problem.

Peano later derived corresponding results for planes cutting a constant volume from an arbitrary given solid, and Ascoli in [1, footnote on p. 128] refers to this to prove that *if* a plane cuts off an extremum volume of all planes through a given point *P* in a given solid, then *P* must be the centroid of that planar section. This result is also established in [14], where Professor Stewart motivates questions like these by suggesting that a gem-cutter might well be interested in removing an impurity at a point *P* in a gem by making a planar cut which leaves a maximum amount of good quality gem.

Because of this work we have these theorems as special cases: given a point *P* in an *n*-hedral angle (or in an elliptical cone) in three-space, with vertex at the origin, then the corner solid with minimum volume determined by a plane through P is the one for which the n-gon (elliptical) outer face has P as its centroid. (In the elliptical case, one can even draw a convincing analogue to Figure 1.)

Day [5] showed that a polyhedron with a given number of faces having minimum volume and containing a given convex solid is tangent to the solid at the centroid of each face. (This involves some work with his Lemma 4.2 and the fact that Day concerns himself with centrally symmetric bodies.) This then gives us a three-dimensional analogue to Theorem 1 in [2]. (See [15, p. 6].)

Finally, I have learned in the meantime that there is in [6] a figure with an element something like what's in my Figure 1, above. In answer to my inquiry Professor Kazarinoff tells me he simply cannot remember where he learned the result, saying that he'd possibly learned it from his father or from Erdös.

References

1. G. Ascoli, Sui baricentri delle sezioni piani di un dominio spaziale connesso, Boll. Un. Mat. Ital., 10 (1931) 123–128.

2. G. D. Chakerian and L. H. Lange, Geometric extremum problems, Math. Mag., 44 (1971) 57–69.

3. M. Chamberlain and S. Moore, Math. Mag., 48 (1975) 238.

4. R. Courant and H. Robbins, What is Mathematics?, Oxford Univ. Press, New York, 1941 pp. 362–363.

5. M. M. Day, Polygons circumscribed about closed convex curves, Trans. Amer. Math. Soc., 62 (1947) 315–319.

6. Nicholas D. Kazarinoff, Geometric inequalities, Random House, New York, 1961, pp. 89 and 122 (Problem #43).

7. L. H. Lange, Several hyperbolic encounters, The Two-Year College Mathematics Journal, 7 (1976) 2–6.

8. G. Pólya, How to solve it, Princeton University Press, 1948, p. 197.

9. ———, Mathematics and plausible reasoning, Volume I, Princeton University Press, 1954, pp. 121, 123, and especially #11 on p. 132.

10. ———, Mathematics and plausible reasoning, Volume II, *ibid.* 1954, 158, 207.

11. ———, Mathematics and plausible reasoning, revised edition, Volume II, *ibid.* 1968, 205–207.

12. ——— and G. Szegö, Isoperimetric inequalities in mathematical physics, Princeton Univ. Press, 1951, pp. 109–111.

13. J. Steiner, Gesammelte Werke, vol. 2, Berlin, 1882, pp. 47, 48.

14. B. M. Stewart, A maximum problem, this MONTHLY, 49 (1942) 454–456.

15. L. Fejes Tóth, Lagerungen in der Ebene, auf der Kugel, und im Raum, Stechert Hafner, Berlin, 1953.

SCHOOL OF SCIENCE, SAN JOSE STATE UNIVERSITY, SAN JOSE, CA 95192.

An Optimization Problem

RICHARD BASSEIN

Department of Mathematics and Computer Science, Mills College, Oakland, CA 94613

Some years ago, I played with a computer game called "Lunar Lander," of which there are several forms in existence. The one I used simulated the following situation. You are piloting a lunar excursion module which has a given amount of fuel and is hurtling towards the surface of the moon. You must decide when and how hard to fire your retrorockets to slow your vehicle to a gentle landing. After much experimentation and consultation with other "simu-astronauts," I found that the best strategy was to wait as long as possible before firing the rockets, then blast away with the maximum force allowed. (That strategy is, in fact, an example of the so called "bang-bang" principle of optimal control; advanced treatments using functional analysis may be found in the references.) This article presents an elementary proof that such a strategy is indeed the best.

To make the central ideas most apparent, let's first consider a simple model in which the change in weight of the remaining fuel is negligible compared to the total weight of the vehicle. Let

$x(t)$ = the height of the vehicle above the surface of the moon

$v(t)$ = the downward velocity of the vehicle; thus $v(t) = -x'(t)$

$f(t)$ = the cumulative amount of fuel consumed; thus $f(0) = 0$

g = the acceleration of gravity near the moon's surface

T = the time at which the vehicle lands; thus $x(T) = 0$.

Our goal is to minimize $f(T)$ while making $v(T) = 0$, for a gentle landing.

Let us assume that the force applied by the retrorockets is proportional to the rate at which fuel is consumed. If $\underline{c}$ is the constant of proportionality and $\underline{m}$ is the mass of the vehicle (including fuel) then Newton's Law says the following about the downward velocity $v(t)$.

$$mv'(t) = mg - cf'(t) \tag{1}$$

By choosing the appropriate units for $f(t)$ we can make $c = m$ and this equation takes the following simpler form.

$$v'(t) = g - f'(t)$$

The use of $f'(t)$ when the force, and therefore the rate of fuel consumption, is changed deserves some discussion. While in reality such changes in $f'(t)$ may be rapid but continuous, it will be simpler, and not terribly inaccurate, to model them as discontinuous. The problem is that $f'(t)$ will be undefined at such moments. The easiest way to deal with this difficulty is to assume that $f(t)$ is continuous and piecewise differentiable and integrate the above equation.

Integrating $v'(t) + f'(t) = g$ gives $v(t) + f(t) = C + gt$, for some constant $\underline{C}$. If the initial velocity of the vehicle is $v_0 = v(0)$, we get $C = v_0$ since $f(0) = 0$. Thus,

$$v(t) + f(t) = v_0 + gt \qquad (2)$$

If $v(T) = 0$, then $f(T) = v_0 + gT$. Our first observation is that to minimize $f(T)$, we need to minimize $\underline{T}$.

Fig. 1 illustrates equation 2 geometrically.

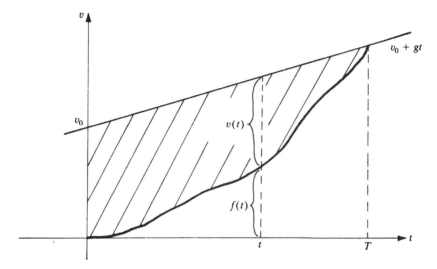

FIG. 1. Meeting a straight line.

Note that the shaded area, between the graph of $f(t)$ and the line $v_0 + gt$, is the integral of $v(t)$ from $t = 0$ to $t = T$, hence is the distance travelled toward the moon. To land, this area must be equal to x_0, the starting height.

Therefore, what we must choose is a graph for $f(t)$ which minimizes $\underline{T}$ and

(1) starts at $f(0) = 0$, since no fuel has been consumed at time $t = 0$,
(2) is nondecreasing, since fuel cannot be unconsumed,
(3) meets the line $v_0 + gt$ at $t = T$ in order to reach $v(T) = 0$, and
(4) encloses, between it and the graph of $v_0 + gt$, an area of x_0.

Let us denote by $\underline{M}$ the maximum rate at which fuel can be consumed. In other words, $f'(t) \le M$ (when $f'(t)$ is defined). Let's see that the graph which minimizes $\underline{T}$ and satisfies conditions (1) through (4) above is $f(t) = 0$ from $t = 0$ to some time $t = T_1$ and then is a straight line of slope $\underline{M}$ from $t = T_1$ to $t = T$, as shown in Fig. 2. This corresponds to waiting as long as possible before firing the rockets, then blasting away at the maximum rate.

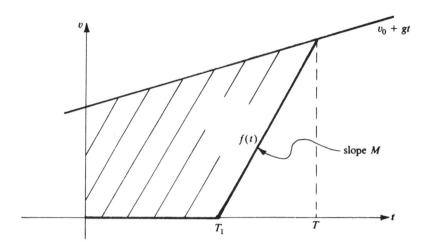

FIG. 2. Holding off, then blasting away.

First note that there is at most one graph of the form described in the previous paragraph which satisfies condition (4), since the area enclosed by such a graph is an increasing function of T_1. (If the choice $T_1 = 0$ already encloses an area larger than x_0, then we don't have enough fuel to slow down to $v(T) = 0$ before we crash into the moon.) Suppose the graph of a different function $g(t)$ also satisfies conditions (1) through (4) with $g'(t) \le M$ and meets $v_0 + gt$ at the same time $\underline{T}$, as shown in Fig. 3. (If it meets the line even earlier, we can just continue it along the line to the point where $t = T$.)

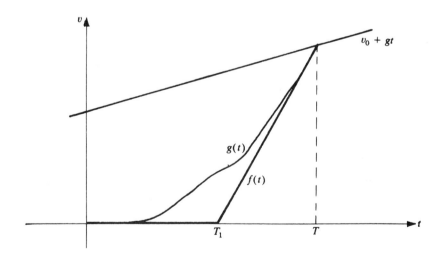

FIG. 3. A different solution?

Now since $g'(t) \le M = f'(t)$ for $T_1 < t < T$, working backwards from $t = T$ shows that $g(t) \ge f(t)$ on that interval. Since $f(t) = 0$ for $0 \le t \le T_1$, we must have $g(t) \ge f(t)$ there, too. But the area enclosed by the graph of $g(t)$ must be x_0, the same as the area enclosed by the graph of $f(t)$. This can only be if $g(t) = f(t)$ for $0 \le t \le T$. Hence we have found the optimal solution.

Finally, let's modify equation 1 to take into account the change in mass due to consumption of fuel. If an amount of fuel $f(t)$ has mass $kf(t)$, then equation 1 becomes

$$(m - kf(t))v'(t) = (m - kf(t))g - cf'(t).$$

Again choosing units for $f(t)$ such that $c = m$, and adjusting the value of k accordingly, we obtain

$$v'(t) = g - \frac{f'(t)}{1 - af(t)},$$

where $a = k/m$. (Note that since $kf(t) < m$, we have $af(t) < 1$.) Isolating the g and integrating this, as we did before, gives

$$v(t) - \frac{1}{a}\log(1 - af(t)) = v_0 + gt.$$

If we let $F(t) = -(1/a)\log(1 - af(t))$, this yields, by analogy with equation 2,

$$v(t) + F(t) = v_0 + gt.$$

The only difference in the remainder of the argument from this point on is that the inequality $f'(t) \le M$ must be put in terms of $F(t)$. From $f(t) = (1/a)$ $(1 - e^{-aF(t)})$ we get $f'(t) = e^{-aF(t)}F'(t)$ which gives us the condition $F'(t) \le Me^{aF(t)}$. The straight line of slope M in the solution of the simplified model comes from the maximum fuel consumption condition $f'(t) = M$. Here we replace that straight line with a solution of $F'(t) = Me^{aF(t)}$. Note that since the right hand side of this first order differential equation is uniformly Lipschitz on a bounded rectangle, it has a unique solution for a given initial value. If T_1 is the moment that firing starts, then $F(T_1) = 0$, and that solution is

$$F(t) = -\frac{1}{a}\log(1 - Ma(t - T_1)).$$

Fig. 4 displays the graph of such a function.

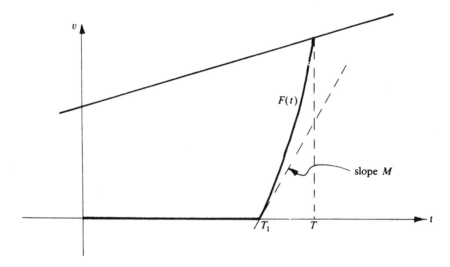

Fig. 4. Maximum firing.

REFERENCES

1. L. Cesari, Optimization Theory and Applications, Springer-Verlag, 1983.
2. R. Holmes, Geometric Functional Analysis and its Applications, Springer-Verlag, 1975.

An Old Max-Min Problem Revisited

MARY EMBRY-WARDROP

Department of Mathematics, Central Michigan University, Mt. Pleasant, MI 48859

Recently, while teaching a beginning calculus class, I introduced the problem of inscribing a rectangle of maximum area in a given right triangle and asked the class how the rectangle should be oriented inside the triangle. I was slightly surprised when a student suggested that one side of the rectangle should be on the hypotenuse of the triangle. In previous classes the suggestions were to fit one corner of the rectangle into the right angle of the triangle. The class soon bogged down in trying to solve the problem with one side of the rectangle on the hypotenuse, but subsequently solved the problem using the other orientation.

I was sufficiently intrigued to solve the problem with both orientations and was surprised to discover that in either case the maximum area was $\frac{1}{4} bh$, where b and h are the base and height of the triangle. Furthermore in each case corners of the rectangle of maximum area lie on the midpoints of the two legs of the triangle.

At this point I was thoroughly intrigued: Could the rectangle be more advantageously oriented and for other orientations did a corner of the rectangle of maximum area lie on the midpoint of a leg of the triangle?

The following assertions describe the results of my investigation. The basic mathematical tools used were the Law of Sines and the technique of maximizing or minimizing a function, usually discussed in a beginning calculus class. I omit the solutions under the assumption that most mathematicians enjoy finding their own solutions more than reading another's.

Let the rectangle be oriented in the triangle as pictured in FIGURE 1. The angle α satisfies $\theta \leqslant \alpha \leqslant \pi/2$. For a given angle α let $A(\alpha)$ be the area of the rectangle of maximum area which can be so inscribed in the triangle.

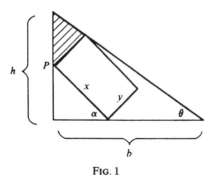

FIG. 1

Assertion 1. Given α the rectangle of largest area is obtained when

$$x = \frac{h}{2 \sin \alpha} \quad \text{and} \quad y = \frac{b \sin \theta}{2 \cos(\alpha - \theta)}.$$

229

In this case

$$A(\alpha) = \frac{bh \sin \theta}{4 \sin \alpha \cos(\alpha - \theta)}$$

and P is the midpoint of the side of the triangle opposite the angle θ.

Assertion 2. The maximum area function $A(\alpha)$, $\theta \leqslant \alpha \leqslant \frac{\pi}{2}$, has largest value at the two endpoints of the interval:

$$A(\theta) = A\left(\frac{\pi}{2}\right) = \frac{1}{4}bh.$$

Assertion 3. The maximum area function $A(\alpha)$, $\theta \leqslant \alpha \leqslant \frac{\pi}{2}$ has smallest value at the midpoint of the interval:

$$A\left(\frac{1}{2}\theta + \frac{\pi}{4}\right) = \frac{bh \sin \theta}{2(1 + \sin \theta)}.$$

Moreover in this case, the shaded triangle in Figure 1 is isosceles with equal sides of length $\frac{1}{2}h$.

There are numerous problems here for a first-semester calculus class and each of the solutions calls upon skills that the students should have developed in geometry, trigonometry, and calculus. One of the interesting facets of Assertions 2 and 3 is that the students would be invited to consider maximizing and minimizing $A(\alpha)$, each value of which is itself a maximum. Few problems of this type seem to be accessible for beginning students. Furthermore, there is a collection of geometric problems only touched upon in Assertions 1–3: Why is P a fixed point as a corner of the rectangles of maximum area? Why is the shaded triangle isosceles when $A(\alpha)$ is minimum? Why is $A(\alpha)$ minimum at the midpoint of the domain of $A(\alpha)$? Is there a geometric proof that $A(\alpha)$ is obtained with sides given by the equation in Assertion 1 or a geometric proof that $A(\theta)$ and $A(\frac{\pi}{2})$ are the maxima of $A(\alpha)$ and $A(\frac{1}{2}\theta + \frac{\pi}{4})$ is the minimum?

It is interesting to note that the problem of inscribing a rectangle of maximum area in an acute triangle is no more difficult than the same problem for a right triangle. Let the rectangle be oriented in an acute triangle, as pictured in FIGURE 2. The angle α satisfies $\theta_2 \leqslant \alpha \leqslant \pi/2$.

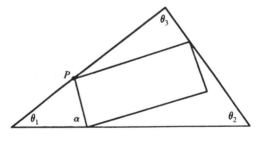

FIG. 2

Once again for a given angle α the rectangle of maximum area, $A(\alpha)$, which can be so inscribed is obtained when P is the midpoint of the side of the triangle opposite θ_2. The maximum of $A(\alpha)$ occurs at the endpoints θ_2 and $\frac{\pi}{2}$ and the minimum occurs at the midpoint $\frac{1}{2}\theta_2 + \frac{\pi}{4}$ of the domain of A. Finally, if the triangle is obtuse, $\theta_3 > \pi/2$ for example, the solution is the same except that $A(\alpha)$ would only be defined for $\frac{\pi}{2} - \theta_1 \leqslant \alpha \leqslant \frac{\pi}{2}$ and in this case the maximum of $A(\alpha)$ would be attained only at $\pi/2$.

Surely the results of this note are known. However, not a single mathematician to whom I have mentioned them admitted having considered the maximum area problem with any orientation other than the most standard one.

Maximize $x(a - x)$

(A small greeting card for Professor George Pólya on his 86th birthday)*

L. H. LANGE

LESTER H. LANGE is Dean of the School of Science at California State University, San Jose (formerly San Jose State College). Before becoming Dean in 1970 he served as Chairman of the Mathematics Department at San Jose for nine years. He has written numerous journal articles, has lectured widely as a Visiting Lecturer of the Mathematical Association of America, and his book, *Elementary Linear Algebra*, was published in 1968. In 1972 he received a Lester R. Ford, Sr. Award from the MAA for distinguished expository writing in mathematics.

If we seek to find the largest rectangle that can be inscribed in a given triangle (as in [2], for example), we encounter the problem stated in the title, where a is some positive constant. The title problem has shown up at various times in mathematical history and it has been solved in several interestingly different ways which we shall discuss below. As a matter of fact, Moritz Cantor [1] says that the solution of this problem in Euclid [3] is the first maximum ever proved in the history of mathematics.

Before giving (what amounts to) the ancient Greek proof and also a proof due to Pierre de Fermat (1601–1665), we record three other proofs (avoiding the developed tools of calculus).

Proof I. Since

$$x(a - x) = (a/2)^2 - (x - a/2)^2,$$

it is clear that we have a maximum if, and only if, $(x - a/2)^2 = 0$, i.e., if and only if $x = a/2$. The maximum value of $x(a - x)$ is then $(a/2)^2$. Of course, the *geometric* interpretation of what we have just accomplished is the proof of this

OLD THEOREM. *Among all rectangles of equal perimeter it is the square which has the largest area. (The number a is the given semiperimeter.)*

Proof II. For this proof, we need as a helping theorem the simplest instance of the (famous) geometric-arithmetic mean inequality involving the two nonnegative numbers x and y:

$$G = \sqrt{xy} \leq \frac{x + y}{2} = A, \tag{α}$$

where equality occurs if and only if $x = y$.

---

*He was born in Budapest on December 13, 1887.

232

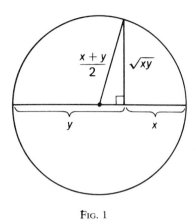

$\dfrac{x+y}{2}$ $\sqrt{xy}$

y x

FIG. 1

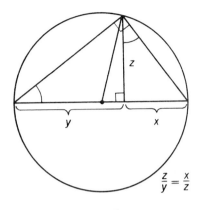

z

y x

$\dfrac{z}{y} = \dfrac{x}{z}$

FIG. 2

A pretty little proof of $G \leq A$ can be read out of FIGURE 1 immediately—where the well-known result in FIGURE 2 has been used, of course. (I wonder when $G \leq A$ first appeared in human history.)

Now, in (α), if A is a *constant*, then the product xy is a *maximum* precisely when $x = y$—and this is the lemma we need in order to maximize $x(a - x)$ in yet another way. We let $(a - x) = y$; notice that $x + y$ is a constant here, and so the product $(x)(a - x)$ will be a maximum if and only if $x = a - x$, i.e., if and only if $x = a/2$ once more.

Proof III. Here is another short way to prove the Old Theorem. The square with perimeter P has sides each equal to $p/4 = m$ and the area of the square is m^2. Any *other* rectangle in this class of rectangles would be such that, if one of its sides is $m + x$ in length, where $0 < x < m$, then an adjacent side necessarily has length $m - x$. The area of this rectangle would then be $(m + x) \times (m - x) = m^2 - x^2 < m^2$, and we've completed the proof once more.

The two proofs that follow are discussed, in varying contexts, in [5] and, to some extent, in [4].

Proof IV. (Ancient Greek version.) This is actually a rendition of the ancient Greek way of establishing Old Theorem, employing their key idea, even though the proof in [3] actually deals with parallelograms rather than rectangles.

The key idea is not unlike the idea used in Proof III: we compare the area of the square with the area of any *other* rectangle having the same perimeter and find that the square wins out. Thus, in FIGURE 3 we have a given rectangle, made up of parts (1) and (2), and a square, made up of parts (2) and (3), of equal perimeter. The proof will then be complete if we show that the area of rectangle (3) is greater than the area of rectangle (1).

Since the appropriate perimeters are equal, we have $2(x + t + s) = 2(t + s + y)$ yielding $x = y$. The area of (1) is $xs \ (= ys)$ and the area of (3) is yt. Then, since $t = s + y$ yields $t > s$, we have $xs = ys < yt$, and the proof is finished.

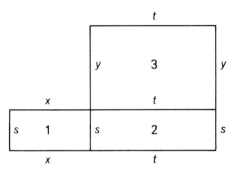

FIG. 3

Proof V. (A rendition of Fermat's version.) If we now consider the collection of all rectangles having the same semiperimeter a, and choose a rectangle which has one side length equal to x, $0 \leq x \leq a$, this rectangle has area $A(x) = (x)(a - x)$, a function of x. Fermat seeks that value of x which maximizes this function $A(x)$, the graph of which is sketched in FIGURE 4.

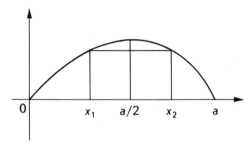

FIG. 4

Since, as x increases from 0 to a, the associated rectangles change continuously from thin tall ones to thin flat ones, we readily see that for each x_1 satisfying $0 < x_1 < a/2$ there exists an x_2, satisfying $a/2 < x_2 < a$, such that $A(x_1) = A(x_2)$. Thus, Fermat writes

$$x_1(a - x_1) = x_2(a - x_2),$$
$$a(x_1 - x_2) = (x_1)^2 - (x_2)^2,$$
$$a = x_1 + x_2,$$

and observes that, hence, as x_1 and x_2 then are chosen closer and closer to each other, we get closer and closer to the situation where $x_1 = x_2 = a/2$, where the highest point of the curve $y = A(x)$ must be located. So, again the square has maximum area.

On page 31 in [4], there is a report of some bold mathematical steps that this sort of thinking led Fermat to take (with respect to finding extrema by looking for repeated roots of certain "expressions," for example).

REFERENCES

1. Mortiz Cantor, *Vorlesungen über Geschichte der Mathematik*, Vol. I, p. 266 of the 1965 (New York) reprint of the 1907 (Teubner) edition.
2. G. D. Chakerian and L. H. Lange, "Geometric Extremum Problems," *Mathematics Magazine*, Vol. 44, No. 2 (March 1971), 57–69.
3. Euclid, *Elements*, Book VI, Proposition 27.
4. Michael S. Mahoney, "Fermat's Mathematics: Proofs and Conjectures," *Science*, vol. 178 (October 1972), 30–36.
5. Otto Toeplitz, *The Calculus: A Genetic Approach* (Chicago: University of Chicago Press), 1963, pp. 80–82.

Construction of an Exercise Involving Minimum Time

ROBERT OWEN ARMSTRONG

ROBERT O. ARMSTRONG has been teaching mathematics at Volunteer State Community College, Gallatin, Tennessee, since 1971. He completed his undergraduate work at Vanderbilt University, received an M.A. in education from Memphis State University, and both his M.A. and D.Ed. in mathematics from the University of Georgia in 1968 and 1971. He has taught junior high school mathematics in Florida, Tennessee and Georgia.

Several investigations have been made into the problem of constructing exercises that require a full understanding of the logic involved in extremum problems [1, 2, 3]. Calculus instructors have a continuing problem of providing good extremum exercises for their students. The technique of exercise construction presented below is intended to produce exercises in which the student must think about the geometry of the problem and the domain of the function. Such exercises require more of the student than the application of algorithms.

Problem: Suppose a man is a feet away from a "straight" shoreline. If he swims straight to the shore, he will reach a point b feet away from his destination on shore. If his swimming velocity is v_1 feet per minute and his running velocity is v_2 feet per minute and $0 < v_1 < v_2$, what is the shortest time in which the man can reach his destination?

The man may be imagined to swim along a line perpendicular to the shoreline a distance of a feet and then run b feet to his destination, or he may swim to a point within the interval of length b feet and then run to his destination, or he may swim straight to his destination. The diagram in FIGURE 1 illustrates his possible routes.

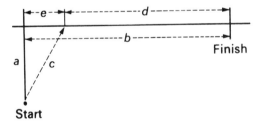

FIG. 1

The following may be called the trigonometric approach. Let α be the acute angle formed by the line segments of lengths a and c feet. The time a given route takes is a function of α where $0 \le \alpha \le \arcsin(b/\sqrt{a^2 + b^2})$. The function is defined by

$$T(\alpha) = (a \sec \alpha)/v_1 + (b - a \tan \alpha)/v_2.$$

Now

$$T'(\alpha) = a(v_1 v_2 \cos^2 \alpha)(v_2 \sin \alpha - v_1) = 0$$

yields the solution $\sin \alpha = v_1/v_2$. If v_1, v_2 are such that $b/\sqrt{a^2 + b^2} \le v_1/v_2$, then

$$\alpha = \arcsin v_1/v_2 > \arcsin\left(b/\sqrt{a^2 + b^2}\right)$$

and α is not in the domain of the function T, or

$$\alpha = \arcsin v_1/v_2 = \arcsin\left(b/\sqrt{a^2 + b^2}\right).$$

So, swimming the straight-line distance of $\sqrt{a^2 + b^2}$ feet is the fastest route. For $T'(0) = -a/v_2 < 0$, and each x_0 in $[0, \arcsin b/\sqrt{a^2 + b^2}]$ is such that $T'(x_0) \ne 0$, since $\arcsin b/\sqrt{a^2 + b^2} < \arcsin v_1/v_2$, and T' is continuous on $[0, \arcsin b/\sqrt{a^2 + b^2}]$. If there were a number x_0 in $[0, \arcsin b/\sqrt{a^2 + b^2}]$ such that $T'(x_0) > 0$, then the Intermediate Value Theorem shows that there exists a number λ such that $0 < \lambda < x_0$ and $T'(\lambda) = 0$, since $T'(0) < 0 < T'(x_0)$.

But then T would have a critical point between 0 and $\arcsin b/\sqrt{a^2 + b^2}$. Hence $T' < 0$ on $[0, \arcsin b/\sqrt{a^2 + b^2}]$. Thus, T is monotone decreasing on $[0, \arcsin b/\sqrt{a^2 + b^2}$. Therefore, if $b/\sqrt{a^2 + b^2} < v_1/v_2$, then $T(\arcsin b/\sqrt{a^2 + b^2})$ is less than $T(x)$ for x in $[0, \arcsin b/\sqrt{a^2 + b^2}]$.

An alternative approach to the problem may be called the Pythagorean approach.

Consider the diagram of the problem in FIGURE 2. If $0 \le v_1/v_2 \le b/\sqrt{a^2 + b^2}$, then $0 < \arcsin v_1/v_2 < \arcsin a/\sqrt{a^2 + b^2}$ and $T'(\arcsin v_1/v_2) = 0$. Now,

$$T''(\alpha) = \frac{a v_1 v_2^2 \cos^2 \alpha + 2 a v_2 \sin^2 \alpha - 2 a v_1 \sin \alpha}{v_1^2 v_2^2 \cos^3 \alpha}.$$

Hence,

$$T''(\arcsin v_1/v_2) = a/v_1 \cos(\arcsin v_1/v_2) > 0$$

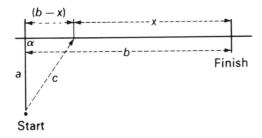

FIG. 2

since a and $v_1 \cos(\arcsin v_1/v_2)$ are positive. Therefore, T has a strict minimum point at $\arcsin v_1/v_2$.

The instructor has the option of choosing a, b such that $b/\sqrt{a^2 + b^2} < v_1/v_2$, which forces the student to think about the geometry of the problem and the fact that T is a monotonically decreasing function on values of the angle α such that $0 < \alpha < 90°$. For example:

(1) $a = 3$, $b = 3$, $v_1 = 4$, $v_2 = 5$.

Answer: $T(\alpha) = T(\arcsin(b/\sqrt{a^2 + b^2}))$

$$= \sqrt{a^2 + b^2}/v_1 = 3\sqrt{2}/4.$$

(2) $a = 5$, $b = 5$, $v_1 = 12$, $v_2 = 13$.

Answer: $T(\alpha) = T(\arcsin(b/\sqrt{a^2 + b^2}))$

$$= \sqrt{a^2 + b^2}/v_1 = 5\sqrt{2}/12.$$

(3) $a = 8$, $b = 6$, $v_1 = 15$, $v_2 = 17$.

Answer: $T(\alpha) = T(\arcsin(b/\sqrt{a^2 + b^2}))$

$$= \sqrt{a^2 + b^2}/v_1 = 10/15$$

$$= 2/3.$$

(4) $a = 7$, $b = 7$, $v_1 = 24$, $v_2 = 25$.

Answer: $T(\alpha) = T(\arcsin(b/\sqrt{a^2 + b^2}))$

$$= \sqrt{a^2 + b^2}/v_1 = 7\sqrt{2}/24.$$

The angle α is clearly a $90°$ angle by the statement of the problem. Hence $c = \sqrt{a^2 + (b - x)^2}$. Time is a function of x in the function f defined by

$$f(x) = \left[\sqrt{a^2 + (b - x)^2}/v_1\right] + x/v_2$$

where $0 \leq x \leq b$. Again v_1, v_2 are the swimming and running velocities, respectively, and $0 < v_1 < v_2$.

Now

$$f'(x) = \left(\frac{1}{v_2} - \frac{1}{v_2}\right)\left(a^2 + (b - x)^2\right)^{-1/2}(b - x).$$

Consider the equation $f'(x) = 0$. Then

$$\frac{v_1}{v_2}\left(a^2 + (b - x)^2\right)^{1/2} = b - x.$$

Every solution to the equation $f'(x) = 0$ is a solution to the equation

$$\left(1 - \frac{v_1^2}{v_2^2}\right)x^2 - 2b\left(1 - \frac{v_1^2}{v_2^2}\right)x + \left(1 + \frac{v_1^2}{v_2^2}\right)b^2 - a^2\left(\frac{v_1^2}{v_2^2}\right) = 0,$$

which results from the usual device of squaring both sides of a radical equation.

Let D be the discriminant of the above quadratic equation:

$$D = 4a^2\left(\frac{v_1^2}{v_2^2}\right)\left(1 - \frac{v_1^2}{v_2^2}\right).$$

Clearly, for $0 < v_1 < v_2$ and $a \neq 0$, $D > 0$. The zeros, r_1, r_2, of the above quadratic equation are

$$r_1 = b + \left(av_1/\sqrt{v_2^2 - v_1^2}\right)$$

and

$$r_2 = b - \left(av_1/\sqrt{v_2^2 - v_1^2}\right).$$

Now $r_1 > b$, so r_1 is not a solution of $f'(x) = 0$. Also, $f'(r_1) > 0$. But $f'(r_2) = 0$.

Also, it is clear that $r_2 < b$. Now if $r_2 < 0$, then there is no critical point in the interval $[0, b]$. But $f'(b) = 1/v_2 > 0$ and the Intermediate Value Theorem insures that $f'(x) > 0$ for x in $[0, b]$. Hence f is monotone increasing on $[0, b]$ and $f(0) = \sqrt{a^2 + b^2}/v_1$ is less than $f(x)$ for x in $[0, b]$.

Also,

$$f''(x) = a^2/v_1\left(a^2 + (b - x)^2\right)^{3/2},$$

and therefore

$$f''(r_2) = f''\left(\frac{b - av_1}{\sqrt{v_2^2 - v_1^2}}\right)$$

$$= \left(\frac{1}{a^2 f_1}\right)\left(1 + \frac{v_1^2}{v_2^2 - v_1^2}\right)^{3/2} > 0,$$

since a, v_1, and v_2 are positive and $v_2 > v_1$. Hence f has a strict minimum at r_2 if $0 < r_2 < b$.

Exercises can be constructed where r_2 is not in $(0, b)$ by choosing a, b, v_1, and v_2 such that $b/a < v_1/\sqrt{v_2^2 - v_1^2}$.

(1) $a = 3$, $b = 3$, $v_1 = 4$, $v_2 = 5$.
   Answer: $f(0) = 3\sqrt{2}/4$.

(2) $a = 5$, $b = 5$, $v_1 = 12$, $v_2 = 13$.
   Answer: $f(0) = 5\sqrt{2}/12$.

(3) $a = 8$, $b = 6$, $v_1 = 15$, $v_2 = 17$.

Answer: $f(0) = 2/3$.

(4) $a = 7$, $b = 7$, $v_1 = 24$, $v_2 = 25$.

Answer: $f(0) = 7\sqrt{2}/24$.

REFERENCES

1. C. S. Ogilvy, "Exceptional Extremum Problems," *American Mathematical Monthly*, Vol. 67 (March 1960), 270–275.
2. Hugh A. Thurston, "So-Called 'Exceptional' Extremum Problems," *American Mathematical Monthly*, Vol. 68 (September 1961), 650–652.
3. J. L. Walsh, "A Regional Treatment of the First Maximum Problem in the Calculus," *American Mathematical Monthly*, Vol. 54 (January 1947), 35–36.

A Bifurcation Problem In First Semester Calculus

W. L. Perry, Texas A & M University, College Station, TX

Although bifurcation phenomena are extremely important in applied mathematics and engineering, it is difficult to present physically motivated examples in courses preceding ordinary differential equations. Our objective is to present an example of bifurcation (i.e., branching of solutions) which is understandable to first semester calculus students. In our problem, a variant of a standard calculus problem, we introduce a real parameter λ and show that the number of solutions changes as λ changes.

Problem: *Given two identical light sources d units apart (Figure 1), find the point on line l where the intensity of illumination is minimum.*

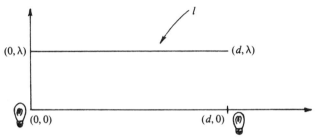

Figure 1.

Most students guess on the basis of symmetry that $x = d/2$ gives minimum illumination for any λ. This guess is incorrect; the (perhaps) surprising answer follows. We begin with the fact that, for a single source, the intensity of illumination at a point is directly proportional to the strength S of the source and inversely proportional to the square of the distance from the source. Therefore, the total intensity of illumination at (x, λ) is

$$I(x) = \frac{kS}{(x^2 + \lambda^2)} + \frac{kS}{(d - x)^2 + \lambda^2}.$$

Note that $I(x)$ is symmetric about $x = d/2$. It is straightforward to verify that

$$I'(x) = -2kS\left[\frac{x}{(x^2 + \lambda^2)^2} - \frac{(d - x)}{((d - x)^2 + \lambda^2)^2}\right]$$

and

$$I''(x) = -2kS\left[\frac{\lambda^2 - 3x^2}{(x^2 + \lambda^2)^3} + \frac{\lambda^2 - 3(d - x)^2}{((d - x)^2 + \lambda^2)^3}\right].$$

241

The problem of solving $I'(x) = 0$ is equivalent to solving

$$x\big((d-x)^2 + \lambda^2\big)^2 - (d-x)(x^2 + \lambda^2)^2 = 0.$$

This may be written as $p(x) = 0$, where $p(x)$ is a fifth degree polynomial. Since $x = d/2$ is a root, we may divide $p(x)$ by $x - (d/2)$ to find that the other roots of $p(x)$ are the roots of

$$q(x) = 2x^4 - 4dx^3 + (4d^2 + 4\lambda^2)x^2 + (-2d^3 - 4d\lambda^2)x + 2\lambda^4.$$

Since $q''(0) > 0$ and $q''(x)$ has no zeros (the discriminant is negative), $q(x)$ is always concave up. Thus, $q(x)$ has at most two real roots (and so $p(x)$ has at most three critical numbers, one of which is $d/2$ and the other two symmetric about $x = d/2$).

Now we also have by direct calculation that $I''(0) < 0 < I'(0)$. Moreover,

$$I''(d/2)\begin{cases} > 0, & \text{if } \lambda < \dfrac{\sqrt{3}\,d}{2} \\[2mm] = 0, & \text{if } \lambda = \dfrac{\sqrt{3}\,d}{2} \\[2mm] < 0, & \text{if } \lambda > \dfrac{\sqrt{3}\,d}{2} \end{cases}$$

and

$$I(0)\begin{cases} > I(d/2), & \text{if } 0 < \lambda < d/\sqrt{2} \\[1mm] = I(d/2), & \text{if } \lambda = d/\sqrt{2} \\[1mm] < I(d/2), & \text{if } \lambda > d/\sqrt{2}. \end{cases}$$

Knowing all this, and using the symmetry of $I(x)$ about $x = d/2$, we can determine the configuration of $I(x)$ on $[0, d]$. Let c_1 and c_2 denote the possible critical values (i.e., solutions of $I'(x) = 0$). Then $I(x)$ has one of the configurations shown in Figure 2.

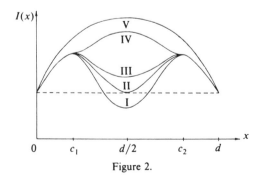

Figure 2.

If $0 \leqslant \lambda < \sqrt{2}\, d/2$, then $I(d/2) < I(0)$ and $I''(d/2) > 0$. Therefore, $I(x)$ is configuration I and $I_{min}(x)$ occurs at $x = d/2$.

If $\lambda = \sqrt{2}\, d/2$, then $I(d/2) = I(0)$ and $I''(d/2) > 0$. Accordingly, $I(x)$ is configuration II and $I_{min}(x)$ occurs at $x = 0, d/2, d$.

If $\lambda > \sqrt{2}\, d/2$, then $I(d/2) > I(0)$. Clearly, $I_{min}(x)$ occurs at $x = 0, d$.

This example frequently generates enough interest among students to lead them to laboratory experiments to test the mathematical results—a healthy exercise in the interplay between mathematics and physics.

To Build a Better Box

Kay Dundas

Kay Dundas is a member of the mathematics department at Hutchinson Community College in Hutchinson, Kansas. He earned B.A. and M.S. degrees at Fort Hays Kansas State University and an Ed.D. with emphasis in mathematics at Oklahoma State University. Besides teaching for seventeen years in his present position, he taught for one year each at Fort Hays Kansas State University, Hays, Kansas and Muskogee High School, Muskogee, Oklahoma.

Most calculus students have encountered the problem of finding the maximum volume of a box that is constructed from a rectangular piece of cardboard by cutting equal squares from each corner and folding up the sides. Have you ever asked your students to actually construct such a box? I have. The students soon discover that the most practical part of this "application of calculus" is the fact that it opens the door to more practical methods of construction. To begin with, removing the corners is ridiculous. If you just cut along one side of each square and use the squares to reinforce the sides, the result is a much stronger box. Another thing they notice, with a little gentle persuasion, is that a box without a top is not very useful. This observation gives me a chance to suggest the construction method shown in Figure 1.

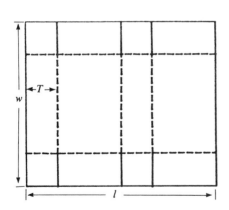
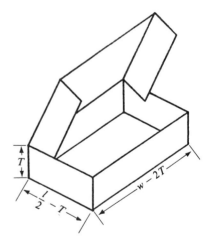

Figure 1.

If you cut along the solid lines and fold along the dotted lines, four well-placed staples will secure a fairly useable box. My students have dubbed this one the *Pizza Box*.

With the no-top construction, cutting out T by T squares from the corners of a rectangle with length l and width w ($w \leq l$), the volume is given by

$$V(T) = T(l - 2T)(w - 2T)$$

for $T < w/2$. With the Pizza Box construction, the volume is given by

$$V(T) = T(l/2 - T)(w - 2T) = T(l - 2T)(w - 2T)/2 \qquad \text{for} \quad T < w/2.$$

Therefore, for any value of T, the volume is half as large using the Pizza Box method, and the maximum occurs at the same value of T in each case.

With a little more prodding, some students will come to the conclusion that restricting the shape of the rectangular piece of cardboard limits the maximum volume of the box. They also can see that this is not a reasonable "real world" restriction. To allow variable dimensions for the rectangle and variable corner sizes would usually require the calculus of several variables. Since this is not available to students when I want to cover this topic, I suggest the following approach.

Suppose A square inches of cardboard is used to construct a box using the Pizza method. Fixing the height at T inches, find the dimensions of the rectangle that will maximize the box's volume.

Taking $w = A/l$ in Figure 1, we have

$$V(l) = T\left(\tfrac{l}{2} - T\right)\left(\tfrac{A}{l} - 2T\right).$$

Then $V'(l) = 0$ when $l = \sqrt{A}$. The cardboard's required dimensions are therefore $l = w = \sqrt{A}$. Using the $\sqrt{A}$ by $\sqrt{A}$ cardboard, we want to find the height T that will maximize this volume. Thus, we begin with

$$V(T) = T\left(\tfrac{1}{2}\sqrt{A} - T\right)\left(\sqrt{A} - 2T\right).$$

Then $V'(T) = 0$ for $T = \sqrt{A}/6$ and the maximum volume of the Pizza Box is $V_{\text{pizza}} = A^{3/2}/27$.

For classroom development, use $A = 144$ square inches because it gives a nice maximum volume of 64 cubic inches when $l = w = 12$ and $T = 2$.

By the time we have solved the Pizza Box problem, some of the students will usually have discovered another commonly used construction method. This method, dubbed the *Popcorn Box*, is shown in Figure 2.

When students first look at this method, they usually choose a box with a square horizontal cross section and the 12 by 12 piece of cardboard that worked for the Pizza Box. Without calculus, they discover that this produces 81 cubic inches of volume—quite an improvement over the previous maximum of 64 cubic inches.

Next, they usually try one of two methods: either they keep the 12 by 12 piece of cardboard and allow the width to vary, or they keep the square base on the box and allow the dimensions of the 144 square inch cardboard to vary. Surprisingly, both methods produce the same maximum volume, $48\sqrt{3} \approx 83.14$. Is this true in general?

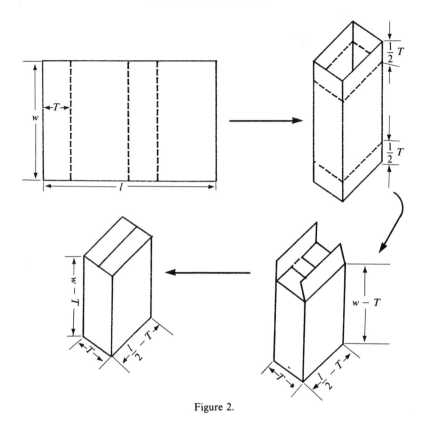

Figure 2.

To check this out for the general case, assume that the cardboard has area A square inches. For $l = w = \sqrt{A}$, we have

$$V(T) = T(\sqrt{A} - T)(\tfrac{1}{2}\sqrt{A} - T)$$

and the maximum volume occurs for $T = (3 - \sqrt{3})\sqrt{A}/6$. For the second alternative, let $T = l/4$ (the box has a square base) giving

$$V(l) = (l/4)^2(\tfrac{A}{l} - \tfrac{l}{4})$$

and maximum volume when $l = 2\sqrt{A/3} = 4w/3$. In both cases, the maximum volume is $V = \sqrt{3}\,A^{3/2}/36$.

The volume of 83.14 isn't much better than the volume of 81 that was obtained without using any calculus. However, when both the shape of the original piece of cardboard and the width T of the box are allowed to vary, the improvement is more dramatic.

To see this, assume again that the area of the cardboard is A square inches. As with the Pizza Box, first fix the box's width at T and let the length of the cardboard vary. This gives

$$V(l) = T(\tfrac{l}{2} - T)(\tfrac{A}{l} - T),$$

and $V'(l) = 0$ when $l = \sqrt{2A}$. Using this $\sqrt{2A}$ by $\sqrt{2A}/2$ cardboard (recall that the area was fixed at A square inches), allow the box width T to vary. Under these conditions,

$$V(T) = T\left(\frac{\sqrt{2A}}{2} - T\right)\left(\frac{A}{\sqrt{2A}} - T\right) = T\left(\tfrac{1}{2}\sqrt{2A} - T\right)^2$$

and $V'(T) = 0$ when $T = \sqrt{2A}/6$. Thus, the Popcorn Box has maximum volume $V_{popcorn} = \sqrt{2}\, A^{3/2}/27$. It is now clear that $V_{popcorn}$ is approximately 41% larger than V_{pizza}.

After spending a class period and a daily assignment on box problems, I like to include a box problem on the next unit test. Usually I give them a specific l by w rectangle, tell them which method of construction to use, and ask them to find the maximum volume. As a test question, I prefer integers for l and w, and rational values for the optimal box dimensions. The following developments show how to choose l and w to accomplish this for the Pizza Box and then for the Popcorn Box.

From Figure 1, we have

$$V(T) = T(l/2 - T)(w - 2T) = (lwT/2) - (l + w)T^2 + 2T^3.$$

Therefore, $V'(T) = 0$ when

$$T = \left(l + w \pm \sqrt{l^2 - lw + w^2}\,\right)/6.$$

The correct T value will be rational when $l^2 - lw + w^2$ is a perfect square. Choosing correct l and w values to accomplish this result is an interesting problem whose solution has been published by the author in an earlier paper "Quasi-Pythagorean Triples for an Oblique Triangle," the TYCMJ 8 (1977), 152–155. The problem is related to the "ambiguous case" triangle pictured in Figure 3.

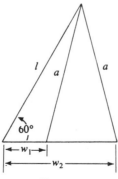

Figure 3.

The cosine law gives $a^2 = l^2 - lw + w^2$ for $w = w_1$ or $w = w_2$. Direct substitution shows that $a = m^2 + mn + n^2$ when

$$l = 2mn + m^2, \quad w_1 = m^2 - n^2 \quad \text{and} \quad w_2 = 2mn + n^2,$$

where $m > n$. It is more difficult to show that all solutions are generated by multiples of these when m and n are relatively prime and do not differ by a multiple of three. The net result is that for $m > n$, the pairs

$$(l, w) = (2mn + m^2, 2mn + n^2) \quad \text{and} \quad (l, w) = (2mn + n^2, m^2 - n^2)$$

generate all the Pizza Box problems one needs.

For the Popcorn Box, referring to Figure 2, we have

$$V(T) = T(l/2 - T)(w - T) = (lwT/2) - (l/2 + w)T^2 + T^3.$$

Thus, $V'(T) = 0$ when

$$T = \left(l + 2w \pm \sqrt{l^2 - 2lw + 4w^2} \right)/6.$$

In this case, $l^2 - l(2w) + (2w)^2$ needs to be a perfect square. This is the same problem as above with w replaced by $2w$. It follows that $(l, 2w) = (2mn + m^2, 2mn + n^2)$ and $(l, 2w) = (2mn + m^2, m^2 - n^2)$ generate the desired dimensions.

$$V(l) = 6(1/2)(l/6)\left(l\sqrt{3}/12\right)\left(144/l - l\sqrt{3}/6\right).$$

Then $V'(l) = 0$ when $l = 4\sqrt[4]{108}$, and the maximum volume is $48\sqrt[4]{12} \approx 89.34$.

This article would have ended here if I had not recently purchased a three way light bulb. It was packaged in an interesting box whose construction is indicated in Figure 4.

The horizontal cross-section of this box is hexagonal, and the ends are folded over just enough to reach the center. This provided a new direction to go in search of a better box.

I assigned an extra credit problem to my class to find the maximum volume using this construction method and 144 square inches of cardboard. Nobody solved the problem, but I'll try again next semester. In my solution, the volume is computed by multiplying the area of six equilateral triangles by the height. This gives the formula

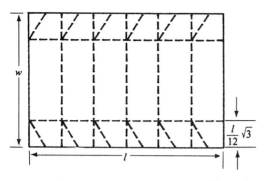 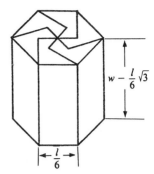

Figure 4.

Many questions could be asked at this point. If we retain the A square inch rectangular piece of cardboard, what is the maximum volume possible using the hexagonal cross section? If more sides are used, will the maximum volume increase? Is some number of sides optimum, or does some smooth curve eventually produce the "best" box?

These questions can be answered under the following restrictions: The polygonal cross-section must be equiangular and have $2n$ sides for some natural number $n > 1$; each of $2n - 2$ sides have length y and the remaining 2 sides have length $(l/2) - (n - 1)y$, where l is the length of the original rectangle. Thus, in the cross-section (see Figure 5), each of the $2n - 2$ isosceles triangles has its vertex angle equal to π/n and its altitude of length $(y/2)\cot(\pi/2n)$.

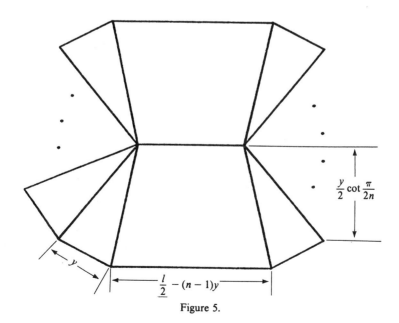

Figure 5.

For the case where $A = 144$, the box has volume

$$V = (y/2)(\cot(\pi/2n))(l - ny)(144/l - y\cot(\pi/2n)).$$

To see this, observe first that the last factor is the height of the box. The cross sectional area is seen when you place the two trapezoids and the $2n - 2$ triangles side by side, with half of the triangles and one trapezoid in the inverted position. This gives a parallelogram with length $l - ny$ and altitude $(y/2)\cot(\pi/2n)$. Taking partial derivatives with respect to l and y, we find that maximum volume

$$V = 128\sqrt{\cot(\pi/2n)} / \sqrt{n}$$

occurs when $l = 12\sqrt{n}\sqrt{\tan(\pi/2n)}$ and $y = 4\sqrt{\cot(\pi/2n)} / \sqrt{n}$.

The hexagonal cross section ($n = 3$) produces a maximum volume of $128\sqrt[4]{27} \approx 97.26$ cubic inches, while an octagonal cross section ($n = 4$) produces a maximum volume of $64\sqrt{\sqrt{2} + 1} \approx 99.44$ cubic inches.

The preceding remarks show the maximum volume V is a function of n. Since $V'(n) > 0$, we see that V is an increasing function. Rewriting $V = 128\sqrt{\cot(\pi/2n)} / \sqrt{n}$ as

$$V = 128\sqrt{\frac{\pi/2n}{\sin(\pi/2n)} \cdot \frac{2\cos(\pi/2n)}{\pi}},$$

we see that $\lim_{n\to\infty} V = 128\sqrt{2/\pi} \approx 102.13$. If we could construct such a box with infinitely many sides, it would have a cross-section in the form of a rectangle with a semicircle on each end. The radius of the semicircles would be $2\sqrt{2/\pi}$, the dimensions of the rectangle would be $4\sqrt{2/\pi}$ and $2\sqrt{\pi/2}$, and the height would be $8\sqrt{2/\pi}$.

A Surprising Max-Min Result

Herbert Bailey, Rose-Hulman Institute of Technology, Terre Haute, IN and University of North Carolina, Chapel Hill, NC

A challenging problem that can be assigned early in the calculus sequence is to *minimize* the area of the triangle formed by the positive coordinate axes and a line tangent to a given curve. For example (Figure 1), consider the curve $y = 1 - x^2$. Then the area of triangle OAB will be a minimum when the point of tangency P has coordinates $(1/\sqrt{3}, 2/3)$.

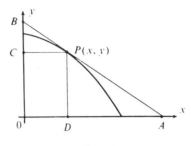

Figure 1.

My students often attempt to solve this problem by letting $x = OA$ and $y = OB$, and considering $A_T = xy/2$ (notwithstanding that (x, y) must be on the curve). They are, in fact, finding the point P that *maximizes* the area of triangle ODP (and rectangle $ODPC$). Unexpectedly, the point that maximizes the rectangular area turns out to be $(1/\sqrt{3}, 2/3)$, the same point that minimizes the triangular area OAB, and so students think they were right after all!

In my search to find a curve for which this difficulty does not occur, I discovered that there are no such curves: every curve which is falling and concave downward in the first quadrant has this max-min property. Before proving and then generalizing this, we first establish the following:

Result 1 (Figure 2). *The area of the triangle OAB formed by the positive coordinate axes and a line through a fixed point P_0 is minimized when P_0 is the midpoint of line segment AB. For any fixed positive intercepts A and B, the rectangle of maximum area that can be inscribed in triangle OAB is $OD_0P_0C_0$, where P_0 is again the midpoint of segment AB.*

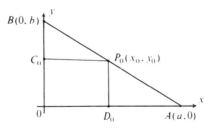

Figure 2.

251

Triangle OAB has area $A_T = ab/2$. Since the hypotenuse AB passes through a fixed point $P_0(x_0, y_0)$, the intercepts a and b satisfy $A_T = ay_0/2[1 - (x_0/a)]$. Setting $dA_T/da = 0$, we find that A_T is minimized for $a = 2x_0$ and $b = 2y_0$. Thus, P_0 is the midpoint of AB. To show that the rectangle $OD_0P_0C_0$ has maximum area, we must maximize $A_R = xy$ with x, y constrained to lie on the fixed line $(x/a) + (y/b) = 1$. Thus, $A_R = xb[1 - (x/a)]$. Setting $dA_R/dx = 0$, we find that A_R is maximized at $x = x_0 = a/2$ and $y = y_0 = b/2$.

Now, let's consider the asserted max-min property for falling, concave downward curves.

Result 2 (Figures 3 and 4). *The point of tangency of the line bounding the first-quadrant curve $y = f(x)$ $(f', f'' < 0$ for $x > 0)$ so as to form a triangle OAB of minimum area with the positive coordinate axes must be the midpoint of segment AB. This same midpoint $P_0(x_0, y_0)$ maximizes the area of the rectangle inscribed in the region bounded by the curve and the positive coordinate axes.*

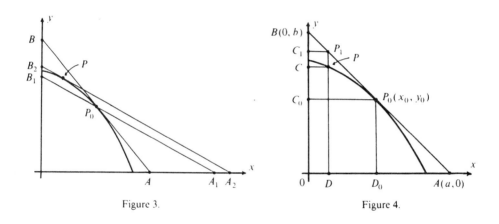

Figure 3.   Figure 4.

That the triangle OAB formed by the tangent line at P_0 has minimum area follows (Figure 3) from

$$OAB < OA_1B_1 < OA_2B_2,$$

where the line A_2B_2 is tangent to the curve at any point $P(x, y)$, and A_1B_1 through $P_0(x_0, y_0)$ is drawn parallel to A_2B_2. The left inequality follows from Result 1, whereas the right inequality follows from the concavity of the curve $y = f(x)$. That the rectangle $OD_0P_0C_0$ has maximum area follows (Figure 4) from

$$OD_0P_0C_0 > ODP_1C_1 > ODPC.$$

The inequality on the right holds because of the downward concavity of the curve $y = f(x)$, which insures that the line segment DP_1 is longer than segment DP. The left inequality follows from Result 1.

Results 1 and 2 are extended to three dimensions by considering volumes of inscribed parallelepipeds (boxes) and circumscribing tetrahedra for a region in the first octant bounded by the coordinate planes and a surface $z = F(x, y)$. We shall assume that the surface is falling ($F_x < 0$ and $F_y < 0$) and concave down.

Result 3 (Figure 5).   *The volume of the tetrahedron formed by the positive coordinate planes and a plane through a fixed point P_0 is minimized when P_0 has coordinates $(a/3, b/3, c/3)$, where a, b, c are the intercepts of the plane with the respective coordinate axes. For any fixed positive values a, b, and c, the box of maximum volume that can be inscribed in tetrahedron $OABC$ will have $P_0(a/3, b/3, c/3)$ as its vertex on the face ABC.*

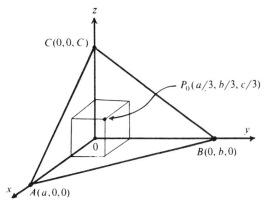

Figure 5.

For fixed $P_0(x_0, y_0, z_0)$, the tetrahedron $OABC$ will have volume $V = (\frac{1}{6})abc$, and this must be minimized subject to the constraint $(x_0/a) + (y_0/b) + (z_0/c) = 1$. Doing this (say, by the method of Lagrange multipliers), we obtain $a = 3x_0$, $b = 3y_0$, and $c = 3z_0$. To show that the inscribed box has maximum volume, we must maximize $V = xyz$ subject to the constraint $(x/a) + (y/b) + (z/c) = 1$. This gives $x = a/3$, $y = b/3$, $z = c/3$, which is precisely P_0.

Result 4 (Figures 6 and 7).   *The point of tangency of the plane bounding the surface $z = F(x, y)$ in the first octant (concave downward, $F_x < 0$ and $F_y < 0$) so as to form a tetrahedron of minimum volume with the positive coordinate planes must be $P_0(a/3, b/3, c/3)$, where a, b, c are the intercepts of the tangent plane with the respective coordinate axes. This same point P_0 maximizes the volume of the box inscribed in the region bounded by the surface $z = F(x, y)$ and the coordinate planes.*

The proof depends on Result 3 and the convexity of the region bounded by the coordinate planes and the surface $z = F(x, y)$. This region is shown in Figures 6 and 7, where P_0 is the point with coordinates $a/3, b/3, c/3$; $P(x, y, z)$ is any point on

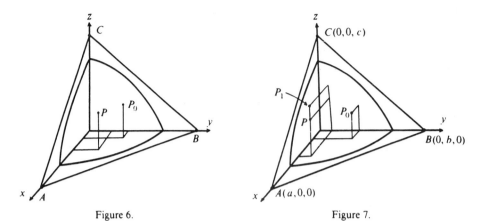

Figure 6.   Figure 7.

the surface $z = F(x, y)$; and $P_1(x, y, z_T)$ is the corresponding point on the triangle ABC.

The tetrahedron containing P_0 has minimum volume since (Figure 6)

$$T_{P_0} < \overline{T}_{P_0} < T_P,$$

where T_{P_0} is the volume of the tetrahedron formed by the tangent plane at $P_0(a/3, b/3, c/3)$, T_p is the volume of the tetrahedron formed by the tangent plane at $P(x, y, z)$ and $\overline{T}_{P_0}$ is the volume of the tetrahedron formed by the plane through P_0 and parallel to the tangent plane at P. The right inequality follows from convexity of $z = F(x, y)$, and the left inequality follows from Result 3. The box at P_0 has maximum volume since (Figure 7)

$$B_{P_0} > B_{P_1} > B_P,$$

where B_{P_i} denotes the volume of the box with diagonal OP_i. The right inequality follows from convexity of the surface $z = F(x, y)$ (convexity insures $z_T > z$), and the left inequality follows from Result 3.

The preceding discussion can now be further enriched by calling students' attention to the following:

> *Let K be a convex region and let $n \geq 3$ be a given integer. If P is a convex n-gon of minimal area containing K, then the midpoints of the sides of P lie on the boundary of K.*

For a geometric proof of this, see "Geometric Extremum Problems" by G. D. Chakerian and L. H. Lange [Math Mag. 44 (1971) 57–69]. In "Polygons Circumscribed About Closed Curves" [Trans. Amer. Math. Soc. 62 (1947) 315–319], M. M. Day shows the above result for two dimensions also holds in three dimensions, with "midpoints of the sides of P" replaced by "centroids of the planes of P."

Acknowledgment. The author would like to thank Henry L. Alder for the above references and for his substantial contributions toward the final version of this capsule.

Hanging a Bird Feeder: Food for Thought

John W. Dawson, Jr., Penn State, York, PA

Calculus instructors who have grown weary of the usual maximum/minimum problems may find the following example of interest. Its context is one familiar to students, yet unlike most geometric optimization problems they will have encountered, the optimal configuration depends in an unexpected way on the numerical values chosen for the parameters. It is a thinly veiled variant of Steiner's problem, a classic problem in geometry which has been unduly neglected by authors of calculus texts.

The Problem.   In the autumn, many people put up feeders for wild birds—and thereby initiate the annual round of "squirrel wars." Seasoned veterans of the combat have learned to thwart the acrobatic rodents by suspending the feeders from wires, which raises the question: What configuration will minimize the length of wire needed?

The wire is strung between two trees a distance D apart and is attached to each of them at a common height above the ground—high enough so a person can walk under the wire near the trees. The feeder is suspended midway between the trees but must be a distance d below the height at which the wires are attached to the trees so a person can reach the feeder easily. There are three configurations to consider, whose shapes resemble the letters T, V, and Y. The first two are special cases of the third. Indeed, if we take the length h of the "tail" of the Y as the independent variable, then $0 \le h \le d$ and the V and T configurations correspond to the endpoints of this interval. Letting L_C denote the length of wire required for configuration C, we have $L_T = D + d$, $L_V = (D^2 + 4d^2)^{1/2}$, and a straightforward calculation shows that $L_Y = h + (D^2 + 4(d - h)^2)^{1/2}$ has but one critical value, namely $h = d - D/2\sqrt{3}$. In order for this quantity to be positive, we must have $D < 2\sqrt{3}\,d$; if so, then we find that $(L_Y)_{\min} = (\sqrt{3}/2)D + d$. Hence $(L_Y)_{\min} < L_T$, and a bit more calculation shows that $(L_Y)_{\min} < L_V$. So, if $D < 2\sqrt{3}\,d$, the Y configuration is best. On the other hand, $L_V < L_T$ whenever $D > \frac{3}{2}d$, so if $D > 2\sqrt{3}\,d$, the V configuration is best.

It is instructive to have students draw the optimal configurations to scale for several different values of the parameters. Better yet, to obtain a physical solution to the minimization problem, wedge three thin pegs between two transparent plates and then dip the apparatus into soapy water. The film makes a configuration that minimizes the total distance connecting the three pegs [Richard Courant and Herbert Robbins, *What is Mathematics?*, Oxford University Press, New York, 1941, p. 392]. By projecting the image of the soap film onto a screen with the aid of an overhead projector, students may then notice, as Steiner did, that the angles between the pegs in the Y configuration are equal. Having made that observation, the students can verify it by computing

$$\frac{D/2}{d - h} = \sqrt{3} = \tan 60°.$$

For an overview of Steiner's problem, see the article by H. W. Kuhn in G. B. Dantzig and B. C. Eaves, *Studies in Optimization*, MAA, Washington, D.C., 1974.

"I don't understand it. Rodney feels he can't use the feeder since he flunked calculus!"

PEAKS, RIDGE, PASSES, VALLEY AND PITS
A Slide Study of $f(x, y) = Ax^2 + By^2$

Cliff Long

The advent of computer graphics is making it possible to pay heed to the suggestion, "a picture is worth a thousand words." As students and teachers of mathematics we should become more aware of the variational approach to certain mathematical concepts, and consider presenting these concepts through a sequence of computer generated pictures.

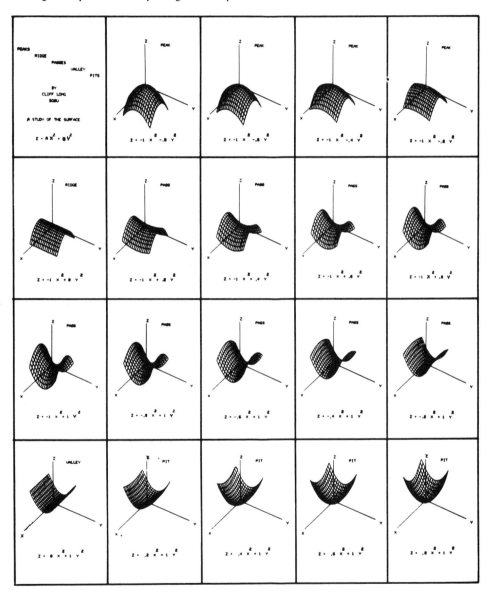

To illustrate this notion, consider the following. In the study of functions of two variables it is usually emphasized that a regular non-planar point on a smooth surface can be classified as one of three distinct types: elliptic, parabolic, hyperbolic [1]. These may be illustrated using the functions $f(x, y) = Ax^2 + By^2$ with the origin being:

(a) elliptic if $A \cdot B > 0$;

(b) parabolic if $A \cdot B = 0$, $A \neq B$ (planar if $A = B = 0$);

(c) hyperbolic if $A \cdot B < 0$.

The slides reproduced here (see page 371) were made at Bowling Green State University from the screen of an Owens-Illinois plasma panel which is an output device for a Data General Nova 800 mini-computer. Many slide sets and super eight movies have been produced by college mathematics teachers under an NSF grant for "Computer Graphics for Learning Mathematics." The institute was held at Carleton College in Northfield, Minnesota, 55057, during the summers of 1973 and 1974. It was under the direction of Dr. Roger B. Kirchner, who, with no small amount of personal effort, has made these slides and movies available at minimum reproduction cost.

It must of course be kept in mind that while one good picture may be worth a thousand words, a thousand poorly chosen ones may be worthless.

Reference

1. D. Hilbert and S. Cohn-Vossen, Geometry and the Imagination, Chelsea, New York, 1952.

DEPARTMENT OF MATHEMATICS, BOWLING GREEN STATE UNIVERSITY, BOWLING GREEN, OH 43403.

A Surface With One Local Minimum

J. Marshall Ash
DePaul University
Chicago, IL 60614

Harlan Sexton
NOSC Code 632B
San Diego, CA 92152

Consider the following statement.

STATEMENT F. *A smooth surface* $(x, y, f(x, y))$ *with one critical point which is a local, but not a global, minimum must have a second critical point.*

Here smooth may be taken to mean that f is infinitely differentiable. A point (a, b) is a critical point if

$$\frac{\partial f}{\partial x}(a, b) = \frac{\partial f}{\partial y}(a, b) = 0.$$

When asked what we thought of this statement, our first reaction was that it ought to be true since a one-dimensional analogue is. After some thought we came up with an "almost rigorous" argument that supported statement F. Somewhat later, assisted by a geometric idea of P. Ash of St. Joseph's University (the first author's brother), we found a counterexample to statement F. Finally, we applied an important principal of mathematical research which A. Zygmund of the University of Chicago has frequently expounded in his seminar: Never be stopped by a counterexample; instead find out what is really happening.

Guided by this maxim, we were able to add a small hypothesis (suggested by William Browder of Princeton University) which did force the conclusion of statement F to follow. More explicitly, we have

THEOREM T. *Let* $f: R^2 \to R$ *be continuously differentiable and have a local, nonglobal minimum. If, further, f is proper* ($f^{-1}(K)$ *is compact whenever K is a compact subset of the range*) *then f must have at least one additional critical point.*

The theorem appears to be unpublished folklore, and we will supply a proof after first presenting some thoughts about statement F, then a counterexample.

In considering statement F, an obvious question to ask is: What happens in one dimension? Let $f: R \to R$ be smooth and have a local minimum at 0, for example, $f(0) = 0$ and $f(x) > 0$ for all $|x| < \delta$. If 0 is not a global minimum then we must have $f(a) < 0$ for some a, say $a > 0$. But then $f(\delta/2) > 0$, $f(a) < 0$ and the intermediate value theorem gives $f(b) = 0$ for some b in $(\delta/2, a)$. But $f(0) = f(b) = 0$ so Rolle's Theorem implies the existence of $c \in (0, b)$ with $f'(c) = 0$. Thus f has a second critical point at c.

Encouraged by this evidence for statement F, let us perform a thought experiment. Suppose we pour water onto the surface from a spout located directly above the critical point. The water will steadily rise. Evidently it must overflow sooner or later if the local minimum is not absolute. The point at which the overflow first occurs must be a second critical point. This argument has strong intuitive appeal, but the idea that the "bowl" around the local minimum has finite volume is implicitly incorporated into the assumption of eventual overflow.

259

The following counterexample shows that statement F is false; although discovered independently, it is somewhat similar to one given by David Smith [1, p. 750]. Define

$$f(x,y) = \frac{-1}{1+x^2} + (2y^2 - y^4)\left(e^x + \frac{1}{1+x^2}\right). \tag{1}$$

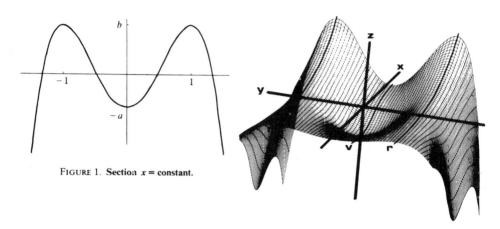

FIGURE 1. Section $x = $ constant.

FIGURE 2

The function f is as differentiable as you like (in fact, it is real-analytic). The point $(0,0)$ is a local, not global minimum, and there are no other critical points.

To find any critical point(s) of f, first consider sections of the form $x = $ constant. Then $f = f(y) = -a + (2y^2 - y^4)(b + a)$ has critical points at $y = 0, 1$, and -1. Now consider sections of the form $y = $ constant. As FIGURE 1 shows, we need only look at the sections $y = 1$, $y = 0$, and $y = -1$, since nowhere else is $\partial f/\partial y = 0$. On the sections $y = -1$, $y = 1$, $f = e^x$ so $\partial f/\partial x = e^x$ is always positive. On the section $y = 0$, $f(x) = -1/(1+x^2)$, so $\partial f/\partial x = 2x/(1+x^2)^2 = 0$ only at $x = 0$. Thus $(0,0)$ is the only critical point.

Since $f(0,0) = -1 > -17 = f(0,2)$, the point $(0,0)$ is not a global minimum. It only remains to show that $(0,0)$ is indeed a local minimum. We have

$$f(x,y) = \left[\frac{-1}{1+x^2}\right] + (2y^2 - y^4)\left(e^x + \frac{1}{1+x^2}\right)$$

$$= \left[-1 + \frac{x^2}{1+x^2}\right] + y^2(2 - y^2)\left(e^x + \frac{1}{1+x^2}\right)$$

$$= f(0,0) + \left\{\frac{x^2}{1+x^2} + y^2(2 - y^2)\left(e^x + \frac{1}{1+x^2}\right)\right\}. \tag{2}$$

If (x, y) is in the unit disc about $(0,0)$, then the quantity in curly brackets in (2) is positive except when $x = y = 0$. This shows $(0,0)$ to be a local minimum.

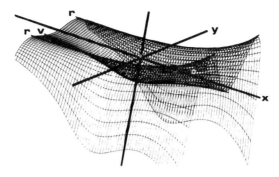

FIGURE 3

We restate the above proof geometrically, using FIGURES 2 and 3. (The 3 coordinate axes are separately scaled here to bring out the salient features.) Each section "parallel" to the y-axis has the shape shown in FIGURE 1 with one dimensional critical points occurring on the ridges labeled r and on the valley bottom labeled v. Since the ridges grow like e^x the tangent plane can be horizontal only at some point of v. The only such point of v is the place where the z axis pierces v, which is the local minimum.

Proof of Theorem T

Assume f satisfies the hypotheses of the theorem. Let f have its local minimum at $A \in R^2$ and let $B \in R^2$ be such that $f(B) < f(A)$. Pour water onto the surface determined by f, above the point A. Then either (i) the water level will rise to arbitrarily great heights, or (ii) the water level will asymptotically approach a finite height h (as actually happens for f defined by (1)), or (iii) the water will overflow. We will proceed to eliminate cases (i) and (ii), in which case our earlier argument will become the proof of the theorem.

In R^2 let $\overline{AB}$ be the line segment joining A and B. Then the continuous function f attains a maximum, say m, on the compact set $\overline{AB}$. The water cannot rise to a height greater than m without spilling so that case (i) is impossible.

If case (ii) were to occur, then the water would be held by a reservoir which would lie over a subset S of the compact set $f^{-1}([f(A), h])$ and whose depth would be everywhere less than $h - f(A)$. Such a reservoir would have finite volume (less than the product of the measure of S with $h - f(A)$). Since we may pour as much water as we like, case (ii) is impossible.

References

[1]   Philip Gillett, Calculus and Analytic Geometry, 2nd ed., D. C. Heath, Lexington, Mass., 1984.

"The Only Critical Point in Town" Test

IRA ROSENHOLTZ
LOWELL SMYLIE
University of Wyoming
Laramie, WY 82071

When searching for absolute extrema of functions of a single variable, it is often convenient to apply the well-known "Only Critical Point in Town" Test: *If f is a continuous function on an interval, which has a local extremum at x_0, and x_0 is the only critical point of f, then f attains an absolute extremum at x_0.* A natural question which arises is "Is the corresponding statement true for functions of two variables (say defined over the entire plane)?" Since our colleagues were evenly split on the question (both halves being quite adamant), and neither a proof nor a counterexample was readily available, we set to work trying to find one.

Progress came slowly at first. Then one bright Monday morning we exchanged pictures of what we thought the level curves of a counterexample might look like. And believe it or not, we both had the same picture!—right down to the location of the mountain, the river bed, and the cliff! It looked something like FIGURE 1. Most of our colleagues were convinced by our picture, but a few remained rightfully skeptical: They wanted a formula. We did too.

Then we noticed something—our level curve picture seemed to have a "saddle point at infinity." So we tried to mimic this.

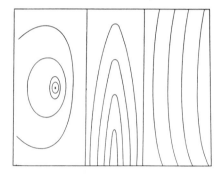

FIGURE 1

Starting with the surface

$$g(x, y) = 3xy - x^3 - y^3,$$

which has a local maximum at $(1,1)$ and a saddle point at $(0,0)$, to "push the saddle point to infinity," we simply took the function

$$f(x, y) = g(x, e^y) = 3xe^y - x^3 - e^{3y}.$$

With this function it is easily seen that the point $(1,0)$ is the only critical point in the plane and that f attains a local maximum there. But it is clearly not an absolute maximum since $f(x,0) \to \infty$ as $x \to -\infty$.

262

In FIGURE 2 you may look at the "entire" graph of f which we have "scrunched" for easy viewing.

We would like to thank Hans Reddinger and some of his University of Wyoming "toys" for the great graphics.

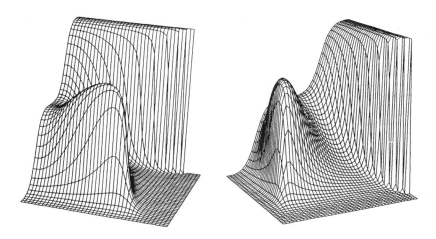

FIGURE 2. Views of the "scrunched" version of $z = 3xe^y - x^3 - e^{3y}$, that is, $z = \arctan[3e^{\tan y} \tan x - \tan^3 x - e^{3\tan y}]$ for $-\pi/2 < x, y < \pi/2$.

BIBLIOGRAPHIC ENTRIES: MAXIMA AND MINIMA

1. *Monthly* Vol. 77, No. 5, pp. 493–501. David Gale, The jeep once more or jeeper by the dozen.

> How to establish refueling depots to cross a desert with minimum amount of fuel.

2. *Monthly* Vol. 78, No. 6, pp. 644–645. David Gale, Correction to "The Jeep once more or Jeeper by the dozen."

3. *Monthly* Vol. 79, No. 2, pp. 160–164. Michel Nicola, Maxima and minima of functions of two variables.

4. *Monthly* Vol. 81, No. 5, pp. 474–480. Joseph B. Keller, Optimal velocity in a race.

5. *Monthly* Vol. 86, No. 9, pp. 767–770. C. H. Heiberg, Extrema for functions of several variables.

6. *Monthly* Vol. 92, No. 9, pp. 631–643. D. Spring, On the second derivative test for constrained local extrema.

7. *Monthly* Vol. 93, No. 7, pp. 558–561. Michael W. Botsko, A first derivative test for functions of several variables.

8. *Monthly* Vol. 96, No. 2, pp. 140–146. James Duemmel, From calculus to number theory.

> Uses number theory to set up extremum problems with integer parameters and rational solutions.

9. *Math. Mag.* Vol. 42, No. 4, pp. 165–174. R. P. Boas, Jr., Inequalities for the derivative of polynomials.

> For each polynomial $p(x)$ of fixed degree with $|p(x)| \leqslant 1$ on $[-1, 1]$, we assign a number, such as the maximum of the derivative, and maximize that number over all $p(x)$.

10. *Math. Mag.* Vol. 60, No. 2, pp. 101–104. William P. Cooke, Building the cheapest sandbox: an allegory with spinoff.

11. TYCMJ Vol. 5, No. 3, pp. 26–30. Norman Schaumberger, A set of trigonometric inequalities with applications to maxima and minima.

12. TYCMJ Vol. 9, No. 1, pp. 16–17. Norman Schaumberger, A calculus proof of the arithmetic-geometric mean inequality.

13. TYCMJ Vol. 14, No. 3, pp. 194–196. Ray C. Shiflett and Harris S. Shultz, When does a square give maximum area?

14. CMJ Vol. 18, No. 3, pp. 219–221. John C. Hegarty, A depreciation model for calculus classes.

15. CMJ Vol. 21, No. 5, pp. 410–414. Robert Lopez and John Mathews, Using a computer algebra system to solve for maxima and minima.

16. CMJ Vol. 21, No. 5, pp. 416–418. David P. Kraines and Vivian Y. Kraines, Extreme and saddle points.

17. CMJ Vol. 22, No. 2, p. 131. Ed Barbeau, The shortest distance from a point to a parabola.

18. CMJ Vol. 22, No. 3, pp. 227–229. Herbert R. Bailey, Hidden boundaries in constrained max-min problems.

9

INTEGRATION
(a)

THEORY

A Fundamental Theorem of Calculus that Applies to All Riemann Integrable Functions

MICHAEL W. BOTSKO
Saint Vincent College
Latrobe, PA 15650

The usual form of the Fundamental Theorem of Calculus is as follows:

THEOREM 1. *Let f be Riemann integrable on* $[a, b]$ *and let g be a function such that* $g'(x) = f(x)$ *on* $[a, b]$. *Then*

$$\int_a^b f(x)\, dx = g(b) - g(a).$$

Unfortunately, this theorem only applies to Riemann integrable functions that are derivatives. Thus it cannot even be used to integrate the following simple function

$$f(x) = \begin{cases} 0 & \text{if } -1 \leqslant x < 0 \\ 1 & \text{if } 0 \leqslant x \leqslant 1. \end{cases}$$

It is the purpose of this note to present a theorem that does apply to every integrable function. In stating our result we will need the following definitions.

Definition 1. The function $f: [a, b] \to R$ satisfies a Lipschitz condition if there exists $M > 0$ such that

$$|f(x) - f(y)| \leqslant M|x - y| \quad \text{for all } x \text{ and } y \text{ in } [a, b].$$

Definition 2. A set E of real numbers has measure zero if for each $\varepsilon > 0$ there is a finite or infinite sequence $\{I_n\}$ of open intervals covering E and satisfying $\sum_n |I_n| \leqslant \varepsilon$

where $|I_n|$ is the length of I_n. If a property holds *except* on a set of measure zero, it is said to hold almost everywhere.

In [2] the author gave an elementary proof of the following result.

LEMMA. *If* $f: [a, b] \to R$ *satisfies a Lipschitz condition and* $f'(x) = 0$ *except on a set of measure zero, then* f *is a constant function on* $[a, b]$.

The proof required no measure theory other than the definition of a set of measure zero. This lemma was then used to prove that a bounded function that is continuous almost everywhere is Riemann integrable. We will use it here to establish our general form of the Fundamental Theorem of Calculus.

THEOREM 2. *Let* f *be Riemann integrable on* $[a, b]$ *and let* g *be a function that satisfies a Lipschitz condition and for which* $g'(x) = f(x)$ *almost everywhere. Then*

$$\int_a^b f(x)\, dx = g(b) - g(a).$$

Proof. Let $F(x) = \int_a^x f(t)\, dt$. Since f is bounded, F satisfies a Lipschitz condition. From the fact that f is continuous except on a set of measure zero (see [3] for an elementary proof), it follows that $F'(x) = f(x)$ almost everywhere. (This shows that every Riemann integrable function is almost everywhere the derivative of a function satisfying a Lipschitz condition.) It follows at once that

$$(F - g)'(x) = F'(x) - g'(x) = f(x) - f(x) = 0$$

almost everywhere. In addition $F - g$ satisfies a Lipschitz condition. By the lemma there exists a real number k such that $F(x) = g(x) + k$ on $[a, b]$. Setting $x = a$ we have $k = -g(a)$. Finally, setting $x = b$, we get

$$\int_a^b f(x)\, dx = F(b) = g(b) - g(a),$$

which completes the proof.

Note that Theorem 2 includes Theorem 1 since any function that has a bounded derivative satisfies a Lipschitz condition.

Let us now integrate the following function. Define

$$f(x) = \begin{cases} -x & \text{if } x \in S = \{1, 1/2, 1/3, \dots\} \\ x^2 + 1 & \text{if } x \in [0, 1] \setminus S. \end{cases}$$

Since f is bounded and continuous except on $S \cup \{0\}$, a set of measure zero, it is Riemann integrable. Let $g(x) = x^3/3 + x$. Then g satisfies a Lipschitz condition and we have that $g'(x) = x^2 + 1 = f(x)$ almost everywhere. Therefore,

$$\int_0^1 f(x)\, dx = g(1) - g(0) = 4/3.$$

In this case $g'(x) \neq f(x)$ on an infinite set and yet Theorem 2 can still be used. In closing, we give a useful corollary of Theorem 2.

COROLLARY. *Let f be Riemann integrable on $[a, b]$ and let g be a continuous function such that $g'(x) = f(x)$ except on a countable set. Then*

$$\int_a^b f(x)\, dx = g(b) - g(a).$$

Proof. To use Theorem 2 we need only show that g satisfies a Lipschitz condition. Since f is integrable there exists $M > 0$ such that $|f(x)| \leq M$ for all x in $[a, b]$. Thus $-M \leq g'(x) \leq M$ except on a countable subset of $[a, b]$. Let $h(x) = Mx - g(x)$. Since h is continuous on $[a, b]$ and $h'(x) = M - g'(x) \geq 0$ except on a countable set, it follows from a result in [1] that h is increasing on $[a, b]$. Thus for c and d in $[a, b]$ with $c < d$ we have $h(c) \leq h(d)$ which gives $g(d) - g(c) \leq M(d - c)$. Similarly, we can show that $-M(d - c) \leq g(d) - g(c)$ and therefore $|g(d) - g(c)| \leq M(d - c)$. Thus g satisfies a Lipschitz condition and the proof follows immediately from Theorem 2.

REFERENCES

1. R. P. Boas, *A Primer of Real Functions*, 3rd edition, Carus Mathematical Monographs of the MAA, No. 13, 1981, pp. 141–142.
2. M. W. Botsko, An elementary proof that a bounded a.e. continuous function is Riemann integrable, *Amer. Math. Monthly* 95 (1988), 249–252.
3. R. R. Goldberg, *Methods of Real Analysis*, Blaisdell Publishing Co., New York, 1964, pp. 163–164.

Finding Bounds for Definite Integrals

W. Vance Underhill, East Texas State University, Commerce, TX

Students in elementary calculus are often dismayed to learn that not every function has an antiderivative, and consequently not every definite integral can be evaluated by the Fundamental Theorem. Although most textbooks discuss such things as Simpson's Rule and the Trapezoid Rule, these methods are usually long and tedious to apply. In many cases, reasonably good bounds for definite integrals can be obtained with little effort by the use of well-known theorems. The fact that techniques for doing this have never been discussed in one place is the motivation for this note.

Except for very specialized and esoteric results, the following three theorems provide methods for obtaining such bounds.

Theorem A. *If f, g, and h are integrable and satisfy $g(x) \leqslant f(x) \leqslant h(x)$ on the interval $[a,b]$, then*

$$\int_a^b g(x)\,dx \leqslant \int_a^b f(x)\,dx \leqslant \int_a^b h(x)\,dx.$$

Theorem B. *On the interval $[a,b]$, suppose that f and g are integrable, g never changes sign, and $m \leqslant f(x) \leqslant M$. Then*

$$m\int_a^b g(x)\,dx \leqslant \int_a^b f(x)g(x)\,dx \leqslant M\int_a^b g(x)\,dx.$$

Theorem C. *If f and g are integrable on $[a,b]$, then*

$$\int_a^b f(x)g(x)\,dx \leqslant \sqrt{\int_a^b f^2(x)\,dx}\ \sqrt{\int_a^b g^2(x)\,dx}\ .$$

Example 1. Find bounds for $\displaystyle\int_1^2 \frac{x\,dx}{\sqrt{x^3+8}}$. Using Theorem B, we choose

$f(x) = \dfrac{1}{\sqrt{x^3+8}}$ and $g(x) = x$. Then $\frac{1}{4} \leqslant f(x) \leqslant \frac{1}{3}$ for $x \in [1,2]$, and $\displaystyle\int_1^2 g(x)\,dx = \frac{3}{2}$. Therefore,

$$.375 \leqslant \int_1^2 \frac{x\,dx}{\sqrt{x^3+8}} \leqslant .500. \tag{1}$$

If we use Theorem C, it is natural to choose $f(x) = \dfrac{x}{\sqrt{x^3+8}}$ and $g(x) = 1$. It follows that

$$\int_1^2 \frac{x\,dx}{\sqrt{x^3+8}} \leqslant \sqrt{\int_1^2 \frac{x^2\,dx}{x^3+8}}\, \sqrt{\int_1^2 1^2\,dx} = \sqrt{\frac{2}{3}\ln\frac{4}{3}} < .438, \qquad (2)$$

a considerable improvement over the upper bound obtained in (1). Since $x^2 \leqslant x^3 \leqslant x^4$ on $[1, 2]$, Theorem A yields

$$\int_1^2 \frac{x\,dx}{\sqrt{x^4+8}} \leqslant \int_1^2 \frac{x\,dx}{\sqrt{x^3+8}} \leqslant \int_1^2 \frac{x\,dx}{\sqrt{x^2+8}}\,.$$

The integral on the right equals $2\sqrt{3}-3$, while the one on the left equals $\frac{1}{2}\ln\left(\dfrac{2+\sqrt{6}}{2}\right)$. Thus,

$$.399 < \int_1^2 \frac{x\,dx}{\sqrt{x^3+8}} < .465. \qquad (3)$$

Combining (1), (2), and (3), we obtain

$$.399 < \int_1^2 \frac{x\,dx}{\sqrt{x^3+8}} < .438. \qquad (4)$$

Note that in each of the three theorems, we were not forced into the choices actually made. Other possibilities exist, resulting in different bounds.

In using Theorem B, one interprets a given integrand as the product of two functions f and g, where g is of one sign and can easily be integrated. The calculation of m and M is usually straightforward. One's inclination, of course, is to let g be something easy to integrate. There is sometimes more than one reasonable "decomposition," as the following example shows.

Example 2. For the integral $\int_0^1 x^2 e^{-x^2}\,dx$, we can let $f(x) = e^{-x^2}$ and $g(x) = x^2$. Then $\dfrac{1}{e} \leqslant f(x) \leqslant 1$ on $[0, 1]$ and $\int_0^1 g(x)\,dx = \frac{1}{3}$. Hence,

$$\frac{1}{3e} \leqslant \int_0^1 x^2 e^{-x^2}\,dx \leqslant \frac{1}{3}\,.$$

Other choices of f and g exist, however. Since xe^{-x^2} can easily be integrated, let $g(x) = xe^{-x^2}$ and $f(x) = x$. Then $0 \leqslant f(x) \leqslant 1$ on $[0, 1]$ and $\int_0^1 g(x)\,dx = (e-1)/2e$ yield

$$0 \leqslant \int_0^1 x^2 e^{-x^2}\,dx \leqslant \frac{e-1}{2e} < .316. \qquad (5)$$

What if $g(x) = x$ and $f(x) = xe^{-x^2}$? Then $\int_0^1 g(x)\,dx = \frac{1}{2}$, and f has its minimum value $m = 0$ and maximum value $M = 1/\sqrt{2e}$. Consequently,

$$0 \leqslant \int_0^1 x^2 e^{-x^2} dx \leqslant \frac{1}{\sqrt{8e}}. \tag{6}$$

A quick check with the calculator shows that the upper bound in (6) is substantially better than in (4) and (5). Combining the best of all cases,

$$.122 < \frac{1}{3e} \leqslant \int_0^1 x^2 e^{-x^2} dx \leqslant \frac{1}{\sqrt{8e}} < .215. \tag{7}$$

If Theorem A is used on this example, we might reason that $x^2 \leqslant x$ on $[0, 1]$, and this leads to the inequality

$$\int_0^1 x^2 e^{-x} dx \leqslant \int_0^1 x^2 e^{-x^2} dx.$$

The integral on the left equals $2 - (5/e) > .160$, giving us a better lower bound than in (7). Consequently,

$$.160 < \int_0^1 x^2 e^{-x^2} dx < .215.$$

Example 3.. Find bounds for $\int_0^{\pi/3} x\, dx/\cos x$. Using Theorem B, the obvious decompositions do not yield particularly good results. Suppose, however, that $f(x) = x/\sin x$ (with $f(0)$ taken to be 1) and $g(x) = \sin x/\cos x$. Then f has $m = 1$ and $M = 2\pi/3\sqrt{3}$, and so

$$.693 < \ln 2 \leqslant \int_0^{\pi/3} \frac{x\, dx}{\cos x} \leqslant \frac{2\pi}{3\sqrt{3}} \ln 2 < .839. \tag{8}$$

Another estimate can also be obtained from Theorem A. Since $1 - x^2/2 \leqslant \cos x \leqslant 1$,

$$\int_0^{\pi/3} x\, dx \leqslant \int_0^{\pi/3} \frac{x\, dx}{\cos x} \leqslant \int_0^{\pi/3} \frac{x\, dx}{1 - \frac{1}{2}x^2}.$$

The lower bound here (.548) is worse than that in (8), but the value of the right-hand integral is $\ln\left(\frac{18}{18 - \pi^2} \right)$. Hence,

$$.693 < \int_0^{\pi/3} \frac{x\, dx}{\cos x} < .795.$$

Students find these methods a welcome change of pace from the routine numerical techniques. They enjoy the challenge to improve on bounds already obtained, and they gain valuable experience in working with inequalities, the heart of analysis.

Average Values and Linear Functions

David E. Dobbs, University of Tennessee, Knoxville, TN

The mean value $M(f; a,b) = \frac{1}{b-a}\int_a^b f(t)\,dt$ of an integrable, real-valued function f defined on the interval $[a,b]$ arises frequently in elementary calculus. For example, when f is the continuous instantaneous rate of change (the derivative) of a function g, then (by the fundamental theorem of calculus)

$$M(f; a,b) = \frac{1}{b-a}\int_a^b g'(t)\,dt = \frac{g(b) - g(a)}{b - a}$$

is just the average rate of change of g over $[a,b]$. Using the mean value theorem for integrals, one can also show for continuous f that $M(f; a,b) = f(c)$ for at least one value of $c \in (a,b)$. As a third illustration, one can express the definite integral of f as the limit of Riemann sums

$$\int_a^b f(t)\,dt = \lim_{n\to\infty} \sum_{i=1}^n f(e_i)\cdot\left(\frac{b-a}{n}\right), \qquad e_i = a + i\left(\frac{b-a}{n}\right)$$

and observe that

$$M(f; a,b) = \lim_{n\to\infty} \frac{1}{n}\cdot\sum_{i=1}^n f(e_i);$$

the mean of the function is a limit of "discrete" means.

The purpose of this note is to demonstrate how the mean value of an integrable function can be used to characterize the function's linearity. Quite apart from their intrinsic interest, the arguments below should be helpful to instructors seeking enrichment material that reinforces many of the topics studied in calculus.

For a linear function $f(x) = Ax + B$, the value $(b - a)M(f; a,b)$ is the area of the shaded trapezoid (Figure 1) with height $b - a$ and bases of length $f(a), f(b)$. Thus,

$$M(f; a,b) = \frac{f(a) + f(b)}{2} \qquad \text{for all} \quad a,b \in \mathbb{R}. \tag{1}$$

As Figures 2 and 3 illustrate, (1) fails for the nonlinear functions g, h. Thus, it seems reasonable to attempt to characterize linearity in terms of (1).

Theorem 1. *A real-valued function f on $[a,b]$ is linear if and only if f is continuous and $M(f; a,x) = \dfrac{f(a) + f(x)}{2}$ for all $x \in (a,b)$.*

Proof. A linear function clearly satisfies the asserted conditions. Suppose, conversely, that f is continuous and satisfies $M(f; a,x) = \dfrac{f(a) + f(x)}{2}$ for all

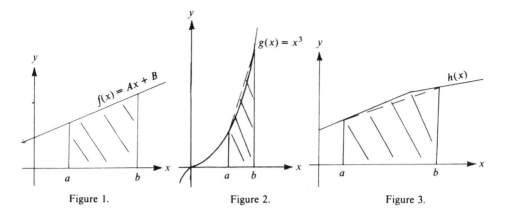

Figure 1.   Figure 2.   Figure 3.

$x \in (a, b)$. Expressing this equality as

$$2 \int_a^x f(t)\,dt = (x - a)\{f(a) + f(x)\} \qquad x \in (a, b) \tag{2}$$

and differentiating both sides of (2) twice, we obtain

$$2f'(x) = (x - a)f''(x) + f'(x) + f'(x).$$

Thus, $(x - a)f''(x) = 0$ for all $x \in (a, b)$. This, of course, means that $f'(x)$ is a constant and $f(x)$ is linear for all $x \in (a, b)$. Because f is continuous on $[a, b]$, it follows that f is linear on all of $[a, b]$.

It is straightforward to verify that a linear function f satisfies

$$\frac{f(a) + f(b)}{2} = f\left(\frac{a + b}{2}\right) \qquad \text{for all} \quad a, b \in \mathbb{R}. \tag{3}$$

This property can also be used to characterize a function's linearity.

Theorem 2. *A real-valued function f on $[a, b]$ is linear if and only if f is differentiable on $[a, b)$ and continuous at b, f' is continuous at a, and $f\left(\frac{a + x}{2}\right) = \frac{f(a) + f(x)}{2}$ for all $x \in (a, b)$.*

Proof. We assume that f has the asserted properties and show that f is linear. Begin (using the chain rule) by differentiating $f\left(\frac{a + x}{2}\right) = \frac{f(a) + f(x)}{2}$ to obtain

$$f'(x) = f'\left(\frac{a + x}{2}\right). \tag{4}$$

Replacing x with $(a + x)/2$ in (4), we get

$$f'\left(\frac{a + x}{2}\right) = f'\left[\frac{a + \left(\frac{a + x}{2}\right)}{2}\right].$$

Iteration produces a sequence

$$x_1 = \frac{a + x}{2} > x_2 = \frac{a + \left(\frac{a + x}{2}\right)}{2} > x_3 > \cdots > x_n = \frac{(2^n - 1)a + x}{2^n} > \cdots$$

such that $f'(x) = f'(x_n)$ for each n and $\lim_{n \to \infty} x_n = a$. Since f' is continuous at a, we have $f'(a) = \lim_{n \to \infty} f'(x_n) = f'(x)$. In other words, $f'(x)$ is the constant value $f'(a)$ for all $x \in (a, b)$. Therefore, f is linear on (a, b). Since f is continuous at a and b, it follows that f is linear on $[a, b]$.

For a linear function f, properties (1) and (3) yield

$$M(f; a, b) = f\left(\frac{a + b}{2}\right) \qquad \text{for all} \quad a, b \in \mathbb{R}. \tag{5}$$

Therefore, we offer another characterization of linearity in terms of (5).

Theorem 3. *A real-valued function f defined on an open interval I is linear if and only if f'' is continuous on I and $M(f; a, x) = f\left(\frac{a + x}{2}\right)$ for all $a, x \in I$ with $x > a$.*

Proof. A function f meeting the asserted conditions must have $f'' = 0$ on I. Suppose not; that is, suppose without loss of generality that $f''(b) > 0$ for some $b \in I$. Then (since f'' is continuous) $f''(c) > 0$ for all c in some open subinterval J of I. Therefore, f' is strictly increasing on J. Now consider $a, x \in J$ with $a < x$. By clearing denominators in $M(f; a, x) = f\left(\frac{a + x}{2}\right)$, differentiating, and rearranging algebraically, one readily obtains

$$\left(\frac{x - a}{2}\right) \cdot f'\left(\frac{a + x}{2}\right) = f(x) - f\left(\frac{a + x}{2}\right). \tag{6}$$

However, by the Mean Value theorem, the right-hand side of (6) equals $f'(d) \cdot \left(\frac{x - a}{2}\right)$ for some $d \in J$ strictly between $\frac{a + x}{2}$ and x. But this means

$$f'\left(\frac{a + x}{2}\right) = f'(d), \tag{7}$$

contradicting the fact that f' is strictly increasing on J. Our proof that $f'' = 0$ on I is thus complete.

Remarks. (a) For a proof of Theorem 3, emphasizing the continuity of f'', assume that $[a, x] \subset I$ and reason as above to obtain (7). Then, by the Mean Value theorem (or Rolle's theorem), $f''(x_1) = 0$ for some x_1 strictly between $(a + x)/2$ and

x. Replace a in the preceding argument by x_1, thus obtaining an x_2 such that

$$\frac{\dfrac{a+x}{2}+x}{2} < \frac{x_1 + x}{2} < x_2 < x$$

and $f''(x_2) = 0$. By iteration, we obtain a sequence $x_1 < x_2 < x_3 < \cdots$ such that all $f''(x_n) = 0$ and $\lim_{n \to \infty} x_n = x$. The continuity of f'' at $x \in I$ therefore yields $f''(x) = \lim_{n \to \infty} f''(x_n) = 0$.

(b)   We state without proof the following companion for Theorem 3. *Let a real-valued function f be given on an open interval I containing 0 by a convergent power series $\sum a_n x^n$. Then f is linear if and only if I contains a positive number r such that $M(f; 0, x) = f\left(\dfrac{x}{2}\right)$ for all $x \in (0, r)$.*

Riemann Integral of cos x

John H. Mathews and Harris S. Shultz, California State University, Fullerton, CA

Lagrange's identity,

$$\sum_{k=0}^{n-1} \cos k\theta = \frac{1}{2} + \frac{\sin\left(n - \frac{1}{2}\right)\theta}{2 \sin \frac{1}{2}\theta},$$

can be verified using mathematical induction and the trigonometric identity,

$$\sin(u + v) - \sin(u - v) = 2 \cos u \sin v.$$

We can use Lagrange's identity to obtain a basic calculus formula. Since

$$\sum_{k=0}^{n-1} \frac{x}{n} \cos k\frac{x}{n} = \frac{x}{n}\left[\frac{1}{2} + \frac{\sin\left(n - \frac{1}{2}\right)\frac{x}{n}}{2 \sin \frac{x}{2n}}\right]$$

is a Riemann sum for the function $f(t) = \cos t$ on the interval $[0, x]$, we have

$$\int_0^x \cos t\, dt = \lim_{n \to \infty}\left[\frac{x}{2n} + \frac{\frac{x}{2n}}{\sin \frac{x}{2n}} \sin\left(x - \frac{x}{2n}\right)\right] = 0 + (1)(\sin x).$$

That is, we have shown that

$$\int_0^x \cos t\, dt = \sin x$$

without using the fundamental theorem of calculus. Compare [James Stewart, *Calculus*, Brooks/Cole, Monterey, CA, pp. 266–267.]

Sums and Differences vs. Integrals and Derivatives

Gilbert Strang

Gilbert Strang studied at MIT and Oxford and UCLA and is now Professor of Mathematics at MIT. His research has been on the borderline between pure and applied mathematics. His teaching led to the textbooks *Linear Algebra and Its Applications* and *Introduction to Applied Mathematics*, and to the organization of the Boston Workshop for Mathematics Faculty. In calculus he teaches mostly freshmen who have not seen derivatives. He was strongly influenced by the 1987 colloquium in Washington and the need for careful change.

This article offers one approach to the understanding, and also to the teaching, of the fundamental theorem of calculus. The ideas expressed here are not new—mathematically they cannot be original, and even pedagogically they might already be in use. If so, that is good. The author has become deeply involved in the effective presentation of calculus—in fact he has gone past the fatal point of no return, beyond which a book has to be written [1]. It is impossible to think about explaining this subject without searching for a fresh way to illuminate the relation of derivatives to integrals.

If the hopeful author of a calculus book is viewed with a lean and lively suspicion —as he certainly is—a small request might be made in return. You know this subject too well, and therefore you must forget what you know. Instead of the "suspension of disbelief" that is required of a playgoer, the reader is asked for something more difficult—a suspension of belief, and of a sure familiarity, in the ideas of calculus. All you are permitted is the knowledge that distance equals (constant) velocity multiplied by time. Of course you are encouraged to learn more.

In spite of the happy spirit in which this article is written, its goal is entirely serious. We all want mathematics to be appreciated. I think this only becomes possible when a part of the subject is understood. It might be unwise for us to transform every student into a full-scale mathematician, but there is no reason to fail in our more limited responsibility—to help students see some of the inspiration behind what they learn.

The Goal (in Two Parts) and the Plan

There are two goals, different but complementary. One is abstract, the other is very concrete. The first is to understand the relation of a function f to its derivative v, and the relation of v to its integral f. Starting cold (from the definitions) this is a

276

substantial challenge. Working with graphs is better. Slope and area are tremendous visual supports. But my experience is that v = velocity and f = distance are the best. The intuition that comes from driving a car, and from ordinary use of the speedometer and odometer, is a free gift to calculus teachers. The relation of v to f is understood implicitly, and our contribution is to make it explicit.

The second goal is more special. The applications of calculus are built on a few functions (amazingly few). The student needs to become familiar with those particular functions. *They come in pairs f and v.* Where the first goal was to know the relation between them, the second goal is to know the functions themselves. Certain special pairs arise constantly in mathematical models, and we might as well connect with those v's and f's from the start.

The first pair is the one that everybody knows: *constant v and linear f.* (I don't skip them because they are known. I explain them because they are known!) The step between function and graph is not automatic in general, but that example is clear. Area and slope get a toehold—they are the area of a rectangle and the slope of a straight line. By no means are those ideas held in a firm grip. There is a long way to go, and the choice of the *next pair* is decisive.

A critical moment comes extremely early (often in the first week). Far ahead is the destination, to understand the key ideas and the key functions of calculus. We absolutely need a bridge. The goal cannot be reached from constant v-linear f in a single step. One approach is to start with *definitions* (of real numbers, functions, limits,...) along with careful examples. I believe that is a seriously wrong start. The approach is taken in good faith, but I am convinced there is a better one. I will concentrate on explaining how a calculus course can go forward from the first pair, to build on the intuition that made it understood before it was expressed.

The second pair is *one step away* from the first pair. It is a piecewise constant v and a piecewise linear f—the best motion seems to be *forward and back.* The velocity is $+V$ up to time T, and then $-V$ from T to $2T$. Every student knows what this motion does, and where it ends up. The distance function f is turned into a graph (Figure 1). The velocity is also expressed by a graph (and negative area for

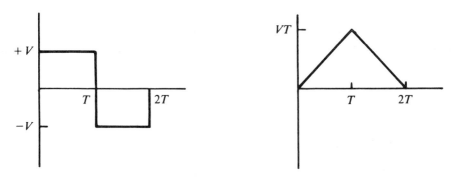

Figure 1
Forward and back: the v-graph and the f-graph.

negative v makes sense). The area under the v-graph should be checked at $T/2$ and $3T/2$. It is essential to take time with visual understanding (including the slope) because the next step will grow out of this one.

Now I can describe the plan. *It is to imitate calculus by algebra.* The velocity-distance relation can be made plain when v changes by finite steps. There is a sequence of velocities $v_1, v_2, v_3, \ldots$. For convenience the jumps occur at evenly spaced times $t = 1, 2, 3, \ldots$. The distances reached at those times are $f_1, f_2, f_3, \ldots$. We are dealing with functions, and at the same time with numbers. A typical case is

$$v = 1, 2, 3, 4, \ldots \quad \text{and} \quad f = 1, 3, 6, 10, \ldots .$$

The one thing every student can do is to fill in the next v and f! The pattern is seen; but it has to be written down. Mathematics is about *discovering* patterns and then *describing* them—and the way to teach mathematics is to do it.

Calculus Before Limits

I will propose four specific v's and f's. They are discrete models of familiar functions. The velocity changes, not continuously but in steps. Calculus will deal with continuous change—that is its central idea—and these models simulate that change by a sequence of jumps.

1. A steadily increasing velocity: $v = 1, 3, 5, 7, \ldots$ imitates $v = 2t$.
2. An oscillating velocity: $v = 1, 0, -1, -1, 0, 1, \ldots$ imitates $v = \cos t$.
3. An exponentially increasing velocity: $v = 1, 2, 4, 8, \ldots$ imitates $v = 2^t$.
4. A burst of speed (then stop): $v = 100, 0, 0, \ldots$ makes f imitate a step function.

The oscillation could come ahead of the others, because it is closest to forward-back motion and has the most attractive graph. But the first case of continuous change, when calculus really starts, will be $v = 2t$. Its discrete form yields a neat pattern—comparable to $v = 1, 2, 3, 4$ but with better f's. The presence of squares is concealed by $f = 1, 3, 6, 10$—now it will stand out.

Example 1. **Increasing velocity.** The four velocities $1, 3, 5, 7$ are displayed on the left of Figure 2. The distance function is on the right. The goal is to uncover, numerically and then algebraically, the relation between

$$v = 1, 3, 5, 7, \ldots \quad \text{and} \quad f = 1, 4, 9, 16, \ldots .$$

At $t = 1$ the distance traveled (starting from zero) is $f_1 = 1$. It equals the velocity (which is 1) multiplied by the time (also 1). At $t = 2$ the distance is $f_2 = 1 + 3$—the velocity in the second time interval is $v_2 = 3$. After three steps f_3 is $v_1 + v_2 + v_3 = 1 + 3 + 5 = 9$. It is a pleasure to prove that the f's are perfect squares, but more important is their relation to the v's.

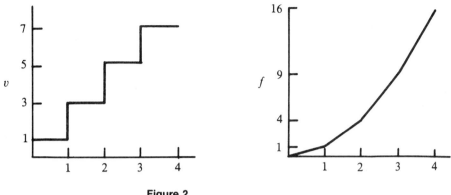

Figure 2
Linear increase in v, squares in f.

The first step begins with the f's. We are given $1, 4, 9, 16, 25$ and we are aiming for $1, 3, 5, 7, 9$. Those v's are the *differences* of the f's:

$$v_j \text{ is the difference } f_j - f_{j-1}. \tag{1}$$

The velocity $v_5 = 9$ is the difference $25 - 16$. It is the slope of the f-graph in the fifth interval. Similarly $v_4 = 7$ is $16 - 9$. I admit to a small difficulty in the first interval (at $j = 1$), from the fact there is no f_0. For this example the natural choice is $f_0 = 0$. The need for a starting point will come back to haunt us (or help us) in calculus.

Remark 1.   The equation $v_j = f_j - f_{j-1}$ asks the student to take a step that we no longer notice. With numbers everything is clear—but now we have switched to letters. There is a risk of muddying up the whole thing by algebra. But we have to do it. The investment in learning algebra is partly justified by our use of it here.

Now comes the crucial step, in the *reverse direction*. How do we recover the f's from the v's? The whole relation between differential calculus and integral calculus rests on the fact that **taking sums is the inverse of taking differences**. We saw the discrete form of differentiation in equation (1), and we will see the discrete form of integration in equation (2).

To repeat: The numbers $1, 3, 5, 7, 9$ are given. The numbers $1, 4, 9, 16, 25$ are wanted. The student sees how to do it, but the professor sees more. This numerical example carries within it—in a limited but absolutely convincing form—the fundamental theorem of calculus. It is still algebra, because v is piecewise constant—but the relation between v and f is the real thing. Why not present it now?

Fundamental Theorem of Calculus (before limits).   *If each* $v_j = f_j - f_{j-1}$ *then*

$$\text{the sum } v_1 + v_2 + \cdots + v_n \text{ equals } f_n - f_0. \tag{2}$$

The area under the graph of v is the change in f.

Proof. We are adding $(f_1 - f_0) + (f_2 - f_1) + (f_3 - f_2) + \cdots + (f_n - f_{n-1})$. The first f_1 is canceled by $-f_1$. Then f_2 is canceled by $-f_2$. At the end only $-f_0$ and $+f_n$ are left. In case $f_0 = 0$, the sum of v's has telescoped into f_n. This is the distance traveled, or the area under v.

Remark 2. It is time for new examples. They provide opportunities to verify the fundamental theorem—but this note will not impose so far on your patience. What the examples offer most of all is something different and more compelling. It is the chance to see simple models of the most crucial functions of mathematics. We are going back to *particular* v's and f's, but not forgetting what they illustrate—a relation that allows us to add v's by discovering f's.

Example 2. **Oscillation.** The six velocities $v = 1, 0, -1, -1, 0, 1$ are on the left of Figure 3. The six distances $f = 1, 1, 0, -1, -1, 0$ are on the right. The f-graph resembles (roughly) a *sine curve*. The v-graph resembles (even more roughly) a *cosine curve*. The f's were formed by the same rule as before, that each f_j equals the sum $v_1 + \cdots + v_j$. It is the total distance, when the velocities are held for one time unit each.

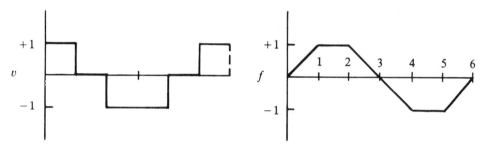

Figure 3
Piecewise constant cosine and piecewise linear sine.

These discrete sines and cosines share three properties (possibly more) with the true waveforms $\sin t$ and $\cos t$:

1. The slope of the discrete sine is the discrete cosine, and the area under the discrete cosine is the discrete sine. (Note: This is true for all v-f pairs—it is not particular to this one.)

2. The f_j are one time step behind the v_j. In other words $v_j = f_{j+1}$—the new distance is the old velocity. Since v_j is always $f_j - f_{j-1}$, that cancels f_{j+1} on the left side of a difference equation:

$$\left(f_{j+1} - f_j\right) - \left(f_j - f_{j-1}\right) = f_{j+1} - 2f_j + f_{j-1} = -f_j. \tag{3}$$

This imitates an important differential equation (second derivative of $\sin t$ equals $-\sin t$). The same is true of the cosines v_j. The time lag $f_{j+1} = v_j$ is a close copy of $\sin(t + \pi/2) = \cos t$—but the delay is $1/6$ of the period instead of $1/4$.

3. The motion returns to the start at $t = 6$ because $v_1 + v_2 + \cdots + v_6 = 0$. The numbers v and f could repeat over the next six intervals, and the graphs would be exactly the same. These are *periodic* waveforms. (*Note:* It would be interesting to take equation (3) as fundamental, and show that the period must be 6.)

The true sine and cosine possess other properties that f and v do not imitate. The outstanding example is $(\cos t)^2 + (\sin t)^2 = 1$. If desired, that could be recovered by the faster oscillation $v = 1, -1, -1, 1$ and $f = 1, 0, -1, 0$ (repeated periodically). However we have to choose the *averages* of v at the jumps, and work with $V_1 = 0$, $V_2 = -1$, $V_3 = 0$, $V_4 = 1$. Then $V_j^2 + f_j^2 = 1$. In many ways the averages are more reasonable anyway.

Example 3.   **Exponential increase** (powers of 2). In this example v doubles at every step: $v = 1, 2, 4, 8, 16, \ldots$. The velocity in the jth time interval is $v = 2^{j-1}$. The novelty comes for the distance f, *which does not begin at zero*. Its starting value must be $f_0 = 2^0 = 1$, if f is to imitate the exponential. Then the next distances are

$$f_1 = 1 + 1 = 2, \qquad f_2 = 1 + 1 + 2 = 4, \qquad f_3 = 1 + 1 + 2 + 4 = 8.$$

The formula is $f_j = 2^j$. It jumps out from Figure 4. On a normal day students won't want a proof. So don't call it a proof! It is a beautiful chance to use the fundamental theorem $\Sigma(\Delta f) = f_n - f_0$ as a way to add up the v's by guessing the answer.

Step 1:   Guess $f_j = 2^j$.

Step 2:   Verify that $f_j - f_{j-1}$ equals v_j.

Conclusion: The sum $v_1 + \cdots + v_n$ is $f_n - f_0$.

In language that comes later, f is an antiderivative. The slope of f is v (piecewise

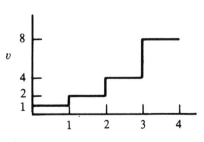
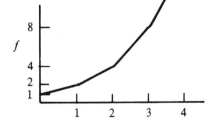

Figure 4
Exponentials $v_j = 2^{j-1}$ and $f_j = 2^j$.

constant), so the area under v must be f (piecewise linear). What matters is an idea that need not be so precise: **The best way to add (or to integrate) is to know the answer.** The answer is correct, provided that $f_j - f_{j-1}$ (or df/dx) brings back v. When calculus writes that answer as $f(b) - f(a)$, the definite integral is not a surprise—it is the natural step from $f_n - f_0$.

The exercise of actually summing the powers of 2 is not to be missed.

Note. The twin example has exponential *decrease*. The velocities $v = -\frac{1}{2}$, $-\frac{1}{4}, -\frac{1}{8}, \ldots$ are negative. The distances $f = \frac{1}{2}, \frac{1}{4}, \frac{1}{8}, \ldots$ are positive (because $f_0 = 1$). That example gives 2^{-j} a meaning beyond pure manipulation of exponents.

Example 4.   **A short burst of speed.** This last example gives a proper importance to the dimension of time. With unit time intervals the slopes were $f_j - f_{j-1}$, and no student is going to notice the division by one. A check on dimensions is the simplest test of any formula, from $f = vt$ to $e = mc^2$, and here are four v's:

$v = 10$	up to	$t = \frac{1}{10}$	(then stop)	$v = 100$	up to	$t = \frac{1}{100}$	(then stop)
$v = \underline{\hspace{1cm}}$	up to	$t = \frac{1}{10,000}$	(stop)	$v = 10^n$	up to	$t = \underline{\hspace{1cm}}$	(stop)

Exercise 1 is to fill in the blanks. Exercise 2 is to draw the graphs of f (with $f_0 = 0$). Exercise 3 (the mathematical one) is to say what those f-graphs have in common, and then *what the v-graphs have in common* (equal areas).

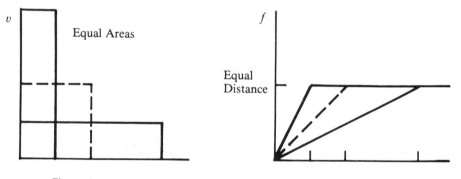

Figure 5
Burst of speed: approach to a step function f and a delta function v.

The areas of these taller and taller rectangles give the same final distance $f = 1$. This seems like the easiest of the four examples, but possibly it will occur to students to *go to the limit*. The f-graph is approaching a step function. Do we ask about the v-graph? I think we should; others may not agree. A step function for f is certainly acceptable (we have been working all along with steps in v). But now the limiting v is *the derivative of a step function.* It is a "*delta function,*" totally concentrated at $t = 0$. Its graph is infinitely tall and infinitely thin. The delta function is zero except at one point, but its integral is the jump in f.

In some respects a delta function is painful (but optional). On the other hand, it is a tremendously important idea and why shouldn't we be the first to teach it? In talking about it I use the word *impulse* and the notation $v(x) = \delta(x)$. Here $v = 10\delta$. I freely admit to the class that no function can be zero except at one point, and still give positive area. (They rather like the idea.) The word distribution is never mentioned (not even a dark hint). My main suggestion is just to do the obvious: Ask the class for the slope of a step function.

Conclusion. The author owes the reader a brief accounting. What is achieved, and what is left? Our purpose was to prepare for one of the central ideas of calculus —the relation between v and f. It is not the *fact* of the fundamental theorem, but the *meaning of that fact* and the *application of that fact*, that our teaching has to aim for.

We proposed two goals. One was general (the v-f relation), the other was particular (special v-f pairs). The examples were limited to piecewise constant velocities, but that limitation brought a reward—the functions are accessible to students at the very start of a course. They give something valuable to do (and new functions to create) at that early time when so few tools are available. Psychologically this time is crucial. Like the rest of us, students form quick opinions. If the course can't get itself started, and the ideas are drowned in an ocean of definitions, those opinions will not be what we hope for.

For college freshmen who studied calculus in high school, the first days are decisive. If nothing is new they stop listening. The models suggested here are likely to be new—and the discrete-continuous analogy deserves all the emphasis we can give it.

A note of caution. I am absolutely not proposing that the whole calculus course be built on this discrete approach. In the end, t^2 and $\sin t$ are easier than $1, 4, 9, 16, \ldots$ and $1, 1, 0, -1, -1, 0$. The key rules of differential calculus (product rule and chain rule) are much more satisfying than their discrete counterparts. It is the limiting operation that makes those rules so terrific.

In integral calculus it is the same. Few of us would hesitate, given the choice between integrating x^p and summing j^p. The beauty of calculus is that it kills off the lower-order corrections in a Riemann sum, to leave areas which are really remarkable. We are grown accustomed to them, and take them so much for granted that we insist on pushing ahead to more complicated and exotic examples. To what purpose, when the ideas and functions that are really needed are so few and so beautiful?

I want to emphasize that the new textbook [1] is aimed at careful change, not revolution. The work of calculus is still there to do, but it need not be drudgery. My effort is to make the book lighter, spiritually as well as materially. I hope others are also writing, to make progress on the problems that the 1987 colloquium presented so forcefully [2].

This short article may bring to mind other ideas and other functions. The velocities $1, r, r^2, r^3, \ldots$ connect f to the sum of a geometric series. But "calculus before limits" doesn't last long. The derivative is introduced very soon, and (separately) the integral appears. Functions are minimized and $y' = cy$ is solved. Still the ideas described here do come back for one short and vital moment. Slope and area have to be brought together into the fundamental theorem of calculus. It is there that the early work with sums and differences, which gave the course something to do at the start, makes a deep theorem look true.

It was a moment of pleasure when I learned from [3] that Leibniz had begun with sums and differences. Perhaps they guided his intuition—until Cauchy arrived, everything was calculus without limits. Whether Newton would approve I don't know. He cared less than others about clear expression (at least on the evidence), but then he left one or two priceless compensations. And our own job is not entirely easy. For teachers who have to explain the ideas of Newton in the notation of Leibniz to pupils who may not be quite so apt as Cauchy, every source of illumination is welcome.

Acknowledgment. This research was supported by Army Research Office grant DAAL 03-86-K 0171 and National Science Foundation grant DMS-87-03313.

References

1. Gilbert Strang, *Calculus*, Wellesley-Cambridge Press, Wellesley, MA (to be published in 1990).
2. Lynn Steen, ed., *Calculus for a New Century*, Mathematical Association of America, Washington, 1988.
3. C. H. Edwards, *The Historical Development of the Calculus*, Springer Verlag, New York, 1979.

BIBLIOGRAPHIC ENTRIES: THEORY

1. *Monthly* Vol. 78, No. 10, pp. 1129–1131. R. C. Metzler, On Riemann integrability.

2. *Monthly* Vol. 80, No. 4, pp. 349–359. E. J. McShane, A unified theory of integration.

3. *Monthly* Vol. 80, No. 6, pp. 615–627. Harley Flanders, Differentiation under the integral sign.

4. *Monthly* Vol. 81, No. 1, p. 145. Harley Flanders, Correction to "Differentiation under the integral sign."

5. *Monthly* Vol. 82, No. 9, pp. 918–919. J. van de Lune, A note on the fundamental theorem for Riemann integrals.

6. *Math. Mag.* Vol. 50, No. 3, pp. 115–122. Bertram Ross, Fractional calculus.

> For a simplified shorter introduction to this topic see Larson, TYCMJ Vol. 5, No. 2, p. 68–70, listed as item 5 in the Bibliography for Sec. 5(a) of this volume.

7. *Math. Mag.* Vol. 60, No. 4, pp. 225–228. S. K. Goel and D. M. Rodrigues, A note on evaluating limits using Riemann sums.

8. TYCMJ Vol. 10, No. 1, pp. 35–37. C. W. Baker, Mean value type theorems of integral calculus.

(b)

TECHNIQUES OF INTEGRATION

INVERSE FUNCTIONS AND INTEGRATION BY PARTS

R. P. BOAS, JR. AND M. B. MARCUS

Students of calculus often have difficulty with inverse functions; some nontrivial exercises with inverse functions may help. We present here a formulation of integration by parts in terms of inverse functions, which in particular makes some elementary integrals rather easier to evaluate. This is by no means new, although we have not found it in standard calculus books. As an application we give a very short proof of Young's inequality.

Let f be a strictly increasing function with (as is appropriate in calculus) a continuous derivative. Integration by parts tells us that

$$(1) \qquad \int_a^b f(x)dx = bf(b) - af(a) - \int_a^b xf'(x)dx.$$

Let $y = f(x)$, $x = f^{-1}(y)$; then (1) can be written

$$(2) \qquad \int_a^b f(x)dx = bf(b) - af(a) - \int_{f(a)}^{f(b)} f^{-1}(y)dy.$$

Now if $u = f(a)$, $v = f(b)$, (2) becomes

$$(3) \qquad \int_{f^{-1}(u)}^{f^{-1}(v)} f(x)dx = vf^{-1}(v) - uf^{-1}(u) - \int_u^v f^{-1}(y)dy.$$

Thus we can always express $\int f^{-1}(y)dy$ in terms of $\int f(x)dx$ (or vice versa).

For example, let $f(x) = \sin x$, $-\pi/2 < x < \pi/2$. Then (3) says that

$$\int_{\sin^{-1}u}^{\sin^{-1}v} \sin x dx = v\sin^{-1}v - u\sin^{-1}u - \int_u^v \sin^{-1}y dy,$$

whence with $u = 0$,

$$\int_0^v \sin^{-1}y dy = v\sin^{-1}v + (1 - v^2)^{\frac{1}{2}} - 1.$$

Those who are familiar with Stieltjes integrals will observe that if we think of (1) as

$$\int_a^b f(x)dx = bf(b) - af(a) - \int_a^b xdf(x),$$

285

it continues to hold as long as f is just strictly increasing and continuous; and (2) is still correct in this case (cf. [7], p. 124).

Young's inequality (in the usual form) says that when f is a strictly increasing continuous function with $f(0) = 0$, and $b > 0$, $t > 0$, then

(4)
$$bt < \int_0^b f(u)du + \int_0^t f^{-1}(y)dy.$$

This is geometrically obvious; some recent articles have been devoted to analytic proofs of it (or of its generalizations: the inequality holds — possibly with weak inequality — even when f is weakly monotonic or discontinuous, provided that f^{-1} is suitably interpreted; see [4], [3], [2]). Applications of (4) are given in [5], pp. 111 ff.; [6], p. 49; [1].

The following proof of (4) is very short, but of course depends on our knowing (2).

For $0 < r < b$ it is obvious that

$$(b - r)f(r) < \int_r^b f(u)du.$$

Write this as

$$bf(r) - \int_0^b f(u)du < rf(r) - \int_0^r f(u)du,$$

and apply (2) to the integral on the right. We get

$$bf(r) - \int_0^b f(u)du < \int_0^{f(r)} f^{-1}(y)dy.$$

If $0 < t < f(b)$, we can take $r = f^{-1}(t)$, and (4) follows.

Some other applications of (2) are given in [1], and a study of various versions of (4) is given in [2].

References

1. R. P. Boas and M. B. Marcus, Inequalities involving a function and its inverse, SIAM J. Math. Analysis, 4 (1973) 585–591.

2. ———, and ———, Generalizations of Young's inequality, J. Math. Analysis Appl., 46 (1974) 36–40.

3. F. Cunningham, Jr., and N. Grossman, On Young's inequality, this MONTHLY, 78 (1971) 781–783.

4. J. B. Diaz and F. T. Metcalf, An analytic proof of Young's inequality, this MONTHLY, 77 (1970) 603–609.

5. G. H. Hardy, J. E. Littlewood and G. Pólya, Inequalities, Cambridge University Press, 1934.

6. D. S. Mitrinović, Analytic Inequalities, Springer-Verlag, New York-Heidelberg-Berlin, 1970.

7. F. Riesz and B. Sz. Nagy, Functional Analysis, Ungar, New York, 1955.

DEPARTMENT OF MATHEMATICS, NORTHWESTERN UNIVERSITY, EVANSTON, IL 60201.

Proof without Words: Integration by Parts

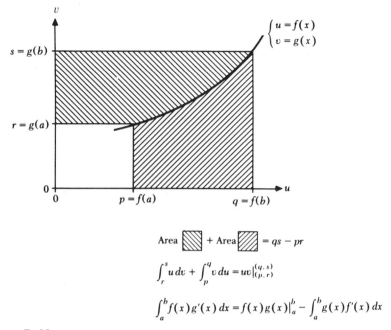

$$\text{Area}\ \boxed{}\ +\ \text{Area}\ \boxed{} = qs - pr$$

$$\int_r^s u\,dv + \int_p^q v\,du = uv\Big|_{(p,r)}^{(q,s)}$$

$$\int_a^b f(x)g'(x)\,dx = f(x)g(x)\Big|_a^b - \int_a^b g(x)f'(x)\,dx$$

—Roger B. Nelsen
Lewis and Clark College
Portland, OR 97219

Ed. Note. This proof appears in some calculus texts, for example, Richard Courant and Fritz John, Introduction to Calculus and Analysis, v. 1, Interscience, New York, 1965, pp. 275–276.

A Discovery Approach to Integration by Parts

John Staib and Howard Anton, Drexel University, Philadelphia, PA

Ideally, a formula for a certain class of problems should arise out of a solution strategy developed for individual problems. For example, the quadratic formula arises out of the complete-the-square strategy for solving a quadratic equation. The usual pedagogy employed in that case is to get the class proficient in that strategy, and then to move naturally into the quadratic formula. It is in this spirit that we offer the following "natural path" to the integration-by-parts formula. We begin with the example

$$\int x \cos x \, dx = ?$$

Well, the derivative of what would produce $x \cos x$? Since the derivative of a product is a sum of two products, maybe the given product is part of the derivative of some product. If so,

$$uv' + u'v = x \cos x + \boxed{}$$

and the missing term must be $(1)(\sin x)$. In other words, to complete $x \cos x$ as the derivative of a product we differentiate one factor (say x) and antidifferentiate the other. So $x \cos x$ is part of $x \cos x + (1)(\sin x)$, which in turn is the derivative of $x \sin x$. Using the "giveth-and-taketh-away" technique, we obtain

$$\int x \cos x \, dx = \int [x \cos x - (1)(\sin x - (1)(\sin x))] \, dx$$

$$= \int \frac{d}{dx}(x \sin x) \, dx - \int \sin x \, dx$$

$$= x \sin x + \cos x + C.$$

The order of u and v can be reversed. Thus,

$$\int x \ln x \, dx = \int \left[x \ln x + \frac{x^2}{2}\left(\frac{1}{x}\right) - \frac{x^2}{2}\left(\frac{1}{x}\right) \right] dx$$

$$= \int \frac{d}{dx}\left(\frac{1}{2} x^2 \ln x \right) dx - \int \frac{1}{2} x \, dx$$

$$= \frac{1}{2} x^2 \ln x - \frac{1}{4} x^2 + C.$$

The strategy becomes clearer with each example: We take the given product and match it up with a complementary term obtained by differentiating one factor of the product and integrating the other. The matched pair is then integrable, leaving us only to worry about the integral of the complementary term which—in some cases—is more easily handled. Thus, this strategy has us first partly integrate the given integral.

Further examples might show all the other aspects of the usual integrate-by-parts lesson. Finally, once the class has gained some proficiency with this strategy, the usual formula may be derived using this strategy.

$$\int f(x)g(x)dx = \int [\,f(x)g(x) + f'(x)G(x) - f'(x)G(x)\,]dx$$

$$= \int \frac{d}{dx}[\,f(x)g(x) + f'(x)G(x)\,]dx - \int G(x)f'(x)dx$$

$$= f(x)G(x) - \int G(x)f'(x)dx + C.$$

Making the substitution $u = f(x)$ and $v = G(x)$ gives the textbook version. However, having arrived at this formula following the path described here, it is not clear that anything is to be gained by introducing the u's and v's. Indeed, it is rather attractive to dispense with the formula by translating it into an "operational chant" as follows:

$$\int f(x)g(x)dx = f(x)G(x) - \int G(x)f'(x)dx$$

$$(\text{1st})(\text{2nd}) \qquad \begin{bmatrix} \text{Hold the 1st} \\ \text{\& anti the 2nd} \end{bmatrix}\begin{bmatrix} \text{Then hold the anti} \\ \text{\& differentiate the 1st} \end{bmatrix}.$$

For example,

$$\int x^2\cos x\,dx = (x^2)(\sin x) - \int (\sin x)(2x)dx$$

$$\underset{\substack{\text{Hold the 1st \&}\\ \text{anti the 2nd}}}{} \quad \underset{\substack{\text{Then hold the anti}\\ \text{\& differentiate the 1st}}}{}$$

$$= x^2\sin x - 2\int x \sin x\,dx$$

$$= x^2\sin x - 2\left\{ \underset{\text{Hold the 1st} \ldots \text{Then hold the} \ldots}{(x)(-\cos x) - \int(-\cos x)(1)dx} \right\}$$

$$= x^2\sin x + 2x\cos x - 2\cos x + C.$$

Formal Integration: Dangers and Suggestions

S. K. STEIN

S. K. STEIN is Professor of Mathematics at the University of California, Davis, He has published in algebra, convex bodies, and topology, and his pedagogical interests are represented by four texts: *Mathematics, The Man-Made Universe*; *Introductory Algebra, A Guided Inquiry* (with C. Crabill); *Geometry, A Guided Inquiry* (with G. D. Chakerian and C. Crabill); and *Calculus and Analytic Geometry*.

When we teach formal integration in calculus, we may tend to forget that we walk a very narrow tightrope. The functions that can be integrated are rather rare, as will be illustrated below. A pause to reflect on this rarity is perhaps long overdue. Let us take a little time to look at some integrals that all of us meet in a calculus course, considering both their mathematical interest and their pedagogical implication.

First, an informal definition. An *elementary function* is any function built up from e^x, $\ln x$, the trigonometric functions and their inverses, and the powers x^a, by the traditional procedures such as addition, multiplication, subtraction, division, or composition of functions. When a student is programmed to differentiate the elementary functions, he is incidentally and painlessly following the steps of a long proof of a theorem which asserts that *the derivative of an elementary function is again an elementary function*. But when the student encounters the antidifferentiation of the elementary functions, he plays a game of a different character: *the antiderivative of an elementary function need not be elementary*. The seemingly simple function e^{x^2} gives trouble, while the frightening $e^{\sqrt{x}}$ has an elementary antiderivative, $2e^{\sqrt{x}}(\sqrt{x} - 1)$. Perhaps we should say that the odds that an elementary function has an elementary antiderivative are slim. To substantiate this claim, let us begin by quoting a theorem of Chebyshev [4, p. 37].

CHEBYSHEV'S THEOREM. *Let p and q be nonzero rational numbers. Then the antiderivative*

$$\int x^p (1 - x)^q \, dx$$

is elementary if and only if at least one of p, q, and p + q is an integer. (The same is true for $\int x^p(1 + x)^q \, dx$.)

A simple illustration of this theorem is $\int \sqrt{x} \sqrt[3]{1 - x} \, dx$, the case $p = 1/2$, $q = 1/3$. Since neither $1/2$, nor $1/3$, nor $1/2 + 1/3$ is an integer, the antiderivative is *not* elementary.

Chebyshev's theorem* has several implications for the typical introductory calculus course. The four theorems that follow concern $\int \sqrt[m]{1 - x^n}\, dx$, $\int \sin^r \theta \cos^s \theta\, d\theta$, arc length, and surface area. All the arguments based on Chebyshev's theorem can easily be presented to a typical calculus class. Indeed, since no steps have been omitted, an interested student should have no trouble reading the proofs himself.

The important integral $\int \sqrt{1 - x^2}\, dx$ is treated in every calculus text. But what about $\int \sqrt[3]{1 - x^2}\, dx$ or $\int \sqrt{1 - x^3}\, dx$, whose integrands differ only slightly from $\sqrt{1 - x^2}$? That these antiderivatives are not elementary is a consequence of the following theorem.

The Integral $\int \sqrt[m]{1 - x^n}\, dx$

THEOREM 1. *Let m and n be positive integers. Then $\int \sqrt[m]{1 - x^n}\, dx$ is elementary if and only if $m = 1$, or $n = 1$, or $m = 2 = n$.*

Proof. In order to make use of Chebyshev's theorem, we introduce the substitution

$$u = x^n, \qquad du = nx^{n-1}\, dx,$$

hence

$$dx = \frac{du}{nx^{n-1}} = \left(\frac{1}{n}\right)\frac{du}{u^{(n-1)/n}} = \left(\frac{1}{n}\right)u^{(1-n)/n}\, du.$$

Then

$$\int \sqrt[m]{1 - x^n}\, dx = \frac{1}{n}\int (1 - u)^{1/m} u^{(1-n)/n}\, du.$$

The question is now reduced to finding out when

$$\int u^{(1-n)/n}(1 - u)^{1/m}\, du \tag{1}$$

is elementary. But (1) is covered by Chebyshev's theorem, which asserts that (1) is

---

*The proof of the easy direction of Chebyshev's theorem runs as follows.

(a) Assume that p is an integer. Let $q = s/t$ where s and t are integers, t positive. Introduce the substitution $u^t = 1 - x$. The integral $\int x^p(1 - x)^q\, dx$ takes the form $\int (1 - u^t)^p \times u^s(-t)u^{t-1}\, du$, the integral of a rational function.

(b) If q is an integer and $p = r/t$, use the substitution $u^t = x$.

(c) Finally, if $p = r/t$, $q = s/t$, and $p + q$ is an integer, rewrite $\int x^p(1 - x)^q\, dx$ as

$$\int x^{p+q}\left(\frac{1 - x}{x}\right)^q dx$$

and use the substitution $u^t = (1 - x)/x$.

elementary if and only if
  (a) $(1 - n)/n$ is an integer, or
  (b) $1/m$ is an integer, or
  (c) $(1 - n)/n + 1/m$ is an integer.
Let us take the three cases in order.

Case (a). Since $(1 - n)/n = (1/n) - 1$, it follows that $(1 - n)/n$ is an integer if and only if $1/n$ is an integer, that is, $n = 1$. (Keep in mind that n is positive.)

Case (b). The fraction $1/m$ is an integer if and only if $m = 1$.

Case (c). The sum
$$\frac{1 - n}{n} + \frac{1}{m}$$
is an integer if and only if
$$\frac{1}{n} + \frac{1}{m} \tag{2}$$
is an integer. Since m and n are positive integers, the only cases in which (2) can be an integer are $m = 1 = n$ and $m = 2 = n$.
  This concludes the proof.

Note that when $m = 1$ the integral $\int \sqrt[m]{1 - x^n}\, dx$ is simply $\int (1 - x^n)\, dx$, which every calculus student finds easy. When $n = 1$,
$$\int \sqrt[m]{1 - x^n}\, dx = \int \sqrt[m]{1 - x}\, dx,$$
which he evaluates with the substitution $u = 1 - x$. The case $m = 2 = n$ stands out as the rare and happy exceptional case, $\int \sqrt{1 - x^2}\, dx$. In no other instance is $\int \sqrt[m]{1 - x^n}\, dx$ elementary. This is the first illustration of how narrow the tightrope is that we take through calculus. The slightest change in the integrand, $\sqrt{1 - x^2}$, and we fall into the abyss of nonelementary antiderivatives.

The Integral $\int \sin^r \theta \cos^s \theta\, d\theta$. We all show the student how to integrate $\sin^m \theta \cos^n \theta$ where m and n are integers. But what if m and n are not integers? Can we, for instance, integrate
$$\sin^{1/2} \theta \cos^{1/2} \theta = \sqrt{\sin \theta \cos \theta}\,?$$
Can we integrate
$$\sin^{1/2} \theta \cos^{3/2} \theta?$$
  Chebyshev's theorem provides the answer to these questions, in the form of Theorem 2.

THEOREM 2. *Let r and s be positive rational numbers. Then the integral*
$$\int \sin^r \theta \cos^s \theta\, d\theta \tag{3}$$

is elementary if and only if r is an odd integer, or s is an odd integer, or r + s is an even integer.

Proof. Introduce the substitution

$$u = \sin \theta,$$

hence

$$\cos \theta = \sqrt{1 - u^2} \quad \text{and} \quad du = \cos \theta \, d\theta.$$

Then (3) becomes

$$\int u^r \left(\sqrt{1 - u^2}\right)^s \frac{du}{\sqrt{1 - u^2}}$$

or

$$\int u^r (1 - u^2)^{(s-1)/2} \, du. \tag{4}$$

The substitution $y = u^2$ will transform (4) to the form considered in Chebyshev's theorem. We let

$$y = u^2, \quad dy = 2u \, du,$$

hence

$$du = \frac{dy}{2\sqrt{y}}.$$

Thus (4) becomes

$$\int y^{r/2} (1 - y)^{(s-1)/2} (1/2) y^{-1/2} \, dy.$$

The question is now, essentially, "When is

$$\int y^{(r-1)/2} (1 - y)^{(s-1)/2} \, dy$$

elementary?"

Again there are three cases:

Case (a). $(r - 1)/2$ is an integer;

Case (b). $(s - 1)/2$ is an integer;

Case (c). $(r - 1)/2 + (s - 1)/2$ is an integer.

In case (a), assume that $(r - 1)/2 = m$, where m is an integer. Then $r = 2m + 1$, that is, r is an odd integer.

Case (b) is similar. In case (c), assume that

$$\frac{r - 1}{2} + \frac{s - 1}{2} = m,$$

where m is an integer. Then

$$r + s = 2(m + 1),$$

which asserts that $r + s$ is an *even* integer. This concludes the proof.

In particular, Theorem 2 tells us that $\int \sin^{1/2} \theta \cos^{1/2} \theta \, d\theta$ is not elementary, but $\int \sin^{1/2} \theta \cos^{3/2} \theta \, d\theta$ is elementary. (Incidentally, since

$$\sin^{1/2} \theta \cos^{1/2} \theta = \sqrt{\sin \theta \cos \theta}$$
$$= (1/\sqrt{2})\sqrt{\sin 2\theta},$$

it follows that $\int \sqrt{\sin \theta} \, d\theta$ is not elementary.)

So far we have considered only formal integration. But even in the usual applications of calculus, such as arc length and area of a surface of revolution, only the cunning and kindness of calculus text authors keep us securely in the realm of elementary antiderivatives.

Arc Length. We all compute the arc length of the curves $y = x^0$, $y = x^1$, and $y = x^2$. The next theorem implies that the curve $y = x^3$ will lead us to an integral that is not elementary.

THEOREM 3. *Let r be a rational number. Then the arc length of the curve $y = x^r$ leads to an elementary function if and only if $r = 1$ or*

$$r = 1 + 1/n$$

for some integer n.

Proof. It will be convenient in the argument to exclude the trivial case $r = 0$ (corresponding to $n = -1$) and $r = 1$. Henceforth assume $r \neq 0, 1$.

The antiderivative for the arc length is

$$\int \sqrt{1 + (rx^{r-1})^2} \, dx$$

or

$$\int \sqrt{1 + r^2 x^{2r-2}} \, dx. \tag{5}$$

A straightforward sequence of two substitutions will transform (5) into the second form in Chebyshev's theorem,

$$\int x^p (1 + x)^q \, dx.$$

First of all, let us dispose of the constant "r^2" in (5) by choosing the function v such that

$$v^{2r-2} = r^2 x^{2r-2},$$

or, what amounts to the same thing,

$$v = r^{2/(2r-2)} x;$$

hence

$$dv = r^{2/(2r-2)}\, dx.$$

Thus (5) becomes

$$\int \sqrt{1 + v^{2r-2}}\, r^{-2/(2r-2)}\, dv.$$

The constant, $r^{-2/(2r-2)}$, goes past the integral sign, and we are faced with the integral

$$\int (1 + v^{2r-2})^{1/2}\, dv. \qquad (6)$$

The substitution $u = v^{2r-2}$ puts (6) into the form covered by Chebyshev's theorem. We have

$$u = v^{2r-2}, \qquad du = (2r-2)v^{2r-3}\, dv, \qquad v = u^{1/(2r-2)}$$

and

$$dv = 1/(2r-2)u^{-(2r-3)/(2r-2)}\, du.$$

The integral (6) becomes

$$[1/(2r-2)] \int (1+u)^{1/2} u^{-(2r-3)/(2r-2)}\, du. \qquad (7)$$

Since $1/2$ has no hope of being an integer, there are only two cases in which (7) will be elementary:

Case (a). $-(2r-3)/(2r-2)$ is an integer, or
Case (b). $1/2 - (2r-3)/(2r-2)$ is an integer.
Consider case (a) first. Let

$$\frac{-(2r-3)}{2r-2} = m, \qquad (8)$$

where m is an integer. Solving (8) for r yields

$$r = 1 + 1/(2m+2), \qquad (9)$$

which says that "r must be 1 plus the reciprocal of an even number."
Next let us turn to case (b),

$$\frac{1}{2} - \frac{2r-3}{2r-2} = n, \qquad (10)$$

where n is an integer. Solving (10) for r yields

$$r = 1 + 1/(2n+1),$$

which says that, "r must be 1 plus the reciprocal of an odd integer."
Cases (b) and (c) coalesce to the single assertion, "r is 1 plus the reciprocal of an integer," and the theorem is proved.

Theorem 3 assures us that the arc length for such curves as $y = x^2$ (correspond-ing to $n = 1$) and $y = x^{2/3}$ (corresponding to $n = -3$) can be computed by the fundamental theorem of calculus, but that for $y = x^3$ or the hyperbola $y = 1/x$ it cannot. Moreover, if the exponent r is larger than 2, the arc length for the curve $y = x^r$ leads to a nonelementary integral. In particular, the only integer values of r for which the arc length of $y = x^r$ is computable are $r = 0$, 1, and 2.

Surface Area. Finding ourselves so restricted in computing arc length, we may wonder what happens when we turn to the area of a surface of revolution. For example, the curve $y = x^3$ leads us to the following integral:

$$\int 2\pi x^3 \sqrt{1 + 9x^4} \, dx,$$

which is easy since x^3 is almost the derivative of $1 + 9x^4$ (suggesting the substitu-tion $u = 1 + 9x^4$). The cases $y = x^1$ and $y = x^2$ are also old chestnuts. But what exponents will work in general? The next theorem answers this question.

THEOREM 4. *Let r be a rational number. Then the area of the surface formed by revolving the curve $y = x^r$ around the x axis leads to an elementary function if and only if $r = 1$ or*

$$r = 1 + 2/n$$

for some integer n.

Proof. As in the proof of Theorem 3, exclude the cases $r = 0$ (corresponding to $n = -2$) and $r = 1$.

The antiderivative for the surface area is

$$\int 2\pi x^r \sqrt{1 + r^2 x^{2r-2}} \, dx, \tag{11}$$

which two substitutions will put into Chebyshev's (second) form.

First choose v such that

$$v^{2r-2} = r^2 x^{2r-2},$$

that is,

$$v = r^{2/(2r-2)} x;$$

hence,

$$dv = r^{2/(2r-2)} \, dx,$$

and

$$x = r^{-2/(2r-2)} v,$$

and

$$dx = r^{-2/(2r-2)} \, dv.$$

Hence (11) becomes

$$\int 2\pi r^{-2r/(2r-2)} v^r \sqrt{1 + v^{2r-2}} \, r^{-2r/(2r-2)} \, dv.$$

Disregarding the constant factor, we are led to consider

$$\int v^r \sqrt{1 + v^{2r-2}} \, dv. \tag{12}$$

Next introduce the substitution $u = v^{2r-2}$, hence

$$du = (2r - 2)v^{2r-3} \, dv, \qquad v = u^{1/(2r-2)},$$

and

$$dv + 1/(2r - 2)u^{-2r-3)/(2r-2)} \, du.$$

Thus (12) becomes

$$\int u^{r/(2r-2)}(1 + u)^{1/2}(1/(2r - 2))u^{-(2r-3)/(2r-2)} \, du,$$

which, except for the constant factor, is

$$\int u^{(3-r)/(2r-2)}(1 + u)^{1/2} \, du. \tag{13}$$

According to the second part of Chebyshev's theorem, there are two cases in which (13) is elementary: either

$$\frac{3 - r}{2r - 2} = m$$

where m is an integer, or

$$\frac{3 - r}{2r - 2} + \frac{1}{2} = n,$$

where n is an integer.

The first case, as a little algebra shows, leads to

$$r = 1 + 2/(2m + 1). \tag{14}$$

The second case leads to

$$r = 1 + 1/n,$$

which we may rewrite as

$$r = 1 + 2/2n. \tag{15}$$

The cases (14) and (15) are summarized in the single assertion

$$r = 1 + 2/p,$$

where p is an integer, and the theorem is proved.

From Theorem 4 it is easy to deduce that if the surface are obtained by rotating the curve $y = x^r$ around the x axis leads to an elementary integral, then when the curve is rotated around the y axis, the surface area again leads to an elementary integral.

Theorems 3 and 4, taken together, assert that "if you can compute the length of $y = x^r$ then you can compute the area of its surface of revolution." (For $r = 1 + 1/n$ can be written as $r = 1 + 2/(2n)$.) However, for every *odd integer* $2n + 1$, the curve $y - x^{1+1/(2n+1)}$ has an "elementary surface area" but not an "elementary arc length." (The hyperbola $y = x^{-1} = 1/x$ and the cubic $y = x^3$ are instances of this phenomenon.)

Pedagogical Reflections. The four theorems shown that as soon as we stray from the "standard" integrals, we will very likely be in trouble. What implication does this possibility have for the calculus teacher? There are several questions he should face:

How much attention should a teacher call to the fact that an elementary function "usually" does not have an elementary integral? Will the student lose faith in the fundamental theorem of calculus?

Conversely, what functions should the student learn to integrate? How much time is formal integration worth? Should the student use the integral tables?

Fortunately, some integrals that occur in mathematics, physics, and engineering fairly often are elementary—for instance, the integrals for the various moments and such specific integrals as $\int e^{ax} dx$ and $\int e^{ax} \sin bx \, dx$. The fact will serve as some guide. On the other hand, we should not present an integration technique simply because it exists (the "I climbed the mountain because it was there" rationale). In particular, I would make these recommendations:

1. Spend at least one day showing the students how to use a table of integrals (not just the short tables in the back of a calculus book).
2. Spend at most one day on integration by partial fractions. A student deserves to be told that any rational function can be integrated, but the only interesting case he is likely to need is $\int 1/(a^2 - x^2) \, dx$, which is in every table of integrals.
3. Give adequate attention to numerical integration. (High-speed computers make numeral integration swift and accurate. In many cases, even if the antiderivative is elementary, the computer is preferred.)
4. Emphasize that a definite integral is the limit of sums. The student will likely go on to use the definite integral over an interval, plane or solid region as a conceptual tool more often than as a computational device.

A Mathematical Epilogue. Some elementary functions have elementary antiderivatives, some do not. (See [1] and [4] for more examples.) Is there a way of deciding? More specifically, is there an algorithm, some fixed general routine, for determining whether an antiderivative is elementary? Recent work by R. H. Risch ([2], [3]) proves that such an algorithm exists. However, I am not suggesting that it can be incorporated in the elementary calculus.

REFERENCES

1. D. G. Mead, "Integration," *American Mathematical Monthly*, *68* (1961), 152–156. (Reprinted in *Selected Papers in Calculus*, ed. by T. M. Apostol, et al., Mathematical Association of America, 1968, 274–278.)
2. R. H. Risch, "The Problem of Integration in Finite Terms," *Transactions of the American Mathematical Society*, *13*^{9} (1969), 167–189.
3. R. H. Risch, "The Solution of the Problem of Integration in Finite Terms," *Bulletin of the American Mathematical Society*, *76*, (1970), 605–608.
4. J. F. Ritt, *Integration in Finite Terms*, New York: Columbia University Press, 1948.

THE EVALUATION OF $\int_a^b x^k dx$

N. Schaumberger, Bronx Community College, City University of New York

The evaluation of $\int_a^b x^k dx$ $(0 \leq a \leq b)$, prior to the introduction of the Fundamental Theorem of Calculus, may be obtained in a manner exhibited in [2] or [3]. Another approach will be presented in which the restrictions on the partitions such as occur in [2] and [3] will be unnecessary. The reader may be interested in comparing the present treatment with that in [1] where the method of telescoping series is also used.

Let $0 \leq a = x_1 < x_2 < \cdots < x_{n+1} = b$ be a partition of $[a, b]$.

Since $x_i^k \leq x_i^{k-j} x_{i+1}^j \leq x_{i+1}^k$ $(j = 0, 1, \cdots, k)$, each subinterval $[x_i, x_{i+1}]$ will contain the point $(x_i^{k-j} x_{i+1}^j)^{1/k}$; and therefore, if $x_{i+1} - x_i$ is denoted by Δx_i,

$$\int_a^b x^k dx = \lim_{\max \Delta x_i \to 0} \sum_{i=1}^n x_i^{k-j} x_{i+1}^j \Delta x_i \quad \text{for any } j \in \{0, 1, \cdots, k\}.$$

Thus

$$(k+1) \int_a^b x^k dx = \lim_{\max \Delta x_i \to 0} \sum_{j=0}^k \left(\sum_{i=1}^n x_i^{k-j} x_{i+1}^j \Delta x_i \right)$$

$$= \lim_{\max \Delta x_i \to 0} \sum_{i=1}^n \left(\sum_{j=0}^k x_i^{k-j} x_{i+1}^j \Delta x_i \right)$$

$$= \lim_{\max \Delta x_i \to 0} \sum_{i=1}^n \left(\sum_{j=0}^k x_i^{k-j} x_{i+1}^{j+1} - \sum_{j=0}^k x_i^{k-j+1} x_{i+1}^j \right)$$

$$= \lim_{\max \Delta x_i \to 0} \sum_{i=1}^n \left(\sum_{j=1}^{k+1} x_i^{k-j+1} x_{i+1}^j - \sum_{j=0}^k x_i^{k-j+1} x_{i+1}^j \right)$$

$$= \lim_{\max \Delta x_i \to 0} \sum_{i=1}^n \left[\left(\sum_{j=1}^k x_i^{k-j+1} x_{i+1}^j + x_{i+1}^{k+1} \right) \right.$$

$$\left. - \left(\sum_{j=1}^k x_i^{k-j+1} x_{i+1}^j + x_i^{k+1} \right) \right]$$

$$= \lim_{\max \Delta x_i \to 0} \sum_{i=1}^n (x_{i+1}^{k+1} - x_i^{k+1}).$$

This last sum telescopes and we have $\int_a^b x^k dx = [1/(k+1)](b^{k+1} - a^{k+1})$.

References

1. T. M. Apostol, Calculus, vol. 1, Blaisdell, New York, 1961, pp. 66–67, p. 35.
2. R. Courant, Differential and Integral Calculus, vol. 1, Interscience, New York, 1947, pp. 84–85.
3. J. M. H. Olmsted, Intermediate Analysis, Appleton-Century-Crofts, New York, 1956. p. 145.

BIBLIOGRAPHIC ENTRIES: TECHNIQUES OF INTEGRATION

1. *Monthly* Vol. 79, No. 9, pp. 963–972. Maxwell Rosenlicht, Integration in finite terms.

2. *Monthly* Vol. 95, No. 2, pp. 126–130. Ashok K. Arora, et al., Special integration techniques for trigonometric integrals.

3. *Math. Mag.* Vol. 45, No. 3, pp. 117–119. Hugh J. Hamilton, The partial fraction decomposition of a rational function.

4. *Math. Mag.* Vol. 53, No. 4, pp. 195–201. Toni Kasper, Integration in finite terms: the Liouville theory.

5. *Math. Mag.* Vol. 62, No. 5, pp. 318–322. A. J. Zajta and S. K. Goel, Parametric integration techniques.

6. TYCMJ Vol. 1, No. 2, pp. 106–107. H. L. Kung, $\int f(x)e^{ax}\,dx$.

Uses undetermined coefficients to integrate when $f(x)$ is a polynomial.

7. TYCMJ Vol. 2, No. 2, pp. 98–100. Louise S. Grinstein, Integration by undetermined coefficients.

8. TYCMJ Vol. 4, No. 2, pp. 91–93. Norman Schaumberger, Evaluating $\int_a^b x^k\,dx$ where k is any negative integer other than -1.

9. TYCMJ Vol. 9, No. 2, pp. 104–105. Alan H. Schoenfeld, A simple antidifferentiation technique.

Advice on the substitution $u = x^a$.

10. TYCMJ Vol. 12, No. 4, pp. 268–270. A. D. Holley, Integration by geometric insight —a student's approach.

11. TYCMJ Vol. 14, No. 1, pp. 60–61, J. E. Nymann, An alternative to partial fractions (part of the time).

12. TYCMJ Vol. 14, No. 2, pp. 110–118. Padmini T. Joshi, Efficient techniques for partial fractions.

13. TYCMJ Vol. 14, No. 2, pp. 168–169. Joseph Wiener, Evaluating integrals by differentiation.

Differentiation with respect to a parameter under the integral sign.

14. CMJ Vol. 16, No. 4, pp. 282–283. Andre L. Yandl, A note on integration by parts.

15. CMJ Vol. 21, No. 4, pp. 307–311. David Horowitz, Tabular integration by parts.

16. CMJ Vol. 22, No. 5, pp. 407–410. Leonard Gillman, More on tabular integration by parts.

17. CMJ Vol. 22, No. 5, pp. 410–413. Leroy F. Meyers, Four crotchets on elementary integration.

18. CMJ Vol. 22, No. 5, pp. 413–415. Xun-Cheng Huang, A shortcut in partial fractions.

Uses the identity $(b - a)/[(u + a)(u + b)] = 1/(u + a) - 1/(u + b)$ when u is a function of x.

19. CMJ Vol. 22, No. 5, pp. 421–429. T. N. Subramaniam and D. E. G. Malm, Reduction formulas revisited.

(c) SPECIAL INTEGRALS

(No papers reproduced in this section)

BIBLIOGRAPHIC ENTRIES: SPECIAL INTEGRALS

1. *Monthly* Vol. 97, No. 1, pp. 39–42. Robert Weinstock, Elementary evaluation of $\int_0^\infty e^{-x^2}\,dx$, $\int_0^\infty \cos x^2\,dx$, $\int_0^\infty \sin x^2\,dx$.

> The method uses differentiation with respect to a parameter under the integral sign. It appears in simplified form in Apostol's *Mathematical Analysis*, (Addison-Wesley), p. 246 of 1st edition (1957), and p. 178 of 2nd edition (1974).

2. *Math. Mag.* Vol. 42, No. 3, p. 113, Clifton T. Whyburn, A different technique for the evaluation of $\int \sec \theta\, d\theta$.

3. *Math. Mag.* Vol. 44, No. 1, pp. 9–11. Kenneth S. Williams, Note on $\int_0^\infty \sin x/x\, dx$.

> An elementary evaluation of $\int_0^\infty (\sin x/x)^n\, dx$ for integer $n \geq 1$ is given in Mathematics Magazine, Vol. 53, No. 3, p. 183.

4. *Math. Mag.* Vol. 62, No. 4, pp. 260–261. John Frohliger and Rick Poss, Just an average integral.

5. TYCMJ Vol. 5, No. 3, p. 58. Norman Schaumberger, Some comments on the exceptional case in a basic integral formula.

> Compare with TYCMJ Vol. 12, No. 1, p. 20–23, reproduced in Ch. 3, Sec. (c) of this volume.

6. TYCMJ Vol. 10, No. 3, p. 202. Norman Schaumberger, Another approach to $\int \sec x\, dx$.

7. CMJ Vol. 20, No. 3, pp. 235–237. Arnold J. Insel, A direct proof of the integral formula for arctangent.

(d)

APPLICATIONS

Disks and Shells Revisited

WALTER CARLIP
Department of Mathematics, Ohio University, Athens, OH 45701

It is a common practice in calculus courses to use the definite integral to *define* the area between the graph of a function and the *x*-axis (see, e.g., [2, p. 252], [3, p. 221], and [4, p. 238]). Soon after, the student is taught two methods to calculate volumes of solids of revolution—the disk method and the shell method—usually with no mention of how *volume* is defined. Most calculus books follow the introduction of disks and shells with several examples in which it is shown that both methods of calculating the volume yield the same answer. The alert student is sure to wonder whether this is always the case.

The equivalence of the disk and shell methods was proven in [1] using integration by parts. We present here a different approach, one that uses only elementary ideas and illustrates an important proof technique.

THEOREM. *Let $f(x)$ be a continuous, invertible function on the interval $[a, b]$, where $a \geqslant 0$. Suppose the region bounded by $y = f(b)$, $x = a$, and the graph of $f(x)$ is rotated about the y-axis. Then the values of the volume of the resulting solid obtained by the disk and shell methods are equal. That is,*

$$\int_{f(a)}^{f(b)} \left(\pi \left[f^{-1}(y) \right]^2 - \pi a^2 \right) dy = \int_a^b 2\pi x [f(b) - f(x)] \, dx.$$

The Ingredients. Early in most calculus curricula, Rolle's Theorem is used to prove the following principle, which is then applied repeatedly.

PRINCIPLE. *If $f(x)$ and $g(x)$ are two functions that satisfy:*
(a) *$f(x)$ and $g(x)$ are differentiable on an interval $[a, b]$,*
(b) *$f'(x) = g'(x)$ for all $x \in [a, b]$, and*
(c) *$f(c) = g(c)$ for one point $c \in [a, b]$,*
then $f(x) = g(x)$ for all $x \in [a, b]$.

Informally, this says that two functions are equal if they are equal at *one* point and have identical derivatives. Although this principle is surprisingly simple, and easily absorbed by students, it has numerous applications and reappears often. Emphasizing this principle helps students see the similarity between proofs that otherwise seem unrelated.

The other ingredients of the proof of our theorem are the Fundamental Theorem of Calculus, the chain rule, and the product rule.

303

Proof. Let $t \in [a, b]$. We define two functions of t as follows:

$$V(t) = \int_{f(a)}^{f(t)} \left(\pi [f^{-1}(y)]^2 - \pi a^2 \right) dy$$

and

$$W(t) = \int_a^t 2\pi x [f(t) - f(x)] \, dx.$$

We need to prove that $V(b) = W(b)$, as these are the two volumes in the theorem. It is easy to show that $V(t) = W(t)$ for all $t \in [a, b]$ by applying the principle given above.

First simplify $W(t)$:

$$W(t) = 2\pi f(t) \int_a^t x \, dx - 2\pi \int_a^t x f(x) \, dx.$$

Now, by the Fundamental Theorem, both $V(t)$ and $W(t)$ are differentiable on the interval $[a, b]$. Furthermore, $V'(t) = (\pi [f^{-1}(f(t))]^2 - \pi a^2) f'(t) = \pi (t^2 - a^2) f'(t)$, and

$$W'(t) = 2\pi f'(t) \int_a^t x \, dx + [2\pi f(t)t - 2\pi t f(t)]$$

$$= 2\pi f'(t) \left[\frac{t^2 - a^2}{2} \right] = \pi (t^2 - a^2) f'(t).$$

Thus, $W'(t) = V'(t)$ for all $t \in [a, b]$. It remains only to observe that $V(a) = W(a) = 0$, by a fundamental property of integrals. $\square$

REFERENCES

1. Charles A. Cable, The disk and shell method, this MONTHLY, 91 (1984) 139.
2. John B. Fraleigh, Calculus of a Single Variable, Addison-Wesley, Reading, Mass., 1985.
3. Al Shenk, Calculus and Analytic Geometry, Goodyear Publ., Santa Monica, Calif., 1979.
4. Michael Spivak, Calculus, Publish or Perish Inc., Berkeley, Calif., 1980.

Ed. Note. This simplifies an earlier paper, *Monthly* Vol. 91, No. 2, p. 139, listed below on p. 334 as Bibliographic entry 4.

UPPER BOUNDS ON ARC LENGTH

RICHARD T. BUMBY

When we feel rigorous about the subject of arc length, we usually insist that the length of a parameterized curve be defined as the least upper bound of the lengths of inscribed polygons. One easily argues that the curve must be longer than the inscribed polygon, for whatever reasonable notion we take of 'length'. The question of why the least upper bound, and not some larger quantity, is never raised. This definition of arc length essentially tells us that we believe that the lengths of curves will be measured with some sort of flexible ruler. But if we use such a tool to estimate distances on a winding road from a road map, we may be in for a surprise.

It is easy to write a program which computes the length of the inscribed polygon of 2^n vertices. Let L_n denote the length with 2^n points equally spaced with respect to the parameter. Then we find that $L_{n+1} - L_n \approx \frac{1}{4}(L_n - L_{n-1})$ experimentally. This suggests that the error in estimating arc length by this method is inversely proportional to the square of the number of points. In order to prove such a result we must find some way of obtaining an upper bound on arc length. Such upper bounds also give further philosophical justification of our definition.

We now consider convex curves and we assert the following axiom:

If a closed convex curve C_1 is contained in an arbitrary closed curve C_2, then the length of C_1 is less than or equal to the length of C_2.

The intuitive justification of this is given by representing C_1 by a block of wood and C_2 by a loop of string. If we pull the loop tight, it will fit snugly around the block, measuring its perimeter. At the same time, the original loop measured the length of C_2. It was made shorter in measuring C_1, completing the justification. From the axiom we easily get that the length of a convex arc is bounded above by the length of a circumscribed polygon. This is, in fact, equivalent to our axiom.

We first assume that we have a convex arc. Even if the arc is not smooth, it can be parameterized by the angle of inclination α of the support (tangent) line. Any other parameter which moved along the curve without backtracking would be a monotonic function of α. We obtain our first upper bounds using the parameterization by α. We choose points $p_0, \cdots, p_n$ on the arc such that p_0 and p_n are the endpoints and the remaining points are determined by given increments of α. These points will be the vertices of the inscribed polygon. From them we construct $p_0^*, \cdots, p_{n+1}^*$, which will be the vertices of the circumscribed polygon, by setting p_i^* equal to the point of intersection of the support lines at p_i and p_{i-1}, with $p_0^* = p_0$, $p_{n+1}^* = p_n$. Note that the direction of the support line at p_i is given by the value of α, which is our basic quantity, and so is well defined here even if there is no well-defined tangent at that point of the curve. For use in the arc length formula we also construct the point

305

p_i' on the arc p_{i-1}, p_i where the support line is parallel to the chord p_{i-1}, p_i. These points are illustrated in Figure 1.

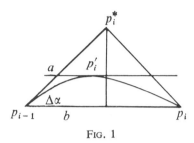

FIG. 1

The difference between the lengths of the inscribed and circumscribed polygons is a sum of quantities of the form $a - b = a(1 - \cos \Delta\alpha) < ca(\Delta\alpha)^2$, for some constant c. Thus, the difference between the lengths of the two polygons is bounded by c times the length of the circumscribed polygon times the square of the largest value of $\Delta\alpha$. Note that $\Delta\alpha$ is the difference in α between p_{i-1} and p_i', or between p_i' and p_i, each of which is bounded by the difference in α between p_{i-1} and p_i. If the points are chosen so that the $\Delta\alpha$ are equal, then the difference between upper and lower approximations to arc length is inversely proportional to the square of the number of points. More generally, if the points are equally spaced with respect to some parameter t such that $d\alpha/dt$ is bounded, the same conclusion holds. For any C^2 parameterization, $d\alpha/dt$ will be continuous. On an interval on which it is never zero, the curve will be (locally) convex. Thus, our empirical observation can be verified in this case.

These upper bounds can also be related to the arc length integral. We illustrate with the special case in which x is the parameter. We use x_i, x_i^*, x_i' for the x coordinates of p_i, p_i^*, p_i'. The length of the inscribed polygon is then

$$\Sigma \sqrt{1 + y'(x_i')^2} \ (x_i - x_{i-1});$$

and the length of the circumscribed polygon is

$$\Sigma \sqrt{1 + y'(x_i)^2} \ (x_{i+1}^* - x_i^*).$$

Each of these is a Riemann sum for $\int \sqrt{1 + y'(x)^2} \, dx$. The assumption of convexity is equivalent to assuming that $y'(x)$, and hence the integrand is a monotonic function of x. The convergence of the Riemann sums in this case is easily established.

There are other related uses of convexity in elementary calculus. In curve sketching, we know that tangents lie on one side and chords on the other side of a convex curve. The author has used an $x - y$ plotter driven by a desk top computer to draw very close inscribed and circumscribed polygons to various curves. The circumscribed polygons were obtained by computing the points of intersection of consecutive tangents.

This bracketing of the curve between tangent and chord can also be applied to Newton's method for finding zeroes of functions. The desired value is shown to lie in a series of nested intervals. This gives estimates on the rate of convergence of the process as part of the calculation. Of course, prior knowledge of convexity is a strong assumption, but one which can frequently be demonstrated. We hope to have shown that this information is valuable in analyzing limiting processes in which convergence is usually from one side.

DEPARTMENT OF MATHEMATICS, RUTGERS COLLEGE, NEW BRUNSWICK, N. J. 08903.

A THEOREM ON ARC LENGTH

JOHN KAUCHER, University of California, Santa Barbara

Many calculus textbooks apply the mean value theorem for integrals to interpret the area under a curve as the area of a related rectangle. The purpose of this note is to interpret the length of arc of a curve as the length of a related line segment.

THEOREM. *If* $f'(x)$ *is continuous for* $a \leq x \leq b$ *and* $t(x)$ *is the length of the segment of the tangent line at* $(x, f(x))$ *intercepted by the lines* $x = a$ *and* $x = b$, *then there exists a number* w, $a \leq w \leq b$, *such that* $t(w)$ *is the length of arc of* $f(x)$ *between* $x = a$ *and* $x = b$.

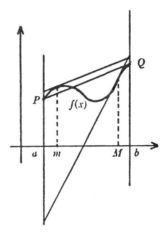

Proof. By the mean value theorem for derivatives, there exists a point m, $a \leq m \leq b$, such that $f'(m) = $ slope of $\overline{PQ}$. Since the line $x = a$ is parallel to $x = b$, then $t(m) = $ length of $\overline{PQ}$. But the length of $\overline{PQ}$ is less than or equal to the length of arc $\overset{\frown}{PQ}$. Therefore $t(m) \leq $ length of $\overset{\frown}{PQ}$.

Since $f'(x)$ is continuous, it has a maximum value at $x = M$, $a \leq M \leq b$. Now

$$t(M) = \sqrt{(b-a)^2 + [f'(M)(b-a)]^2} = (b-a)\sqrt{1 + [f'(M)]^2}.$$

But since $f'(M) \geq f'(x)$,

$$\sqrt{1 + [f'(M)]^2} \geq \sqrt{1 + [f'(x)]^2}.$$

Integrating from $x = a$ to $x = b$, we get $t(M) \geq $ length of $\overset{\frown}{PQ}$.

Since $f'(x)$ is continuous, $t(x)$ is continuous. We proved that $t(m) \leq $ length of $\overset{\frown}{PQ} \leq t(M)$. By the intermediate value theorem there must exist a number w such that $t(w) = $ length of $\overset{\frown}{PQ}$.

A NOTE ON ARC LENGTH

JOHN T. WHITE, Texas Technological College

Upon reading an article by Kaucher [1] one is led to consider motivating arc length of a curve described by the graph of f on $[a, b]$ by approximating the curve by line segments which are tangent to the curve and adding these up.

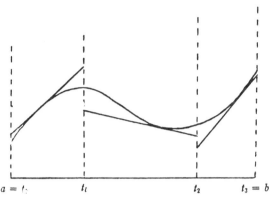

$$a = t_0 \qquad t_1 \qquad t_2 \qquad t_3 = b$$

FIG. 1.

More precisely we have

DEFINITION. *Let f' be continuous on $[a, b]$; then the curve described by the graph on f on $[a, b]$ has upper length L if L is the greatest lower bound of*

$$(1) \qquad \sum_{i=0}^{n-1} \sqrt{[f'(\eta_i)\Delta t_i]^2 + [\Delta t_i]^2}$$

taken over all partitions $P = \{t_0, t_1, \cdots, t_n\}$ of $[a, b]$, where η_i is a point in $[t_i, t_{i+1}]$ where $|f'|$ assumes its maximum value on $[t_i, t_{i+1}]$.

Thus, each term in (1) is the length of the tangent line segment between t_i and t_{i+1} having maximum absolute slope.

One then defines the lower length l in an analogous fashion and defines the curve to have length if both l and L exist and are equal, in which case the length is taken to be this common value.

THEOREM. *If f' is continuous on $[a, b]$, then the curve described by the graph of f has length.*

Proof. It is seen immediately from integration theory that L is the upper integral of $\sqrt{[f'(t)]^2 + 1}$ over $[a, b]$ and l is the lower integral. Hence, the length exists and is

$$\int_a^b \sqrt{[f'(t)]^2 + 1} \, dt.$$

Reference

1. John Kaucher, A theorem on arc length, this MAGAZINE, 42 (1969) 132–133.

"Mean Distance" in Kepler's Third Law

Sherman K. Stein

University of California, Davis

Kepler's third law relates the time required for a planet to complete one revolution around the sun to the "mean distance" from the planet to the sun. For Kepler this "mean distance" referred to the radius of the large circle which carries the Copernican epicycles that approximate the planet's orbit. This is implicit in Kepler's statement of the third law in *Harmony of the World*, Book V, Chapt. 3:

> *Res est certissima exactissimaque quod proportionis quae est inter binorum quorumque planetarum tempora periodica, sit praecise sesquialtera proportionis mediarum distantiarum, id est orbium ipsorum.*

But in modern expositions "mean distance" is usually defined as the average of the smallest and largest distances from the planet to the sun. Kepler would probably have accepted this definition as equivalent to his. Since it coincides with the semimajor axis of the elliptical orbit, some texts use the semimajor axis as the definition of mean distance, without any explanation.

The term "mean distance", when first encountered by a student who has studied calculus, suggests a different view: the *average distance* from the planet to the sun *throughout* its orbit. Let us see how this type of average is related to Kepler's mean distance.

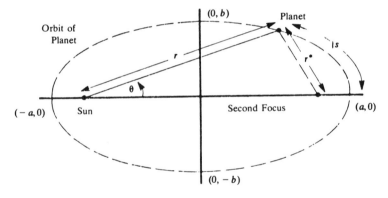

FIGURE 1.

Consider an ellipse with semimajor axis a and semiminor axis $b < a$. Place the pole of a polar coordinate system at the sun and the polar axis along the major axis of the ellipse (see FIGURE 1). The orbit may be assumed to have the polar equation $r = pe/(1 - e \cos \theta)$, where e, $0 < e < 1$, is the eccentricity of the ellipse, p is a constant related to a by the equation $pe = a(1 - e^2)$, and $b^2 = a^2(1 - e^2)$. Three averages come to mind: the average of r with respect to angle θ, or time t, or arc length s. Let us examine each in turn, and see if any coincide with Kepler's mean distance.

The average of r with respect to angle θ is defined as $(\int_0^{2\pi} r \, d\theta)/2\pi$. Now

$$\int_0^{2\pi} r \, d\theta = \int_0^{2\pi} \frac{pe}{1 - e \cos \theta} \, d\theta = \frac{2\pi pe}{\sqrt{1 - e^2}},$$

310

a value that can be obtained by first rationalizing the integrand, as in freshman calculus, or by using a table of integrals. Thus

$$\frac{1}{2\pi} \int_0^{2\pi} r \, d\theta = \frac{2\pi p e}{2\pi \sqrt{(1-e^2)}} = \frac{a(1-e^2)}{\sqrt{1-e^2}} = a\sqrt{1-e^2} = b.$$

So the average of r with respect to θ is simply b, the semiminor axis. It is not influenced by a at all.

The average of r with respect to time t is defined as $(\int_0^T r \, dt)/T$ where T is the duration of one revolution. Since the radial arm of a given planet sweeps out area at a constant rate, $\frac{1}{2} r^2 (d\theta/dt)$ is a constant which we will denote by h. It follows that $Th = \pi ab$, the area of the ellipse. Thus

$$\frac{1}{T} \int_0^T r \, dt = \frac{h}{\pi ab} \int_0^{2\pi} r \frac{dt}{d\theta} \, d\theta = \frac{h}{\pi ab} \int_0^{2\pi} (r^3/(2h)) \, d\theta = \frac{1}{2\pi ab} \int_0^{2\pi} r^3 \, d\theta,$$

where

$$\int_0^{2\pi} r^3 \, d\theta = \frac{p^3 e^3 (2+e^2)\pi}{(1-e^2)^{5/2}} = b(3a^2 - b^2)\pi.$$

Hence

$$\frac{1}{T} \int_0^T r \, dt = \frac{b(3a^2 - b^2)\pi}{2\pi ab} = \frac{3a}{2} - \frac{b^2}{2a} = \frac{3a}{2} - \frac{a^2(1-e^2)}{2a} = a\left(1 + \frac{e^2}{2}\right).$$

Therefore the average of r with respect to t is $a(1 + e^2/2)$, which is larger than a and is influenced by both a and b.

Finally, we compute the average of r with respect to arc length. Since the integral for the arc length of an ellipse cannot be evaluated in finite terms, we will proceed indirectly, utilizing the defining property of an ellipse that the sum of the distances from any point on the ellipse to the two foci is constant. To begin, we introduce the radius r^* from a typical point on the orbit to the second focus (see FIGURE 1). Let L be the arc length of the orbit. Then by symmetry, $\int_0^L r \, ds = \int_0^L r^* \, ds$. But by the definition of an ellipse, $r + r^* = 2a$; hence, $\int_0^L (r + r^*) \, ds = 2aL$. Consequently, $\int_0^L r \, ds$ is half of $2aL$; hence $(\int_0^L r \, ds)/L = a$. Thus the average of r with respect to arc length is a, which is precisely Kepler's "mean distance".

Would Kepler feel comfortable with the basic idea of this proof, though calculus had not been available in his time? Probably "yes". On at least two occasions he paired off distances whose sum is twice the semimajor axis; see [2, p. 315 and p. 334]. In the first instance he was almost integrating.

Incidentally, the average distances with respect to time and angle are discussed in [3, p. 304], but no mention is made there of average with respect to arc length. Also the three averages are defined in [1, p. 750–752] but only the averages with respect to time and angle are computed. It might be of interest to obtain a synthetic evaluation of the average with respect to angle, and an analytic (calculus) evaluation of the average with respect to arc length.

I would like to thank J. H. Heilbron of the Center for History of Science and Technology at the University of California at Berkeley for advice concerning Kepler's perspective, and G. D. Chakerian and Murray Klamkin for calling my attention to references [1] and [3].

References

[1]   J. Edwards, A Treatise on the Integral Calculus, vol. 2, Chelsea, New York, 1954.

[2]   A. Koyre, The Astronomical Revolution, Cornell University Press, Ithaca, New York, 1973.

[3]   W. P. MacMillan, Statics and the Dynamics of a Particle, McGraw-Hill, New York, 1927.

Some Problems of Utmost Gravity

WILLIAM C. STRETTON, *College of DuPage*

In high school and two-year college mathematics and physics classes, we often encounter problems relating to the force due to gravity. These problems sometimes contain such assertions as, "One can consider the mass of the earth concentrated at its center," or, "The earth's gravitational force on an object beneath its surface is directly proportional to the distance of the object from the center." I wonder how many secondary mathematics and physics teachers have seen demonstrations of these facts? Proofs can sometimes be found in physics or astronomy texts, but these proofs often contain some rather dubious mathematics.

The objective of this paper is threefold. First, to show that, with respect to gravitational attraction, the entire mass of a uniformly dense sphere does act as if it were concentrated at its center. Second, to show that the gravitational attraction of a sphere on point mass inside the sphere is indeed directly proportional to the distance of the point mass from the center of the sphere. (This leads to an interesting corollary.) And, third, to show that for a body of nonspherical shape, the mass does not necessarily act as though it were concentrated at its center of gravity. This last objective will be attained by means of a simple counterexample.

Let's first look at the counterexample. Consider a thin straight wire of length $2a$ and mass per unit length w. Its total mass is $M = 2aw$. Let a point mass m be located at a distance b from the midpoint of the wire. If the mass of the wire could be considered to be concentrated at its center of gravity, then the gravitational force of attraction between the wire of mass M and the point mass m would be, by Newton's Law,

$$F = \frac{GMm}{b^2}. \tag{1}$$

Let us actually calculate this force in the simplest case. Let the wire extend from $-a$ to $+a$ on the x-axis and let point mass m be located at $(0, b)$ on the y-axis as shown in FIGURE 1.

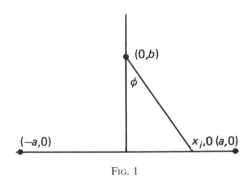

FIG. 1

If the wire is divided into n very small equal segments of length Δx, then the vertical component of the force between the ith segment centered at x_i and m is, by Newton's law, approximately

$$\Delta F_i \approx \frac{Gmw\Delta x_i}{x_i^2 + b^2} \cos \phi. \tag{2}$$

The horizontal component is canceled out by an equal but opposite force on the segment at $-x_i$ and so the horizontal components can be neglected. From the figure, we see that

$$\cos \phi = \frac{b}{\sqrt{x_i^2 + b^2}}.$$

The total force is then found by taking the limit of the sum of the vertical components as $n \to \infty$ and $\Delta x \to 0$. That is,

$$F = \lim_{\Delta x \to 0} \sum_{i=1}^{n} \Delta F_I = \int_{-a}^{a} \frac{Gmwb\, dx}{\left(x^2 + b^2\right)^{3/2}}. \tag{3}$$

Evaluating this integral, we have

$$F = \frac{Gmw \cdot 2a}{b\sqrt{a^2 + b^2}} = \frac{GmM}{b\sqrt{a^2 + b^2}}$$

where $M = 2aw$. Note that this result is not the same as (1). However, if b is very large compared to a, then $\sqrt{a^2 + b^2} \approx b$, and (1) would give a close approximation to the exact result (3). This is the case when calculating the gravitational attraction between the earth and small objects a great distance from its center. The radius is so large compared to the dimensions of the objects that (1) gives correct results for all practical purposes.

Now consider a sphere of density w, radius a, and a point mass m located b units from the center of the sphere, $b > a$. For simplicity, take the center of the sphere as origin and let the point mass be on the z-axis at $(0, 0, b)$ as shown in FIGURE 2. Then the equation of the sphere is $x^2 + y^2 + z^2 = a^2$. The problem is to calculate the force of attraction between the sphere and m. To this end, let the interior of the sphere be divided into rectangular solid elements by passing planes through the sphere perpendicular to the coordinate axes at intervals of Δx on the x-axis, Δy on the y-axis, and Δz on the z-axis. Let $P(x_i, y_i, z_i)$ be the center of the ith element with volume $\Delta V_i = \Delta x \Delta y \Delta z$ and mass $\Delta M_i = w\Delta x \Delta y \Delta z$. Then the vertical component of the force of attraction between m and ΔM_i is given approximately by Newton's law and is $\Delta F_i \approx Gm\,\Delta M_i/s^2 \sin \phi$ where s is the distance between m and P,

$$s = \sqrt{(b - z_i)^2 + x_i^2 + y_i^2}$$

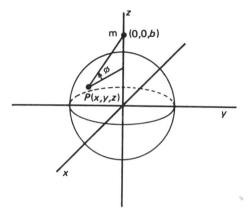

FIG. 2

and

$$\sin \phi = \frac{b - z_i}{\sqrt{(b - z_i)^2 + x_i^2 + y_i^2}}.$$

The horizontal component of the force is canceled by an equal but opposite component due to the element symmetrical to ΔM_i with respect to the z-axis. Hence the total force between m and the sphere can be found by taking the limit of the sum of the vertical components for each ΔM_i as Δx, Δy, and Δz all approach zero and the number of elements n approaches ∞.

$$F = \lim_{n \to \infty} \sum_{i=1}^{n} \frac{Gm\Delta M_i}{s^2} \sin \phi$$

$$= \lim_{n \to \infty} \sum_{i=1}^{n} \frac{Gmw(b - z_i)\Delta z \Delta y \Delta x}{\left[(b - z_i)^2 + x_i^2 + y_i^2\right]^{3/2}}. \tag{4}$$

From (4) we get

$$F = Gmw \int_{-a}^{a} \int_{\sqrt{a^2-x^2}}^{\sqrt{a^2-x^2}} \int_{-\sqrt{a^2-x^2-y^2}}^{\sqrt{a^2-x^2-y^2}} \frac{(b - z)\, dz\, dy\, dx}{\left[(b - z)^2 + x^2 + y^2\right]^{3/2}}. \tag{5}$$

It is convenient to change to polar coordinates in the xy plane. Then the equation of the circular section of the sphere in the xy plane becomes $r = a$, $x^2 + y^2$ can be replaced by r^2, and the area element $dy\, dx$ becomes $r\, dr\, d\theta$. Now the integral with appropriate limits is

$$F = Gmw \int_{0}^{2\pi} \int_{0}^{a} \int_{-\sqrt{a^2-r^2}}^{\sqrt{a^2-r^2}} \frac{(b - z)\, dz\, r\, dr\, d\theta}{\left[(b - z)^2 + r^2\right]^{3/2}}. \tag{6}$$

The first integration, with respect to z, can be accomplished by using the standard formula

Thus: $\mu^n \, du = \dfrac{\mu^{n+1}}{n+1}$.

$$\int_{-\sqrt{a^2-r^2}}^{\sqrt{a^2-r^2}} \frac{(b-z)\,dz}{\left[(b-z)^2+r^2\right]^{3/2}}$$

$$= -\frac{1}{2}\int_{-\sqrt{a^2-r^2}}^{\sqrt{a^2-r^2}} \left[(b-z)^2+r^2\right]^{-3/2}(-2)(b-z)\,dz$$

$$= \left. \frac{1}{\left[(b-z)^2+r^2\right]^{1/2}} \right|_{-\sqrt{a^2-r^2}}^{\sqrt{a^2-r^2}}$$

$$= \frac{1}{\left[a^2+b^2-2b\sqrt{a^2-r^2}\right]^{1/2}} - \frac{1}{\left[a^2+b^2+2b\sqrt{a^2-r^2}\right]^{1/2}}. \quad (7)$$

Therefore we have

$$F = Gmw\int_0^{2\pi}\int_0^a\left[\frac{1}{\left[a^2+b^2-2b\sqrt{a^2-r^2}\right]^{1/2}}\right.$$

$$\left. - \frac{1}{\left[a^2+b^2+2b\sqrt{a^2-r^2}\right]^{1/2}}\right]r\,dr\,d\theta. \quad (8)$$

The next integration, with respect to r, can be done by an appropriate substitution. Let $u = \sqrt{a^2-r^2}$ and let $a^2+b^2 = c^2$. Then $u^2 = a^2-r^2$, $u\,du = -r\,dr$ and u decreases from a to 0 when r increases from 0 to a. Then

$$\int_0^a \frac{r\,dr}{\left[a^2+b^2-2b\sqrt{a^2-r^2}\right]^{1/2}} = \int_a^0 \frac{-u\,du}{\sqrt{c^2-2bu}},$$

$$\int_0^a \frac{-r\,dr}{\left[a^2+b^2+2b\sqrt{a^2-r^2}\right]^{1/2}} = \int_a^0 \frac{u\,du}{\sqrt{c^2+2bu}}. \quad (9)$$

These integrals can now be found in a standard table of integrals or worked out with the aid of another substitution,

$$v = \sqrt{c^2 \pm 2bu}\ .$$

In either case we get

$$\int_a^0 \frac{-u\,du}{\sqrt{c^2-2bu}} = \frac{2c^3 - 2c^2\sqrt{c^2-2ab} - 2ab\sqrt{c^2-2ab}}{6b^2} \quad (10a)$$

and

$$\int_a^0 \frac{u\,du}{\sqrt{c^2 + 2bu}} = \frac{-2c^3 + 2c^2\sqrt{c^2 + 2ab} - 2ab\sqrt{c^2 + 2ab}}{6b^2}. \tag{10b}$$

Combining these results and using the facts that $c^2 = a^2 + b^2$, $b > a$, and therefore $\sqrt{c^2 - 2ab} = \sqrt{a^2 - 2ab + b^2} = b - a$ we get

$$\int_0^a \left[\frac{1}{\left[a^2 + b^2 - 2b\sqrt{a^2 - r^2}\right]^{1/2}} - \frac{1}{\left[a^2 + b^2 + 2b\sqrt{a^2 - r^2}\right]^{1/2}} \right] r\,dr = \frac{2a^3}{3b^2}. \tag{11}$$

Therefore, after the second integration, we have

$$F = Gmw \int_0^{2\pi} \frac{2a^3}{3b^2}\,d\theta. \tag{12}$$

The third integration is very easy and so we have

$$F = Gmw \cdot \frac{2a^3}{3b^2}\theta \Big|_0^{2\pi} = Gmw \cdot \frac{4\pi a^3}{3b^2}. \tag{13}$$

Now the volume of the sphere is $4/3\,\pi a^3$ and the mass M of the sphere is $4/3\,\pi a^3 w$. Hence (13) can be written as

$$F = \frac{GmM}{b^2}. \tag{14}$$

This shows that the gravitational attraction of the sphere on the point mass m acts as if the entire mass of the sphere were concentrated at its center, a distance b from m.

Now suppose the point mass m is located inside the sphere. Nothing is changed in the above derivation except that now $b < a$ and so $\sqrt{a^2 - 2ab + b^2} = a - b$ instead of $b - a$. This causes the integral (11) to have the value $2/3b$. Then instead of (14) our final result is (15): $F = Gmw \cdot 4/3\,\pi b$. This shows that the gravitational attraction of a sphere for a point mass inside the sphere is directly proportional to the distance b between the point mass and the center of the sphere.

Finally, consider a point mass on the surface of the sphere. Then $b = a$ and the integral (11) has the value $2/3a$. Then the force becomes (16): $F = Gmw \cdot 4/3\,\pi a$. Now returning to a point inside the sphere, $b < a$, it can be considered as lying on the surface of an inner sphere, radius b, surrounded by a spherical shell of thickness $a - b$. The force of attraction of the inner sphere can now be found by (16) to be $F = Gmw \cdot 4/3\pi b$. But this is precisely the result given by (15) for the attraction of the whole sphere on m. Hence we have the remarkable result that the outer shell contributes absolutely nothing to the gravitational attraction of a sphere for an inner point mass.

A New Look at an Old Work Problem

BERT K. WAITS AND JERRY L. SILVER

BERT K. WAITS is Assistant Professor of Mathematics at The Ohio State University, Columbus, and is the Director of the CRIMEL program designed to individualize freshman mathematics instruction. Professor Waits is also the Regional Campus Mathematics Coordinator for the Mathematics Department. He has had several articles published in mathematical journals and is the co-author of several textbooks.

JERRY L. SILVER, Assistant Professor of Mathematics at The Ohio State University, is also actively involved in the CRIMEL program developing innovative instructional aides. He is the co-author of several textbooks, and is currently interested in the area of operations research and general optimization theory.

Many popular texts approach the concept of work in elementary integral calculus in a manner similar to the following.

DEFINITION. *The amount of work, W, done on an object in moving it from a to b along a coordinate line is given by*

$$W = \int_a^b F(x)\, dx,$$

where $F(x)$ is the force applied to the object at position x [1].

The first example after the development of the definition is invariably the classic spring problem using Hooke's law.

EXAMPLE. *Find the amount of work done in stretching a spring 6″ from its rest position if it takes a force of 5 lbs. to hold the spring 2″ from its rest position.*

SOLUTION According to Hooke's law there is a constant k so that $F(x) = kx$ is the force necessary to hold the spring x units from its rest position. Hence $F(2) = 5 = k \cdot 2$. So $k = 5/2$. Therefore

$$W = \int_0^6 \tfrac{5}{2}x\, dx = 45 \text{ inch-lbs.}$$

The next type of example in most texts is slightly more difficult. It typically involves pumping liquid out of a conical or parabolic container.

EXAMPLE. *Find the amount of work required to pump a liquid, weighting ρ lbs. per cubic foot, over the rim of a parabolic tank 4 feet deep with top radius 2 feet.*

Solution (Disk Method). Consider the interval $[0, 4]$ partitioned by points $y_0 = 0, y_1, y_2, \ldots, y_{n-1}, y_n = 4$ and the resulting "disks" of liquid with volume $\pi x_i^2 \Delta y_i$

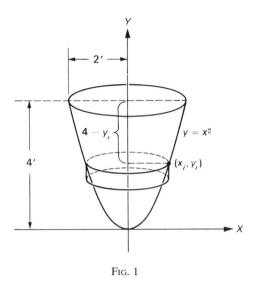

Fig. 1

$(\Delta y_i = y_i - y_{i-1})$. The amount of work required to move such a disk would be the weight of the liquid times the distance required to remove it to the top of the tank,

$$W_i = (4 - y_i)\rho\pi x_i^2 \Delta y_i.$$

An approximation to the total work done is then

$$W \approx \sum_{i=1}^{n} (4 - y_i)\rho\pi y_i \Delta y_i.$$

This expression leads to an integral

$$W = \int_0^4 \rho\pi(4 - y) y \, dy$$

$$= \frac{32\rho\pi}{3} \text{ foot-lbs. of work.}$$

In this example the usual definition of finding work directly as the integral of a force function over a certain interval is not applied. Some authors even claim that the work definition "... *cannot* be applied to the *total* (problem)..." [2], or they assert that this problem "... *requires* a somewhat different principle" [3] (emphasis added). Others use the "disk method" without mentioning that it is not a direct application of their work definition [4].

We do not want to imply that there is something incorrect or even pedagogically unsound about their approach. However, we feel the tank-type work problem can be solved in a direct, natural manner applying the usual definition of work as the integral of force over a certain distance. Unfortunately, it is easy to go astray, as did one student in the following solution.

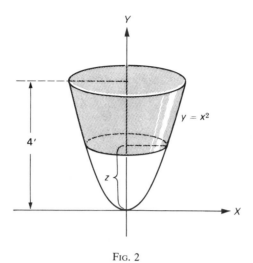

$$y = x^2$$

4′

z

FIG. 2

Student's Solution (Incorrect). To remove the liquid we push it up so that the level is z feet from bottom (FIGURE 2). For $z \in [0, 4]$, we take the force function to be the weight of the liquid remaining. Hence $F(z) = \rho V(z)$ where $V(z)$ is the shaded volume.

$$F(z) = \rho \int_z^4 \pi x^2 \, dy$$

$$= \rho \pi \int_z^4 y \, dy$$

$$= \frac{\rho \pi}{2} (16 - z^2).$$

So

$$W = \int_0^4 \frac{\rho \pi}{2} (16 - z^2) \, dz$$

$$= \frac{64 \rho \pi}{3} \qquad \left(not \; \frac{32 \rho \pi}{3} \right).$$

To alter the student's solution so as to arrive at the correct answer, we proceed as follows.

Correct Solution (Work Definition Method). We define a force function representing the weight of the remaining liquid at any given level. Let $z \in [0, 4]$. If $V(z)$ is the volume shaded below in FIGURE 3, then the force required to hold the liquid in the tank when the level is z feet from the bottom is simply $F(z) = \rho V(z)$.

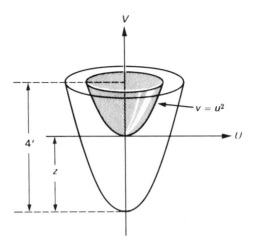

Fig. 3

It is easy to see that

$$V(z) = \int_0^{4-z} \pi u^2 \, dv$$

$$= \int_0^{4-z} \pi v \, dv$$

$$= \pi \frac{(4-z)^2}{2}.$$

Therefore,

$$W = \int_0^4 \rho V(z) \, dz$$

$$= \rho \pi \int_0^4 \frac{(4-z)^2}{2} \, dz$$

$$= \frac{32 \rho \pi}{3}.$$

Can you find the fallacy in using the volume described in the incorrect solution? Perhaps this topic would be an interesting problem for first-year calculus students.

REFERENCES

1. R. E. Johnson and F. L. Kiokemeister, *Calculus With Analytic Geometry*, 4th ed., Boston: Allyn and Bacon, 1969, p. 261.
2. G. B. Thomas, *Calculus and Analytic Geometry*, 4th ed., Reading, Mass.: Addison-Wesley Publishing Co., 1968, p. 225.
3. M. H. Protter and C. B. Morrey, *College Calculus With Analytic Geometry*, Reading, Mass.: Addison-Wesley Publishing Co., 1964, p. 245.
4. Louis Leithold, *The Calculus With Analytic Geometry*, 2nd ed., New York: Harper & Row, Publishers, 1972, p. 354.

Some Surprising Volumes of Revolution

G. L. Alexanderson
L. F. Klosinski

G. L. Alexanderson, Professor of Mathematics and Chairman of the Department of Mathematics at the University of Santa Clara, Santa Clara, California, is the co-author of several textbooks, including A First Undergraduate Course in Abstract Algebra, 1973.

L. F. Klosinski, Assistant Professor of Mathematics at the University of Santa Clara, is Director of the M.A.A. Visiting Lecturer Program in Northern California. He also directs the Santa Clara Invitational High School Mathematics Contest.

George Polya in his *Mathematics and Plausible Reasoning* calculates the volume of a "bead", a sphere with a right circular cylindrical hole (the axis of the cylinder passes through the center of the sphere). He goes even further to consider the cases where the hole in the bead may be conical or, indeed, in the form of a paraboloid of revolution. In each of these cases, the volume remarkably does not depend on the radius of the sphere from which the bead is made, but only on other parameters. The purpose of this note is to point out that similar striking formulas can be obtained for other combinations of conics.

We make use of the following integral to evaluate these volumes:

$$
\begin{aligned}
I &= \int_{x_1}^{x_2} (x - x_1)(x - x_2)\, dx \\
&= \left[(x - x_1)\frac{(x - x_2)^2}{2} \right]_{x_1}^{x_2} - \int_{x_1}^{x_2} \frac{(x - x_2)^2}{2}\, dx \\
&= \frac{(x_1 - x_2)^3}{6}.
\end{aligned}
$$

For the first problem described above, let $x = x_1$ and $x = x_2$ be the points of intersection of the line $y = k$ and $x^2 + y^2 = r^2$, $0 < k < r$. Then the volume of revolution (see Figure 1) formed by rotating the shaded region about the x-axis is

$$
\begin{aligned}
V &= \int_{x_1}^{x_2} \pi\big[(r^2 - x^2) - k^2\big]\, dx \\
&= -\pi \int_{x_1}^{x_2} (x - x_1)(x - x_2)\, dx = -\pi \frac{(x_1 - x_2)^3}{6} = \frac{\pi h^3}{6}
\end{aligned}
$$

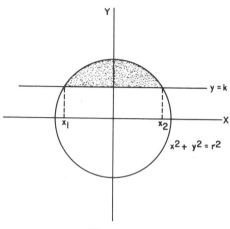

Figure 1.

where $h = (x_2 - x_1)$ is the "height" of the bead. (In the above calculation, we have made use of the fact that two quadratics are identically equal if they have the same roots and the same leading coefficient.)

We note that the volume of this bead is a function only of the height, h, and does not depend on r, the radius of the sphere. Consider then two beads both of height 2 inches: one formed from a sphere the size of an orange and the other formed from a sphere the size of the earth. Remarkable as it may appear, both of these beads have exactly the same volume! This would seem to defy the intuition, although it is made more plausible by considering the thickness of each bead.

Similar striking results appear in the case of the rotation of areas between other conics. Let us consider the volume of rotation we get by rotating about the x-axis the region between the parabola $y^2 = ax + b$ and the line $y = mx + c$ (Figure 2).

$$V = \int_{x_1}^{x_2} \pi[ax + b - (mx + c)^2]\, dx$$

$$= -\pi m^2 I = \frac{\pi h^3}{6}\, m^2.$$

The volume depends on m, the slope of the line, and again on h. But intercepts of the line and the location of the focus of the parabola, parameters which one might reasonably expect in the volume formula, do not appear, at least explicitly.

Other volumes for various combinations of conics appear below, with implicit assumptions on the values of the parameters so the two curves in each case will intersect in two points in order for the curves to form a region, which is then rotated about the x-axis to form a solid of revolution.

The volume formulas, thus, involve only h, the "height" of the solid formed, and three other parameters, where appropriate: m, the slope of a line, and the ratios a/b and c/d in the cases of hyperbolas and ellipses. In case (1), if the ellipse is a circle, the volume is $\pi h^3(1 + m^2)/6$ which can be simplified to $\pi h p^2/6$ where

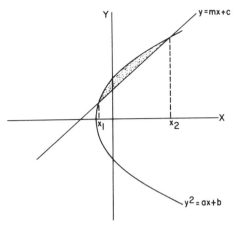

Figure 2.

	Upper Curve	Lower Curve	Volume
(1)	$\dfrac{(x-k)^2}{b^2} + \dfrac{y^2}{a^2} = 1$	$y = mx + n$	$\dfrac{\pi h^3}{6}\left(\dfrac{a^2}{b^2} + m^2\right)$
(2)	$\dfrac{(x-k)^2}{d^2} + \dfrac{y^2}{c^2} = 1$	$\dfrac{(x-j)^2}{b^2} + \dfrac{y^2}{a^2} = 1$	$\dfrac{\pi h^3}{6}\left(\dfrac{c^2}{d^2} - \dfrac{a^2}{b^2}\right)$
(3)	$\dfrac{(x-k)^2}{b^2} + \dfrac{y^2}{a^2} = 1$	$y^2 = ex + f$	$\dfrac{\pi h^3}{6}\dfrac{a^2}{b^2}$
(4)	$\dfrac{(x-k)^2}{b^2} + \dfrac{y^2}{a^2} = 1$	$\dfrac{y^2}{c^2} - \dfrac{(x-r)^2}{d^2} = 1$	$\dfrac{\pi h^3}{6}\left(\dfrac{c^2}{d^2} + \dfrac{a^2}{b^2}\right)$
(5)	$y = mx + n$	$\dfrac{y^2}{a^2} - \dfrac{(x-r)^2}{b^2} = 1$	$\dfrac{\pi h^3}{6}\left(\dfrac{a^2}{b^2} - m^2\right)$
(6)	$\dfrac{y^2}{c^2} - \dfrac{(x-r)^2}{d^2} = 1$	$\dfrac{y^2}{a^2} - \dfrac{(x-s)^2}{b^2} = 1$	$\dfrac{\pi h^3}{6}\left(\dfrac{a^2}{b^2} - \dfrac{c^2}{d^2}\right)$
(7)	$y^2 = ex + f$	$\dfrac{y^2}{a^2} - \dfrac{(x-r)^2}{b^2} = 1$	$\dfrac{\pi h^3}{6}\dfrac{a^2}{b^2}$
(8)	$y^2 = ex + f$	$y = mx + n$	$\dfrac{\pi h^3}{6} m^2$

p is the length of the cut. Of course, if $m = 0$, we have the first case examined, the volume of an ordinary bead.

The volume problems described above provide some curious examples for classes covering the calculation of volumes of revolution. Students, when asked what parameters they expect to appear in an answer, will surely be surprised with some of the results above and a lesson can be learned from these: intuition, while essential, is not always reliable.

Surface Area and the Cylinder Area Paradox

Frieda Zames

Frieda Zames is Assistant Professor of Mathematics at New Jersey Institute of Technology where she has been employed since 1966. For the past three years, she has also been working with a federally funded program devising curricula and teaching remedial mathematics to educationally deprived adults in Newark. She earned the Ph.D. in Mathematics Education at New York University in 1972.

Introduction. In 1890, H. A. Schwartz astounded the mathematical world by publishing an example which invalidated the accepted definition of surface area [4]. The "cylinder area paradox" is the name given to this insight that opened the floodgates of mathematical inquiry to investigations into surfaces and surface areas [1].

The primary reason that mathematicians were led astray was the assumption that there is a direct analogy between arc length and surface area. In 1948, Tibor Rado [2] stated: "The present status of the theory supports the view that far-reaching analogies do exist. But the analogies lie deep while the discrepancies are conspicuous . . . ." It seems preferable to indicate the incorrect analogy assumed before illustrating the paradox that pointed up the inconsistencies of an uncritical acceptance.

The Assumption. The length $L(C)$ of a finite curve C will be defined in the usual way. Assume C_i is any sequence of inscribed polygonal arcs such that $C_i \to C$ uniformly. If P is an inscribed polygonal curve, then $L(P) \leqslant L(C)$. Also, if $\epsilon > 0$ is given, there exists a polygonal curve P_ϵ such that $L(P_\epsilon) > L(C) - \epsilon$. These two intuitive ideas show that

$$L(C) = \sup_P L(P)$$

where P is taken over all inscribed polygons. It follows from this that there exists a sequence of polygonal approximations C_i where $C_i \to C$ uniformly such that

$$L(C_i) \to L(C).$$

It was this definition that was confounded when an attempt was made to generalize to areas.

The surface area $A(S)$ of a bounded surface was defined in a manner analogous to the definition of the length of a curve given above. As an example, the following definition of the area of a surface S bounded by a curve is similar to that given by J. A. Serrat in 1880 [5]. Let S_i be any sequence of inscribed polyhedral surfaces which converge uniformly to the given surface S, i.e., $S_i \to S$ uniformly. If P is an inscribed polyhedral surface of surface S, then $A(P) \leqslant A(S)$. In addition, if $\epsilon > 0$

is given, there exists a polyhedral surface P_ϵ such that $A(P_\epsilon) > A(P) - \epsilon$. Therefore,

$$A(S) = \sup_P A(P),$$

where P is taken over all inscribed polyhedral surfaces. Thus, it was incorrectly
concluded that there exists a sequence of inscribed polyhedrons S_i where $S_i \to S$
uniformly such that

$$A(S_i) \to A(S).$$

This definition was shown to be untenable by H. A. Schwartz in a letter dated
December 25, 1880, to A. Genocchi and independently by G. Peano in a class
lecture of May 1882 [1].

The Cylinder Area Paradox. Until Schwartz published this paradox, the definition of surface areas just described was the generally accepted one. The importance
of this example is that the paradox arose from using the assumed definition on an
extremely simple surface, a cylinder [6].

Let S be the lateral surface of a right circular cylinder of height h and radius r.
Divide S into m bands by circles lying in planes parallel to the base such that each
band has altitude h/m. Select two adjacent bands, as in the figure below, and
divide each of the three circles into n congruent arcs such that the endpoints of
these arcs are the vertices of the inscribed triangles. The arcs on the top and the
bottom will be vertically aligned but the arcs in the middle circle are all displaced
through one-half an arc-length. Assume ΔBAC is any one of the congruent
inscribed triangles and let D and E be the midpoints of segment BC and arc BC,
respectively. Let O be the center of the circle containing arc BC, so that ΔBOC is
parallel to the base of the cylinder and assume $\angle BOC = 2\theta$. Thus, $\theta = \pi/n$ and

$$|BC| = 2r \sin \theta = 2r \sin \frac{\pi}{n}.$$

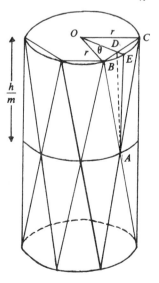

Figure 1.

To find the area of $\triangle BAC$, it is first necessary to apply the Pythagorean theorem to $\triangle ADE$ in order to calculate the altitude $|AD|$:

$$|DE| = |OE| - |OD| = r - r\cos\theta = r\left(1 - \cos\frac{\pi}{n}\right),$$

and

$$|AD|^2 = |AE|^2 + |DE|^2 = \left(\frac{h}{m}\right)^2 + r^2\left(1 - \cos\frac{\pi}{n}\right)^2,$$

so that

$$\text{area }(\triangle BAC) = \frac{1}{2}|BC|\,|AD|,$$

$$= \frac{1}{2}\left(2r\sin\frac{\pi}{n}\right)\sqrt{\left(\frac{h}{m}\right)^2 + r^2\left(1 - \cos\frac{\pi}{n}\right)^2},$$

$$= \left(r\sin\frac{\pi}{n}\right)\sqrt{\left(\frac{h}{m}\right)^2 + r^2\left(1 - \cos\frac{\pi}{n}\right)^2}.$$

In one band of the cylinder between two consecutive parallel circles there are $2n$ of these congruent triangles. Thus, for the m bands, there are $2mn$ such congruent triangles and the sum of their areas is

$$A(m, n) = 2mn\left(r\sin\frac{\pi}{n}\right)\sqrt{\left(\frac{h}{m}\right)^2 + r^2\left(1 - \cos\frac{\pi}{n}\right)^2},$$

$$= 2r\left(n\sin\frac{\pi}{n}\right)\sqrt{h^2 + (mr)^2\left(1 - \cos\frac{\pi}{n}\right)^2}.$$

It will now be seen that the way in which the limits are taken determine the value of

$$\lim_{m, n\to\infty} A(m, n).$$

Case 1. Let $n \to \infty$ first with m held fixed and then let $m \to \infty$.

$$\lim_{m\to\infty}\left[\lim_{n\to\infty} A(m, n)\right]$$

$$= \lim_{m\to\infty}\left[\lim_{n\to\infty} 2r\left(n\sin\frac{\pi}{n}\right)\sqrt{h^2 + (mr)^2\left(1 - \cos\frac{\pi}{n}\right)^2}\right].$$

Each of the following functions can be approximated by the first few terms of its Taylor series expansion, so that

$$\sin\frac{\pi}{n} \approx \frac{\pi}{n} \qquad \text{and} \qquad \cos\frac{\pi}{n} \approx 1 - \frac{\pi^2}{2n^2}.$$

These approximations are sufficiently accurate for evaluating the following limits:

$$\lim_{n\to\infty}\left(n\sin\frac{\pi}{n}\right) = \lim_{n\to\infty} n\left(\frac{\pi}{n}\right) = \pi,$$

and

$$\lim_{n\to\infty}\left(1 - \cos\frac{\pi}{n}\right) = \lim_{n\to\infty}\frac{\pi^2}{2n^2} = 0.$$

These two results imply that

$$\lim_{m\to\infty}\left[\lim_{n\to\infty} A(m, n)\right] = \lim_{m\to\infty} 2r\pi h = 2\pi rh.$$

It is useful to examine this limit geometrically. When $n \to \infty$ first and m remains fixed, the number of triangles in a band increases indefinitely and approaches the surface area of the band. In this case, since the number of bands is considered fixed at this point, the final answer turns out to be independent of m. Thus, the sum of the areas of these triangles approaches the expected value for the lateral surface area.

Case 2. Let $m \to \infty$ first with n held fixed and then let $n \to \infty$.

$$\lim_{n\to\infty}\left[\lim_{m\to\infty} A(m, n)\right]$$

$$= \lim_{n\to\infty}\left[\lim_{m\to\infty} 2r\left(n \sin \frac{\pi}{n}\right)\sqrt{h^2 + (mr)^2\left(1 - \cos \frac{\pi}{n}\right)^2}\,\right]$$

$$= 2r \lim_{n\to\infty}\left[\left(n \sin \frac{\pi}{n}\right)\lim_{m\to\infty}\sqrt{h^2 + (mr)^2\left(1 - \cos \frac{\pi}{n}\right)^2}\,\right]$$

$$= 2r\left[\lim_{n\to\infty}\left(n \sin \frac{\pi}{n}\right)\cdot \infty\right] = \infty.$$

As in Case 1, this limit is interesting to look at geometrically. When $m \to \infty$ first and n is held fixed, the number of bands increase indefinitely while the number of triangles in any band is constant. Also, as $m \to \infty$, for fixed n, $A \to E$ which implies that ΔBAC becomes practically perpendicular to the surface! It is no wonder that the area of a sequence of sums of such triangles approaches infinity.

Case 3. Let m and n approach infinity simultaneously so that $m/n^2 = c$ where $c \geqslant 0$ is a constant. Clearly, $A(m, n) = A(cn^2, n)$.

$$\lim_{n\to\infty} A(cn^2, n) = \lim_{n\to\infty}\left[2r\left(n \sin \frac{\pi}{n}\right)\sqrt{h^2 + (rcn^2)^2\left(1 - \cos \frac{\pi}{n}\right)^2}\,\right]$$

$$= 2r \lim_{n\to\infty}\left(n \sin \frac{\pi}{n}\right)\cdot\left[\lim_{n\to\infty}\sqrt{h^2 + (rc)^2\left(n^2\left(1 - \cos \frac{\pi}{n}\right)\right)^2}\,\right]$$

Using the Taylor series approximation for $1 - \cos(\pi/n)$ as $\pi^2/2n^2$,

$$\lim_{n\to\infty}\left[n^2\left(1 - \cos \frac{\pi}{n}\right)\right] = \frac{\pi^2}{2}.$$

By combining these results, the calculation can now be completed:

$$\lim_{n\to\infty} A(cn^2, n) = 2\pi r\sqrt{h^2 + \frac{r^2\pi^4 c^2}{4}}\,.$$

Since $c \geqslant 0$, all answers greater than or equal to $2\pi rh$ are possible and thus

$$\lim_{n \to \infty} A(cn^2, n) \geqslant 2\pi rh.$$

To illustrate this, let $c = 2\sqrt{3} \, h/\pi^2 r$. Consequently,

$$\lim_{n \to \infty} A(cn^2, n) = 2\pi r \sqrt{h^2 + \frac{r^2 \pi^4}{4} \cdot \frac{12h^2}{\pi^4 r^2}} = 2\pi r \sqrt{h^2 + 3h^2} = 4\pi rh.$$

Using the original evaluation of

$$A(m, n) = 2r\left(n \sin \frac{\pi}{n}\right)\sqrt{h^2 + (rm)^2\left(1 - \cos \frac{\pi}{n}\right)^2},$$

it is clear that $2\pi rh$ is the minimum value for $\lim_{m, n \to \infty} A(m, n)$. The essence of the paradox is the realization that there is no unique answer for $\lim_{m, n \to \infty} A(m, n)$. However, $2\pi rh$ is a lower bound for all these limits.

It might be interesting for the reader to prove:

1. $\lim_{n \to \infty} A(cn, n) = 2\pi rh$,
2. $\lim_{n \to \infty} A(cn^3, n) = \infty$,
3. $\lim_{n \to \infty} A(c\sqrt{n}, n) = 2\pi rh$.

Conclusion. Since the publication of Schwartz's paradox became known, a great number of new interpretations for a theory of surface areas have been proposed. Rado [3] points out that "In most cases the idea of approximating the given surface by polyhedrons has been altogether dropped." Lebesgue's definition of surface area devised in 1902 is an exception. According to Lebesgue, the area of a surface is defined as

$$A(S) = \inf\left[\lim_{i \to \infty} \inf A(S_i)\right]$$

where the infimum of $A(S_i)$ is taken over all sequences of polyhedral surfaces that converge uniformly to S. This definition of surface area created by Lebesgue circumvented the contradictions posed by the cylinder area paradox and also maintained the surface area concept as basically geometrical. However, the theory of surface area is still surprisingly incomplete although very satisfactory results have been obtained for certain special classes of surfaces [3].

REFERENCES

1. Lamberto Cesari, Surface Area, Princeton University Press, New York, 1956, 24–26.
2. Tibor Rado, Length and Area, Amer. Math. Soc. Colloq. Publication, New York, 1948, 6.
3. ———, On the Problem of Plateau, Chelsea, New York, 1951, 2–18.
4. H. A. Schwartz, Sur une définition erronée de l'aire surface courte, Gesammelte Mathematische Abhandlungen, Berlin, 1890, II, 309–311. This example cited in a letter from H. A. Schwartz was first published in the second mimeographed edition of the Cours by C. Hermite, 1882–1883, 35–36.
5. J. A. Serrat, Cours de Calcul Différentiel et Intégral (2nd ed.), Paris, 1880, 293.
6. G. B. Thomas, Jr., Calculus and Analytic Geometry (4th ed.), Addison-Wesley, Reading Mass., 1968, 568–571.

Using Integrals to Evaluate Voting Power

Philip D. Straffin, Jr.

Philip Straffin, Jr., received his Ph.D. from the University of California at Berkeley in 1971 and is now Associate Professor of Mathematics at Beloit College. For the past several years he has taught a course on game theory and the social sciences in Beloit's interdisciplinary division.

When we teach integration in an elementary calculus course, we owe it to our students to present some realistic samples of how integrals can be used in areas outside mathematics and mathematical physics. Computing volumes of solids and doing work problems is not enough. In this article, I would like to present an application of the integration of polynomials which has proved popular with my Calculus I students.

The problem is one from political science: how to numerically evaluate the power of voters in a *weighted voting body*. A weighted voting body can be represented by a symbol

$$[q; w_1, w_2, \ldots, w_n],$$

where there are n voters, the ith voter casts w_i votes, and a quota of q votes is necessary to pass a bill. For example, the symbol

$$\text{(1)} \qquad \begin{array}{c} [7; 4, 3, 2, 1] \\ \text{A B C D} \end{array}$$

represents a body in which there are four voters (call them A, B, C, D) casting 4, 3, 2, 1 votes respectively, and 7 votes are necessary to pass a bill.

Weighted voting bodies are fairly common in political situations. Classic examples include voting by shareholders in a corporation, several United Nations organizations, the World Bank, the European Economic Community, the New Mexico legislature, and many county governments in New York State. A multitude of other examples can be found in [1]. A legislature in which each member casts only one vote, but where members belong to different political parties and vote under strict party discipline, can also be thought of as a weighted voting body. In this interpretation, example (1) might represent a legislature of 10 members, with 4 belonging to party A, 3 to party B, 2 to party C, and 1 to party D.

The naive way to think of the distribution of power in a weighted voting body like that of example (1) is to suppose that power is in strict proportion to number of votes. Thus, A has 40% of the votes and hence should have 40% of the power. A little reflection should convince students of the naivete of this supposition. For instance, note that in (1) A has *veto power*: even if B, C, and D all favor a bill, it cannot pass without A's approval. This observation should lead us to believe that A might have considerably more than 40% of the power. Two other compelling examples are:

(2) [6; 7, 1, 1, 1] Here A has 70% of the vote, but she clearly has all the power. A is a *dictator*, in the sense that a bill passes if and only if A votes for it.

(3) [6; 3, 3, 3, 1] Here D has 10% of the vote, but no power. D's vote can never make any difference to the outcome, and D is called a *dummy*.

To obtain a more realistic measure of power than just proportion of votes, consider the following model. Let us suppose that each bill which comes before the voting body can be assigned a number $p \in [0, 1]$, where p is the probability that any member of the body will vote for the bill. We can think of p as the "level of acceptability" of the bill. Some bills will be generally unacceptable (p near 0) and will usually (but not always) be overwhelmingly defeated; others will be generally acceptable (p near 1) and will usually pass; and others will be controversial (p near $1/2$).

We think of each member of the voting body as being concerned with the question "What is the probability that my vote will *make a difference* to the outcome on a bill?" This probability will, in general, vary as a function of p. We define power as follows:

Definition. The *power* of a member in a weighted voting body is the probability that that member's vote will make a difference to the outcome on a bill with acceptability p, averaged over all p between 0 and 1.

The notion of integration comes in, of course, when we take the average on $[0, 1]$. An example should make clear how this works. Consider

(4)
$$\frac{[3; 2, 1, 1]}{\text{A B C}}.$$

Suppose a bill with acceptability p comes before this body. Thus each member votes for the bill with probability p. Notice that A's vote will make a difference to the outcome if either B or C, or both, vote for the bill. (If they both vote against it, it will fail regardless of what A does.) Thus the probability that A's vote will make a difference is

$$f_A(p) = \underset{\text{B yes, C no}}{p(1-p)} + \underset{\text{B no, C yes}}{(1-p)p} + \underset{\text{B, C yes}}{p^2} = 2p - p^2.$$

Similarly, B's vote will make a difference if A votes "yes" and C votes "no". (If A votes no, the bill will fail regardless of what B does; if A and C both vote yes, the bill will pass regardless of what B does.) Thus

$$f_B(p) = \underset{\text{A yes, C no}}{p(1-p)} = p - p^2.$$

By symmetry, we also have

$$f_C(p) = p - p^2.$$

Our measure of power is traditionally denoted by the letter ϕ. Thus we have, according to our definition,

$$\phi_A = \int_0^1 f_A(p) \, dp = \int_0^1 (2p - p^2) \, dp = 2/3$$

$$\phi_B = \phi_C = \int_0^1 (p - p^2) \, dp = 1/6.$$

A has two-thirds of the power, with B and C each having one-sixth. This seems reasonable if we note that A has veto power in this example.

Students should enjoy checking that in example (1),

$$f_A(p) = p + p^2 - p^3 \qquad \phi_A = 7/12$$

$$f_B(p) = p - p^3 \qquad \phi_B = 3/12$$

$$f_C(p) = f_D(p) = p^2 - p^3 \qquad \phi_C = \phi_D = 1/12.$$

Thus A, with 40% of the vote, has 58% of the power. C and D, with different proportions of the vote, have the same power. This simply reflects the fact that C's vote is no more likely to be decisive to the outcome on a bill than is D's.

Students should also check that this measure of power gives the intuitive answer for examples like (2) and (3). More complicated examples than (1) or (4) are good exercises in elementary probability manipulations. Power calculations for all "essentially different" voting bodies with four or fewer members are given in the appendix of [5].

The power index ϕ calculated according to our definition above has an interesting history in political science. It was first defined by Lloyd Shapley, a mathematical game theorist at the RAND Corporation, and Martin Shubik, an economist at Yale, in a classic paper in 1954 [4]. It has been known since as the Shapley-Shubik power index and has become the most widely accepted measure of voting power in political science. See [3], for example. However, Shapley and Shubik's original definition was completely combinatorial, and seemingly unrelated to the definition given above. The equivalence of the two definitions was first shown in a complicated context in [2], and more simply in [6]. References [1] and [5] include general discussions of this and other power indices.

REFERENCES

1. William Lucas, Measuring Power in Weighted Voting Systems, in Case Studies in Applied Mathematics, CUPM, 1976, 42–106.
2. Guillermo Owen, Multilinear Extensions of Games, Management Science, Series A, 18 (1972) P64–P79.
3. W. H. Riker and P. Ordeshook, An Introduction to Positive Political Theory, Prentice-Hall, Englewood Cliffs, N. J., 1973, chapter 6.
4. L. S. Shapley and M. Shubik, A Method for Evaluating the Distribution of Power in a Committee System, American Political Science Review, 48 (1954) 787–792.
5. P. D. Straffin, Jr., Power Indices in Politics, MAA Modules in Applied Mathematics, Ithaca, N. Y., 1976.
6. P. D. Straffin, Jr., Homogeneity, Independence and Power Indices, Public Choice, 30 (1977) 107–18.

Area of a Parabolic Region

R. Rozen and A. Sofo, Royal Melbourne Institute of Technology, Melbourne, Australia

Although students can quickly recall that the area of a circle is πr^2 and the area of an ellipse is πab, there does not appear to be a standard formula that they can recall when dealing with areas of parabolic regions. Thus it may be instructive to prove the following:

The area bounded by the parabola $y = ax^2 + bx + c$ and the x-axis is

$$A = \frac{\Delta^{3/2}}{6a^2},$$ (1)

when $\Delta = b^2 - 4ac$ is positive. In particular,

$$A = \frac{2}{3} BH,$$ (2)

where B and H, respectively, denote the base length and height of the enclosed area.

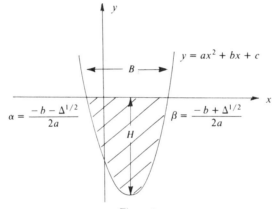

Figure 1.

The area of the parabolic region is given by:

$$A = \left| \int_{\alpha}^{\beta} (ax^2 + bx + c)\, dx \right| = \frac{a}{3}(\beta^3 - \alpha^3) + \frac{b}{2}(\beta^2 - \alpha^2) + c(\beta - \alpha).$$ (3)

Now from

$$\beta + \alpha = \frac{-b}{a}, \quad \beta - \alpha = \frac{\Delta^{1/2}}{a}, \quad \text{and} \quad \alpha\beta = \frac{c}{a},$$

we compute:

$$\beta^2 - \alpha^2 = \frac{-b\Delta^{1/2}}{a^2} \quad \text{and} \quad \beta^3 - \alpha^3 = (\beta - \alpha)\left[(\beta + \alpha)^2 - \alpha\beta\right] = \frac{(b^2 - ac)\Delta^{1/2}}{a^3}.$$

332

Substitution of these expressions into (3) yields (1). We can now easily verify (2) by completing the square

$$y = a\left(x + \frac{b}{2a}\right)^2 - \frac{\Delta}{4a}$$

and observing that y has extremal value $\frac{-\Delta}{4a}$ when $x = \frac{-b}{2a}$. Thus, $H = \left|\frac{\Delta}{4a}\right|$. And since $B = \beta - \alpha = \frac{\Delta^{1/2}}{a}$, it is now clear that

$$A = \frac{\Delta^{3/2}}{6a^2} = \frac{2}{3}\left(\frac{\Delta^{1/2}}{a}\right)\left(\frac{\Delta}{4a}\right) = \frac{2}{3}BH.$$

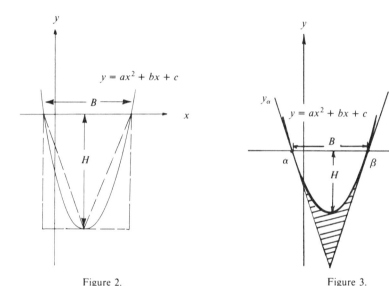

Figure 2.   Figure 3.

Example 1.   (Figure 2)   Comparing the parabolic area $A_p = \frac{2}{3}BH$ with the area $A_T = \frac{1}{2}BH$ of the inscribed triangle and the area $A_R = BH$ of the circumscribed rectangle, we see that

$$A_T = \frac{3}{4}A_p, \quad A_p = \frac{2}{3}A_R, \quad \text{and} \quad A_T = \frac{1}{2}A_R.$$

Example 2.   (Figure 3)   Find the area between the parabola $y = ax^2 + bx + c$ and the tangents to the parabola at its roots. Since $2a\alpha = -b - \Delta^{1/2}$ and $2a\beta = -b + \Delta^{1/2}$, we see that

$$y'(\alpha) = 2a\alpha + b = -\Delta^{1/2} \quad \text{and} \quad y'(\beta) = 2a\beta + b = \Delta^{1/2}.$$

Thus, the equations of the tangent lines at $x = \alpha$ and at $x = \beta$ are, respectively,

$$y_\alpha = -\Delta^{1/2}(x - \alpha) \quad \text{and} \quad y_\beta = \Delta^{1/2}(x - \beta).$$

Since these lines intersect when $x = \dfrac{\alpha + \beta}{2}$, the triangle formed by these tangents has height $|\Delta^{1/2}(\dfrac{\alpha - \beta}{2})| = \dfrac{\Delta}{a} = 2H$. Accordingly, the shaded area is

$$A = \tfrac{1}{2}(B)(2H) - \tfrac{2}{3}BH = \tfrac{1}{3}BH$$

$$= \frac{\Delta^{3/2}}{12a^2}.$$

BIBLIOGRAPHIC ENTRIES: APPLICATIONS

1. *Monthly* Vol. 76, No. 4, pp. 355–366. A. W. Goodman and Gary Goodman, Generalizations of the theorem of Pappus.

2. *Monthly* Vol. 83, No. 1, pp. 26–30. M. S. Klamkin and D. J. Newman, Inequalities and identities for sums and integrals.

3. *Monthly* Vol. 84, No. 7, pp. 534–541. Sherwood F. Ebey and John J. Beauchamp, Larval fish, power plants, and Buffon's needle problem.

4. *Monthly* Vol. 91, No. 2, p. 139. Charles A. Cable, The disk and shell method.

5. *Monthly* Vol. 98, No. 2, pp. 139–143. James P. Butler, The perimeter of a rose.

The width of a rose petal is treated in *Monthly* vol. 97, No. 10, p. 907–911, reproduced in Ch. 5, Sec. (b) of this volume, p. 128; for the area see the next entry.

6. *Math. Mag.* Vol. 43, No. 3, pp. 156–157. A. A. Aucoin, On the leaf curves.

7. *Math. Mag.* Vol. 53, No. 1, pp. 36–39. Scott E. Brodie, Archimedes' axioms for arc-length and area.

8. *Math. Mag.* Vol. 54, No. 5, pp. 261–269. P. W. Kuchel and R. J. Vaughn, Average lengths of chords in a square.

9. *Math. Mag.* Vol. 56, No. 2, pp. 104–110. Andrew M. Rockett, Arc length, area, and the arcsine function.

10. *Math. Mag.* Vol. 64, No. 1, pp. 32–34. James E. McKenney, Finding the volume of an ellipsoid using cross-sectional slices.

Shows how to find the volume with 3 equally-spaced slices. No calculus, but good calculus-related material.

11. TYCMJ Vol. 7, No. 4, pp. 43–45. John W. Dawson, Jr., Some ridge-length problems.

Gives three examples of finding arc lengths of cross-sections of solids.

12. TYCMJ Vol. 8, No. 4, pp. 203–206. James A. Burns, Some effects of rationing.

13. CMJ Vol. 17, No. 5, pp. 414–415. Frances W. Lewis, Defining area in polar coordinates.

14. CMJ Vol. 21, No. 3, pp. 225–227. Douglass L. Grant, Moments on a rose petal.

15. CMJ Vol. 21, No. 5, pp. 384–389. Daniel Cass and Gerald Wildenberg, Relations between surface area and volume in lakes.

16. CMJ Vol. 22, No. 4, pp. 318–321. Anthony Lo Bello, The volume and centroid of the step pyramid of Zoser.

(e)

MULTIPLE INTEGRALS AND LINE INTEGRALS

Change of Variables in Multiple Integrals: Euler to Cartan

From formalism to analysis and back; methods of proof come full circle.

VICTOR J. KATZ

University of the District of Columbia
Washington, DC 20008

Leonhard Euler first developed the notion of a double integral in 1769 [7]. As part of his discussion of the meaning of a double integral and his calculations of such an integral, he posed the obvious question: what happens to a double integral if we change variables? In other words, what happens to $\int\int_A f(x, y)\, dx\, dy$ if we let $x = x(t, v)$ and $y = y(t, v)$ and attempt to integrate with respect to t and v? The answer is provided by the change-of-variable theorem, which states that

$$\int\int_A f(x, y)\, dx\, dy = \int\int_B f(x(t, v), y(t, v)) \left| \frac{\partial x}{\partial t} \frac{\partial y}{\partial v} - \frac{\partial x}{\partial v} \frac{\partial y}{\partial t} \right| dt\, dv \qquad (1)$$

where the regions **A** and **B** are related by the given functional relationship between (x, y) and (t, v). This result, and its generalization to n variables, are extremely important in allowing one to transform complicated integrals expressed in one set of coordinates to much simpler ones expressed in a different set of coordinates. Every modern text in advanced calculus contains a discussion and proof of the theorem. (For example, see [5], [1], [18].)

Leonhard Euler 1707–1783

Euler interpreted this result formally; namely, he considered $dx\,dy$ as an "area element" of the plane. So his aim was to show that his area element transformed into a new "area element" $\left|\dfrac{\partial x}{\partial t}\dfrac{\partial y}{\partial v}-\dfrac{\partial x}{\partial t}\dfrac{\partial y}{\partial v}\right|\,dt\,dv$ under the given change of variables. Obviously, if we merely change coordinates by a translation, rotation, and/or reflection, the area element is transformed into a congruent one. So Euler noted that if t and v are new orthogonal coordinates related to x and y by a translation through constants a and b, a clockwise rotation through the angle θ whose cosine is m, and a reflection through the x-axis, i.e.,

$$x = a + mt + v\sqrt{1 - m^2}$$

$$y = b + t\sqrt{1 - m^2} - mv,$$

then $dx\,dy$ should be equal to $dt\,dv$. Unfortunately, when he performed the obvious formal calculation

$$dx = m\,dt + dv\sqrt{1 - m^2}\,,$$

$$dy = dt\sqrt{1 - m^2} - m\,dv$$

and multiplied the two equations, he arrived at

$$dx\,dy = m\sqrt{1 - m^2}\,dt^2 + (1 - 2m^2)\,dt\,dv - m\sqrt{1 - m^2}\,dv^2,$$

which, he noted, was obviously wrong and even meaningless (see FIGURE 1). Even more so, then, would a similar calculation be wrong if t and v were related to x and y by more complicated transformations. It was thus necessary for Euler to develop a workable method; i.e., one that in the above situation gives $dx\,dy = dt\,dv$ and, in general, gives $dx\,dy = Z\,dt\,dv$, where Z is a function of t and v.

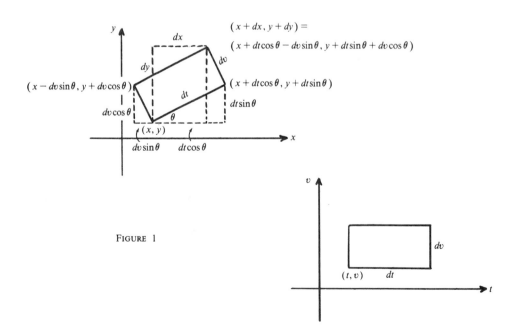

FIGURE 1

To see how he arrived at his method, we must first consider his definition and calculation of double integrals. After noting that $\int\int Z\,dx\,dy$ means an "indefinite" double integral, i.e., a function of x and y which when differentiated first with respect to x and with respect to y gives $Z\,dx\,dy$, Euler proceeded to calculate "definite" integrals over specified planar regions $\mathbf{A}$ in the way familiar to calculus students. Thus, he wrote the integral as $\int dx \int Z\,dy$ and holding x constant, he integrated with respect to y between the functions $y=f_1(x)$ and $y=f_2(x)$ which bounded the region $\mathbf{A}$; finally he integrated with respect to x between its minimum and maximum values in $\mathbf{A}$. He interpreted this integral in the obvious way as a volume. In particular, he integrated $\int\int\sqrt{c^2-x^2-y^2}\,dx\,dy$ over various regions to calculate volumes of portions of a sphere. Finally, he noted that $\int\int_{\mathbf{A}} dx\,dy$ is precisely the area of $\mathbf{A}$ and explicitly calculated the area of the circle given by $(x-a)^2+(y-b)^2=c^2$ to be πc^2.

Since the method of double integration involves leaving one variable fixed while dealing with the other, Euler proposed a similar method for the change-of-variable problem: change variables one at a time. First he introduced the new variable v and assumed that y could be represented as a function of x and v. So $dy=P\,dx+Q\,dv$ where P and Q are the appropriate partial derivatives. Now by assuming x fixed, he obtained $dy=Q\,dv$ and $\int\int dx\,dy=\int\int Q\,dx\,dv=\int dv\int Q\,dx$ (FIGURE 2). Next, he let x be a function of t and v and put $dx=R\,dt+S\,dv$. So by holding v constant, he calculated $\int dv\int Q\,dx=\int dv\int QR\,dt=\int\int QR\,dt\,dv$. This gave Euler the first solution to his problem: $dx\,dy=QR\,dt\,dv$.

Obviously, this was not completely satisfactory, since Q may well depend on x, and, in addition, the method was not symmetric. So Euler continued, now representing y as a function of t and v, hence $dy=T\,dt+V\,dv$. Then, formally, $dy=P\,dx+Q\,dv=P(R\,dt+S\,dv)+Q\,dv=PR\,dt+(PS+Q)\,dv$. So $PR=T$ and $PS+Q=V$, which gives $QR=VR-ST$. Euler's final answer was that $dx\,dy=(VR-ST)\,dt\,dv$. He noted again that simply multiplying the expressions for dx and dy together and rejecting the terms in dt^2 and dv^2 gives $(RV+ST)\,dt\,dv$, which differs by a sign from the correct answer. After a further note that one must always take the absolute value of the expression $VR-ST$ (since area is positive) he proceeded to confirm the correctness of his result through several increasingly complex examples.

This "proof" was typical of Euler's use of formal methods in many parts of his vast mathematical work. As a developer of algorithms to solve problems of various sorts, Euler has never been surpassed. (We can see that Euler's method, in modern notation, amounts to first factoring the transformation $x=x(t,v)$, $y=y(t,v)$ into two transformations, the first being $x=x(t,v)$, $v=v$ and the second $x=x$, $y=y(x,v)$. This can be done by "solving" $x=x(t,v)$ for t in the form $t=h(x,v)$ and then writing $y=y(h(x,v),v)$. Then $P=y_1\dfrac{\partial h}{\partial x}$, $Q=y_1\dfrac{\partial h}{\partial v}+y_2$,

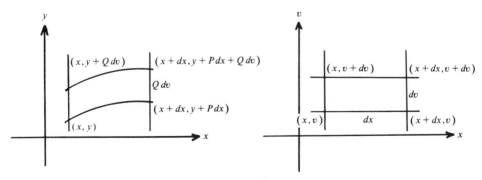

FIGURE 2

Joseph-Louis Lagrange 1736–1813

$R = x_1$, $S = x_2$, $T = y_1$ and $V = y_2$, where subscripts denote partial derivatives. Since $x(h(x,v),v)$ $= x$ and $h(x(t,v),v) = t$, we calculate that $\dfrac{\partial h}{\partial x} x_1 = 1$ and $\dfrac{\partial h}{\partial x} x_2 + \dfrac{\partial h}{\partial v} = 0$, so $PR = T$ and $PS + Q = V$.)

In 1773 J. L. Lagrange also had need of a change-of-variable formula—this time for triple integrals [12]. He was interested in determining the attraction which an elliptical spheroid exercised on any point placed on its surface or in the interior. Since the general expression for attraction at any point was well known, the difficulty lay in integrating over the entire body. Even though the problem had already been solved geometrically, Lagrange, as part of his general philosophy of treating mathematics analytically, attempted a different solution.

To solve his problem, Lagrange had to calculate a triple integral. Since, following Euler's method, this had to be done by first holding two variables constant, integrating with respect to the third from one surface of the body to another, then evaluating the ensuing double integrals, he was quickly led to very complicated integrands. He realized that new coordinates were needed to replace the rectangular ones in order to make the integration tractable. Thus he proceeded to develop a general formula for changing variables in a triple integral. Lagrange's method was similar to Euler's in that he let vary only one variable at a time, but the details differed.

Given, then, x, y, and z as functions of new variables p, q, r, Lagrange wrote

$$dx = A\,dp + B\,dq + C\,dr$$
$$dy = D\,dp + E\,dq + F\,dr \tag{2}$$
$$dz = G\,dp + H\,dq + I\,dr$$

where $A, B, \ldots, I$ are, of course, the appropriate partial derivatives. His aim was to calculate the volume of the infinitesimal parallelepiped $dx\,dy\,dz$ (the "volume element") in terms of $dp\,dq\,dr$. To do this, he calculated each "difference" (i.e., edge of the parallelepiped) separately, regarding the other two variables as constant. First x and y are held constant; thus $dx = 0$ and $dy = 0$; the first two equations in (2) become

$$A\,dp + B\,dq + C\,dr = 0$$
$$D\,dp + E\,dq + F\,dr = 0.$$

Lagrange solved these two equations for dp and dq in terms of dr and substituted in the expression for dz in (2) to get

$$dz = \frac{G(BF - CE) + H(CD - AF) + I(AE - BD)}{AE - BD}\,dr.$$

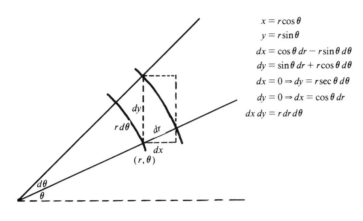

$$x = r\cos\theta$$
$$y = r\sin\theta$$
$$dx = \cos\theta\, dr - r\sin\theta\, d\theta$$
$$dy = \sin\theta\, dr + r\cos\theta\, d\theta$$
$$dx = 0 \Rightarrow dy = r\sec\theta\, d\theta$$
$$dy = 0 \Rightarrow dx = \cos\theta\, dr$$
$$dx\, dy = r\, dr\, d\theta$$

FIGURE 3

Next, x and z are assumed constant and only y varies; so $dx = 0$ and $dz = 0$. It follows immediately that $dr = 0$ and $A\, dp + B\, dq = 0$; therefore, $dp = -(B/A)\, dq$ and

$$dy = \frac{AE - BD}{A}\, dq.$$

Finally, y and z are taken as constant, so $dy = 0$ and $dz = 0$. Thus $dr = 0$ and $dq = 0$, which implies that $dx = A\, dp$. By multiplying together the expressions obtained for dx, dy, and dz, Lagrange calculated his result:

$$dx\, dy\, dz = (AEI + BFG + CDH - AFH - BDI - CEG)\, dp\, dq\, dr. \tag{3}$$

This is, of course, our standard formula. The result for three-dimensional integrals is analogous to (1), and in modern notation, is written as

$$\iiint_{A} f(x, y, z)\, dx\, dy\, dz = \iiint_{B} f(x(p,q,r), y(p,q,r), z(p,q,r)) \left| \frac{\partial(x,y,z)}{\partial(p,q,r)} \right| dp\, dq\, dr$$

where $\dfrac{\partial(x,y,z)}{\partial(p,q,r)}$ is the functional determinant of x, y, z with respect to p, q, r. (FIGURE 3 illustrates Lagrange's idea for the case of two variables and polar coordinates.)

We note that Lagrange, like Euler, dealt with the differential forms formally; there is absolutely no infinitesimal approximation that we would require in a similar proof today. But this formalism is typical of some of Lagrange's other work, in particular, his attempt to develop the calculus without limits by the use of algebra and infinite series [11], [13]. Also like Euler, Lagrange noted that the most obvious thing to do to try to obtain the change-of-variable formula would be to multiply together the original expressions (2) for dx, dy, and dz. However, he wrote, this product would contain squares and cubes of dp, dq, and dr and so would not be valid in an expression of a triple integral. Hence he had to use the step-by-step formal approach already outlined.

Lagrange applied his result to the case of spherical coordinates and was then able to perform the integrations he needed. Similarly, A. Legendre [15] and Pierre S. Laplace [14] soon after used essentially the same method to get similar results. These men were also interested in the change-of-variable formula in order to determine the attraction exercised by solids of various shapes, for which they needed to compute complicated integrals.

Carl Friedrich Gauss 1777–1855

In 1813 Carl F. Gauss gave a geometric argument for a special case of the change-of-variable theorem for two variables, although in a somewhat different context [8]. Gauss' method of proof contrasts sharply with that of Euler. Gauss was developing the idea of a surface integral in connection with studying attractions. As part of this he gave a method for finding the element of surface in three-space so that he could integrate over such a surface. He started by parametrizing the surface using three functions x, y, z of the two variables p, q. He then noted that given an infinitesimal rectangle in the p-q plane whose vertices were (p,q), $(p+dp,q)$, $(p,q+dq)$, $(p+dp, q+dq)$, there was a corresponding "parallelogram" element in the surface whose vertices were (x,y,z), $(x+\lambda\,dp, y+\mu\,dp, z+\nu\,dp)$, $(x+\lambda'\,dq, y+\mu'\,dq, z+\nu'\,dq)$, and $(x+\lambda\,dp+\lambda'\,dq, y+\mu\,dp+\mu'\,dq, z+\nu\,dp+\nu'\,dq)$, where

$$dx = \lambda\,dp + \lambda'\,dq$$
$$dy = \mu\,dp + \mu'\,dq \qquad (4)$$
$$dz = \nu\,dp + \nu'\,dq.$$

(One can easily calculate the above result from the definitions and properties of the relevant partial derivatives.) It follows that the projection of the infinitesimal parallelogram onto the x-y plane is the parallelogram whose vertices are (x, y), $(x+\lambda\,dp, y+\mu\,dp)$, $(x+\lambda'\,dq, y+\mu'\,dq)$, $(x+\lambda\,dp+\lambda'\,dq, y+\mu\,dp+\mu'\,dq)$ and whose area is clearly $\pm(\lambda\mu' - \mu\lambda')\,dp\,dq$. (See FIGURE 4.) Gauss was therefore able to compute the element of surface as $dp\,dq((\mu\nu' - \nu\mu')^2(\nu\lambda' - \lambda\nu')^2(\lambda\mu' - \mu\lambda')^2)^{1/2}$ and thus to integrate this over the p-q region corresponding to his surface. (In this paper, Gauss used his special cases of the divergence theorem and his parametric method for calculating a surface element to evaluate certain "surface integrals" for the case of an ellipsoid given by $x = A\cos(p), y = B\sin(p)\cos(q), z = C\sin(p)\sin(q)$ for $0 \leqslant p \leqslant \pi, 0 \leqslant q \leqslant 2\pi$.)

If we let $z = 0$ so that the "surface" is part of the x-y plane, then Gauss' argument shows that the new "area element" is $|\lambda\mu' - \mu\lambda'|\,dp\,dq$, hence that $\int\int dx\,dy = \int\int |\lambda\mu' - \mu\lambda'|\,dp\,dq$, a special case of the change-of-variable theorem from which the general case may easily be derived. Gauss' argument differs considerably from those of Euler and Lagrange. He essentially made use of analytic and geometric methods instead of using the formal approach of his predecessors. But as was typical of Gauss, he did not provide all the steps necessary to complete his analytic argument, especially since he was dealing with infinitesimals. The missing parts can, however, be readily supplied.

The next mathematician to break new ground in this field was Mikhail Ostrogradskii, in 1836. A Russian mathematician who studied in France in the 1820's, he later returned to St. Petersburg where he produced many works in applied mathematics. Unfortunately, some of his most important discoveries appear to have been totally ignored, at least in Western Europe. Not only

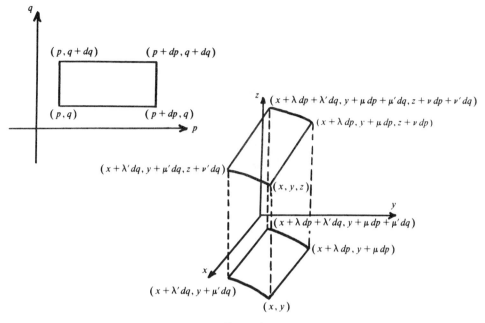

FIGURE 4

did he give the first generalization of the change-of-variable theorem to n variables, but he also first proved and later generalized the divergence theorem [10], wrote integrals of n-forms over n-dimensional "hypersurfaces," and, as we shall see below, gave the first proof of the change-of-variable theorem for double integrals using infinitesimal concepts. All of these results were eventually repeated by other mathematicians with no credit to Ostrogradskii.

In his 1836 paper [16], Ostrogradskii generalized to n dimensions the change-of-variable theorem and Lagrange's proof of it. Given that $X, Y, Z, \ldots$ are all functions of $x, y, z \ldots$, Ostrogradskii first calculated $dX, dY, dZ, \ldots$ in terms of $dx, dy, dz, \ldots$. Then by holding all variables except X constant, he had $dY = dZ = \cdots = 0$, so he could solve for dX in terms of dx by using determinants; continuing with each variable in turn he calculated expressions for $dY, dZ, \ldots$ in terms of $dy, dz, \ldots$ and by multiplying showed that $dX \, dY \, dZ \ldots = \Delta \, dx \, dy \, dz \ldots$ where Δ is the functional determinant of $X, Y, Z, \ldots$ with respect to $x, y, z, \ldots$. Ostrogradskii did not state this result as a formula for transforming multiple integrals, but he did apply it to convert a hypersurface integral with $n + 1$ terms of the form $dx \, dy \ldots$, to an ordinary n-dimensional integral in n new variables.

Both Carl Jacobi [9] and Eugene Catalan [4] published papers in 1841 giving clearly the general change-of-variable theorem for n-dimensional integrals. Catalan's proof was also similar to Lagrange's in its use of formal manipulations on one variable at a time. Jacobi's paper was the culmination of a series of articles concerning this theorem; it contained additional results such as the multiplication rule for the composition of several changes of variable. Jacobi's work was referred to shortly thereafter by Cauchy and soon his name became tied to the theorem. In fact, the functional determinant Δ is now known as the Jacobian rather than the "Ostrogradskian."

Two years after his 1836 paper, Ostrogradskii published in [17] a proof of the change-of-variable formula in two variables which used the same basic idea as had Gauss. He first criticized the proofs of Euler and Lagrange, and, by implication, his own earlier proof. He claimed that,

Mikhail Ostrogradskii 1801–1861

assuming that x and y were functions of u and v, if one first used $dx = 0$ to solve for dy in terms of du (that is, to evaluate one side of the differential rectangle) one could not then assume that du would be 0 when one tried to evaluate dx by setting $dy = 0$ (to find the other side of the rectangle). In fact, he wrote, you would have to use a new set of differentials, δu and δv, in evaluating the other side, and, once you did that, you came up with an incorrect result.

So Ostrogradskii returned to the meaning of $\int\int V\,dx\,dy$ as a sum of differential elements. Using a method similar to that of Gauss, although staying strictly in two dimensions, he proceeded to recalculate the area of these elements. He carefully chose each element to be bounded by two curves where u was constant and two curves where v was constant. If ω denotes the area of such an element, he noted that by the definition of the definite integral, $\int\int V\,dx\,dy = \int\int V\omega$. It is easy to calculate ω (see FIGURE 5) since the four vertices have coordinates $(x, y), \left(x + \dfrac{\partial x}{\partial u}du,\, y + \dfrac{\partial y}{\partial u}du \right)$,

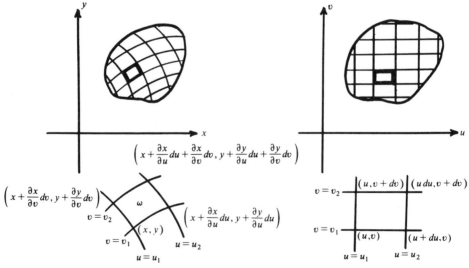

FIGURE 5

$\left(x + \dfrac{\partial x}{\partial v} dv, y + \dfrac{\partial y}{\partial v} dv \right)$, and $\left(x + \dfrac{\partial x}{\partial u} du + \dfrac{\partial x}{\partial v} dv, y + \dfrac{\partial y}{\partial u} du + \dfrac{\partial y}{\partial v} dv \right)$. By elementary geometry, the area of this parallelogram is $\pm \left(\dfrac{\partial x}{\partial u} \dfrac{\partial y}{\partial v} - \dfrac{\partial x}{\partial v} \dfrac{\partial y}{\partial u} \right) du\, dv$ and so the integral formula becomes

$$\int \int V\, dx\, dy = \pm \int \int V \left(\frac{\partial x}{\partial u} \frac{\partial y}{\partial v} - \frac{\partial x}{\partial v} \frac{\partial y}{\partial u} \right) du\, dv.$$

Ostrogradskii further noted that this method could be easily extended to three dimensions but not more, since there is not a corresponding geometrical result in four dimensions. We must note, of course, that Ostrogradskii had not explicitly justified using the standard formula for the area of a parallelogram when, in fact, the area is actually that of a "curvilinear" parallelogram. However, it was common practice in that time (as we noted also about Gauss' proof), to ignore explicit arguments about infinitesimal approximation.

Only four years later, a proof similar to that of Ostrogradskii appeared in Augustus DeMorgan's text *Differential and Integral Calculus* [6], one of the first "analytic" textbooks to appear in English. It is doubtful that DeMorgan had read Ostrogradskii's work, for his approach is somewhat different; he was considering how to calculate a double integral over a plane region bounded by four curves, where the standard method of integrating, first with respect to one variable between two functions of the other and then with respect to the second between constant limits, will not work. But his method of attack, via the definition of the double integral as a limit, the division of the given region into subregions bounded by curves where u was constant and where v was constant, and the calculation of areas of curvilinear quadrilaterals, is very close to that of Ostrogradskii. DeMorgan went even further, however, to provide detailed reasoning as to why the errors of approximation—third order infinitesimals—may be safely ignored.

It is also interesting that DeMorgan prefaced his results by stating that Legendre's proof (which was identical to that of Lagrange) was "so obscure in its logic as to be nearly unintelligible, if not dubious."

Ostrogradskii and DeMorgan, then, had moved away from the formal symbolic approach of Euler and Lagrange. But we should emphasize that the former had not justified equating the "elements of area" $dx\, dy$ and $\left| \dfrac{\partial(x,y)}{\partial(u,v)} \right| du\, dv$ themselves, as the latter had attempted to do. They had only showed the equality of the integrals over the appropriate regions. A new justification for the formal symbolic approach only came with Elie Cartan and his theory of differential forms.

Beginning in the mid 1890's, Cartan wrote a series of papers in which he formalized the subject of differential forms, namely the expressions which appear under the integral sign in line and surface integrals. As part of this formalization, he used the Grassmann rules of exterior algebra for calculations with such forms. In a paper of 1896 [2], as an example of such a calculation, he was able to do what Euler could not; namely, if $x = x(t, v)$ and $y = y(t, v)$, he could multiply $dx = \dfrac{\partial x}{\partial t} dt + \dfrac{\partial x}{\partial v} dv$ and $dy = \dfrac{\partial y}{\partial t} dt + \dfrac{\partial y}{\partial v} dv$ using the rules $dt\, dt = dv\, dv = 0$ and $dt\, dv = - dv\, dt$ to show that

$$dx\, dy = \left(\frac{\partial x}{\partial t} \frac{\partial y}{\partial v} - \frac{\partial x}{\partial v} \frac{\partial y}{\partial t} \right) dt\, dv.$$

In 1899 [3], Cartan went into much more detail on the rules for operating with these differential forms. And again, one of his first examples was the change-of-variable formula.

As a final point, we note that proofs using the methods of Euler, Lagrange, and Ostrogradskii all appeared in textbooks through the first third of the twentieth century. There were, naturally, attempts to make all three methods more rigorous. A readily available example of this (for the proofs of Euler and Ostrogradskii) occurs in Courant's *Differential and Integral Calculus* [5]. Most current textbooks, on the other hand, use an entirely different proof based on Green's theorem.

References

[1] R. C. Buck, Advanced Calculus, 2nd ed., McGraw-Hill, New York, 1965, p. 296 ff.

[2] E. Cartan, Le principe de dualité et certaines intégrales multiples de l'espace tangentiel et de l'espace réglé, Bull. Soc. Math. France, 24 (1896) 140–177; Oeuvres (2) 1, 265–302.

[3] ———, Sur certaines expressions différentielles et sur le problème de Pfaff, Annales École Normale, 16 (1899) 239–332; Oeuvres (2) 1, 303–397.

[4] E. Catalan, Sur la transformation des variables dans les intégrales multiples, Mémoires couronnés par l'Académie Royale des Sciences et Belles-Lettres de Bruxelles, 14 (1841) 1–48.

[5] R. Courant, Differential and Integral Calculus, vol. 2, Interscience, New York, 1959, p. 247 ff.

[6] A. DeMorgan, The Differential and Integral Calculus, Robert Baldwin, London, 1842.

[7] L. Euler, De formulis integralibus duplicatis, Novi comm. acad. scient. Petropolitanae, 14 (1769) 72–103; Opera (1) 17, 289–315.

[8] C. Gauss, Theoria attractionis corporum sphaeroidicorum ellipticorum homogeneorum methodo novo tractata, Comm. soc. reg. scient. Gottingensis, 2 (1813): Werke 5, 1–22.

[9] C. Jacobi, De determinantibus functionalibus, Crelle, 22 (1841) 319–359; Werke 3, 393–438.

[10] V. Katz, The history of Stokes' theorem, this MAGAZINE, 52 (1979) 146–156.

[11] J. Lagrange, Théorie des Fonctions Analytiques, Paris, 1797; Oeuvres 9.

[12] ———, Sur l'attraction des sphéroïdes elliptiques, Nouveaux Mémoires de l'Académie Royale des Sciences et Belles-Lettres de Berlin, (1773); Oeuvres 3, 619–658.

[13] ———, Leçons sur le Calcul des Fonctions, Paris, 1801.

[14] P. Laplace, Traité de Mécanique Céleste, Paris, 1799; Oeuvres 2.

[15] A. Legendre, Mémoire sur les intégrales doubles, Histoire de l'Académie Royale des Sciences avec les Mémoires de Mathématique et de Physique, (1788) 454–486.

[16] M. Ostrogradskii, Mémoire sur le calcul des variations des intégrales multiples, Mémoires de l'Académie Impériale des Sciences de St. Petersbourg, (6) 3 (part 1) (1836), 36–58; also in Crelle, 15 (1836) 332–354.

[17] ———, Sur la transformation des variables dans les intégrales multiples, Mémoires de l'Académie Impériale des Sciences de St. Petersbourg, cl. math., (1838) 401–407.

[18] A. Taylor and W. R. Mann, Advanced Calculus, 2nd ed., Xerox, Lexington, 1972, p. 490 ff.

Three Aspects of Fubini's Theorem

J. Chris Fisher
University of Regina
Regina, Canada S4S 0A2

J. Shilleto
6 Locksley Avenue, #5B
San Francisco, CA 94122

Which of the three propositions in the box—(1), (2) or (3)—would you consider to be the most palpably true? Our first choice is (1), while (3) is second, and (2) is a close third. This is because

---

Let $f(x, y), \dfrac{\partial}{\partial x} g(x, y)$, and $\dfrac{\partial^2}{\partial y \partial x} h(x, y)$ be continuous real-valued functions in the rectangle $\{(x, y): a \leqslant x \leqslant b, c \leqslant y \leqslant d\}$. Then:

$$(1) \quad \int_a^x \int_c^y f(u, v) \, dv \, du = \int_c^y \int_a^x f(u, v) \, du \, dv,$$

$$(2) \quad \frac{\partial}{\partial x} \int_c^y g(x, v) \, dv = \int_c^y \frac{\partial}{\partial x} g(x, v) \, dv,$$

$$(3) \quad \frac{\partial^2}{\partial y \partial x} h(x, y) = \frac{\partial^2}{\partial x \partial y} h(x, y).$$

---

the geometrical evidence for (1) provides a more compelling argument than the naturalness and sense of order of (2) and (3). In fact, (3)'s interpretation using velocities actually *detracts* from its believability (as we shall see)!

These statements are surprising in light of the fact that *using only the fundamental theorem of calculus and some routine manipulations, any one of these propositions can be derived from any other.*

Many of our observations can be found in [2], and some of the ideas are suggested by exercises in [1, p. 793], [3, p. 61], and [4, pp. 464–465]. Nevertheless, they are missing from contemporary calculus texts and deserve occasional airings. In addition to bringing [2] back to light, our goal here is to emphasize the intuitive content of this circle of ideas.

Statement (1), a special case of Fubini's theorem, can be interpreted as follows:

> One gets just as much tomato to eat if he slices it from left to right or from back to front.

Compare this with the mental gymnastics required to untangle the interpretation of (3):

> A person walks on a hillside and points a flashlight along a tangent to the hill; then the rate at which the beam's direction changes when walking south and pointing east equals its rate of change when walking east and pointing south.

We leave the interpretation of (2) to the reader. (Hint: The left side of (2) is the rate of change of the cross-sectional area of the tomato slices mentioned above. Does your interpretation of (2) convince you of its validity?)

Proofs that the statement (i) implies (i ± 1) are readily found in textbooks (or see [2]). As a typical example, here is the standard proof that (1) implies (2). We assume (1) and define

$f(x, y) = \dfrac{\partial}{\partial x} g(x, y)$. That is,

$$\int_a^x f(u, y)\, du = g(x, y) - g(a, y).$$

Then

$$\frac{\partial}{\partial x} \int_c^y g(x, v)\, dv = \frac{\partial}{\partial x} \int_c^y \left(\int_a^x f(u, v)\, du + g(a, v) \right) dv$$

$$= \frac{\partial}{\partial x} \int_c^y \int_a^x f(u, v)\, du\, dv + \frac{\partial}{\partial x} \int_c^y g(a, v)\, dv.$$

Since $\int_c^y g(a, v)\, dv$ is a function of y only, its partial derivative with respect to x is zero, and (having assumed (1))

$$\frac{\partial}{\partial x} \int_c^y g(x, v)\, dv = \frac{\partial}{\partial x} \int_a^x \int_c^y f(u, v)\, dv\, du$$

$$= \int_c^y f(x, v)\, dv$$

$$= \int_c^y \frac{\partial}{\partial x} g(x, v)\, dv.$$

The proof's only nontrivial steps use the fundamental theorem of calculus. Indeed, one rather undesirable feature of this proof is that the details make it seem as if something more is involved. Let us therefore change our notation to one of operators to bring out the essence of the above argument. Define

$$D_x f := \frac{\partial f}{\partial x}, \qquad D_x^{-1} f := \int_a^x f(u, y)\, du,$$

$$D_y f := \frac{\partial f}{\partial y}, \quad \text{and} \quad D_y^{-1} f := \int_c^y f(x, v)\, dv.$$

In this notation, statements (1), (2), (3) become

(1) $$D_x^{-1} D_y^{-1} = D_y^{-1} D_x^{-1},$$

(2) $$D_x D_y^{-1} = D_y^{-1} D_x,$$

and

(3) $$D_x D_y = D_y D_x.$$

The fundamental theorem of calculus for $f = f(z)$ is *essentially* $D_z D_z^{-1} f = D_z^{-1} D_z f = f$, where "essentially" means that $D_z^{-1} D_z f$ should have a constant of integration. Of course, in the present context that constant eventually disappears (much as it did in the detailed proof), a fact that can conveniently be left as an exercise. With this warning, the proof that (1) implies (2) now reads

$$D_x D_y^{-1} \underset{\text{F.T.}}{=} D_x D_y^{-1} \left(D_x^{-1} D_x \right) \underset{(1)}{=} D_x \left(D_x^{-1} D_y^{-1} \right) D_x \underset{\text{F.T.}}{=} D_y^{-1} D_x$$

Here is (2) implies (3):

$$D_y D_x \underset{\text{F.T.}}{=} D_y D_x D_y^{-1} D_y \underset{(2)}{=} D_y D_y^{-1} D_x D_y \underset{\text{F.T.}}{=} D_x D_y.$$

The proofs that (3) implies (2) and (2) implies (1) can be obtained by interchanging D with D^{-1} in the lines above.

We should emphasize that because $D^{-1}Df$ differs from f by a constant, the above argument does not constitute a rigorous proof that (i) implies (i − 1). It is, however, an amusing exercise to decode such a symbolic argument to check that each constant of integration really does disappear. Here, for example, is a proof that (3) implies (2) (by decoding $D_y^{-1} D_x = D_y^{-1} D_x D_y D_y^{-1} = D_y^{-1} D_y D_x D_y^{-1} = D_x D_y^{-1}$):

$$\int_c^y \frac{\partial}{\partial x} g(x, v)\, dv \underset{\text{F.T.}}{=} \int_c^y \frac{\partial}{\partial x} \left(\frac{\partial}{\partial v} \int_c^v g(x, t)\, dt \right) dv$$

$$\underset{(3)}{=} \int_c^y \frac{\partial}{\partial v} \frac{\partial}{\partial x} \int_c^v g(x, t)\, dt\, dv$$

$$\underset{\text{F.T.}}{=} \frac{\partial}{\partial x} \int_c^y g(x, t)\, dt - \frac{\partial}{\partial x} \int_c^c g(x, t)\, dt$$

$$= \frac{\partial}{\partial x} \int_c^y g(x, v)\, dv.$$

The ideas touched upon in this note seem to be appropriate for any calculus course, rigorous or not. At one level they provide an attractive way of proving (3): merely explain how it follows quickly from (1). At any level they provide the opportunity to stress normally unseen connections while providing one more chance to show (and show off) the power of the fundamental theorem of calculus.

Note finally that one can easily avoid the intermediate proposition (2), since (3) follows *directly* from (1):

$$D_y D_x = D_y D_x \left(D_y^{-1} D_x^{-1} \right) D_x D_y = D_y D_x \left(D_x^{-1} D_y^{-1} \right) D_x D_y = D_x D_y.$$

We would like to thank Jerry Marsden and John Wilker for their helpful comments and references.

References

[1]   Jerrold Marsden and Alan Weinstein, Calculus, Benjamin/Cummings, 1980.
[2]   R. T. Seeley, Fubini implies Leibniz implies $F_{yx} = F_{xy}$, Amer. Math. Monthly, 68 (1961) 56–57.
[3]   Michael Spivak, Calculus on Manifolds, W. A. Benjamin, New York, 1965.
[4]   R. E. Williamson, R. H. Crowell and H. Trotter, Calculus of Vector Functions, 3rd ed., Prentice-Hall, 1972.

The Largest Unit Ball in Any Euclidean Space

JEFFREY NUNEMACHER
Oberlin College
Oberlin, OH 44074

In what dimensional Euclidean space does the unit ball have greatest volume? greatest surface area? The usual approach to this problem is to find explicit formulas for the volume and surface area and then to analyze their behavior. The standard derivations of these formulas are based on recurrence relations and typically involve some advanced calculus. For various approaches see, for example, [1, p. 411]; [2, p. 302]; [3, p. 220]; [4, p. 502]; [5, p. 324]. This note solves the problem by working directly from the recurrence relations. This approach is pleasingly simple and makes the argument accessible to a multivariable calculus class.

Let $B_n(r)$ denote the open ball of radius r in R^n, i.e.,

$$B_n(r) = \left\{ (x_1, x_2, \ldots, x_n) \,\middle|\, \sum_{i=1}^{n} x_i^2 < r^2 \right\},$$

and let $V_n(r)$ denote its volume. Since balls are similar n-dimensional objects, it is not surprising that there are constants a_n so that $V_n(r) = a_n r^n$. This statement can be proved using the change of variables formula (see, e.g., [4, p. 500]). A more elementary argument can be carried out based on approximation by Riemann sums, using the basic observation that if all sides of an n-box are magnified by r, then the volume is magnified by r^n.

By definition we have

$$V_n(r) = \underset{B^n(r)}{\iint} \cdots \int 1 \, dx_1 \cdot dx_2 \cdots dx_n = \int_{-r}^{r} \left(\underset{\Sigma_{i=1}^{n-1} x_i^2 < r^2 - x_n^2}{\iint \cdots \int} 1 \, dx_1 \, dx_2 \cdots dx_{n-1} \right) dx_n.$$

Since the value of the inner integral is $V_{n-1}\left(\sqrt{r^2 - x_n^2} \right)$, we find that

$$V_n(r) = \int_{-r}^{r} a_{n-1} \left(\sqrt{r^2 - x_n^2} \right)^{n-1} dx_n.$$

A standard trigonometric substitution now shows that

$$V_n(r) = 2 a_{n-1} r^n \int_0^{\pi/2} \cos^n\theta \, d\theta.$$

Thus the volumes $V_n \equiv V_n(1)$ of the unit balls are related by the first-order recurrence relation

$$V_n = 2 V_{n-1} \int_0^{\pi/2} \cos^n\theta \, d\theta,$$

where $V_1 = 2$, the length of the interval $(-1, 1)$ in R^1.

For $n \geq 0$ let b_n denote the coefficient $2 \int_0^{\pi/2} \cos^n\theta \, d\theta$. A standard integration by parts yields the second-order recurrence relation $b_n = ((n-1)/n) b_{n-2}$, which together with the initial conditions $b_0 = \pi$ and $b_1 = 2$ allows the calculation of any particular b_n. It is clear that $\{b_n\}$ is decreasing, and since for any $\varepsilon > 0$, $\lim_{n \to \infty} \cos^n\theta = 0$ uniformly for $\varepsilon \leq \theta \leq \pi/2$, we see that $\lim_{n \to \infty} b_n = 0$. We have $V_n = b_n V_{n-1}$, so that V_n increases until $b_n < 1$, then it decreases to 0. Calculation of the first several terms shows that $b_5 > 1$ but $b_6 < 1$; thus the greatest volume occurs in dimension $n = 5$.

A similar method can be used to show that S^{n-1}, the unit sphere of R^n, has greatest surface area when $n = 7$. Let $A_n(r)$ denote the surface area of S^{n-1}. The crucial ingredient is the believable but not-quite-obvious fact that $d/dr\, V_n(r) = A_n(r)$. This formula can be obtained by using the change of variables formula with generalized polar coordinates (see, e.g., [3, p. 340]). Here is an elementary proof.

Let $0 = r_0 < r_1 < \cdots < r_n = r$ be a partition of $[0, r]$ and let $\Delta r_n = r_n - r_{n-1}$. Then by elementary geometric considerations

$$\sum_{i=1}^{n} A_n(r_{i-1})\, \Delta r_i < V_n(r) < \sum_{i=1}^{n} A_n(r_i)\, \Delta r_i.$$

It follows that $V_n(r) = \int_0^r A_n(s)\, ds$, whence $(d/dr)\, V_n(r) = A_n(r)$.

Now since $V_n(r) = a_n r^n$, we have $A_n(r) = n a_n r^{n-1}$, so that $A_n \equiv A_n(1) = n V_n$. Thus

$$A_n = n b_n V_{n-1} = \frac{n}{n-1}\, b_n A_{n-1},$$

which gives a one-step recurrence relation for A_n. Again the sequence of coefficients is decreasing towards 0, so that A_n increases until $(n/(n-1))\, b_n < 1$, then decreases towards 0. Since $(n/(n-1))\, b_n = b_{n-2}$ and b_5 is the last b_n to exceed 1, the surface area of the unit sphere is greatest in R^7.

It is an interesting consequence of this argument that both V_n and A_n approach 0 as n gets large.

References

[1]   T. Apostol, Calculus, Vol. 2, 2nd ed., John Wiley & Sons, New York, 1969.
[2]   R. Courant, Differential and Integral Calculus, Vol. 2, Wiley-Interscience, New York, 1936.
[3]   W. Fleming, Functions of Several Variables, 2nd. ed., Springer-Verlag, New York, 1977.
[4]   S. Lang, Undergraduate Analysis, Springer-Verlag, New York, 1983.
[5]   J. Marsden, Elementary Classical Analysis, Freeman, San Francisco, 1974.

Ed. Note. See related papers listed as entries 7 and 8 in the Bibliography for this section, p. 353.

Volumes of Cones, Paraboloids, and Other "Vertex Solids"

PAUL B. MASSELL
United States Naval Academy
Annapolis, MD 21402

While performing some calculations involving the volume of a solid circular paraboloid $z(r) = h(1 - (r/a)^2)$ (with $h > 0$) as illustrations of the Divergence Theorem in vector calculus, the author noticed that the ratio of the volume of the portion of the solid paraboloid above the polar plane to that of the solid cylinder with the same base and height h (its associated solid cylinder) is equal to $1/2$ for all values of the radius a. A natural question is whether this ratio holds for elliptical paraboloids or for paraboloids with any simple curve as a base. Another question is whether there is a similar ratio that is independent of the shape and size of the base for exponents other than 2 in the formula for $z(r)$. Our theorem answers these questions for a class of solids we call vertex solids; a class that includes cones and paraboloids.

We will now define vertex solids. Let $r = g(\theta)$ describe a simple closed curve in the polar plane such that $0 \leqslant g(\theta)$ for $0 \leqslant \theta \leqslant 2\pi$. Let the point V be on the positive z-axis with $z = h$ (this will be the top vertex for the vertex solid). For each fixed θ in $[0, 2\pi]$, consider the curves $z_k(r) = h(1 - (r/g(\theta))^k)$ where k is a positive constant (if $g(\theta) = 0$, let $z_k(r) = h$). If k is an integer, then $z_k(r)$ is clearly the unique curve of kth degree that goes through V and the point $(\theta, g(\theta), 0)$ in the polar plane and that has the property $d^i z/dr^i = 0$ at V for $i = 1, 2, \ldots, k - 1$ (for fixed θ). Now consider r and θ as independent variables and view the above expression for $z_k = z_k(r, \theta)$ as representing a surface. If $r = g(\theta)$ describes an ellipse, then z_1 represents an elliptical cone and z_2 represents an elliptical paraboloid. For all $k > 0$, we call the solid defined by the set of points (r, θ, z) satisfying $0 \leqslant r \leqslant g(\theta)$, $0 \leqslant \theta \leqslant 2\pi$, $0 \leqslant z \leqslant z_k(r, \theta)$ a *vertex solid*. Its *associated solid cylinder* is the set of points (r, θ, z) satisfying $0 \leqslant r \leqslant g(\theta)$, $0 \leqslant \theta \leqslant 2\pi$, $0 \leqslant z \leqslant h$.

THEOREM. *The ratio of the volume of the vertex solid of degree k to that of its associated solid cylinder is $k/(k + 2)$. (Thus for an elliptical cone the volume is $1/3$ that of its associated cylinder's volume πabh; for an elliptical paraboloid its volume is $1/2$ of πabh. Here, a and b are the minor and major radii of the ellipse. The ratios $1/3$ and $1/2$ hold for cones and paraboloids (respectively) with any base that is describable by a simple closed curve.)*

Proof.

$$\text{vol(vertex solid)} = \int_0^{2\pi} \int_0^{g(\theta)} z_k(r, \theta) \cdot r \, dr \, d\theta$$

$$\text{vol(solid cylinder)} = \int_0^{2\pi} \int_0^{g(\theta)} h \cdot r \, dr \, d\theta = A.$$

Substitution of the expression for z_k and a fairly simple integration reveals that the ratio of the volume of the vertex solid to that of its associated solid cylinder is

$$\frac{A - (2/(k+2))A}{A} = \frac{k}{k+2}.$$

Notes:

(1) This result easily can be extended to the case where the base of the vertex solid does not lie below the vertex V. In this case, the vertex solid is not entirely contained in its solid cylinder.

(2) As k increases, the vertex solid occupies more and more of its associated solid cylinder, and in the limit occupies all of it.

(3) Consider the cross sections of the vertex solid and its solid cylinder generated by the plane $\theta = c$ (constant). (Assume $g(c) > 0$.) Denoting them by C_v and C_c, it's easy to see that area(C_v)/area$(C_c) = k/(k+1)$.

I would like to thank my colleague Tom Mahar for a very helpful discussion of this result and Bruce Richter for encouraging me to publish it. Thanks are also due to two referees for several helpful suggestions for improving the readability of the paper.

Interchanging the Order of Integration

STEWART VENIT

STEWART VENIT is Assistant Professor of Mathematics at California State University, Los Angeles, where he has been teaching since 1971. He received his Ph.D. from the University of California at Berkeley in 1971. His special area of interest is numerical analysis.

Most calculus texts, after introducing the subject of double integrals, give examples to show that interchanging the given order of integration can sometimes be a time-saving convenience. For instance, if an integration is performed in the order indicated, it may perhaps be necessary to subdivide the region and/or use special techniques to evaluate the integral, while reversing the order of integration may result in a particularly simple computation. We can provide the student with much better motivation for this topic by demonstrating that the interchange of order process can be more than just a convenience—sometimes it is a necessity!

We must first convince the student that certain indefinite integrals cannot be evaluated in closed form. This can be done by asking him to "swallow" a relatively reasonable assumption, which is a special case of Liouville's theorem.

THEOREM. *Let f and g be rational functions with the degree of $g > 0$. If $\int f \cdot \exp(g)\, dx$ can be integrated in closed form, then there exists a rational function R such that $\int f \cdot \exp(g)\, dx = R \cdot \exp(g)$.*

In [1], Mead uses this theorem to give a technique whereby students can demonstrate for themselves that certain integrals cannot be evaluated in closed form.*

Examples of these are $\int \exp(x^2)\, dx$, $\int (\sin x / x)\, dx$, and $\int (1/\ln x)\, dx$.

We are now in a position to show the student the possible necessity for the interchange of order operation.

Example 1. $\int_R \int \exp(y^2)\, dy\, dx$, where R is the region bounded by the curves $y = x$ and $y = x^{1/3}$. As it stands, the integration cannot be done. However, interchanging the order, we get

$$\int_0^1 \int_{y^3}^{y} \exp(y^2)\, dx\, dy,$$

which can be done with the aid of integration by parts.

---

*In the Fall, 1970 issue of this Journal, Kung uses a simpler version of this theorem to provide an alternative to integration by parts for computing $\int f \cdot \exp(ax)\, dx$, where f is a polynomial in x.

Example 2. $\int_S\int (\sin y/y)\,dy\,dx$, where S is the region bounded by the curves $y = 1$, $y = 2$, $y = x/2$, and $y = x$. Interchange of order yields the simple integral

$$\int_1^2 \int_y^{2y} \left(\frac{\sin y}{y} \right) dx\,dy.$$

Example 3. $\int_T\int (1/\ln y)\,dy\,dx$, where T is the region bounded by the curves $y = \exp(x)$, $y = \exp(\sqrt{x})$, $y = 3$, and $y = 4$. Here, reversing the order yields

$$\int_3^4 \int_{\ln y}^{(\ln y)^2} \left(\frac{1}{\ln y} \right) dx\,dy.$$

The integration with respect to y involves finding $\int \ln y\,dy$, a good review exercise in itself!

REFERENCE

1.   D. Mead, Integration, Amer. Math. Monthly, 68 (1961) 152–156.

BIBLIOGRAPHIC ENTRIES: MULTIPLE INTEGRALS AND LINE INTEGRALS

1. *Monthly* Vol. 78, No. 1, pp. 42–45. D. E. Varberg, Change of variables in multiple integrals.
2. *Monthly* Vol. 84, No. 3, pp. 201–204. John M. Karon and James V. Ralston, Double integrals as initial value problems.
3. *Monthly* Vol. 88, No. 9, pp. 701–704. Ronald W. Gatterdam, The planimeter as an example of Green's theorem.
4. *Math. Mag.* Vol. 43, No. 2, pp. 85–89. Peter A. Lindstrom, Evaluation of double integrals by means of the definition.
5. *Math. Mag.* Vol. 52, No. 3, pp. 146–156. Victor J. Katz, The history of Stokes' theorem.
6. *Math. Mag.* Vol. 61, No. 3, pp. 164–169. Anthony J. Lo Bello, The volumes and centroids of some famous domes.
7. *Math. Mag.* Vol. 62, No. 2, pp. 101–107. David J. Smith and Mavina K. Vamanamurthy, How small is a unit ball?

The volume V_n of a unit ball increases for $n \le 5$ then decreases for $n \ge 5$.

8. CMJ Vol. 15, No. 2, pp. 126–134. Marshall Fraser, The grazing goat in n dimensions.
9. CMJ Vol. 17, No. 4, pp. 326–337. Bart Braden, The surveyor's area formula.

10

NUMERICAL, GRAPHICAL, AND MECHANICAL METHODS (INCLUDING USE OF COMPUTERS)

NUMERICAL DIFFERENTIATION FOR CALCULUS STUDENTS

DAVID A. SMITH

It is becoming fashionable to make the standard calculus course more "relevant" or "applied" by the addition of numerical methods, perhaps at the expense of more traditional topics, or in a supplementary or "laboratory" course attached to the regular course. Of course, the student must use these methods to solve problems with the aid of a computer, or else they will be meaningless to him. The first trickles of an anticipated flood of books presenting supplementary numerical material for calculus courses have already appeared. In such books the subject of integration is handled in predictable fashion: upper and lower sums, the trapezoidal rule, Simpson's rule, and perhaps some discussion of error estimates. Indeed, these subjects are also in most of the standard calculus books, but the problems posed there for the student are necessarily contrived or trivial when it cannot be assumed that there is a computer handy with which to do the arithmetic.

In the same manner, the method proposed in supplementary texts for numerical differentiation is usually the method first presented in the standard text: compute a limit of difference quotients. The real objective of such material in computer-oriented texts is to *illustrate* the concepts of limit and derivative, but nevertheless this is usually the only differentiation method presented. Difference quotients have a disturbing tendency to turn out to be 0 when Δx gets small enough that the computer cannot tell the difference between $f(x)$ and $f(x + \Delta x)$. The instructor must then explain that the terminal zeros in the computed sequence are to be ignored, and the "real" limit is the last non-zero entry (unless $f'(x) = 0$, of course). This procedure may lead the perceptive student to conclude that a computer is not a very useful tool for differentiating a function.

The purpose of this note is to point out that simple, effective numerical differentiation formulas are readily accessible at the first-year calculus level, and that the derivation of these formulas *and* their error estimates can reveal a useful application of a very elementary result, namely Rolle's Theorem.

Let us suppose that our problem is to compute $f'(0)$. This choice is a matter of convenience for the derivation, and the result is easily translated elsewhere on the x-axis later. We shall approximate $f'(0)$ by first fitting a polynomial $p(x)$ to conveniently chosen points on the graph of f, and then computing $p'(0)$, a trivial task. In deriving Simpson's rule, the instructor will have already carried out the fit of a quadratic polynomial to the points at $-h, 0, h$. Without repeating the details, we simply note that one finds easily:

$$(1) \qquad p'(0) = \frac{1}{2h}[f(h) - f(-h)].$$

(This is also the slope of the chord joining the points at $-h$ and h.) Similarly, one may fit a quartic polynomial $p(x)$ to five points at $x = 0, \pm h, \pm 2h$. This leads to five linear equations in the coefficients of p. Only the coefficient of x (i.e., $p'(0)$) is needed, and one easily solves for this coefficient:

$$(2) \qquad p'(0) = \frac{1}{12h}[f(-2h) - 8f(-h) + 8f(h) - f(2h)].$$

Formula (2) is presumably "better" than formula (1), but how much better? How does one choose h in each case to get an acceptable approximation for $f'(0)$? The answers to such questions depend on having a reasonable error estimate, which we now provide in a form for mathematicians. For students, one would want to consider only the cases $k = 1$ and 2 of the following theorem. Even in the $k = 1$ case, the theorem says that quite good answers for $f'(0)$ can be obtained from formula (1) for values of h that are only moderately small (on the order of, say, 10^{-4}) provided f is reasonably "nice". Such an h is large enough to avoid the numerical anomalies inherent in computation of $f'(0)$ via a "limit" of difference quotients.

THEOREM. *Let f be a function with a continuous $(2k + 1)$-th derivative on the interval $[-kh, kh]$, where k is a positive integer. Let p be a polynomial of degree $2k$ that agrees with f at the $2k + 1$ points $0, \pm h, \cdots, \pm kh$. Then*

$$(3) \qquad |f'(0) - p'(0)| < Kh^{2k} \Big/ \binom{2k}{k} (2k + 1),$$

where K is any bound for $|f^{(2k+1)}(x)|$ on $[-kh, kh]$.

Proof. Let $E(x) = f(x) - p(x)$, and note that we seek a bound for $|E'(0)|$. E has as roots at least the $2k + 1$ numbers $0, \pm h, \cdots, \pm kh$, and so does the polynomial $P(x) = \prod_{j=-k}^{k}(x-jh)$. Hence these numbers are also roots of any function of the form

(4) $$F(x) = E(x) - cP(x),$$

where c is a constant. It follows from Rolle's Theorem, for any such F, that $F'(x)$ has $2k$ roots other than 0. The choice of $c = E'(0)/P'(0)$ makes 0 a root of F' as well, so that F' has $2k + 1$ distinct roots in $(-kh, kh)$. Note that $P'(0)$ is the coefficient of x in $P(x)$, i.e., $\prod_{j \neq 0}(-jh)$, so that

(5) $$c = (-1)^{k}E'(0)/(k!)^{2}h^{2k}.$$

By another (multiple) application of Rolle's Theorem to F', $F^{(2k+1)}$ must have at least one root in $(-kh, kh)$, say t. Now one may compute directly from (4) and the definitions of E and P that

(6) $$F^{(2k+1)}(x) = f^{(2k+1)}(x) - c(2k + 1)!.$$

Setting $x = t$, and using (5), we obtain from (6) the equation

(7) $$0 = f^{(2k+1)}(t) \pm E'(0)(2k + 1)!/(k!)^{2}h^{2k}.$$

Finally, solving (7) for $E'(0)$ and using the assumed bound K, we conclude that

$$|E'(0)| \leq K(k!)^{2}h^{2k}/(2k + 1)!,$$

which is equivalent to (3).

The questions raised earlier may now be answered. The error in (1) is roughly "proportional" to h^2 while that in (2) is "proportional" to h^4. The proportionality constants cannot reasonably be evaluated, of course, since they depend on bounds for the third and fifth derivatives, while we are still computing a single value of the first derivative. Nevertheless, the theorem provides strong evidence that values of h large enough to avoid numerical difficulties due to cancellation will produce highly accurate values of the derivative. The usual pragmatic approach to determining how small h should be is to successively halve h, recomputing (1) or (2) until halving h no longer makes a difference in the desired number of decimal places.

The following example illustrates this via a short *PL/I* program (see p. 357) that uses formula (2) and tabulates f and f' for

$$f(x) = \frac{[\cos(x^2 - 1) + x^2]^{\frac{1}{4}}}{(3x^2 + 17)^{1/3}}.$$

The example is artificially complicated, of course, as textbook examples are often artificially simple. By calculating $f'(x)$, the diligent student will discover $f'(0) = 0$, but not much else about this function. The tabulation shows the existence of at least seven relative extrema on $[0, 5]$ and also provides evidence of an absolute minimum at 0 (note symmetry). A closer tabulation would reveal another pair of extrema in $[4.5, 5]$, plus approximate locations of extrema and inflection points.

Remarks and acknowledgments. It has been pointed out by Allen Ziebur and by the referee that formulas (1) and (2) and error estimates of the form $O(h^{2k})$ can be derived easily from Taylor's formula. We prefer the approach above for freshmen because the error estimate is more explicit and the proof requires only Rolle's Theorem.

Thanks are due to David Hayes for suggestions that led to several improvements in an earlier version of this note, including a somewhat cleaner proof of the Theorem.

```
  10.         LET  F(X)=SQRT(COS(X*X-1)+X*X)/(3*X*X+17)**(1/3);
  20.         LET  DF(X,H)=(F(X-2*H)-8*F(X-H)+8*F(X+H)-F(X+2*H))/(12*H);
  30.         GET LIST(A,B,S);
  40.         PUT EDIT('X','F(X)','DF(X)','H')(X(3),A,(3) (X(8),A));
  50.         DO X=A TO B BY S;
  60.         H=.25;
  70.         Y=DF(X,H);
  80.         DO  WHILE(H>.00000001);
  90.         H=H/2;
 100.         Z=DF(X,H);
 110.         IF ABS(Y-Z)>=.000001 THEN Y=Z; ELSE GO TO OUT;
 120.         END ;
 130.         PUT EDIT(X,F(X))(F(6,2),F(12,5));
 140.         GO TO END;
 150.    OUT: PUT EDIT(X,F(X),Y,2*H)(F(6,2),(2) F(12,5),E(13,3));
 160.    END: END ;
 170.         STOP ;
?EXECUTE
A
?0
B
?5
S
?.5
     X          F(X)          DF(X)          H
   0.00        0.28587       0.00000       2.500E-01
   0.50        0.37983       0.30393       6.250E-02
   1.00        0.52100       0.20840       3.125E-02
   1.50        0.55721      -0.05376       3.125E-02
   2.00        0.56470       0.24438       3.125E-02
   2.50        0.78937       0.43210       1.562E-02
   3.00        0.84288      -0.11190       1.562E-02
   3.50        0.93690       0.39413       1.562E-02
   4.00        0.97094      -0.03038       7.812E-03
   4.50        1.07804       0.01503       1.562E-02
   5.00        1.11693       0.29717       7.812E-03
```

MATHEMATICS DEPARTMENT, DUKE UNIVERSITY, DURHAM, N.C. 27706.

THE ERROR OF THE TRAPEZOIDAL METHOD FOR A CONCAVE CURVE

S. K. STEIN

Consider the area of the region between the curve $y = \log x$ and the polygonal approximation formed by using as vertices the points $(n, \log n)$, $n = 1, 2, 3, \ldots$. This (unbounded) region is indicated by the shading in the accompanying diagram.

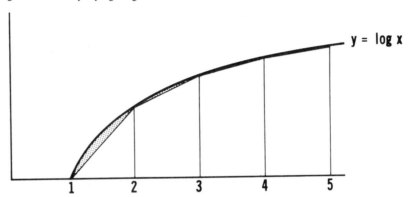

That this region has finite area can provide a basis for establishing Stirling's estimate of $n!$, as in ([2]: 361–363) and shown below in Theorem 3. This note presents a short geometric proof that the area is finite; this proof uses only the concavity of the function $\log x$. This result, stated in Theorem 1, is equivalent to ([1]: 170). Theorem 2 is a generalization.

THEOREM 1. *Let f be a function defined for $x \geq 1$ such that $f(x) \geq 0$, $f^{(1)}(x) \geq 0$, and $f^{(2)}(x) \leq 0$ for all $x \geq 1$. Let n be a positive integer and let T_n be the trapezoidal estimate of $\int_1^n f(x)dx$ based on the numbers $1, 2, \ldots, n$. Then*

$$\int_1^n f(x)dx - T_n \leq \frac{f(2) - f(1)}{2}.$$

Proof. For any integer $i = 2, 3, \ldots, n - 1$, consider the following diagram, which is the key to the proof.

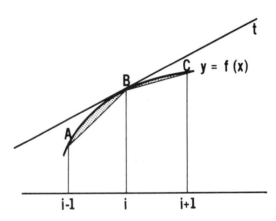

358

The line t is tangent to the graph of f at $B = (i, f(i))$. The areas of the shaded regions represent the errors of two adjacent summands in the trapezoidal approximation.

When the shaded region from B to C is translated by the vector $B - C$, no point of the resulting region R lies below t. Thus R and the shaded region from A to B, which have the common right-hand point B, share no interior points. Now for each $i = 2, 3, \ldots, n - 1$ translate the corresponding region BC so that the right-hand end point coincides with the point $(2, f(2))$. Observe that no two of the resulting translates share an interior point with each other or with the shaded region AB associated with the case $i = 2$. Thus the total error of the trapezoidal approximation is at most the area of the right triangle with vertices, $(1, f(1))$, $(1, f(2))$, and $(2, f(2))$, which has area $(f(2) - f(1))/2$. This proves the theorem.

Theorem 1 has two immediate consequences, which deserve mention. First of all, it shows that the infinite series which describes the error of the trapezoidal method over the interval $[1, \infty)$ using the numbers $1, 2, 3, 4, \ldots$, is convergent:

$$\sum_{n=1}^{\infty} \left(\int_n^{n+1} f(x)dx - \frac{f(n+1) + f(n)}{2} \right) \leq \frac{f(2) - f(1)}{2}.$$

Also, with the aid of Theorem 1 and the fact that $\int_1^n f(x)dx - T_n$ is nonnegative, it is easy to establish the inequalities

$$-\frac{f(1)}{2} - \frac{f(n)}{2} \leq \int_1^n f(x)dx - \sum_{i=1}^{n} f(i) \leq \frac{f(2)}{2} - f(1) - \frac{f(n)}{2},$$

which, as in ([1]: 172), can be used to estimate $\sum_{i=1}^{n} f(i)$.

The same type of argument that proved Theorem 1 establishes the following more general theorem.

THEOREM 2. *Let f be a function defined for $x \geq 1$ such that $f(x) \geq 0$, $f^{(1)}(x) \geq 0$, and $f^{(2)}(x) \leq 0$ for all $x \geq 1$. Let $1 = a_1 < a_2 < a_3 < \cdots < a_n < \cdots$ be an infinite sequence such that the sequence of differences $a_2 - a_1, a_3 - a_2, \ldots$ is bounded, say by b. Then the series*

$$\sum_{n=1}^{\infty} \left[\int_{a_n}^{a_{n+1}} f(x)dx - \frac{f(a_{n+1}) + f(a_n)}{2} (a_{n+1} - a_n) \right]$$

is convergent, and its sum is at most $b^2 f'(1)/2$.

If the sequence of differences, $a_{n+1} - a_n$, is unbounded, the series of Theorem 2 need not be convergent. For example, let $f(x) = \sqrt{x}$ and $a_n = n^2$. In this case

$$\int_{a_n}^{a_{n+1}} f(x)dx - \frac{f(a_{n+1}) + f(a_n)}{2} (a_{n+1} - a_n) = 1/6$$

for all n.

THEOREM 3. *There is a constant c such that $n! \sim c\sqrt{n}(n/e)^n$.*

Proof. Let a_n denote the difference between $\int_1^n \log x \, dx$ and the trapezoidal approximation, T_n, based on the values $1, 2, \cdots, n$. By Theorem 1, $\lim_{n \to \infty} a_n$ exists. Let a denote this limit.

Now, $\int_1^n \log x \, dx - T_n = a_n$ or

$$x \log x - x \Big|_1^n - \sum_{k=1}^{n-1} \frac{\log(k) + \log(k+1)}{2} = a_n.$$

or

$$n \log n - n + 1 - (\log 2 + \log 3 + \cdots + \log n) + \frac{\log n}{2} = a_n.$$

Application of the exponential function to both sides yields

$$\frac{e\sqrt{n}(n/e)^n}{n!} = e^{a_n}.$$

Thus, $n! \sim e^{1-a}\sqrt{n}(n/e)^n$, and the theorem is established.

The fact that the constant c in Theorem 3 equals $\sqrt{2\pi}$ is usually established by utilizing Wallis's theorem:

$$\lim_{n \to \infty} \frac{(n!)^2 2^{2n}}{(2n)!\sqrt{n}} = \sqrt{\pi}.$$

This limit is determined by evaluating $I(n) = \int_0^{\pi/2} \sin^n\theta\, d\theta$ and exploiting the inequalities $I(2n+1) < I(2n) < I(2n-1)$.

Incidentally, Stirling's formula implies that the shaded region described in the opening paragraph has area $1 - \log\sqrt{2\pi}$.

References

1. R. C. Buck, Advanced Calculus, McGraw-Hill, New York, 1965.
2. R. Courant, Differential and Integral Calculus, Vol. 1, Blackie, London, 1963.

DEPARTMENT OF MATHEMATICS, UNIVERSITY OF CALIFORNIA, DAVIS, CA 95616.

A NON-SIMPSONIAN USE OF PARABOLAS IN NUMERICAL INTEGRATION

ARTHUR RICHERT

Department of Mathematics, Southern College, Collegedale, TN 37315

This paper presents a numerical integration technique which uses parabolas differently than does Simpson's rule. The degree of precision of the technique is three, as is the case for Simpson's rule, while the error bound is an improvement over Simpson's rule by a factor of almost eleven. The technique provides an example of the usefulness of Taylor polynomials in numerical analysis which may be presented to beginning calculus students as soon as Taylor's theorem has been covered.

In Simpson's rule, the interval of integration, $[a, b]$, is divided into an even number, n, of equal subintervals each of length $h = (b - a)/n$. The approximation is then given by

$$(1) \qquad \int_a^b f(x)\, dx \approx \frac{h}{3}\left[f(x_0) + 4f(x_1) + 2f(x_2) + \cdots + 4f(x_{n-1}) + f(x_n) \right],$$

with an error bound of

$$(2) \qquad |\text{Error}| \leqslant h^4 M_4 (b - a)/180,$$

where M_4 is an upper bound on the magnitude of $f^{(4)}$ over the interval $[a, b]$.

We begin in the same manner with the exception that n is not required to be even. As an approximation to f over the ith subinterval, $[x_{i-1}, x_i]$, we use the second degree Taylor polynomial for f expanded about X_i, the midpoint of the ith subinterval. This polynomial is given by

$$T_i(x) = f(X_i) + f^{(1)}(X_i)(x - X_i) + f^{(2)}(X_i)(x - X_i)^2/2.$$

Direct integration yields

$$(3) \qquad \int_{x_{i-1}}^{x_i} T_i(x)\, dx = h \cdot f(X_i) + \frac{h^3}{24} \cdot f^{(2)}(X_i).$$

The approximation analogous to (1) is obtained by summing (3) over the n subintervals to get

$$\int_a^b f(x)\, dx \approx h \sum_{k=1}^{n} f(X_k) + \frac{h^3}{24} \sum_{k=1}^{n} f^{(2)}(X_k).$$

This formula is of interest in its lack of dependence on $f^{(1)}$ and also in that it is merely the midpoint rule plus a "correction" term involving the second derivative.

To determine an error bound we assume $f \in C^4[a, b]$ and express f over the ith subinterval in the form

$$f(x) = \sum_{k=0}^{3} \frac{f^{(k)}(X_i)}{k!}(x - X_i)^k + \frac{f^{(4)}(c_x)}{24}(x - X_i)^4,$$

where for each $x \in [x_{i-1}, x_i]$ an appropriate c_x between x and X_i is guaranteed by Taylor's theorem. Direct integration utilizing the weighted mean value theorem for integrals yields

$$\int_{x_{i-1}}^{x_i} f(x)\, dx = h \cdot f(X_i) + \frac{h^3}{24} \cdot f^{(2)}(X_i) + \frac{h^5}{1920} \cdot f^{(4)}(d_i),$$

where $d_i \in (x_{i-1}, x_i)$. Thus the absolute error over the ith subinterval is bounded by $h^5 M_4/1920$ and the error expression analogous to (2) is

$$|\text{Error}| \leqslant nh^5 M_4/1920 = h^4 M_4(b - a)/1920.$$

The degree of precision remains three while the error bound is reduced by a factor of nearly eleven. The cost of this improvement is the availability of $f^{(2)}$ and an additional $n - 1$ evaluations.

Reconsidering Area Approximations

George P. Richardson

The Rockefeller College of Public Affairs and Policy, State University of New York at Albany,
Albany, NY 12222

A recent survey of calculus texts for adoption revealed that all contained presentations of the familiar method of approximating a definite integral by summing the areas of n trapezoids of equal width h,

$$A \doteq \sum_{i=1}^{n} \frac{f(x_{i-1}) + f(x_i)}{2} h.$$

Only one text contained any mention of the simpler approximation obtained by summing the areas of rectangles computed at midpoints,

$$A \doteq \sum_{i=1}^{n} f\left[\frac{x_{i-1} + x_i}{2}\right] h,$$

and there the midpoint-rectangle method was quickly dismissed in an example. The overwhelming tendency to ignore the midpoint-rectangle method in modern texts is unfortunate, and somewhat puzzling, because it is conceptually simpler than the trapezoid approximation and is actually more accurate. A good case can be made for the claim that the midpoint-rectangle method is the first numerical approximation to the definite integral that students should see.

Although the greater accuracy of the midpoint-rectangle approximation is well known, it is nonetheless surprising to many, perhaps partly because we are so used to seeing the trapezoid method emphasized in introductory texts. Yet, the result is easy to see geometrically (see FIGURE 1). The comparison of the area approximations is facilitated by noting that the area of the midpoint-rectangle is equal to the area of another trapezoid determined by the tangent to the graph of f at the midpoint. From the figure it is then evident that as long as $f''(x)$ is continuous and nonzero over $[x_{i-1}, x_i]$, so that the direction of curvature does not change, the area representing the midpoint-rectangle error will always be less than the area representing the trapezoid error.

Most calculus texts state without proof a theorem equivalent to the following. If f has a continuous second derivative on $[a, b]$ then the error e_T in approximating

$$\int_b^a f(x)\, dx$$

using n trapezoids of width $h = (b - a)/n$ satisfies

$$0 \leqslant e_T \leqslant \frac{1}{12} h^2 (b - a) \left[\max_{a \leqslant x \leqslant b} |f''(x)| \right].$$

The corresponding error bound e_R for n rectangles, which is not commonly

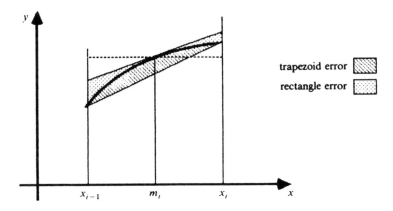

FIG. 1. Geometric comparison of the midpoint-rectangle and trapeziod approximations for $\int_{x_{i-1}}^{x_i} f(x)\,dx$.

mentioned in calculus texts, is

$$0 \leqslant e_R \leqslant \frac{1}{24}h^2(b-a)\left[\max_{a \leqslant x \leqslant b} |f''(x)|\right].$$

Proofs can be found in most numerical analysis texts, e.g. [1], [2], [3], [6, pp. 361–400]. While the inequalities here do not support a precise conclusion, the denominators in these error bounds certainly hint that the midpoint-rectangle approximation is likely to be about twice as accurate as the trapezoid method.

   The following derivation of series expressions for the approximation errors of the two methods (stimulated by [4, pp. 80–84]) is of interest because it is accessible to first-year calculus students, is apparently not well known, and gives more explicit statements of the approximation errors.

   Suppose f satisfies the hypotheses of Taylor's Theorem over $[a, b]$. Partition $[a, b]$ into n subintervals $[x_{i-1}, x_i]$ each of length h. Let m_i represent the midpoint of $[x_{i-1}, x_i]$, and for any x in $[x_{i-1}, x_i]$ expand $f(x)$ in a Taylor series about m_i. On integrating over $[x_{i-1}, x_i]$, the even powers in the resulting series cancel, leaving

$$\int_{x_{i-1}}^{x_i} f(x)\,dx = f(m_i)h + \frac{h^3}{2^2 3!}f''(m_i) + \frac{h^5}{2^4 5!}f^{(4)}(m_i) + \cdots . \tag{1}$$

Summing over the n subintervals yields the rectangle approximation and a series expression for its error:

$$\int_a^b f(x)\,dx = \sum_{i=1}^n f(m_i)h + \frac{h^3}{2^2 3!}\sum_{i=1}^n f''(m_i) + \frac{h^5}{2^4 5!}\sum_{i=1}^n f^{(4)}(m_i) + \cdots . \tag{2}$$

   To derive the comparable result for the trapezoid rule, use the same Taylor series to express $f(x_{i-1})$ and $f(x_i)$ in terms of $f(m_i)$ and its derivatives. Averaging

$f(x_{i-1})$ and $f(x_i)$ and then solving for $f(m_i)$ produces

$$f(m_i) = \frac{f(x_{i-1}) + f(x_i)}{2} - \frac{h^2}{2^2 2!} f''(m_i) - \frac{h^4}{2^4 4!} f^{(4)}(m_i) - \cdots . \tag{3}$$

Substituting the right side of (3) into (1) in place of $f(m_i)$ and collecting like powers of h yields

$$\int_{x_{i-1}}^{x_i} f(x)\, dx = \frac{f(x_{i-1}) + f(x_i)}{2} h - \frac{2h^3}{2^2 3!} f''(m_i) - \frac{4h^5}{2^4 5!} f^{(4)}(m_i) - \cdots ,$$

which gives the following result for the trapezoid rule, comparable to (2):

$$\int_a^b f(x)\, dx = \sum_{i=1}^n \frac{f(x_{i-1}) + f(x_i)}{2} h - \frac{2h^3}{2^2 3!} \sum_{i=1}^n f''(m_i)$$
$$- \frac{4h^5}{2^4 5!} \sum_{i=1}^n f^{(4)}(m_i) - \cdots . \tag{4}$$

From the series error terms in (2) and (4) it is again evident that the rectangle approximation is of the same order as the trapezoid approximation but is likely to be about twice as accurate. More precisely, the error terms show that both methods are exact for linear functions, while the midpoint-rectangle approximation has exactly half the error of the trapezoid approximation for quadratic and cubic polynomials. For functions with nonzero higher derivatives over $[a, b]$, the smaller error inherent in the midpoint-rectangle approximation will theoretically become evident once h is sufficiently small.

There is both puzzle and promise in these results. The puzzle is that the midpoint-rectangle method is almost uniformly ignored in introductory calculus texts, and the trapezoid rule universally emphasized, although the midpoint rule is likely to be more accurate. Moreover, the midpoint-rectangle method is conceptually simpler and is closer to the fundamental notion of a Riemann sum. For building intuition about the definite integral, probably every student should experiment with a simple computer implementation of the method.

Besides their pedagogical implications, the series expansions for the approximation errors in (2) and (4) provide a further benefit. As suggested by Figure 1, the errors have opposite signs. Taking advantage of that fact, one can easily note that doubling (2) and adding it to (4) eliminates the terms involving h^3 [4, pp. 80–84]. Thus on dividing by 3 we obtain an approximation of the definite integral that is accurate to an error proportional to h^5:

$$\int_a^b f(x)\, dx = \frac{1}{6} \sum_{i=1}^n [f(x_{i-1}) + 4f(m_i) + f(x_i)] h - \frac{h^5}{2^3 \cdot 5!} \sum_{i=1}^n f^{(4)}(m_i)$$
$$- \frac{2h^7}{2^5 3 \cdot 7!} \sum_{i=1}^n f^{(6)}(m_i) - \cdots .$$

This substantial improvement is, of course, Simpson's rule, and equations (2) and (4) provide an elegant derivation. (One introductory calculus-with-applications text has a nicely intuitive development along these lines without the detailed error terms [3, pp. 454–457].) The same technique done with trapezoid approximations with n and $2n$ trapezoids also yields Simpson's rule and can be efficiently iterated in Romberg's method [5] to produce much more accurate, higher-order approximations.

REFERENCES

1. Ake Bjorck and Germund Dahlquist, Numerical Methods, Prentice-Hall, Englewood Cliffs, N.J., 1974.
2. Richard L. Burden and J. Douglas Faires, Numerical Analysis, 3rd ed., Prindle, Weber & Schmidt, Boston, 1985.
3. Larry J. Goldstein, David C. Lay, David I. Schneider, Calculus and its Applications, 4th ed., Prentice-Hall, 1987.
4. Robert Sedgewick, Algorithms, Addison-Wesley, 1984.
5. Stanley Wagon, Evaluating Definite Integrals on a Computer: Theory and Practice, UMAP unit 432, EDC/UMAP, Newton, Mass., 1980.
6. David M. Young and Robert Todd Gregory, A Survey of Numerical Mathematics, vol. 1, Addison-Wesley, Reading, Mass., 1972.

An Interpolation Question Resolved by Calculus

MARTIN D. LANDAU AND WILLIAM R. JONES

MARTIN D. LANDAU is Associate Professor of Mathematics at Lafayette College, Easton, Pennsylvania, where he has been teaching since 1965. He received his Ph.D. in mathematics from Lehigh University in 1967, and his special areas of interest are real analysis and topology.

WILLIAM R. JONES received his Ph.D. in mathematics from Rutgers University in 1963, and since then has been teaching at Lafayette University where he is Associate Professor of Mathematics. His areas of interest include real analysis, complex analysis and differential equations.

In high school, students learn linear interpolation in tables of functions such as the trigonometric, logarithmic, power, root, exponential, and hyperbolic functions. The functional value sought is approximated by means of a straight line segment joining the two points whose coordinates are the nearest tabular entries. It is natural to ask between which entries in a table the process of interpolation yields the largest error. This question affords students taking an introductory course in the calculus an opportunity to appreciate the power of the calculus.

A few results of the differential calculus are employed in this paper. First, $f'(x) > 0$ in an interval implies f increases in this interval; $f'(x) < 0$ in an interval implies f decreases in the interval. Second, if $f''(x) > 0$ in an interval, the graph of f is concave up in this interval; $f''(x) < 0$ in an interval implies the graph is concave down in the interval. If the graph is concave up on an interval, then the line segment joining two points of the graph lies above the graph, and hence the approximation obtained by linear interpolation between these points is too large; similarly, if the graph is concave down, the line segment is below the graph and the approximation is too small.

Let us call the tabulated function $f(x)$ and suppose that it is tabulated from a to b with fixed increments h. If x is a tabulated point (but $x \neq b$), we may wish to interpolate between x and $x + h$. We are then interested in $f(x + rh)$ where r lies between 0 and 1. The linear interpolation method leads to writing, for $f(x + h) \neq f(x)$,

$$\frac{f(x + rh) - f(x)}{f(x + h) - f(x)} \approx \frac{(x + rh) - x}{(x + h) - x} = r.$$

Thus,

$$f(x + rh) - f(x) \approx r[f(x + h) - f(x)],$$

and, finally,

$$f(x + rh) \approx (1 - r)f(x) + rf(x + h).$$

The error in this approximation depends on both x and r. Since we are only concerned with the magnitude of the error and not its sign, we will consider the error function

$$E(x,r) = |(1 - r)f(x) + rf(x + h) - f(x + rh)|.$$

From our prior remarks, however, we see that

$$E(x,r) = (1 - r)f(x) + rf(x + h) - f(x + rh)$$

if f is concave up on $[x, x + h]$, and

$$E(x,r) = f(x + rh) - (1 - r)f(x) - rf(x + h)$$

if f is concave down on $[x, x + rh]$. Although we only defined $E(x,r)$ for $0 < r < 1$, we can permit $r = 0$ and $r = 1$; in such cases there is no interpolation error and we set $E(x,0) = E(x,1) = 0$.

Now let us suppose that f is continuous on the closed interval $[a, b]$ and has derivatives up to the third order on the open interval (a, b). The tabulated functions we shall discuss later all satisfy these conditions on the intervals on which they are tabulated. Let us temporarily think of x as fixed, so that $E(x, r)$ is a function of r where $0 \leq r \leq 1$. Since f is continuous, this function of r is also continuous on the closed interval $[0, 1]$, and hence assumes its maximum by the Extreme Value Theorem of the calculus. Let us denote this by

$$M(x) = \max\{E(x,r)|0 \leq r \leq 1\}.$$

Since f is tabulated on the discrete set of points $\{a, a + h, \ldots, b - h, b\}$, the function $M(x)$ is defined for all points of this set except b. We are interested in whether $M(x)$ increases or decreases as x increases; that is, does the worst interpolation error in an interval get worse or better as we move to the next interval on the right? We will string together three elementary propositions that permit us to answer this question.

PROPOSITION 1. *If for each r in $(0, 1)$ the function $E(x, r)$ is a strictly increasing function of x, then $M(x)$ is also a strictly increasing function. (Similarly, E strictly decreasing implies M strictly decreasing.)*

Proof. Suppose $x_1 < x_2 \neq b$, where both x_1 and x_2 are tabular points. Then $M(x_1) = \max\{E(x_1, r)|0 \leq r \leq 1\}$ and there is an r_1 between 0 and 1 such that $M(x_1) = E(x_1, r_1)$. Similarly, there is an r_2 between 0 and 1 such that $M(x_2) = E(x_2, r_2)$. Since $E(x, r)$ is a strictly increasing function of x for fixed r, we know that $E(x_1, r_1) < E(x_2, r_1)$. And since $E(x_2, r_2)$ is the maximum of $E(x_2, r)$ in $[0, 1]$, we know that $E(x_2, r_1) \leq E(x_2, r_2)$. Combining, we see that $M(x_1) < M(x_2)$ as we wished to show.

Now we can establish the behavior of the function M if we know that $E(x, r)$ is increasing (or decreasing) in x for each r in $(0, 1)$. To ascertain when this condition is satisfied, it is natural to think now of r as fixed and x as the variable, and then to consider the sign of the derivative of E with respect to x. This presents a problem, however, for x is a discrete variable instead of a continuous variable. We can get around this be defining a new function that agrees with E on its domain but with a continuous independent variable. To this end, we define

$$G(x;r) = |(1 - r)f(x) + rf(x + h) - f(x + rh)|$$

for *all* real x in $[a, b - h]$. The semicolon between x and r is to indicate that r is a parameter. Thus we are defining a family of functions of x with the parameter r restricted to lie in $(0, 1)$. Now we can obtain our second result.

PROPOSITION 2. *Suppose for each r in $(0, 1)$ that $G'(x; r) > 0$ on $(a, b - h)$, then M is a strictly increasing function of x. (Similarly, $G' < 0$ implies that M is a strictly decreasing function.)*

Proof. As we know, $G' > 0$ on $(a, b - h)$ for each r, combined with the continuity of G on $[a, b - h]$, implies that G is strictly increasing on $[a, b - h]$; that is, $a \le x_1 < x_2 \le b - h$ implies that $G(x_1; r) < G(x_2; r)$ for any r in $(0, 1)$. Now suppose that x_1 and x_2 are tabulated values; that is, they are in the domain of E. Then $a \le x_1 < x_2 \le b - h$ ensures that $E(x_1, r) = G(x_1; r) < G(x_2; r) = E(x_2, r)$, and we see that $E(x, r)$ is a strictly increasing function of x for each r in $(0, 1)$. Hence, Proposition 1 yields the conclusion that M is strictly increasing in x.

Now we will establish conditions which imply the hypothesis of Proposition 2.

PROPOSITION 3. *Suppose each of f'' and f''' has fixed sign (non-zero) on (a, b). Then if $f''(x) \cdot f'''(x) > 0$ on (a, b), for each r in $(0, 1)$ $G'(x; r) > 0$ on $(a, b - h)$, and hence M is a strictly increasing function. (Similarly, $f'' \cdot f''' < 0$ implies $G' < 0$, and hence M is strictly decreasing.)*

Proof. Suppose $f''(x) > 0$ and $f'''(x) > 0$ on (a, b). Then for each r in $(0, 1)$,

$$G(x;r) = (1 - r)f(x) + rf(x + h) - f(x + rh)$$

since the graph of f is concave up. Differentiating,

$$G'(x;r) = (1 - r)f'(x) + rf'(x + h) - f'(x + rh).$$

But since $f'''(x) > 0$ (and f''' is the second derivative of f'), we know that the graph of f' is also concave up, and hence the interpolated value for f', $(1 - r)f'(x) + rf'(x + h)$, lies above the true functional value $f'(x + rh)$. Therefore $G'(x; r) > 0$. By Proposition 2, this yields the conclusion that M is strictly increasing in x. Similar arguments apply to the three remaining cases.

These three propositions comprise a proof of our main result.

THEOREM. *If f is continuous on $[a, b]$ and each of $f''(x)$ and $f'''(x)$ has constant nonzero sign on (a, b) and if $f''(x) \cdot f'''(x)$ is positive (resp. negative) on (a, b), then the function $M(x)$, i.e., the maximum absolute linear interpolation error associated with a table for f, is increasing (resp. decreasing) on its domain.*

Geometrically this says essentially that the interpolation error increases with x in an interval in which the graphs of f and f' agree in concavity, but decreases if f and f' differ in concavity. This is true whether we consider interpolating a fixed percentage of the way between neighboring tabulated points (fixed r) or whether we consider the maximum error in each subinterval. Applying this to the trigonometric functions, we see that the sine, tangent and secant interpolation errors increase as x increases through the first quadrant, whereas their confunctions have decreasing errors. For x positive we see that e^x, $\sinh x$, $\cosh x$ and x^3 have increasing errors as x increases, whereas $\ln x$, e^{-x}, $\sqrt{x}$, $\sqrt[3]{x}$ and $1/x$ have decreasing errors. Two commonly tabulated functions which fall into neither of the above categories are x^2 and $\tanh x$. For $f(x) = x^2$ we have $f''' = 0$, and it is easy to verify that for fixed r the interpolation error is a constant throughout the table and that the maximum error occurs at the midpoint of each subinterval. For $\tanh x$, the second derivative is negative for all $x > 0$ whereas the third derivative changes sign from negative to positive as x passes through the value x_0 satisfying $2\sinh^2 x = 1$. From tables, $x_0 \approx 0.66$, so as we interpolate in the table for increasing values of x, the error increases as we approach x_0 and decreases beyond x_0. For the unique interval containing x_0 (we would not expect x_0 to be an entry in the table), the theorem does not determine how the error compares with those obtained by interpolating in the preceding and following intervals of the table. Thus for $\tanh x$, the maximum error will occur in one of these three adjacent intervals.

Another commonly tabulated function whose behavior with respect to interpolation error is similar to that of the hyperbolic tangent is

$$f(x) = \int_{-\infty}^{x} \frac{1}{\sqrt{2\pi}} e^{-t^2/2} \, dt$$

tabulated for $x \geqslant 0$. This yields the area under the standard normal curve. By the fundamental theorem of calculus

$$f'(x) = \frac{1}{\sqrt{2\pi}} e^{-x^2/2},$$

and we can easily verify that the graph of f is concave down for $x > 0$ whereas that of f' changes from concave down to concave up for x positive as it passes through an inflection point at $x_0 = 1$. Thus once again the error increases as we approach x_0 and decreases thereafter. In this case, x_0 will always appear as an entry in the table, so the maximum error occurs in either of the two adjacent intervals having 1 as an endpoint.

Generalized Cycloids: Discovery via Computer Graphics

Sheldon P. Gordon

Sheldon Gordon is an Associate Professor of Mathematics at Suffolk County Community College, where he has been on the faculty since 1974. For the last three years, Dr. Gordon served as the project director of an NSF CAUSE grant project intended to promote the use of the computer into almost all course offerings in mathematics, the sciences, engineering and economics. The present paper is an outgrowth of this project. Dr. Gordon is currently involved in the newly formed National Consortium of Computers in Mathematical Sciences Education, an organization dedicated to encourage the educational use of the computer at all levels of mathematics instruction throughout the nation.

Introduction. One of the most significant roles for the computer in mathematics is that it provides the opportunity to discover new mathematical relationships. By eliminating the need to actually perform computations, the computer allows the user to consider a large number of otherwise inaccessible results and so find previously hidden patterns. This is particularly true using the graphical capabilities of the computer.

In this regard, the author has developed a program which will draw the graph of virtually any pair of parametric equations

$$x = f(t), \quad y = g(t),$$

where the functions f and g are continuous. This program is being used for demonstrations in our Calculus courses and is used on an individual basis by the students outside of class. The present paper describes one particular outcome of the use of this program.

When treating the topic of parametric equations in the traditional calculus sequence, there are a limited number of standard curves which are usually covered. One of the most prominent is the cycloid, which is given by the pair of parametric equations

$$x = at - a\sin t$$
$$y = a - a\cos t. \tag{1}$$

This represents the locus of points described by a fixed point on a rolling circle of radius a. The graph of this function, as drawn by the computer, is shown in Figure 1 with $a = 5$.

The author gratefully acknowledges the support provided by the National Science Foundation under grant #SER77-05817. The views expressed in this article are those of the author and do not necessarily reflect the views of the Foundation.

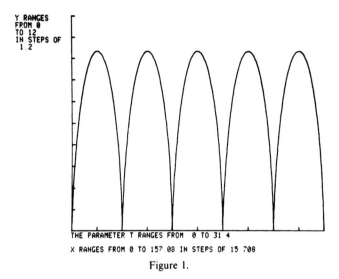

Figure 1.

However, since the computer does all the work once the user selects the desired function, it is very simple and natural to begin to wonder "what happens if the above expressions are modified?" For example, suppose the following variation is tried:

$$x = at - a \sin t$$
$$y = at - a \cos t. \tag{2}$$

The resulting graph is shown in Figure 2 with $a = 5$ again. It appears to be related to the original cycloid, but is inclined at an angle rather than remaining horizontal.

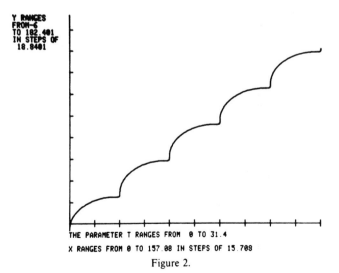

Figure 2.

Moreover, if we try a further variation, such as

$$x = at - a \sin t$$

$$y = a\sqrt{t} - a \cos t, \tag{3}$$

then we get the result in Figure 3, which is still another version of a cycloid.

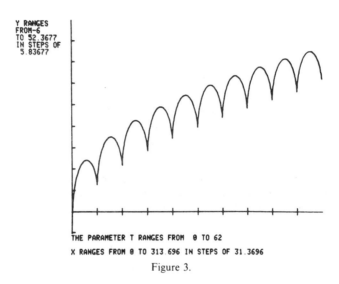

Figure 3.

By this stage, it is clear that what is being generated is a series of generalized cycloids, all of which depend on the first term in the expression for y. If we now examine the equations (1) for the basic cycloid, we notice that the terms $a \cdot t$ and a, in the expressions for x and y, represent precisely the coordinates of the center of the rolling circle, (at, a). Using this interpretation applied to the system (2), we see that the first terms here, at and at, apparently represent the coordinates of the center of a circle which is rolling along a line whose slope is 1.

Similarly, system (3) would seem to represent a cycloid formed when a circle rolls in such a manner that its center remains on the curve $y = \sqrt{ax}$.

In actuality, this interpretation is not quite correct. We must first analyze what is meant by saying that "a circle rolls along the horizontal" as in the definition of the cycloid. In fact, such a situation involves two distinct, simultaneous motions: first, there is the linear motion of the strictly horizontal translation of the circle at a constant rate, a; second, there is the rotational motion of the circle about its center with constant angular velocity of 1. The net effect is the motion of a rotating circle of radius a as it moves horizontally such that the center always lies on the line $y = a$.

The results in systems (2) and (3) therefore involve a generalization of the concept of cycloid to cover the case where the rotating circle moves in such a way that its center lies on curves other than the line $y = a$. Furthermore, if the physical aspect of motion, "rolling," is replaced by the simultaneous mathematical operations of translation and rotation, then a variety of different generalizations of the cycloid become immediately evident.

Generalized Cycloids along the Horizontal. We now consider the situation where a circle of radius a is moving horizontally to the right with constant velocity a. Moreover, we assume that the circle is rotating with constant angular velocity $d\theta/dt = \omega$. Therefore, the angle through which a fixed point on the circle rotates in time t is given by $\theta = \omega t$ and the resulting arc length is $s = a\omega t$. When $\omega = 1$, the arc length exactly matches the horizontal displacement of the circle, at, which is the result of "rolling."

When $\omega < 1$, the result is akin to "skidding" where the forward motion is faster than mere rotation would account for. When $\omega > 1$, the result is comparable to a car's tire spinning on ice with minimal forward movement.

The corresponding parametrization of the coordinates of a fixed point on the circle is then given by

$$x = at - a\sin(\omega t)$$

$$y = a - a\cos(\omega t). \tag{4}$$

In Figures 4 and 5, the resulting curves are shown for values of $\omega = 4$ and $\frac{1}{2}$ respectively.

Moreover, it is clear that it should also be possible to vary the horizontal velocity of the circle. If we denote this by b, then the corresponding parametric equations are

$$x = bt - a\sin(\omega t)$$

$$y = a - a\cos(\omega t). \tag{5}$$

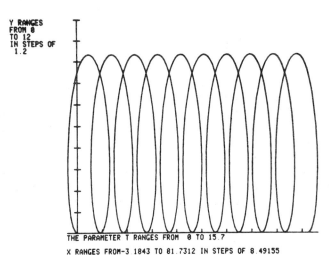

Y RANGES FROM 0 TO 12 IN STEPS OF 1.2

THE PARAMETER T RANGES FROM 0 TO 15.7
X RANGES FROM -3.1843 TO 81.7312 IN STEPS OF 8.49155

Figure 4.

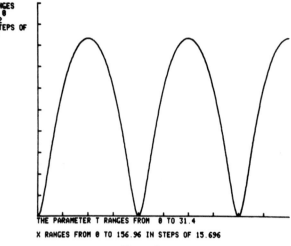

Figure 5.

The corresponding graphs, predictably enough, are very similar to the ones shown in Figures 4 and 5 and therefore no specific illustrations are included.

Generalized Cycloids along Arbitrary Curves. We now consider a different generalization of the concept of cycloid which encompasses the examples in systems (2) and (3). Suppose that we have a circle of radius a which is rotating with constant angular velocity ω. Moreover, suppose that the circle is moving in such a way that its center always lies on a curve $v = h(x)$. Finally, suppose that the circle moves in such a way that its horizontal velocity remains constant and equal to b. The resulting parametric equations for a fixed point on this circle are given by

$$x = bt - a \sin \omega t$$

$$y = h(bt) - a \cos \omega t. \tag{6}$$

In view of this, equations (2) represent the generalized cycloid formed when the circle of radius a rotates with angular velocity $\omega = 1$ and constant horizontal velocity $b = a$ such that its center lies on the line $y = x$. Furthermore, equations (3) represent the generalized cycloid, using $a = b$, and $\omega = 1$ so that the center of the circle lies on the curve $y = \sqrt{ax}$.

Incidentally, it is worth noting that the case $\omega = 1$ in the present context does not reduce to the situation of a rolling circle as it did in the horizontal case. In particular, the actual parametrization for a circle of radius a which rolls up an incline at an angle α (so that the center lies on the line $y = (\tan \alpha)x$ through the origin) is given by

$$x = at \cos \alpha - a \sin(t - \alpha)$$

$$y = at \sin \alpha - a \cos(t - \alpha).$$

As a final illustration, Figure 6 shows the generalized cycloid which results from a circle rotating along a sine curve.

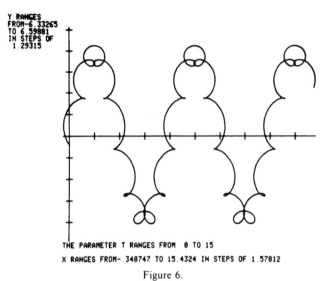

Y RANGES
FROM-6.33265
TO 6.59881
IN STEPS OF
1.29315

THE PARAMETER T RANGES FROM 0 TO 15

X RANGES FROM-.348747 TO 15.4324 IN STEPS OF 1.57812

Figure 6.

Finally, it is worth remarking that the computer program used for the graphs was written using a BASIC version of the PLOT 10 graphics routines, designed for use on a Tektronics 4006 terminal supported by a PDP 11/34 computer system. This program, as well as a series of other graphics programs for calculus, is available to anyone who can use them.

Behold! The Midpoint Rule is Better Than the Trapezoidal Rule for Concave Functions

Frank Burk, California State University, Chico, CA

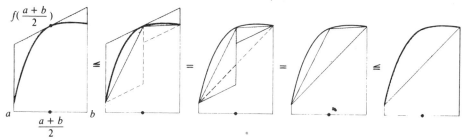

Applications of Transformations to Numerical Integration
Chris W. Avery and Frank P. Soler, De Anza College, Cupertino, CA

The method of substitution to produce a closed-form solution for an integral is well known to the beginning calculus student. This capsule addresses the application of transformations to numerical integration, in particular, to the evaluation of improper integrals. We use Simpson's rule as the primary tool, as it is the most powerful numerical integration technique available to mathematics students at the freshman/sophomore levels. The algorithm provides adequate precision for the evaluation of definite integrals as long as the error formula can be relied on to provide useful information; that is, if the integrand has four continuous derivatives. What happens if a derivative or the integrand itself is discontinuous?

Consider first the integral

$$\int_1^4 \sqrt{x}\, dx = 14/3. \tag{1}$$

Simpson's rule with 20 subintervals (or 10 "double subintervals," conventionally denoted by S_{20}) yields 4.666666, the correct answer to machine accuracy. On the other hand, for the integral

$$\int_0^4 \sqrt{x}\, dx = 16/3, \tag{2}$$

$S_{100} = 5.332683$ gives less than four significant digits of the correct answer.

We seek a way to overcome this lack of precision. The dependence of the error bound on the derivative of the integrand, which approaches infinity at $x = 0$, provides a clue. To speed up convergence, we may transform the integrand to one whose derivatives are defined throughout the interval: if $u^2 = x$, then $2u\, du = dx$, and (2) becomes

$$\int_0^2 2u^2\, du. \tag{3}$$

Of course, Simpson's rule (or any integration method of order greater than 2) is exact on this integral. Thus, a reasonable strategy when using Simpson's rule for integrands that have discontinuous derivatives is to find a transformation to remove the discontinuities. The same strategy often works if the integrand itself is discontinuous (that is, if the integral is improper).

Consider the integral

$$\int_0^{\pi/2} \sqrt{1 + \tan x}\, dx. \tag{4}$$

Even with a large number of subintervals, Simpson's rule cannot come close to evaluating this integral. We start by separating the zero of $\tan x$ from its singularity by rewriting (4) as

$$\int_0^{\pi/4} \sqrt{1 + \tan x}\, dx + \int_{\pi/4}^{\pi/2} \sqrt{1 + \tan x}\, dx.$$

The first integral is now easily evaluated using Simpson's rule: $S_{12} = 0.938449$, which is accurate to the 6 decimals shown. The second integral may be transformed by $\tan x = 1/u^2$ to

$$2 \int_0^1 \frac{\sqrt{u^2 + 1}}{u^4 + 1} \, du. \tag{5}$$

For this integral, $S_{20} = 1.952222$, which is also accurate to 6 decimal places. Hence, the value of (4) is $.938449 + 1.952222 = 2.890671$, accurate to at least 5 decimal places.

An interesting family of functions with applications in probability theory (after minor modifications of no interest to us here) is given by $f(t) = t^{(n/2)-1}e^{-t/2}$ for small positive integers n. The integral

$$\int_0^1 f(t) \, dt \tag{6}$$

converges for $n \geq 1$, but there are obvious problems for several odd values of n. For $n = 3$, we obtain the proper integral

$$\int_0^1 t^{1/2}e^{-t/2} \, dt. \tag{7}$$

We may remove the discontinuity in the derivative of the integrand exactly as we did for (2). The substitution $t = u^2$ transforms (7) to

$$\int_0^1 2u^2 e^{-u^2/2} \, du. \tag{8}$$

For this integral: $S_{16} = 0.498187$, which is accurate to 6 decimal places.

Similarly, setting $n = 1$ in (6) yields the improper integral

$$\int_0^1 t^{-1/2}e^{-t/2} \, dt. \tag{9}$$

The same transformation used to obtain (8) yields

$$\int_0^1 2e^{-u^2/2} \, du. \tag{10}$$

For this integral: $S_{12} = 1.711249$, which is accurate to at least 5 decimal places.

Transformations may also be used to evaluate integrals with infinite limits of integration. For example, consider

$$\int_0^\infty \frac{dx}{\sqrt{x^3 + 1}}. \tag{11}$$

We want to find a transformation that will turn the interval $[0, \infty)$ into a finite interval. In this case, a transformation of the form $x = 1/u^n$ is a likely candidate. However, as with (4), we cannot use this transformation because the lower bound is 0. Thus, we first split the interval at $x = 1$ to obtain

$$\int_0^1 f(x) \, dx + \int_1^\infty \frac{dx}{\sqrt{x^3 + 1}}. \tag{12}$$

The first integral is readily evaluated using Simpson's rule: $S_{24} = 0.909604$, accurate to 6 decimal places. The second integral is transformed by $x = 1/u^2$ to

$$2\int_0^1 \frac{du}{\sqrt{u^6 + 1}}.$$ (13)

Here $S_{24} = 1.89476$. Therefore, the value of (11) is $.909604 + 1.89476 = 2.80436$.

Remarks.   At De Anza College, the teaching of the beginning calculus sequence is undergoing major revisions, stimulated by the advent of powerful interactive software that allows for the exploration of the calculus from numerical and graphical standpoints. Computer assignments allow teacher and student to interact with concepts without experiencing restrictions inherent with algebraic manipulations. Visual and numerical verifications of theoretical results previously not well understood by the students have led to new insights in the classroom. The use of substitution explored in this note is a direct result of the mathematical stimulation generated by use of the computer. We believe the microcomputer will drastically alter the teaching of mathematics in ways that will demand new pedagogical strategies, of which this note provides one example.

Circumference of a Circle—The Hard Way

David P. Kraines, Duke University, Durham, NC, Vivian Y. Kraines, Meredith College, Raleigh, NC, and David A. Smith, Duke University, Durham, NC

The purpose of this capsule is to develop divide-and-conquer methods for numerical integration of improper integrals. The software we will use is *MicroCalc, ver. 3.01*, Harley Flanders, MathCalcEduc, Ann Arbor, MI, 1987. This package was designated "Distinguished" in the 1987 EDUCOM/NCRIPTAL Software Awards competition.

This lesson brings together concepts of arc length, area, numerical integration, improper integrals, inverse functions, and error estimation to solve a problem whose answer is already known; the techniques are applicable to important problems whose answers *cannot* be known in closed form. We assume that numerical integration has been introduced along with the integral and the fundamental theorem of calculus, and that this lesson is part of "applications of the integral." We do not assume exposure to transcendental functions or improper integrals. The former are not needed, and this can be the first exposure to the latter.

MicroCalc is a collection of graphical, numerical, and symbolic units for single- and multi-variable calculus. All selections of units and options are made from menus by single keypresses. Functions and constants can be entered in almost standard mathematical notation, and up to six functions (inputs or the results of symbolic calculations) are retained in memory for transfer to other units. Among its useful features are a Function Editor, which enables function algebra, substitution, composition, and differentiation, and a Scratch Pad, which is a full-function, four-register calculator.

Lesson. Pose the problem of finding the length of the circle $x^2 + y^2 = 1$. We all know the answer to this problem should be 2π—how can we see that this answer follows from our definitions?

First, replace the problem by that of finding the length of the upper semicircle; the semicircle is the graph of a function, and the length π will be familiar when we see it. Then, using the blackboard, calculate the integral to be evaluated:

$$\int_{-1}^{1} \sqrt{1 + (y')^2}\, dx = \int_{-1}^{1} \frac{dx}{\sqrt{1 - x^2}} = \int_{0}^{1} \frac{2\, dx}{\sqrt{1 - x^2}}.$$

Using *MicroCalc*'s graphing unit, draw the integrand to give students a visual sense of the problem *as area* under the curve $y = 2/\sqrt{1 - x^2}$. Figure 1 shows *MicroCalc*'s graph in the window $[-2, 2] \times [0, 10]$.

Ask students to suggest a strategy for finding the area under the graph from 0 to 1. (Ignore for now the question of whether the area is finite; both the arc length interpretation and the calculation to follow are convincing arguments.) Students will almost certainly suggest the equivalent of integrating up to $1 - \varepsilon$ for smaller and

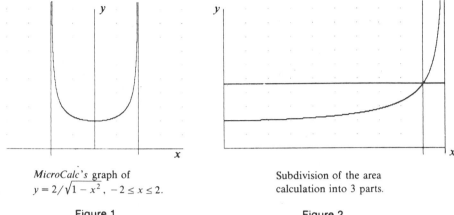

MicroCalc's graph of
$y = 2/\sqrt{1 - x^2}$, $-2 \leq x \leq 2$.

Subdivision of the area
calculation into 3 parts.

Figure 1   **Figure 2**

smaller ε. Accept the suggestion (for the time being), and switch to *MicroCalc's* Approximate Integration unit, which does fast evaluations of midpoint, trapezoidal, and Simpson's rules. For simplicity, we assume that only the Simpson results are of interest, and we follow the numerical analyst's rule of thumb: *When we double N and run again, the digits that don't change will never change*; i.e., *they already agree with the exact answer*. (For proper integrals, but not for this one, you may prefer to use the known error bound; if so, *MicroCalc* can calculate the necessary derivatives and allow you to quickly determine bounds for the fourth derivative, either by tabulation or by graphing.)

Before doing too many calculations, explain that one should separate off a part of the problem that is easy for Simpson's rule, say

$$\int_0^{0.9} \frac{2\,dx}{\sqrt{1 - x^2}},$$

and then concentrate on the remaining "strip" of area; Simpson's rule works better on shorter intervals. You can quickly find, with $N = 100$ and $N = 200$, that this (proper) integral has the value 2.2395390, correct to the number of digits shown.

If you now follow the students' suggestion, say, by integrating from 0.9 to $1 - 10^{-n}$ for $n = 2, 3, \ldots$, you will work harder and longer to get less information at each step. Furthermore, it will not be at all clear what the sequence of integral values is converging to (if anything).

Let's draw another picture to show more clearly where the difficulty lies. Switch back to the graphing unit, and draw the integrand on the interval $[0, 1]$, then add vertical lines at $x = 0.9$ and $x = 1$ (Figure 2). Our best piece of information to date is the highly accurate value of the integral on the interval $[0, 0.9]$. Select *MicroCalc's* Scratch Pad; enter $x = 0.9$, and calculate y by pressing the function key that represents the integrand. That gives $f(x) = 4.5883146774\ldots$, which we abbreviate

p. Add to the graph the horizontal line $y = p$ (also shown in Figure 2). We have now divided the area problem into three parts, two of which are the integral up to 0.9 and a rectangle of area $p/10$. It is clear from Figure 2 that direct calculation of the third area is asking Simpson's rule to fit a nearly *vertical* segment of graph with segments of parabolas, a near impossibility. On the other hand, this would not be much of a problem if the "bad" part of the graph were nearly *horizontal*. How can we make it so?

Perceptive students will suggest either inverting the function or replacing vertical elements of area by horizontal ones. Either way, you can quickly calculate that the remaining problem is equivalent to evaluating $\int_p^\infty [1 - \sqrt{1 - 4/x^2}]\,dx$. Use the graphing unit to show the graph of this new integrand with the lines $y = 0.1$ and $x = p$ superimposed. Now explain (or let the students suggest) a strategy for integrating on an infinite interval: First integrate on a finite interval that includes most of the area, say $[p, 10]$. Then add integrals of smaller and smaller function values on longer and longer intervals until doing so no longer makes a difference.

Select the integration unit again, and generate the following table of integrals on $[a, b]$:

a	b	Integral
p	10	0.242547343
10	100	0.18067415
100	1000	0.0180007
1000	10000	0.00180000

Let the students guess the next integral entry—then confirm it. The pattern of decrease by a factor of 10 continues indefinitely.

Now select the Scratch Pad again to sum the results to 7 places (corresponding to our least accurate entry): 2.2395390 (for the integral from 0 to 0.9), 0.4588315 (for the rectangular area), 0.2425473, 0.1806742, 0.0180007 (from the last table), and 0.002 (for all the remaining terms—challenge the students to explain this entry). The result is 3.1415927, and we *know* all these digits are correct, with the possible exception of the last; as it happens, it is correct too. No need to look it up—you can ask the Scratch Pad for a value of π (the important Greek letters are assigned to function keys).

As follow-ups, ask the students to use the ideas developed here to find the length of an ellipse and to evaluate $\int_0^{\pi/2}\sqrt{1 + \tan x}\,dx$. The second problem appears in Chris W. Avery and Frank P. Soler, Applications of transformations to numerical integration, *College Mathematics Journal* 19 (1988) 166–168 with a different, but equally interesting, mode of solution.

NOTE ON SIMPSON'S RULE

ANON, Erewhon-upon-Wabash

We derive the remainder in Simpson's Rule by using an integral formula.

Suppose the interval of integration is $[-h, h]$, centered at the origin, and the given integrand $F(x)$ satisfies

(i) F has a continuous third derivative on $[-h, h]$,

(ii) the restrictions of F''' to the intervals $[-h, 0]$ and $[0, h]$ are differentiable, and the derivatives F^{iv} are integrable (but do not necessarily satisfy $F^{iv}(0-) = F^{iv}(0+)$).

Let $P_2(x)$ be the quadratic interpolating $F(x)$ at $-h$, 0, h. The error in Simpson's Rule is obtained by integrating the difference $f(x) = F(x) - P_2(x)$. This difference satisfies (i) and (ii), and in addition $f(-h) = f(0) = f(h) = 0$. The following formula is proved by integration by parts:

$$\int_{-h}^{0} (x + h)^3(3x - h)f^{iv}(x)dx + \int_{0}^{h} (x - h)^3(3x + h)f^{iv}(x)dx = 72\int_{-h}^{h} f(x)dx.$$

Now assume $F(x)$ satisfies (i), (ii), and

(iii) $|F^{iv}(x)| \leq M$ on $[-h, h]$.

Then $f(x)$ also satisfies (iii) and the error can be estimated:

$$72\left|\int_{-h}^{h} f(x)dx\right| \leq M\int_{-h}^{0} (x + h)^3(h - 3x)dx + M\int_{0}^{h} (h - x)^3(3x + h)dx$$

$$= 2M\int_{0}^{h} (h - x)^3(3x + h)dx = \tfrac{4}{5}Mh^5.$$

Finally,

$$\left|\int_{-h}^{h} F(x)dx - \int_{-h}^{h} P_2(x)dx\right| \leq \tfrac{1}{90}Mh^5.$$

Reference

John Todd, Introduction to the Constructive Theory of Functions, Academic Press, New York, 1963, p. 122.

BIBLIOGRAPHIC ENTRIES: NUMERICAL, GRAPHICAL, AND MECHANICAL METHODS AND APPROXIMATIONS (INCLUDING USE OF COMPUTERS)

1. *Monthly* Vol. 77, No. 9, pp. 999–1001. G. J. Porter, A computer assisted approach to integral calculus via Jordan content.

2. *Monthly* Vol. 79, No. 3, pp. 282–290. Solomon Garfunkel, A laboratory and computer based approach to calculus.

3. *Monthly* Vol. 79, No. 3, pp. 290–293. H. W. Hethcote and A. J. Schaeffer, A computer laboratory course for calculus and linear algebra.

4. *Monthly* Vol. 81, No. 2, pp. 163–168. G. L. Thesing and C. A. Wood, Using the computer as a discovery tool in calculus.

5. *Monthly* Vol. 84, No. 9, pp. 726–728. George H. Grown, Jr., On Halley's variation of Newton's method.

6. *Monthly* Vol. 86, No. 5, pp. 386–391. Sheldon P. Gordon, A discrete approach to computer-oriented calculus.

7. *Monthly* Vol. 87, No. 2, pp. 124–128. Edward Rozema, Estimating the error in the trapezoidal rule.

8. *Monthly* Vol. 87, No. 4, pp. 243–251. George Miel, Calculator calculus and roundoff errors.

9. *Monthly* Vol. 93, No. 6, pp. 476–478. P. J. Rippon, Convergence with pictures.

The pictures illustrate convergence of series

10. TYCMJ Vol. 5, No. 4, pp. 4–11. Arne Broman, Computer calculation of integrals.

11. TYCMJ Vol. 7, No. 3, pp. 10–14. Burt M. Rosenbaum, An integral approximation exact for fifth-degree polynomials.

12. TYCMJ Vol. 7, No. 4, pp. 1–3. Stewart M. Venit, Remarks concerning the delta method for approximating roots.

13. CMJ Vol. 17, No. 2, pp. 172–181. Michael E. Frantz, Interactive graphics for multivariate calculus.

14. CMJ Vol. 17, No. 5, pp. 418–422. Frank Burk, Numerical integration via integration by parts.

15. CMJ Vol. 18, No. 3, pp. 222–223. Frank Burk, Archimedes' quadrature and Simpson's rule.

16. CMJ Vol. 19, No. 1, pp. 43–52. James A. Uetrecht, A clamped Simpson's rule.

17. CMJ Vol. 29. No. 3, pp. 238–251. Richard D. Neidinger, Automatic differentiation and APL.

18. CMJ Vol. 21, No. 1, pp. 51–55. John H. Mathews, Teaching Riemann sums using computer symbolic algebra systems.

19. CMJ Vol. 21, No. 4, pp. 314–322. Steven Schonefeld, Some examples illustrating Richardson's improvement.

20. CMJ Vol. 22, No. 1, pp. 3–12. Gilbert Strang, A chaotic search for i.

21. CMJ Vol. 22, No. 4, pp. 327–331. Paul B. Massell, Using computer graphics to help analyze complicated functions.

22. CMJ Vol. 22, No. 5, pp. 415–417. Russell Jay Hendel, Differentiation via partial fractions: A case against CAS.

11

INFINITE SEQUENCES
AND SERIES
(a)

THEORY

THE INTERVALS OF CONVERGENCE OF SOME POWER SERIES

EUGENE SCHENKMAN, Purdue University

In this note we deduce the exact radius of convergence for some Maclaurin expansions by methods available to students in a first-year calculus course. We begin by sketching a proof that if $f(x) = \sum_{i=0}^{\infty} a_i x^i$ with $a_0 = 1$ and $\sum_{i=1}^{\infty} |a_i \lambda^i|$ < 1 for some $\lambda > 0$, then the Maclaurin expansion of $1/f(x)$ has a radius of convergence at least λ. For if $0 \leq |x| \leq \lambda$, then $f(x) > 0$ and

$$\log[f(x)] = \log\{1 - [1 - f(x)]\}$$

$$= -\sum_{n=1}^{\infty} \frac{1}{n} [1 - f(x)]^n = -\sum_{n=1}^{\infty} \frac{1}{n} \left(-\sum_{i=1}^{\infty} a_i x^i\right)^n.$$

This double series converges absolutely for $|x| \leq \lambda$, so its terms can be arranged to form a power series $\sum_{j=1}^{\infty} b_j x^i$ which converges to $\log f(x)$ for $|x| \leq \lambda$. Consequently

$$\frac{1}{f(x)} = e^{-\log f(x)} = \sum_{n=0}^{\infty} \frac{1}{n!} \left(-\sum_{j=1}^{\infty} b_j x^j\right)^n,$$

which can also be written as a power series in x, convergent for $|x| \leq \lambda$. This power series is the Maclaurin expansion of $1/f(x)$, and its radius of convergence is thus at least λ.

The interval of convergence of the function under consideration will be henceforth denoted by ρ.

THEOREM 1. *The interval of convergence for* $\sec x$ *is* $\rho = \frac{1}{2}\pi$.

Proof. It is clear that $\rho < \frac{1}{2}\pi$ since $\sec \frac{1}{2}\pi = \infty$.

In the following we shall need to know that all the coefficients of the Maclaurin expansion of $\sec x$ are nonnegative. To see this we show inductively that the nth derivative of sec x is sec x times a polynomial P_n (tan x) in tan x with nonnegative coefficients: Indeed, $d(\sec x)/dx = \sec x \tan x$, and if d^n (sec x)/dx^n were sec x P_n(tan x), then

$$\frac{d^{n+1}\sec x}{dx^{n+1}} = \sec x\big[P'_n(\tan x)(\tan^2 x + 1) + P_n(\tan x)\tan x\big]$$

$$= \sec x P_{n+1}(\tan x),$$

where P_{n+1} is a polynomial with nonnegative coefficients. Now

$$\sec x = \frac{1}{\cos x} = \frac{1}{2\cos^2(\tfrac{1}{2}x) - 1} = \tfrac{1}{2}\sec^2(\tfrac{1}{2}x)\frac{1}{1 - \tfrac{1}{2}\sec^2(\tfrac{1}{2}x)}$$

$$= \tfrac{1}{2}\sec^2(\tfrac{1}{2}x)\sum_{n=0}^{\infty}\big[\tfrac{1}{2}\sec^2(\tfrac{1}{2}x)\big]^n,$$

with the last equality valid if $\tfrac{1}{2}\sec^2(\tfrac{1}{2}x) < 1$, that is, $\tfrac{1}{2}x < \tfrac{1}{4}\pi$. The last sum can be written as a double sum in terms of the expansion of $\sec(\tfrac{1}{2}x)$, and since this expansion has non-negative terms, the whole expression on the right is a power series in $\tfrac{1}{2}x$. Hence from the validity of the expansion of secant for $\tfrac{1}{2}x < \lambda$, follows its validity for $x < 2\lambda$ as long as $\lambda < \tfrac{1}{4}\pi$, so $\rho = \tfrac{1}{2}\pi$.

THEOREM 2. *The interval of convergence for* tan x *and* sech x *is* $\rho = \tfrac{1}{2}\pi$.

This is clear since tan $x = \sin x \sec x$, and sech $x = \sec ix$.

THEOREM 3. *The interval of convergence for* $x/\sinh x$ *is* $\rho = \pi$.

Proof. Since $\sinh x = 2\sinh(\tfrac{1}{2}x)\cosh(\tfrac{1}{2}x)$, it follows that

$$\frac{x}{\sinh x} = \operatorname{sech}(\tfrac{1}{2}x)\frac{\tfrac{1}{2}x}{\sinh(\tfrac{1}{2}x)}.$$

Thus the expansion of $x/\sinh x$ is valid for $\tfrac{1}{2}x < \lambda$ provided $\lambda < \tfrac{1}{2}\pi$. Hence $\rho = \pi$.

THEOREM 4. *The interval of convergence for* $x/(e^x - 1)$ *is* $\rho = 2\pi$.

Proof. We have

$$\frac{e^x - 1}{x} = e^{x/2}\frac{e^{x/2} - e^{-x/2}}{x} = e^{x/2}\frac{\sinh(\tfrac{1}{2}x)}{\tfrac{1}{2}x},$$

hence

$$\frac{x}{e^x - 1} = \frac{\tfrac{1}{2}x}{\sinh(\tfrac{1}{2}x)}e^{-x/2}.$$

It follows that $\rho = 2\pi$. Cf. [1, p. 139].

The author is grateful to Professor G. R. MacLane for some illuminating discussions during the preparation of this note.

Reference

1. I. I. Hirschman, Jr., Infinite Series, Holt, Rinehart and Winston, New York, 1962.

AN ALTERNATIVE TO THE INTEGRAL TEST FOR INFINITE SERIES

G. J. PORTER, University of Pennsylvania

Infinite series are usually studied in a calculus course following the development of the integral. One reason for this placement is the desire to have the integral test available. An earlier study of infinite series might be desired to complement the study of sequences or to study Taylor series as an immediate application of the derivative. In these cases an alternative to the integral test is needed.

One alternative is the Cauchy Condensation Test; this method seems to be well known in Europe and Latin America, but not in the United States. Many calculus teachers are aware that this test (perhaps not by this name) may be used to prove that the series $\sum 1/n$ diverges. I suspect a smaller number are aware that it may be used for all the series which are usually studied by the integral test. It is the point of this note to recall the test and give several examples of its use.

THEOREM (Cauchy Condensation Test). *Let $\sum_{n=1}^{\infty} a_n$ be a series of positive terms such that $a_{n+1} \leq a_n$ for all n. Then $\sum_{n=1}^{\infty} a_n$ converges if and only if the condensed series $\sum_{j=1}^{\infty} 2^j a_{2^j}$ converges.*

Proof. Since

$$2^{j-1} a_{2^j} \leq a_{2^{j-1}+1} + a_{2^{j-1}+2} + \cdots + a_{2^j} \leq 2^{j-1} a_{2^{j-1}}$$

we have

$$\sum_{j=1}^{\infty} 2^{j-1} a_{2^j} \leq \sum_{n=2}^{\infty} a_n \leq \sum_{j=1}^{\infty} 2^{j-1} a_{2^{j-1}} \leq 2 \sum_{j=1}^{\infty} 2^{j-1} a_{2^j}$$

and the theorem follows.

Example 1: $\sum_{n=1}^{\infty} 1/n^{\alpha}$. The condensed series is

$$\sum_{j} \frac{2^j}{(2^j)^{\alpha}} = \sum_{j} \frac{1}{(2^j)^{\alpha-1}} = \sum_{j} \frac{1}{(2^{\alpha-1})^j}.$$

This is a geometric series and converges if and only if $2^{1-\alpha} < 1$, i.e., $\alpha > 1$. Thus the given series converges if and only if $\alpha > 1$.

Example 2: $\sum_{n=2}^{\infty} 1/n(\log n)^{\alpha}$. The condensed series is

$$\sum_{j} \frac{2^j}{2^j (\log(2^j))^{\alpha}} = \sum_{j} \frac{1}{(\log(2^j))^{\alpha}} = \sum_{j} \frac{1}{(j \log 2)^{\alpha}} = \frac{1}{(\log 2)^{\alpha}} \sum_{j} \frac{1}{j^{\alpha}}$$

which converges if and only if $\alpha > 1$ by Example 1.

Example 3. $\sum_{n=2}^{\infty} 1/n \log n (\log(\log n))^{\alpha}$. The condensed series is

$$\sum_j \frac{2^j}{2^j \log(2^j)(\log(\log 2^j))^{\alpha}} = \sum_j \frac{1}{j(\log 2)(\log(j \log 2))^{\alpha}}$$

$$= \frac{1}{\log 2} \sum \frac{1}{j(\log j + \log 2)^{\alpha}}$$

which converges if and only if $\alpha > 1$ by comparison with Example 2.

A SEQUENCE OF CONVERGENCE TESTS

J. R. NURCOMBE

1. Introduction. In recent years, various notes (e.g., [3] and [4]) have appeared purporting to contain some new extension of Cauchy's root test, although the material involved is essentially well known, being found in different forms in [1, p. 44] and [2, p. 282]. One possible reason for this is that the treatments given in [1] and [2] fail to establish any clear formal similarity between Cauchy's test and succeeding tests of the logarithmic scale of convergence. It is therefore perhaps worthwhile to remedy this defect. An attempt to do this is made in Theorem 1, and the remainder of this note endeavors to clarify some of the points arising in the references.

2. It is familiar that the convergence or divergence of a series of positive terms Σa_n may be deduced by comparing it with the standard series of the logarithmic scale, which converge for $p > 1$ and diverge for $p \leqslant 1$:

$$\sum_{n=1}^{\infty} 1/p^n, \qquad \sum_{n=1}^{\infty} 1/n^p, \qquad \sum_{n=2}^{\infty} 1/n(\log n)^p, \qquad \sum_{n=3}^{\infty} 1/n\log n(\log\log n)^p,$$

etc. This can be expressed in a form similar to the root test.

THEOREM 1. *Let Σa_r be a series of positive terms. If* (i) $\overline{\lim}\, a_n^{1/n} < 1$ *or* (ii) $\overline{\lim}\, a_n^{1/\log n} < 1/e$, *or* (iii) $\overline{\lim}(na_n)^{1/\log_2 n} < 1/e$, *or* (iv) $\overline{\lim}(n\log n \cdot a_n)^{1/\log_3 n} < 1/e$, *then in each case, Σa_r converges, where $\log_2 n = \log\log n$, etc.*

Proof. The argument is the same in all cases and (iii) is chosen as typical. For all $n \geqslant N$, $(na_n)^{1/\log_2 n} \leqslant 1/e^p, p > 1$. Thus $na_n \leqslant 1/e^{p\log_2 n}$ or $a_n \leqslant 1/n(\log n)^p$, implying Σa_r is convergent.

REMARKS. (1) The tests for divergence can be given in a similar manner, but since the limit forms are weaker than the non-limit forms in this case, they are omitted.

(2) (i) is Cauchy's test. (ii) and (iii) are, respectively, the comparison test equivalents of Raabe's and Gauss's tests. (iv) is the next comparison test of the logarithmic scale, and succeeding tests can be formulated in an analogous and obvious way. It is (ii) and (iii) which are given in [3].

(3) Substantially the same relationship exists between these tests and the corresponding ratio tests as exists between Cauchy's test and d'Alembert's test.

(4) The above method of formulating the comparison tests could be called the "exponential form." The ratio tests can also be given in this form. D'Alembert's test is unchanged. Raabe's test becomes: if $\lim(a_{n+1}/a_n)^n < 1/e$, then Σa_r converges, and Gauss's test is:

$$\text{if } \lim n(a_{n+1}/a_n)^{n\log n} < 1/e, \text{ then } \Sigma a_r \text{ converges.}$$

3. Although the comparison tests are more powerful than the ratio tests, the latter can be strengthened and generalized in the following way.

THEOREM 2. *Let Σa_r be a series of positive terms, and k, a fixed positive integer. If* $\lim a_{n+k}/a_n \lessgtr 1$, *then $\Sigma a_r \begin{cases} \text{converges} \\ \text{diverges.} \end{cases}$*

Proof. Consider the k subseries,

$$\sum_{r=0}^{\infty} a_{1+rk}, \sum_{r=0}^{\infty} a_{2+rk}, \ldots, \sum_{r=1}^{\infty} a_{rk},$$

which collectively comprise all the terms of $\Sigma_{r=1}^{\infty} a_r$. Every subseries converges for the first hypothesis by d'Alembert's test and, since the sum of a series of positive terms is invariant under alterations in the order of the terms, Σa_r converges. For the second hypothesis, every subseries diverges by d'Alembert's test, so Σa_r cannot converge.

REMARKS. (1) Series can be readily constructed to show that $\lim a_{n+k}/a_n <$ $\lim a_n^{1/n} < \overline{\lim} a_n^{1/n} \leqslant \overline{\lim} a_{n+k}/a_n$, which is known to be true if $k=1$, is false if $k>1$. For example, if $\Sigma a_r = u + v^2 + u^3 + v^4 + \cdots$, take $u=1/2$, $v=2/3$. Then $\underline{\lim} a_{n+2}/a_n = 1/4 < \overline{\lim} a_{n+2}/a_n = 4/9 < \underline{\lim} a_n^{1/n} = 1/2 < \overline{\lim} a_n^{1/n} = 2/3$. Thus it is no longer necessarily true that Cauchy's test includes the extended forms of d'Alembert's test. Furthermore, for this example, the whole scale of ratio tests will fail when $k=1$, since the upper and lower limits of (a_{n+1}/a_n) include unity, although when $k=2$, the convergence of the series is easily deduced from theorem 2.

(2) Naturally, if there are infinitely many zero terms, the ratio tests are inapplicable.

(3) The same extensions apply to Raabe's, Gauss's, and the other tests of the logarithmic scale. For example, the extended form of Raabe's test when $k=2$ shows that the example in remark (1) of [3] converges. To prove this, we must show that $\lim n(a_{n+2}/a_n - 1) < -2$, when $a_{n-1} = a_n = (1-3/n)^{n\log n}$ for n even.

Now

$$\frac{a_{n+2}}{a_n} = \frac{\left(1 - \dfrac{3}{n+2}\right)^{(n+2)\log(n+2)}}{\left(1 - \dfrac{3}{n}\right)^{n\log n}}.$$

Also

$$(n+2)\log(n+2) = (n+2)\left(\log n + \log\left(1 + \frac{2}{n}\right)\right)$$

$$= n\log n + 2\log n + 2 + O\left(\frac{1}{n}\right),$$

so

$$\frac{a_{n+2}}{a_n} = \left(1 + \frac{3}{n-3}\right)^{n\log n} \cdot \left(1 - \frac{3}{n+2}\right)^{n\log n + 2\log n + 2 + O\left(\frac{1}{n}\right)}$$

$$= \left\{\left(1 - \frac{3}{n+2}\right)\left(1 + \frac{3}{n-3}\right)\right\}^{n\log n} \cdot \left(1 - \frac{3}{n+2}\right)^{2\log n + 2 + O\left(\frac{1}{n}\right)}$$

$$= \left(1 + \frac{6}{(n+2)(n-3)}\right)^{n\log n} \cdot \left(1 - \frac{6\log n}{n+2} - \frac{6}{n+2} + O\left(\frac{\log n}{n}\right)^2\right)$$

$$= \left(1 + \frac{6n\log n}{(n+2)(n-3)} + O\left(\frac{\log n}{n}\right)^2\right)\left(1 - \frac{6(1+\log n)}{n+2} + O\left(\frac{\log n}{n}\right)^2\right)$$

$$= 1 - \frac{6\left(1 + \log n - \dfrac{n\log n}{n-3}\right)}{n+2} + O\left(\frac{\log n}{n}\right)^2.$$

Thus

$$n\left(\frac{a_{n+2}}{a_n} - 1\right) = \frac{-6n\left(1 - \dfrac{3\log n}{n-3}\right)}{n+2} + O\left(\frac{\log n}{n}\right)^2$$

and the required limit is -6, which implies convergence.

References

1. T. J. I'A. Bromwich, An Introduction to the Theory of Infinite Series, 2nd ed., Macmillan, 1931.
2. K. Knopp, Theory and Application of Infinite Series, 2nd ed., Blackie, 1947.
3. I. S. Murphy, Glasgow Math. J., 17 (1976) 151–154.
4. S. W. Reyner, this MONTHLY, 73 (1966) 998–1000.

HALL GREEN TECHNICAL COLLEGE, BIRMINGHAM, ENGLAND.

EVERY POWER SERIES IS A TAYLOR SERIES

Mark D. Meyerson

Department of Mathematics, U.S. Naval Academy, Annapolis, MD 21402

For every function f, defined and with all its derivatives defined at the origin, we define its Taylor series at the origin as $y = a_0 + a_1 x + a_2 x^2 + \cdots$, where $a_n = f^{(n)}(0)/n!$. (Then one studies convergence. The standard example $y = \exp(-1/x^2)$ shows that the Taylor series may fail to converge to the original function.) The following question naturally arises: Is a given power series the Taylor series for some function? We show the answer is yes. Our construction follows that of Gelbaum and Olmsted ([2, pp. 69–70]) who show that $\sum_{n=0}^{\infty} n! x^n$ is a Taylor series. My thanks to John Kalme for showing me that example.

THEOREM. *Given any sequence* $\{a_n\}_{n=0}^{\infty}$ *of real numbers, there is a function with Taylor series* $\sum_{n=0}^{\infty} a_n x^n$.

Proof. For $n = 0, 1, 2, \ldots$, let the function g_n be defined by

$$g_n(x) = \begin{cases} a_n n! & \text{if} \quad |x| \leqslant 1/(2|a_n|n! + 1) \\ 0 & \text{if} \quad |x| \geqslant 2/(2|a_n|n! + 1) \end{cases}$$

and elsewhere as a "smoothing function" which is monotonic in each interval where it is defined and which makes all derivatives of g_n exist (see [2, p. 40]). Let $f_0 = g_0$ and for $n \geqslant 1$ let

$$f_n(x_n) = \int_0^{x_n} \int_0^{x_{n-1}} \cdots \int_0^{x_2} \int_0^{x_1} g_n(x_0)\, dx_0\, dx_1 \cdots dx_{n-2}\, dx_{n-1}.$$

Then for $n \geqslant 1$, $|f_n^{(n-1)}(x)| = |\int_0^x g_n(t)\, dt| \leqslant 1$; so integrating $n - k - 1$ times we see that $|f_n^{(k)}(x)| \leqslant |x|^{n-k-1}/(n-k-1)!$, where $0 \leqslant k \leqslant n - 1$ and $f_n^{(0)} = f_n$. By the Weierstrass M-test, $\sum_{n=0}^{\infty} f_n^{(k)}(x)$ converges uniformly on every bounded interval ($k \geqslant 0$). Hence $\sum_{n=0}^{\infty} f_n(x)$ converges to some $f(x)$ and $f^{(k)}(x) = \sum_{n=0}^{\infty} f_n^{(k)}(x)$ for $k \geqslant 1$ (see [6, p. 140]). But $f_n^{(k)}(0) = \delta_{nk} a_n n!$, so $f^{(n)}(0) = a_n n!$ ($n, k \geqslant 0$). □

COROLLARY. *There are uncountably many functions with any given Taylor series.*

Proof. For f as above and λ real, consider the function h defined by

$$h(x) = f(x) + \lambda \exp(-1/x^2)$$

for $x \neq 0$ and $h(0) = f(0)$. □

Note that we could have centered the Taylor series at any point, not necessarily the origin.

There are many earlier proofs of this result. It seems to have been proved first by E. Borel in 1895 ([1, see p. 44]). He made careful use of the harmonic series to choose coefficients for a power series defining a function with derivatives close to the desired ones. The error is small enough to be accounted for by an analytic function, and so the sum of these two functions is the desired function. Then in 1938 G. Pólya ([4, see p. 244]) gave a proof using machinery he developed for handling systems of equations with infinitely many unknowns. More recently A. Rosenthal ([5]) gave a proof expressing the desired function as an infinite sum of monomials with explicitly calculated exponential coefficients. And closest to the method here is the proof of H. Mirkil ([3]), which is done in n dimensions.

This work was supported by the Naval Academy Research Council Fund.

References

1. Émile Borel, Sur quelques points de la théorie des fonctions, Ann. Sci. École Norm. Sup., (3), no. 12 (1895) 9–55.

2. Bernard R. Gelbaum and John M. H. Olmsted, Counterexamples in Analysis, Holden-Day, 1964.

3. H. Mirkil, Differentiable functions, formal power series, and moments, Proc. Amer. Math. Soc., 7 (1956) 650–652.

4. G. Pólya, Eine einfache, mit funktionentheoretischen Aufgaben verknüpfte, hinreichende Bedingung für die Auflösbarkeit eines Systems unendlich vieler linearer Gleichungen, Comment. Math. Helv., 11 (1938–1939) 234–252.

5. Arthur Rosenthal, On functions with infinitely many derivatives, Proc. Amer. Math. Soc., 4 (1953) 600–602.

6. Walter Rudin, Principles of Mathematical Analysis, 2nd ed., McGraw-Hill, New York, 1964.

POWER SERIES WITHOUT TAYLOR'S THEOREM

WELLS JOHNSON

Department of Mathematics, Bowdoin College, Brunswick, ME 04011

It is typical in today's calculus texts to develop the full theory of sequences and infinite series before deriving the power series expansions for any of the standard transcendental functions. The Taylor polynomials and Taylor's theorem are usually presented for arbitrary functions, and only after considerable development do students ever see, for example, the application of computing the number e to five decimal places. We suggest here reversing the approach by first deriving the power series for e^x, $\sin x$, and $\cos x$, giving some numerical applications, and then using the examples to motivate the general theory behind Taylor's theorem.

The usual infinite series representation for e, for example, can be derived quite easily as a wonderful application of integration by parts. For $n \geqslant 0$, let $E_n = \int_0^1 t^n e^{-t} \, dt$. The integrands are easily graphed and the definite integrals E_n can be interpreted as areas. Integration by parts gives the reduction formula

$$(1) \qquad E_n = -e^{-1} + nE_{n-1}, \qquad n \geqslant 1,$$

and E_0 satisfies the equation

$$(2) \qquad e = 1 + eE_0.$$

Multiplication by $e/n!$ transforms (1) into the nice recursion relation

$$(3) \qquad \frac{eE_{n-1}}{(n-1)!} = \frac{1}{n!} + \frac{eE_n}{n!}, \qquad n \geqslant 1,$$

and repeated substitution of (3) into (2) gives

$$(4) \qquad e = 1 + \frac{1}{1!} + \frac{1}{2!} + \cdots + \frac{1}{n!} + R_n, \qquad n \geqslant 1,$$

where the remainder term $R_n = eE_n/n!$.

We are thus led directly to the notion of convergence of an infinite sum. Since the areas represented by E_n approach zero as n gets large, the R_n's approach zero very rapidly because of the presence of the $n!$ in the denominator. Hence the long sums give very good approximations to e, and these approximations get dramatically better as we take n larger.

More precisely, for $0 \leqslant t \leqslant 1$, we have $e^{-1} \leqslant e^{-t} \leqslant 1$, and so

$$\frac{1}{e(n+1)} \leqslant E_n \leqslant \frac{1}{n+1}.$$

Hence R_n satisfies the inequality

$$(5) \qquad \frac{1}{(n+1)!} \leqslant R_n \leqslant \frac{e}{(n+1)!}, \qquad n \geqslant 1,$$

and the error term, while not less than the next term in the sum (which makes perfectly good sense), is no larger than e times the next term in the sum (which is a nontrivial result). We can now compute e to five decimal places if we wish, but more importantly, the concept of convergence of an infinite sum to a particular number is beautifully and simply illustrated with a nontrivial example.

To derive the power series for e^x, we simply replace e^{-t} by e^{-xt} in the integrand for E_n. The reduction formula

$$(1') \qquad xE_n = -e^{-x} + nE_{n-1}, \qquad n \geqslant 1,$$

394

can be put into the nicer form

(3')
$$\frac{x^n e^x E_{n-1}}{(n-1)!} = \frac{x^n}{n!} + \frac{x^{n+1} e^x E_n}{n!}, \qquad n \geq 1.$$

Repeatedly applying (3') to the equation $e^x = 1 + xe^x E_0$, we arrive at the power series expansion for e^x and an exact expression for the remainder term of the finite partial sums. Convergence follows by fixing x, bounding the integrals E_n, and letting n go to infinity.

The power series for the functions $\sin x$ and $\cos x$ can also be derived directly. One way is to consider the definite integrals

(6)
$$S_n = \int_0^1 (1 - t)^n \cos xt \, dt \quad \text{and} \quad C_n = \int_0^1 (1 - t)^n \sin xt, \qquad n \geq 0.$$

Direct integration gives

(7)
$$\sin x = xS_0 \quad \text{and} \quad \cos x = 1 - xC_0.$$

Integration by parts gives the reduction formulas

(8)
$$xS_n = nC_{n-1} \quad \text{and} \quad xC_n = 1 - nS_{n-1}, \qquad n \geq 1.$$

Combining these two reduction formulas, we obtain

(9)
$$x^2 S_n = n - n(n-1)S_{n-2} \quad \text{and} \quad x^2 C_n = x - n(n-1)C_{n-2}, \qquad n \geq 2,$$

which can be put into the forms

(10)
$$\frac{x^{n-1} S_{n-2}}{(n-2)!} = \frac{x^{n-1}}{(n-1)!} - \frac{x^{n+1} S_n}{n!} \quad \text{and} \quad \frac{x^{n-1} C_{n-2}}{(n-2)!} = \frac{x^n}{n!} - \frac{x^{n+1} C_n}{n!}, \qquad n \geq 2.$$

Repeated substitution of (10) into (7) yields the power series for $\sin x$ and $\cos x$, respectively, and an easy analysis of the error terms proves convergence.

At this point it can be pointed out that, by substitution, e^{kx} also has a power series development, and that the coefficient of x^n in this series is of the form $f^{(n)}(0)/n!$. Also, the Taylor series for e^x, $\sin x$, and $\cos x$ about the point a, say, can be derived from the relations

$$e^x = e^a \cdot e^{x-a},$$
$$\sin x = \sin a \cos(x - a) + \cos a \sin(x - a),$$
$$\cos x = \cos a \cos(x - a) - \sin a \sin(x - a),$$

and it is easily seen that the coefficient of $(x - a)^n$ in these expansions always has the proper form: $f^{(n)}(a)/n!$.

Students are now prepared for Taylor's theorem and its proof. Integration by parts leads to the integral form of the remainder term, which, by bounding the integral as we did in inequality (5), is all that is needed to prove convergence for many specific functions.

MORE—AND MOORE—POWER SERIES WITHOUT TAYLOR'S THEOREM

I. E. LEONARD

Department of Mathematical Sciences, Northern Illinois University, DeKalb, IL 60115

JAMES DUEMMEL

Department of Mathematics, Western Washington University, Bellingham, WA 98225

Inspired in part by the article "Power series without Taylor's theorem" by Wells Johnson in the June-July 1984 issue of the MONTHLY, we present here an extremely simple method for obtaining the power series expansions for the functions $\sin x$, $\cos x$, e^{-x}, e^x. The approach we take uses no "sophisticated" concepts such as integration by parts, reduction formulas, or recursion relations; it uses only the monotonicity of the definite integral, and the basic inequalities for the functions in question. We make no claim of originality: the first author learned of the technique from Professor R. A. Moore, at Carnegie-Mellon University, while the second author encountered it in a calculus text (see [2], p. 78).

Starting with the inequality, $\cos t \leqslant 1$, and integrating repeatedly over the interval $[0, x]$, we obtain inequalities involving the successive Taylor polynomials for sine and cosine. For example, four integrations yield, for $x \geqslant 0$,

$$x - \frac{x^3}{3!} \leqslant \sin x \leqslant x$$

and

$$1 - \frac{x^2}{2!} \leqslant \cos x \leqslant 1 - \frac{x^2}{2!} + \frac{x^4}{4!}.$$

Students are quickly convinced that $\sin x$ and $\cos x$ are trapped between successive partial sums of their usual Taylor series expansions. The error estimates

$$\left| \cos x - \sum_{k=0}^{n} (-1)^k \frac{x^{2k}}{(2k)!} \right| \leqslant \frac{x^{2n+2}}{(2n+2)!}$$

and

$$\left| \sin x - \sum_{k=0}^{n} (-1)^k \frac{x^{2k+1}}{(2k+1)!} \right| \leqslant \frac{x^{2n+3}}{(2n+3)!}$$

are evident, and a little more argument will convince even the most skeptical students that the series actually converge to $\sin x$ and $\cos x$.

As soon as the derivative of the exponential function has been covered, this approach will also yield the Taylor series for e^{-x} from the inequality $e^{-t} \leqslant 1$, for $t \geqslant 0$. This series can be used to estimate e^{-1} and hence e. In particular, the estimate $(24/9) < e < 3$ is easily obtained in this way. This could be used below for students who find 3^b more acceptable than e^b.

The power series expansion for e^x, $x \geqslant 0$, can be obtained from the inequality $1 \leqslant e^t \leqslant e^b$, for $0 \leqslant t \leqslant b$ (b fixed), by integrating repeatedly, and we obtain, after rearranging,

$$1 + x + \frac{x^2}{2!} + \cdots + \frac{x^{n+1}}{(n+1)!} \leqslant e^x \leqslant 1 + x + \frac{x^2}{2!} + \cdots + \frac{x^{n+1}}{(n+1)!} e^b$$

for $0 \leqslant x \leqslant b$. This inequality gives the error estimate

$$0 \leqslant e^x - \sum_{k=0}^{n+1} \frac{x^k}{k!} \leqslant \frac{x^{n+1}}{(n+1)!} (e^b - 1)$$

for $0 \leqslant x \leqslant b$. Now a simple argument showing that $\lim_{n \to \infty} (x^n/n!) = 0$, for fixed x, yields the power series expansions for $\sin x$, $\cos x$, e^{-x}, e^x, valid for $x \geqslant 0$. Note that taking $x = b = 1$ in the above inequality gives

$$\frac{1}{(n+1)!} \leqslant e - \sum_{k=0}^{n} \frac{1}{k!} \leqslant \frac{e}{(n+1)!} .$$

References

1. W. Johnson, Power series without Taylor's theorem, this MONTHLY, 91 (1984) 367–369.
2. J. F. Randolph, Calculus, Macmillan, New York, 1952.

Ed. Note. See also the related paper by Deng Bo reproduced on p. 207.

N! AND THE ROOT TEST

Charles C. Mumma II

Department of Mathematics, University of Washington, Seattle, WA 98195

I would be willing to bet that 99.98% of all freshman calculus students (perhaps even more!) attempting to determine the convergence of $\sum_{n=1}^{+\infty} \dfrac{2^n}{n!}$ would *not* use the root test. This is not surprising, considering that most texts do not evaluate $\lim_{n \to +\infty} \sqrt[n]{n!}$. However, armed with the knowledge that this limit is $+\infty$, the root test becomes more versatile and accessible.

LEMMA. $\lim_{n \to +\infty} \sqrt[n]{n!} = +\infty$.

Proof. First notice that $(2n)! \geqslant \prod_{k=n}^{2n} k \geqslant n^{n+1}$. Consequently

$$\sqrt[2n]{(2n)!} \geqslant \sqrt[2n]{n^{n+1}} \geqslant \sqrt{n} \quad \text{and} \quad \sqrt[2n+1]{(2n+1)!} \geqslant \sqrt[2n+1]{n^{n+1}} \geqslant \sqrt{n}.$$

A more accurate analysis of this limit yields a rather interesting result.

LEMMA. $\lim_{n \to +\infty} \dfrac{\sqrt[n]{n!}}{n} = \dfrac{1}{e}$.

Proof. Taking logarithms, we get

$$\ln\left(\frac{\sqrt[n]{n!}}{n}\right) = \frac{1}{n}\ln(n!) - \ln(n) = \frac{1}{n}\sum_{k=1}^{n}\ln(k) - \frac{1}{n}\sum_{k=1}^{n}\ln(n) = \sum_{k=1}^{n}\ln\left(\frac{k}{n}\right)\frac{1}{n}.$$

As $n \to +\infty$, this becomes $\int_{0}^{1}\ln(x)\,dx = -1$.

This result is usually proved using the fact that the root test is stronger than the ratio test, that is,

$$\liminf \frac{a_{n+1}}{a_n} \leqslant \liminf \sqrt[n]{a_n} \leqslant \limsup \sqrt[n]{a_n} \leqslant \limsup \frac{a_{n+1}}{a_n},$$

where $(a_n)_{n \in \mathbf{N}}$ is an arbitrary sequence of positive real numbers. The beauty of the present proof is its simplicity and directness and its use of methods readily available to freshman calculus students.

A Simple Test for the nth term of a Series to Approach Zero

JONATHAN LEWIN
Department of Mathematics, Kennesaw College, Marietta, GA 30061

MYRTLE LEWIN
Department of Mathematics, Agnes Scott College, Decatur, GA 30030

Using Stirling's formula, one may see at once that if $a_n = (2n)!/4^n(n!)^2$, then a_n is of the order of $1/\sqrt{n}$, and one may conclude from the alternating series test that the series $\Sigma(-1)^n a_n$ is conditionally convergent. At an elementary level, however, the convergence of the latter series may be a little more difficult to obtain. Since $a_{n+1}/a_n = (2n+1)/(2n+2) < 1$ for each n, it is clear that the sequence (a_n) is decreasing, but it is not immediately obvious within the environment of a typical calculus course that $a_n \to 0$ as $n \to \infty$. For this purpose, one might use the following simple result which takes a leaf out of the theory of infinite products:

THEOREM. *Suppose* (a_n) *is a decreasing sequence of positive numbers and for each natural number* n, *define* $b_n = 1 - a_{n+1}/a_n$. *Then the sequence* (a_n) *converges to zero if and only if the series* Σb_n *diverges.*

Proof. We note first that unless $b_n \to 0$ as $n \to \infty$, both of the series Σb_n and $\Sigma \log(1 - b_n)$ diverge. On the other hand, if $b_n \to 0$, then $b_n/(-\log(1 - b_n)) \to 1$ as $n \to \infty$, and it follows from the limit comparison test that Σb_n diverges if and only if $\Sigma \log(1 - b_n)$ diverges. We note also that since $0 < b_n < 1$, we have $\log(1 - b_n) < 0$ for each n.

Now since $1 - b_n = a_{n+1}/a_n$ for every n, it is clear that $a_n = a_1(1 - b_1)(1 - b_2)(1 - b_3) \cdots (1 - b_{n-1})$ for each $n \geq 2$, and we therefore conclude that $a_n \to 0$ iff $\log a_n \to -\infty$ iff $\log a_1 + \Sigma_{i=1}^n \log(1 - b_i) \to -\infty$ iff $\Sigma \log(1 - b_n)$ diverges iff Σb_n diverges.

Returning now to the above example, we see that $b_n = 1/(2n + 2)$ for each n, and the obvious divergence of Σb_n implies that $a_n \to 0$. The same technique gives an easy proof of the convergence of such series as $\Sigma((-1)^n n^n/e^n n!)$, and the series $\Sigma\binom{\alpha}{n}$ of binomial coefficients with $\alpha > -1$.

399

Convergence and Divergence of $\sum_{n=1}^{\infty} 1/n^p$

TERESA COHEN
Pennsylvania State University
University Park, PA 16802

WILLIAM J. KNIGHT
Indiana University at South Bend
South Bend, IN 46615

There is a little known proof of the divergence of the harmonic series that goes as follows. Suppose $1/1 + 1/2 + 1/3 + 1/4 + \cdots$ converges to a number S. Then the even numbered terms clearly converge to $\frac{1}{2}S$. But this means that the odd numbered terms must converge to the other half of S, which is impossible because

$$\tfrac{1}{1} > \tfrac{1}{2}, \ \tfrac{1}{3} > \tfrac{1}{4}, \ \tfrac{1}{5} > \tfrac{1}{6}, \ldots.$$

Thus the series must diverge.

This proof is so simple that it can be used with high school students. The purpose of this note is to show how the idea of the proof can be modified to establish the convergence of $\sum_{n=1}^{\infty} 1/n^p$ when $p > 1$.

Write S_N for the Nth partial sum of the series. Then

$$S_{2N+1} = 1 + \left[\frac{1}{2^p} + \frac{1}{4^p} + \cdots + \frac{1}{(2N)^p} \right] + \left[\frac{1}{3^p} + \frac{1}{5^p} + \cdots + \frac{1}{(2N+1)^p} \right]$$

$$< 1 + \left[\frac{1}{2^p} + \frac{1}{4^p} + \cdots + \frac{1}{(2N)^p} \right] + \left[\frac{1}{2^p} + \frac{1}{4^p} + \cdots + \frac{1}{(2N)^p} \right]$$

$$= 1 + \frac{1}{2^p} S_N + \frac{1}{2^p} S_N < 1 + 2^{1-p} S_{2N+1}$$

because $S_N < S_{2N+1}$. Thus $(1 - 2^{1-p}) S_{2N+1} < 1$. Since $p > 1$ the factor $1 - 2^{1-p}$ is positive, and so we have $S_{2N+1} < (1 - 2^{1-p})^{-1}$ for all N. Since $S_{2N} < S_{2N+1}$ we see that the increasing sequence $\{S_N\}$ is bounded above by $(1 - 2^{1-p})^{-1}$. Hence it converges.

Finally, note that when $p < 1$ the series $\sum 1/n^p$ diverges by comparison with the harmonic series.

A Differentiation Test for Absolute Convergence

Yaser S. Abu-Mostafa

California Institute of Technology
Pasadena, CA 91125

In this note, we describe a new test which provides a necessary and sufficient condition for absolute convergence of infinite series. The test is based solely on differentiation and is very easy to apply. It also provides a pictorial illustration for absolute convergence and divergence.

The discovery was made a few years ago when I was asked to give a lecture on infinite series to my classmates. The subject was new to us and some basic ideas were not quite appreciated at that early stage. I tried to give an informal or pictorial illustration for any concept that sounded abstract. When I mentioned that an infinite series would converge absolutely if its *far away* terms became *small enough*, I had to explain to the students (and to myself as it turned out) *how* far and *how* small.

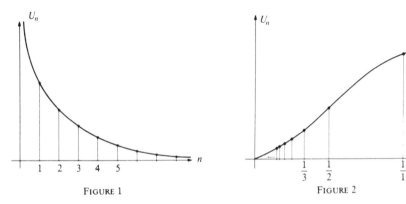

FIGURE 1   FIGURE 2

Plotting the series term by term with the summation index running along the x-axis was not very successful (FIGURE 1). I never seemed to get far enough and the terms looked quite small even with the harmonic series which, I had recently learned, was divergent. I had to come up with a picture that would show what a small term is and how it remains small if one multiplies the whole series by 10^{100} and why no matter how far away a point is, it may still not be far enough.

Remembering the duality of zero and infinity, I thought it might be nice to plot the series term by term with the *inverse* of the summation index running along the x-axis (FIGURE 2) so that we see the whole thing crowded near zero. First of all, the curve of any convergent series had to "hit" the origin since the terms go to zero as $n \to \infty$. I felt that the shape of the curve near the origin should also be related to convergence.

When I plotted divergent series like $\Sigma 1/n$ and $\Sigma 1/\sqrt{n}$, I ended up with positive or infinite slopes at the origin, but the convergent series $\Sigma 1/n^2$ had slope zero at the origin (FIGURE 3). The correspondence between the zero slope and the idea of "small terms" was appealing since a zero slope multiplied by 10^{100} is still a zero slope. The fact that the slope of a curve is a limit concept was in accordance with the "far away" idea. After the lecture, I rushed to check whether my particular illustration might generalize. After some scribbling, it gave rise to a valid criterion for absolute convergence which I call the **differentiation test**.

You probably have guessed the mechanism of the test by now. Roughly speaking, you take the infinite series in question, ΣU_n, and construct the function f defined by $f(1/n) = U_n$. First check that $f(0) = 0$. Now differentiate f and check the value of $f'(0)$: if this is also zero, the series is convergent.

Let us state this formally with the proper qualifying conditions:

DIFFERENTIATION TEST. *Let $\Sigma_{n=1}^{\infty} U_n$ be an infinite series with real terms. Let $f(x)$ be any real function such that $f(1/n) = U_n$ for all positive integers n and d^2f/dx^2 exists at $x = 0$. Then $\Sigma_{n=1}^{\infty} U_n$ converges absolutely if $f(0) = f'(0) = 0$ and diverges otherwise.*

Notice that there is a requirement that $f''(0)$ exists for the test to apply. When this requirement is satisfied, the test is guaranteed to determine whether the series is absolutely convergent or divergent. We will say more about relaxing the requirement of existence for $f''(0)$ later on, but first we present some examples to see how the test works.

We start with simple examples in which we know whether or not the series is convergent (after all, we haven't proved anything yet). Consider the two series

$$\sum_{n=1}^{\infty} \sin\frac{1}{n} \quad \text{and} \quad \sum_{n=1}^{\infty} 1 - \cos\frac{1}{n}.$$

There are many ways to verify that the first series is divergent while the second is convergent. Applying the differentiation test, we examine the functions $\sin x$ and $1 - \cos x$, both of which have second derivatives at $x = 0$. The test now tells us that the first series is divergent ($f'(0) = 1 \neq 0$) and the second is absolutely convergent ($f(0) = f'(0) = 0$).

Another interesting example is the geometric series $\Sigma_{n=1}^{\infty} a^n$ where $0 < a < 1$. When we substitute x for $1/n$, we get the function $f(x) = a^{1/x} = e^{(\ln a)/x}$ (for $x > 0$ and zero otherwise). Since $\ln a < 0$, $f(x)$ goes to zero very quickly as $x \to 0$ (FIGURE 4). In fact, $f(x)$ has the derivatives of *all* orders at $x = 0$ equal to zero. To see this, differentiate $f(x)$ any number of times and you will always get a (finite) polynomial in $1/x$ multiplied by $f(x)$ itself. Since the exponential is "stronger" than any polynomial, all the derivatives will go to zero as $x \to 0$. This suggests that the geometric series with $0 < a < 1$ is *very* convergent, which is indeed the case.

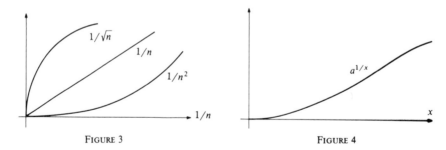

FIGURE 3   FIGURE 4

Now we show some examples where you can determine convergence or divergence right away using the differentiation test while others will require effort to get the result. In fact, for the following examples, using other techniques to determine convergence is practically the same as writing the proof for the differentiation test in the general case.

Consider the infinite series

$$\sum_{n=1}^{\infty} \int_{0}^{1/n} g(t)\, dt$$

where $g(t)$ is any function that has a derivative at $t = 0$. You are required to determine the conditions on g for the series to converge absolutely. How long does it take you to conclude that it will converge absolutely if, and only if, $g(0) = 0$? To verify the result, try substituting simple functions like t^2 or e^{-t} for $g(t)$ and carry out the integration, then test the resulting series for convergence or divergence using standard methods.

Now consider:

$$\sum_{n=1}^{\infty} \sinh\left(\tanh\frac{1}{n} - \tan\frac{1}{n} + \sec\frac{1}{n^2} - \cosh\frac{1}{n}\right).$$

Since $\sinh(\tanh x - \tan x + \sec x^2 - \cosh x)$ is analytic and has zero value and zero derivative at $x = 0$, the series is absolutely convergent. You can try other compositions of simple functions like these and see that the differentiation test is equivalent to expanding the composite function $f(x)$ in a Taylor series about $x = 0$ and checking that the lowest power in the expansion is at least x^2.

If you would like to see other techniques for dealing with these examples, go through the following proof of the differentiation test which depends on such techniques.

Proof. Our proof of the differentiation test depends on L'Hospital's rule, the limit comparison test [1], and the integral test.

Since d^2f/dx^2 is assumed to exist at $x = 0$, we are guaranteed (among other things) that $f(x)$ is continuous at $x = 0$ and is differentiable in a neighborhood of $x = 0$ (we will need the latter to apply L'Hospital's rule). We thus have the following steps relating the conditions on $f(x)$ to the absolute convergence of $\sum_{n=1}^{\infty}U_n = \sum_{n=1}^{\infty}f(1/n)$:

(1) $f(0) = 0$ is necessary for any convergence, since $\lim_{n \to \infty}U_n = \lim_{x \to 0}f(x) = f(0)$ and if this is non-zero, the series must diverge.

(2) Suppose that $f(0)$ does equal zero, but $f'(0) = a \neq 0$. Then $\lim_{x \to 0}f(x)/x = \lim_{x \to 0}(f(x) - f(0))/(x - 0) = a$. Consequently, $\lim_{n \to \infty}|U_n|/(1/n) = |a| \neq 0$. By the limit comparison test, $\sum_{n=1}^{\infty}U_n$ diverges absolutely since the harmonic series also does.

(3) We have determined that $f(0) = f'(0) = 0$ is necessary for convergence. We now assume that this condition holds and prove sufficiency. Take $0 < u < 1$ and consider the limit

$$\lim_{x \to 0^+} \frac{f(x)}{x^{1+u}} = \lim_{x \to 0^+} \frac{f'(x)}{(1+u)x^u}$$

$$= \frac{1}{1+u} \lim_{x \to 0^+} \left(\frac{f'(x) - f'(0)}{x - 0}\right)x^{1-u} = \frac{f''(0)}{1+u} \lim_{x \to 0^+} x^{1-u} = 0,$$

where the first equality is an application of L'Hospital's rule. Therefore, $\lim_{n \to \infty}|U_n|/(1/n)^{1+u} = 0$ and again by the limit comparison test, $\sum_{n=1}^{\infty}U_n$ must converge absolutely since $\sum_{n=1}^{\infty}1/n^{1+u}$ converges absolutely by the integral test.

Steps (1), (2), (3) complete the proof.

Perhaps you noticed in part (3) of the proof that the convergence did not depend critically on the existence of $f''(0)$. This is indeed the case and the existence of $f''(0)$ can be replaced by a weaker condition. We note that the condition cannot be completely removed since $\sum_{n=2}^{\infty}1/n \ln n$, which is absolutely divergent by the integral test, has terms $f(1/n)$ where $f(x) = -x/\ln x$ (for $x > 0$ and zero otherwise); this function $f(x)$ has zero value and zero derivative at $x = 0$, but a non-existent second derivative.

The existence of $f''(0)$ in the differentiation test can be replaced, for example, by the existence of $\lim_{x \to 0^+}f'(x)/x^u$ or $d^2|x|^u f(x)/dx^2|_{x=0}$ for some $0 < u < 1$ (both conditions are implied by the existence of $f''(0)$ when $f(0) = f'(0) = 0$). Very minor modification of part (3) of the proof above is needed in these cases. Certain weaker conditions will also work; their discovery is left as a simple exercise.

It is also obvious that only the existence of $f'(0)$ is needed to conclude absolute divergence of a series using the test. Since divergence is seldom good news, I choose to leave the test in its simple symmetric form. Finally, one can apply the test with $f(1/n) = |U_n|$ instead of U_n. This covers complex series as well.

I should like to acknowledge Dr. Brent Smith and one of the referees for their assistance.

References

[1]   T. M. Apostol, Mathematical Analysis, Addison-Wesley, 1974, pp. 183–193.
[2]   A. E. Taylor & W. R. Mann, Advanced Calculus, John Wiley, 1972, pp. 598–630.

A Note on Infinite Series

Louise S. Grinstein

Louise S. Grinstein is Professor of Mathematics at Kingsborough Community College of the City University of New York, where she has been teaching since 1966. Her Ph.D. in mathematics education was received from Columbia University in 1965. In addition to pedagogical experience, Professor Grinstein has worked in industry as a computer programmer and systems analyst. She has also contributed articles to various professional journals.

Recently, a student of mine came across the following problem: "Does the series $3 + (3/4) + (11/27) + (9/32) + \cdots$ converge or diverge?" The given answer was "diverges".[1] He attempted to reconstruct the general term but was stymied. The class discussed the problem noting that the denominators $1 = 1$, $4 = 2^2$, $27 = 3^3$ are of the form k^k. The fourth denominator conforms to this pattern if the term $9/32$ is written as $72/4^4$. The remaining difficulty centered about the numerators 3, 3, 11, 72, ... Since no obvious answer was readily forthcoming, I used this opportunity to point out that an infinite number of infinite series exist with the given as first four terms. Some of these converge while others diverge.

The convergent case. A cubic polynomial $f(n) = an^3 + bn^2 + cn + d$ can be found satisfying the conditions that $f(1) = 3$; $f(2) = 3$; $f(3) = 11$; $f(4) = 72$. The general term is therefore

$$a_n = (7.5n^3 - 41n^2 + 70.5n - 34)/n^n.$$

The ratio test yields:

$$t = \lim_{n \to \infty} \left| \frac{a_{n+1}}{a_n} \right| = \lim_{n \to \infty} \left[\left(\frac{n}{n+1} \right)^{n-3} \cdot \frac{1}{n+1} \cdot \frac{\left(7.5 - \frac{41}{n+1} + \frac{70.5}{(n+1)^2} - \frac{34}{(n+1)^3} \right)}{\left(7.5 - \frac{41}{n} + \frac{70.5}{n^2} - \frac{34}{n^3} \right)} \right]$$

$$= \lim_{n \to \infty} \frac{1}{e} \cdot \frac{1}{n+1} = 0 < 1,$$

and thus the series converges absolutely.

The divergent case. Disregarding the first two terms, a general term valid for $n \geqslant 3$ is:

$$a_n = \frac{n^{n-1} + 2^{2n-5}}{n^n} = \frac{1}{n} + \frac{2^{2n-5}}{n^n}.$$

---

[1] Frank Ayres, Jr., "Theory and Problems of Differential and Integral Calculus", Schaum's Outline Series, Schaum Publishing Company, second ed., p. 229, New York, 1964.

For all integers $n \geqslant 3$, $a_n > 1/n$. Thus, the given series, from the third term on, dominates the divergent harmonic series and so diverges.

This problem points out the ambiguity inherent in such frequently posed questions as "Given the numerical values of a finite number of terms at the beginning of a series, state whether the series converges or diverges". Where the formation of the general term is obvious, students tend to forget that the obvious answer is not the only one.

Power Series for Practical Purposes

"Take a bone from a dog—what remains?"—The White Queen

Ralph Boas

The author retired from Northwestern University in 1980 with the title of Professor Emeritus of Mathematics, a title which, he likes to point out, has nothing to do with merit. He has also retired as editor of the "American Mathematical Monthly;" the December 1981 issue has a symbolic portrait of him on the cover. This article is an outgrowth of his hobby of trying to encourage people to look at apparently complicated things in a simple way.

1. Why Taylor series? We hear a great deal about teaching mathematics that is applicable to "the real world." Consequently I was nonplussed at being told recently, in all seriousness, that most teachers will reject a calculus text out of hand if it doesn't discuss the formula for the remainder in a Taylor series.

This claim may well be true; but, if so, most teachers are trying to teach the wrong course. The Taylor remainder is an important piece of mathematics associated with calculus—important for proving theorems, that is. So are uniform convergence and Lebesgue integration, but most teachers realize that these topics will not be appreciated by the mass of freshmen and sophomores. Why is the Taylor remainder conceived to be of such central importance?

The conventional answer is that with the remainder formula we can estimate remainders and so be sure that power series represent the functions we got them from. This is indeed true for the exponential, sine and cosine functions, for the binomial series—and not for much of anything else. Indeed, these are practically the only elementary functions whose successive derivatives are simple enough to calculate in the first place and then to estimate. Just try to calculate enough derivatives to estimate the remainder after eight terms for $\cos(x - \sin x)$. (This function occurs in the theory of frequency modulation.) You could, of course, use a computer to evaluate the derivatives, but in that case you probably wouldn't need the series anyway.

Given that the remainder formula is useful primarily for theoretical purposes, can we justify the introduction of Taylor series at all? Isn't it now only of academic interest that

$$e^x = 1 + x + \frac{x^2}{2!} + \frac{x^3}{3!} + \cdots ?$$

Not long ago, if you wanted a fairly precise approximation to $e^{0.01982}$, you couldn't do much better than to calculate a few terms of the series with $x = 0.01982$. Nowadays, of course, you push seven buttons on the little pocket calculator and get 1.02001772, a result more precise than anything you are likely to need. You can obtain values of $\cos(x - \sin x)$ almost as easily.

Nevertheless, many people who apply mathematics do value the ability to write down several terms of the series of such functions. The usual technique for $\cos(x - \sin x)$ would be to form the Maclaurin series for $x - \sin x$, substitute this series for y in the Maclaurin series for $\cos y$, and rearrange the result as a power series in x. Authors of calculus books seem to be largely unaware that this is a completely legitimate process. But, given the existence of calculators and computers, why should anyone bother to work out the Maclaurin series by any method?

In the first place there are things that calculators don't yet do, or do only after special programming. One of these is the evaluation of nonelementary functions. I have seen a calculator that has built-in gamma functions, but, as far as I know, there is as yet no calculator with built-in Bessel functions or elliptic integrals. Another operation that is not easy on a calculator is the calculation of a table of values of an indefinite integral. Still another is making calculations of higher precision than was built in. You cannot, at least at present, appeal directly to a calculator for values of

(1) $J_0(\sin x)$,

(2) $\displaystyle\int_0^x \frac{dt}{\sqrt{t^3 + 2t + 1}}$,

or

(3) $e^{-x^2/2} - \dfrac{1}{x} \displaystyle\int_0^x te^{-t^2/2}\, dt$.

Here J_0 is the Bessel function

(4) $J_0(x) = \displaystyle\sum_{k=0}^{\infty} (-1)^k (x/2)^{2k} / (k!)^2$;

(2) is an elliptic integral, and (3) is a solution of [1] $x^2 y'' + x^3 y' + (x^2 - 2)y = 0$.

It is relatively easy to start from well-known power series and work out the first few coefficients of the Maclaurin series of (1), (2) or (3). If one of these functions turns up in a practical problem, it will probably have to be entered into a computer to be used for further computation. It is much easier to enter a small number of coefficients than a whole table of values.

2. Why it works.   To get power series for functions like (1), (2) or (3), we need to know that power series can be differentiated or integrated, multiplied or divided, and (most importantly) substituted into each other; then, if necessary, rearranged as power series again, with the same result as if we had worked out the coefficients by differentiation. Some of the proofs are beyond the scope of a calculus course, but at least correct statements of the relevant theorems are easily understood. (The same can be said for other theorems that are used in calculus courses.)

It is most convenient to state the theorems for complex series; but they can be specialized (although not all the proofs can) to real series by reading "interval of convergence" for "disk of convergence."

A. Differentiation and integration are permissible in the interior of the disk of convergence. This is straightforward to prove and is done in many textbooks.

B. Multiplication. If the Maclaurin series of f and g both converge for $|z| < r$, then their formal product, arranged as a series of ascending powers, has radius of convergence at least r and represents fg. (Here r is not necessarily the radius of convergence of either series.)

This is a special case of the theorem that the formal product of two absolutely convergent series converges (absolutely) to the product of the sums of the series. In fact, we have

$$\sum_{n=0}^{\infty} a_n z^n \cdot \sum_{m=0}^{\infty} b_m z^m = \sum_{k=0}^{\infty} c_k z^k, \quad \text{where } c_k = \sum_{j=0}^{k} a_j b_{k-j}.$$

Hence we can calculate the coefficients in the product series or program them for a computer to calculate.

C. Division. If the Maclaurin series of f and g converge for $|z| < r$ and $g(z) \neq 0$ for $0 \leqslant |z| < r$, then if the Maclaurin series for f is divided by the Maclaurin series for g by long division (as if the series were polynomials), the resulting series represents f/g for $|z| < r$.

Again, r is not necessarily the radius of convergence of either series.

If $h = f/g$ then $f = gh$. Write $h(z) = \sum_{n=0}^{\infty} c_n z^n$, multiply h by g by using the formula under B, and solve for the c_n recursively. We can then see by induction that these c_n are exactly what one gets by long division.

For example, C does not allow us to divide the series for $\cos z$ by the series for $\sin z$ near $z = 0$; fortunately, since $\cot z$ does not have a Maclaurin series.

D. Substitution. The following theorem, although not the most general possible, covers many cases that are likely to arise in practice. Curiously enough, it is absent from most calculus books and is not discussed adequately in most modern textbooks on complex analysis. It was rather difficult to locate a formal proof (see, however, [2] or [3]); I give one in §3 in case anyone wants to see it. It involves nothing deeper than the theorem that a uniformly convergent series of analytic functions can be differentiated term by term.

Substitution theorem. *Let $f(w)$ be represented by the series $\sum_{n=0}^{\infty} a_n w^n$ for $|w| < s$; let $g(z)$ be represented by $\sum_{k=0}^{\infty} b_k z^k$ for $|z| < r$, and let $|g(0)| < s$. Then the Maclaurin series of $F(z) = f(g(z))$ can be obtained by substituting $w = \sum_{k=0}^{\infty} b_k z^k$ into $\sum_{n=0}^{\infty} a_n w^n$ and rearranging the result as a power series in t.*

The resulting series represents $F(z)$ in any disk $|z| < t$ in which F is analytic; there is such a disk because F is analytic at 0. In practice, f and g are often elementary functions and the radius of convergence of the Maclaurin series of F can be found by inspection.

Notice particularly what the theorem does *not* permit. For example, we cannot get a Maclaurin series for $\ln(1 - \cos z)$ by substituting the Maclaurin series for $\cos z$ into the series for $\ln(1 - w)$, because the composite function is not even defined at

$z = 0$. The theorem does not let us try because the Maclaurin series for $\ln(1 - w)$ has radius of convergence $s = 1$ but $\cos 0 = 1$ is not less than s. We must also be careful to avoid the mistake of substituting the Maclaurin series of g into the Maclaurin series of f when g is analytic at $z = a$ but $g(a)$ is outside the disk of convergence of the Maclaurin series of f. For example, consider a branch of $(1 - 2\cos z)^{1/2}$. The Maclaurin series of $(1 - 2w)^{1/2}$ converges only for $|w| < \frac{1}{2}$, but $\cos 0 = 1$; even though $(1 - 2\cos z)^{1/2}$ is analytic at $z = 1$, we cannot substitute $w = \cos z$ into a divergent Maclaurin series and expect to get a meaningful result when z is near 0.

3. Proof of the substitution theorem. Since $F(z)$ is analytic at 0 it has a Maclaurin series $\sum \lambda_n z^n$. On the other hand, when $|g(z)| < s$ we have

$$F(z) = f(g(z)) = \sum_{m=0}^{\infty} \phi_m [g(z)]^m = \sum_{m=0}^{\infty} \phi_m \sum_{n=0}^{\infty} c_{m,n} z^n$$

by the multiplication theorem. What we want to show is that

(5) $\lambda_n = \sum_{m=0}^{\infty} \phi_m c_{m,n}.$

Now $\sum \lambda_n z^n$ is uniformly convergent in any closed disk $|z| < r$ where F is analytic and hence can be differentiated repeatedly in a neighborhood of 0. Setting $z = 0$, we get

$$\lambda_0 = \sum_{m=0}^{\infty} \phi_m c_{m,0}.$$

Now $\sum \phi_m w^m$ converges uniformly for $|w| \leq s_1 < s$ and so $\sum \phi_m [g(z)]^m$ converges uniformly provided $|g(z)| \leq s_1$. This will be the case if $|z|$ is sufficiently small and s_1 is sufficiently close to s, since $|g(0)| < s$. Hence $\sum \phi_m [g(z)]^m$, although not a power series, can also be differentiated term by term; consequently

$$F'(z) = \sum_{m=0}^{\infty} \phi_m \sum_{n=0}^{\infty} n c_{m,n} z^{n-1}$$

$$= \sum_{n=0}^{\infty} n \lambda_n z^{n-1};$$

setting $z = 0$ we get

$$\lambda_1 = \sum_{n=0}^{\infty} c_{1,n}.$$

This process can be continued to yield (5).

This shows that the series for $f(g(z))$ can be rearranged into a power series when z is sufficiently close to 0; this series then represents $F(z)$ in the largest disk, center at 0, in which F is analytic.

REFERENCES

1. E. Kamke, Differentialgleichungen, Lösungsmethoden und Lösungen, vol. 1, 3rd ed., Leipzig, Akademische Verlagsgesellschaft, 1944, p. 450, (2.208).
2. A. I. Markushevich, Theory of Functions of a Complex Variable, vol. 1 (translated and edited by R. A. Silverman), Prentice-Hall, Englewood Cliffs, N.J., 1965, p. 433.
3. W. F. Osgood, Lehrbuch der Funktionentheorie, vol. 1, 5th ed., Teubner, Leipzig, 1928, p. 362.

BIBLIOGRAPHIC ENTRIES: THEORY

1. *Monthly* Vol. 78, No. 2, pp. 164–170. O. E. Stanaitis, Integral tests for infinite series.

2. *Monthly* Vol. 87, No. 1, pp. 36–38. Richard Johnsonbaugh, The alternating series defined by an increasing function.

3. *Monthly* Vol. 96, No. 1, pp. 41–42. J. M. Patim, A very short proof of Stirling's formula.

A very short nonelementary proof based on the Lebesgue dominated convergence theorem.

4. *Math. Mag.* Vol. 51, No. 2, pp. 83–89. R. P. Boas, Estimating remainders.

5. *Math. Mag.* Vol. 51, No. 4, pp. 235–238. Mark A. Pinsky, Averaging an alternating series.

6. *Math. Mag.* Vol. 57, No. 4, pp. 209–214. J. B. Wilker, Stirling ideas for freshman calculus.

7. TYCMJ Vol. 4, No. 2, pp. 16–29. G. Baley Price. Telescoping sums and the summation of sequences.

Many examples of partial sums that can be evaluated explicitly using the telescoping property: arithmetic and geometric progressions, sums of bionomial coefficients, factorial polynomials, and power sums.

8. TYCMJ Vol. 9, p. 191. Thomas W. Shilgalis. Flow chart for infinite series.

9. TYCMJ Vol. 12, No. 1, pp. 54–56. G. C. Schmidt, Uniqueness of power series representations.

10. TYCMJ Vol. 14, No. 3, pp. 253–256. Lenny K. Jones, Sequences, series, and Pascal's triangle.

11. CMJ Vol. 17, No. 2, pp. 165–166. J. Richard Morris, Counterexamples to a comparison test for alternating series.

12. CMJ Vol. 17, No. 5, p. 417. Norman Schaumberger, Another approach to a class of slowly diverging series.

(b)

SERIES RELATED TO THE

HARMONIC SERIES

AN INTERESTING SUBSERIES OF THE HARMONIC SERIES

A. D. WADHWA

The series $\Sigma\, 1/n$ diverges, but it is known that if we omit the terms corresponding to the integers whose decimal representations contain a specified digit at least once (the digit 0, for example), the resulting series converges ([1], [2], [3], [4], [5]).

This seems surprising at first sight; but for very large n the integers that do not have a particular digit in their decimal representations are quite scarce. Indeed, there are $9 \cdot 10^{n-1}$ integers with n-digit decimal representations, but only 9^n of them omit 0, and $9^n/9 \cdot 10^{n-1} \to 0$ as $n \to \infty$. Since few people seem to be aware of the convergence of the harmonic series thinned out in this way, we give a brief proof, and show that the sum of the series is between 20.2 and 28.3. The series with some other digit omitted can be treated similarly. We also show that the series is considerably more than merely convergent: in fact $\Sigma\, n^{-\alpha}$ converges, for the integers n in question, if $\alpha > \log_{10} 9 = 0.95 +$. There are corresponding results if integers are represented in other bases. In particular, if we use base 100 we see that, if we delete from the harmonic series only the reciprocals of integers whose decimal representations contain at least two zeros, the resulting series also converges (similarly for any fixed number of zeros).

The n-digit integers which do not have any digit 0 can be written $a_1 \cdots a_n$ (understood to mean, as usual, $10^{n-1}a_1 + 10^{n-2}a_2 + \cdots + a_n$), where $0 < a_k \leq 9$. Since all the terms are positive, the convergence of $\Sigma\, 1/k$ over all k of this form is equivalent to the convergence of

$$S = \sum_{n=1}^{\infty} \sum \frac{1}{a_1 \cdots a_n} .$$

For a given n there are 9^n terms in the inner sum, and each exceeds 10^{n-1}, so

$$S \leq \sum_{n=1}^{\infty} 9^n/10^{n-1} = 90.$$

To go further, we notice that of the n-digit integers, one-ninth have leading digit 1, one-ninth have leading digit 2, and so on; so of the 9^n n-digit integers, 9^{n-1} are between 10^{n-1} and $2 \cdot 10^{n-1}$, and so on. These integers then contribute at most

$$10^{-n+1} \cdot 9^{n-1}(1 + \tfrac{1}{2} + \cdots + \tfrac{1}{9}) = (0.9)^{n-1}(1 + \tfrac{1}{2} + \cdots + \tfrac{1}{9})$$

to S, and consequently

$$S < (1 + \tfrac{1}{2} + \cdots + \tfrac{1}{9}) \sum_{n=1}^{\infty} (0.9)^{n-1} = 10(1 + \tfrac{1}{2} + \cdots + \tfrac{1}{9}) < 28.3.$$

On the other hand, the integers between 10^{n-1} and $2 \cdot 10^{n-1}$ are less than $2 \cdot 10^{n-1}$, and so on, so that for $n \geqq 2$ the n-digit integers contribute at least

$$(0.9)^{n-1}(\tfrac{1}{2} + \cdots + \tfrac{1}{10})$$

to S. Hence

$$S > 1 + \tfrac{1}{2} + \cdots + \tfrac{1}{9} + \sum_{n=2}^{\infty} (0.9)^{n-1}(\tfrac{1}{2} + \cdots + \tfrac{1}{10}) > 20.189.$$

Closer bounds are given in [2].

The remainder after the reciprocals of the k-digit integers have been added is

$$R_n = \sum_{n=k+1}^{\infty} \sum \frac{1}{a_1 \cdots a_n},$$

(where again the denominator means $10^{n-1}a_1 + 10^{n-2}a_2 + \cdots + a_n$, $0 < a_j \leqq 9$). Since each denominator is less than 10^n and the inner sum contains 9^n terms,

$$R_n > \sum_{n=k+1}^{\infty} (0.9)^n = 9 \cdot (0.9)^k.$$

Thus $R_n \to 0$ quite slowly (for $k = 20$ it is still greater than 1), and so (even though $R_n \to 0$) there is no possibility of calculating S by adding up the successive terms. However, S can be calculated by first estimating the remainder more carefully; a machine calculation gave $S = 23.10$, correct to two decimal places.

If we write integers in base k instead of base 10, we see that the corresponding thinned-out harmonic series has a sum between

$$k\left\{1 + \tfrac{1}{2} + \cdots + \frac{1}{k-1}\right\} - (k-1)^2/k \quad \text{and} \quad k\left\{1 + \tfrac{1}{2} + \cdots + \frac{1}{k-1}\right\}.$$

Similarly, it follows that $\Sigma \, 1/n^\alpha$ over the integers whose base k representations contain no 0 is convergent if $\alpha > [\log(k-1)]/\log k$, and its sum is between

$$\frac{k^\alpha}{k^\alpha - k + 1}\{2^{-\alpha} + \cdots + k^{-\alpha}\} \quad \text{and} \quad \frac{k^\alpha}{k^\alpha - k + 1}\{1 + 2^{-\alpha} + \cdots + (k-1)^{-\alpha}\}.$$

I wish to thank Professor S. D. Chopra for bringing the problem to my attention and for his interest in the solution. I am indebted to R. A. Honsberger for references and to the referee for some numerical results.

References

1. T. M. Apostol, Mathematical Analysis, Addison-Wesley, Reading, Mass., 1960, p. 384, Ex. 12–16.
2. F. Irwin, A curious convergent series, this MONTHLY. 23 (1916) 149–152.
3. A. J. Kempner, A curious convergent series, this MONTHLY. 21 (1914) 48–50.
4. E. J. Moulton, Solution of Problem 453, this MONTHLY. 23 (1916) 302–303.
5. G. Pólya and G. Szegő, Aufgaben und Lehrsätze aus der Analysis, Springer, 1925, Problem I 124.

DEPARTMENT OF MATHEMATICS, KURUKSHETRA UNIVERSITY, KURUKSHETRA, INDIA.

A PROOF OF THE DIVERGENCE OF $\Sigma\, 1/p$

Ivan Niven, University of Oregon

First we prove that $\Sigma'\, 1/k$ diverges, where Σ' denotes the sum over the squarefree positive integers. Each positive integer is uniquely expressible as a product of a squarefree positive integer and a square, so for any positive integer n,

$$\left(\sum_{k<n}' 1/k \right)\left(\sum_{j<n} 1/j^2 \right) \geqq \sum_{m<n} 1/m.$$

Here the second sum is bounded but the third sum is unbounded as n increases, so the first sum must be unbounded. Next suppose that $\Sigma\, 1/p$ converges to β, the sum taken over all primes p. By dropping all terms beyond x in the series expansion of e^x or exp (x), we see that exp $(x) > 1+x$ for $x>0$. Hence for each positive integer n

$$\exp(\beta) > \exp\left(\sum_{p<n} 1/p \right) = \prod_{p<n} \exp(1/p) > \prod_{p<n} (1 + 1/p) \geqq \sum_{k<n}' 1/k.$$

But this contradicts the unboundedness of the last sum, so $\Sigma\, 1/p$ diverges.

PROOFS THAT $\Sigma 1/p$ DIVERGES

The theorem of the title, in which p runs through the primes, was deduced by Euler in 1737 from the "equation"

$$\sum_{n=1}^{\infty} \frac{1}{n} = \prod \left(1 - \frac{1}{p} \right)^{-1}.$$

Kronecker later fixed up this argument by replacing n and p by n^A and p^A, $A > 1$, and letting $A \to 1$ [1].

It is my purpose in this paper to examine some fairly recent (compared to the age of the theorem, anyway) easy proofs of this theorem, where the word "easy" needs some definition. Of course any theorem may be proved very compactly if enough preliminary knowledge is assumed, and many number theory texts derive Euler's theorem from much stronger results. Here we are concerned with direct proofs appropriate for an undergraduate number theory class.

Our presumed audience will know something about infinite series (otherwise the statement of the theorem will not even make sense), but only from calculus, along with some elements of number theory.

We may not assume, for example, familiarity with the connection between the convergence of the series Σa_n and the infinite product $\prod(1 \pm a_n)$. In fact, although our audience knows what it means for a series to converge absolutely, it does not know that rearranging such a series is justified. This means that some arguments must be complicated by replacing series with finite sums.

The proofs below have been left in more or less their original form, except that the notation is standardized. The letter p always represents a prime, and all lower-case letters except e represent positive integers. For given positive integers a and b with $a < b$, let P denote the set of primes p satisfying $a < p \leq b$, let M be those integers n all of whose prime divisors are in P, and for a given integer x let M_x be all elements of M not exceeding x. Let $|S|$ denote the number of elements in the set S.

Erdős's Proof. Paul Erdős published the following proof in 1938 [2]. If $\Sigma 1/p$ converges we can choose b so that $\Sigma_{p>b} 1/p < \frac{1}{2}$. Take $a = 1$. Suppose $n \in M_x$, and write $n = k^2 m$, where m is square-free. Since $m = \prod_{S} p$, where S is some subset of P, m can assume at most $2^{|P|}$ values. Also $k < \sqrt{n} < \sqrt{x}$. Thus $|M_x| < 2^{|P|} \sqrt{x}$.

Now the number of positive integers $< x$ divisible by a fixed p does not exceed x/p. Thus $x - |M_x|$, the number of such integers divisible by some prime greater than b, satisfies

$$x - |M_x| \leq \sum_{p>b} \frac{x}{p} < \frac{x}{2}.$$

We see

$$\frac{x}{2} < |M_x| < 2^{|P|} \sqrt{x},$$

or $\sqrt{x} < 2^{|P|+1}$, which is clearly false for x sufficiently large.

Comments. The proof above, which is notable for its lack of series manipulations, is given in the classic book by Hardy and Wright [3], as well as by Calvin Long's text [4].

Bellman's and Moser's Proofs. The details of Richard Bellman's 1943 proof [5] and Leo Moser's 1958 proof [6] will be omitted, since both appeared in this MONTHLY. Bellman assumed an a sufficiently large so that $\Sigma_P 1/p < 1$ with $b = \infty$, and from this derived the convergence of first $\Sigma_M 1/n$ and then the harmonic series.

Moser derived from the false assumption that $\Sigma 1/p$ converges the true conclusion that $\pi(x)/x \to 0$ as $x \to \infty$, where $\pi(x)$ denotes (as usual) the number of primes $\leqslant x$. A contradiction was then produced from this and the assumption that $\Sigma_P 1/p < \frac{1}{2}$ for large enough a.

Bellman used the rearrangement of positive series several times in the proof, and Moser used the result that a convergent series is Cesàro summable to the same limit [7], which our hypothetical audience is unlikely to have seen.

Dux's Proof. In 1956 Erich Dux [8] gave a proof that began and ended similarly to Bellman's but also made use of the rearrangement of positive series. Let $a = 1$ and, assuming $\Sigma 1/p$ converges, choose b so that $\Sigma_{p>b} 1/p = A < 1$. Define M' to be all $n' > 1$ divisible by primes only exceeding b, and M'' to be all positive integers not in M or M'. (Note that 1 is in M.)

Then, since P is finite,

$$\sum_M \frac{1}{n} = \prod_P \left(1 + \frac{1}{p} + \frac{1}{p^2} + \cdots\right) = \prod_P \left(1 - \frac{1}{p}\right)^{-1} < \infty,$$

and

$$\sum_{M'} \frac{1}{n'} < \sum_{p>b} \frac{1}{p} + \left(\sum_{p>b} \frac{1}{p}\right)^2 + \cdots = \frac{A}{1-A} < \infty;$$

so

$$\sum_{M''} \frac{1}{n''} = \left(-1 + \sum_M \frac{1}{n}\right) \sum_{M'} \frac{1}{n'} < \infty.$$

This contradicts the divergence of the harmonic series.

Clarkson's Proof. This 1966 proof by James A. Clarkson [9] calls to mind Euclid's proof that the number of primes is infinite. If $\Sigma 1/p$ converges, we can choose a so that $\Sigma_P 1/p < \frac{1}{2}$ for all b. Let $Q = \prod_{p<a} p$. For fixed r, it is possible to choose b large enough so that all the factors of the numbers $1 + iQ$, $1 \leqslant i \leqslant r$, are in P, since if $p < a$ then $p \nmid 1 + iQ$.

Now each term of the sum $\Sigma_{i=1}^r 1/(1 + iQ)$ whose denominator is a product of j primes (not necessarily distinct) occurs at least once in the expansion of

$$\left(\sum_P \frac{1}{p}\right)^j < 2^{-j}. \tag{1}$$

Thus

$$\sum_{i=1}^r \frac{1}{1+iQ} < \sum_{j>1} 2^{-j} < 1.$$

But, since r was arbitrary, this implies that the harmonic series converges.

Comments. The expansion of $(\Sigma_P 1/p)^j$ recalls the proofs of Bellman and Dux. Getting the inequality after (1) requires rearrangement of infinite series. This proof (with $b = \infty$) is reproduced in Apostol's *Introduction to Analytic Number Theory* [10]. The last sentence of the proof is most easily justified by comparison with $\Sigma 1/2Qi$.

Two More Proofs. A very simple identity forms the basis for two more proofs of my own design, namely,

$$\left(1 + \frac{1}{p}\right)\left(1 + \frac{1}{p^2} + \frac{1}{p^4} + \cdots + \frac{1}{p^{2k}}\right) = 1 + \frac{1}{p} + \frac{1}{p^2} + \cdots + \frac{1}{p^{2k+1}}.$$

Taking the product over P and letting $k \to \infty$ yields

$$\prod_P \left(1 + \frac{1}{p}\right) \sum_M \frac{1}{n^2} = \sum_M \frac{1}{n}. \tag{2}$$

Since $\Sigma 1/n^2$ converges and $\Sigma 1/n$ diverges, this means,

$$\text{for } a = 1, \prod_P \left(1 + \frac{1}{p}\right) \to \infty \quad \text{as } b \to \infty. \tag{3}$$

Continuation A. Since for $C > 0, e^C = 1 + C + C^2/2! + \cdots > 1 + C$, we have

$$\prod_P \left(1 + \frac{1}{p}\right) < \prod_P e^{1/p} = \exp\left(\sum_P \frac{1}{p}\right).$$

This, with (3), shows, that $\Sigma 1/p$ diverges.

Continuation B. By (3) $\prod_P (1 + 1/p) \to \infty$ as $b \to \infty$ for any fixed a, and so the same is true for $\Sigma_M 1/n$ by (2). If $\Sigma 1/p$ converges, we can choose a so that $\Sigma_P 1/p < \frac{1}{2}$ for all b, then choose b and x large enough so that $\Sigma_{M_x} 1/n > 2$. Since every n in M_x except 1 is of the form pn for $p \in P$ and $n \in M_x$, we have

$$\sum_P \frac{1}{p} \sum_{M_x} \frac{1}{n} > \sum_{M_x} \frac{1}{n} - 1.$$

Then

$$\frac{1}{2} > \sum_P \frac{1}{p} > 1 - \left(\sum_{M_x} \frac{1}{n}\right)^{-1} > 1 - \frac{1}{2},$$

a contradiction.

Comments. The proof using Continuation A is so simple (and the theorem is so old) that it would be foolhardy to call it new, although I have not found the arrangement anywhere. Of course $e^C > 1 + C$ may also be proved without recourse to the series for e^C, but I believe most of our hypothetical audience will remember this expansion. Continuation B has a better chance of being a novelty.

The reader should note a very recent proof by Frank Gilfeather and Gary Meisters [11].

References

1. L. E. Dickson, History of the Theory of Numbers, vol. 1, Chelsea, New York, 1952, p. 413.
2. P. Erdös, Uber die Reihe $\Sigma 1/p$, Mathematica, Zutphen. B., 7 (1938) 1–2.
3. G. H. Hardy and E. M. Wright, The Theory of Numbers, Oxford, 1954, pp. 16–17.
4. Calvin T. Long, Elementary Introduction to Number Theory, 2nd ed., Heath, Lexington, 1972, pp. 63–64.
5. R. Bellman, A note on the divergence of a series, this MONTHLY, 50 (1943) 318–319.
6. Leo Moser, On the series $\Sigma 1/p$, this MONTHLY, 65 (1958) 104–105.
7. R. R. Goldberg, Methods of Real Analysis, Blaisdell, New York, 1964, p. 91.
8. Erich Dux, Ein kürzer Beweis der Divergenz der unendlichen Reihe $\Sigma_{r=1} 1/p_r$, Elem. Math., 11 (1956) 50–51.
9. James A. Clarkson, On the series of prime reciprocals, Proc. Amer. Math. Soc., 17 (1966) 541.
10. Tom M. Apostol, Introduction to Analytic Number Theory, Springer-Verlag, New York, 1976, pp. 18–19.
11. W. G. Leavitt, The sum of the reciprocals of the primes, Two-Year College Math. J., 10 (1979) 198–199.

MATHEMATICS DEPARTMENT, ILLINOIS STATE UNIVERSITY, NORMAL, IL 61761.

The Sum of the Reciprocals of the Primes

W. G. Leavitt, University of Nebraska, Lincoln, NE

For a long time it has been known that the sum of the reciprocals of the primes $\sum 1/p = 1/2 + 1/3 + 1/5 + 1/7 + 1/11 + \cdots$ forms a divergent series. Recently a very neat, new proof has been devised by Frank Gilfeather and Gary Meisters, who are to be thanked for their permission to present it here.

For a given integer $n \geqslant 2$, we take the set of all primes $p \leqslant n$ and consider the product

$$\prod_{p \leqslant n} \left(\frac{p}{p-1} \right) = \prod_{p \leqslant n} \left(\frac{1}{1 - 1/p} \right) = \prod_{p \leqslant n} \left(1 + \frac{1}{p} + \frac{1}{p^2} + \cdots \right). \tag{1}$$

Since each $k \leqslant n$ is a product of various powers of certain primes less than or equal to n, we are certain to have $1/k$ as one of the terms in the above multiplication

$$\left(1 + \frac{1}{2} + \frac{1}{2^2} + \cdots \right)\left(1 + \frac{1}{3} + \frac{1}{3^2} + \cdots \right)\cdots .$$

This is true for all $k \leqslant n$. Therefore, it follows from (1) that

$$\prod_{p \leqslant n} \left(\frac{p}{p-1} \right) > \sum_{k=1}^{n} \frac{1}{k} .$$

Since the logarithm function is monotonic (preserves inequalities), this gives

$$\sum_{p \leqslant n} (\log p - \log(p-1)) > \log\left(\sum_{k=1}^{n} \frac{1}{k} \right). \tag{2}$$

On the other hand,

$$\sum_{p \leqslant n} (\log p - \log(p-1)) = \sum_{p \leqslant n} \left(\int_{p-1}^{p} \frac{1}{x} \, dx \right) < \sum_{p \leqslant n} \left(\frac{1}{p-1} \right) \leqslant \sum_{p \leqslant n} \frac{2}{p} . \tag{3}$$

Combining inequalities (2) and (3), we arrive at

$$\sum_{p \leqslant n} \frac{1}{p} > \frac{1}{2} \log\left(\sum_{k=1}^{n} \frac{1}{k} \right). \tag{4}$$

The right-hand side of (4) increases without limit as n increases. Therefore $\sum 1/p$ diverges.

Remark. It may be surprising to note that the sum of the reciprocals of the twin primes (i.e., pairs of primes which differ by 2, such as 11, 13 or 17, 19) is a convergent series. [See, for example, E. Landau, *Elementary Number Theory*, 2nd ed., Chelsea (1966), pp. 94–103.]

The Bernoullis and the Harmonic Series

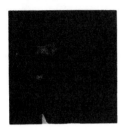

After receiving a B.S. in mathematics from the University of Pittsburgh in 1969, William Dunham moved westward to complete a 1974 Ph.D. at the Ohio State University, where he did his thesis in general topology under Professor Norman Levine. Since graduate school, Dr. Dunham has taught mathematics at Hanover College, an institution where faculty must be generalists and where the liberal arts are taken seriously. In that environment, his interests shifted toward the history of mathematics. Professor Dunham received a grant, in 1983, from the Lilly Endowment, Inc., to pursue these historical interests, and it was while engaged in such pursuits that he stumbled upon the 300-year old mathematical morsel described in this paper.

Any introduction to the topic of infinite series soon must address that first great counterexample of a divergent series whose general term goes to zero—the harmonic series $\sum_{k=1}^{\infty} 1/k$. Modern texts employ a standard argument, traceable back to the great 14th Century Frenchman Nicole Oresme (see [3], p. 92), which establishes divergence by grouping the partial sums:

$$1 + \frac{1}{2} > \frac{1}{2} + \frac{1}{2} = \frac{2}{2}$$

$$1 + \frac{1}{2} + \left(\frac{1}{3} + \frac{1}{4}\right) > \frac{2}{2} + \left(\frac{1}{4} + \frac{1}{4}\right) = \frac{3}{2}$$

$$1 + \frac{1}{2} + \frac{1}{3} + \frac{1}{4} + \left(\frac{1}{5} + \frac{1}{6} + \frac{1}{7} + \frac{1}{8}\right) > \frac{3}{2} + \left(\frac{1}{8} + \frac{1}{8} + \frac{1}{8} + \frac{1}{8}\right) = \frac{4}{2},$$

and in general

$$1 + \frac{1}{2} + \frac{1}{3} + \cdots + \frac{1}{2^n} > \frac{n+1}{2},$$

from which it follows that the partial sums grow arbitrarily large as n goes to infinity.

It is possible that seasoned mathematicians tend to forget how surprising this phenomenon appears to the uninitiated student—that, by adding ever more negligible terms, we nonetheless reach a sum greater than any preassigned quantity. Historian of mathematics Morris Kline ([5], p. 443) reminds us that this feature of the harmonic series seemed troubling, if not pathological, when first discovered.

So unusual a series could not help but attract the interest of the preeminent mathematical family of the 17th Century, the Bernoullis. Indeed, in his 1689 treatise

"Tractatus de Seriebus Infinitis," Jakob Bernoulli provided an entirely different, yet equally ingenious proof of the divergence of the harmonic series. In "Tractatus," which is now most readily found as an appendix to his posthumous 1713 masterpiece *Ars Conjectandi*, Jakob generously attributed the proof to his brother ("Id primus deprehendit Frater"), the reference being to his full-time sibling and part-time rival Johann. While this "Bernoullian" argument is sketched in such mathematics history texts as Kline ([5], p. 444) and Struik ([6], p. 321), it is little enough known to warrant a quick reexamination.

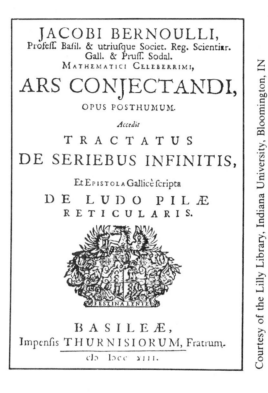

The proof rested, quite unexpectedly, upon the *convergent* series

$$\frac{1}{2} + \frac{1}{6} + \frac{1}{12} + \frac{1}{20} + \cdots = \sum_{k=1}^{\infty} \frac{1}{k(k+1)}.$$

The modern reader can easily establish, via mathematical induction, that

$$\sum_{k=1}^{n} \frac{1}{k(k+1)} = \frac{n}{n+1},$$

and then let n go to infinity to conclude that

$$\sum_{k=1}^{\infty} \frac{1}{k(k+1)} = 1.$$

Jakob Bernoulli, however, approached the problem quite differently. In Section XV of *Tractatus*, he considered the infinite series

$$N = \frac{a}{c} + \frac{a}{2c} + \frac{a}{3c} + \frac{a}{4c} + \cdots,$$

then introduced

$$P = N - \frac{a}{c} = \frac{a}{2c} + \frac{a}{3c} + \frac{a}{4c} + \frac{a}{5c} + \cdots,$$

and subtracted termwise to get

$$\frac{a}{c} = N - P = \left(\frac{a}{c} - \frac{a}{2c} \right) + \left(\frac{a}{2c} - \frac{a}{3c} \right) + \left(\frac{a}{3c} - \frac{a}{4c} \right) + \cdots$$

$$= \frac{a}{2c} + \frac{a}{6c} + \frac{a}{12c} + \frac{a}{20c} + \cdots. \tag{1}$$

Thus, for $a = c$, he concluded that

$$\frac{1}{2} + \frac{1}{6} + \frac{1}{12} + \frac{1}{20} + \cdots = \frac{1}{1} = 1. \tag{2}$$

Unfortunately, Bernoulli's "proof" required the subtraction of two divergent series, N and P. To his credit, Bernoulli recognized the inherent dangers in his argument, and he advised that this procedure must not be used without caution ("non sine cautela"). To illustrate his point, he applied the previous reasoning to the series

$$S = \frac{2a}{c} + \frac{3a}{2c} + \frac{4a}{3c} + \cdots$$

and

$$T = S - \frac{2a}{c} = \frac{3a}{2c} + \frac{4a}{3c} + \frac{5a}{4c} + \cdots.$$

Upon subtracting termwise, he got

$$\frac{2a}{c} = S - T = \frac{a}{2c} + \frac{a}{6c} + \frac{a}{12c} + \frac{a}{20c} + \cdots, \tag{3}$$

which provided a clear contradiction to (1).

Bernoulli analyzed and resolved this contradiction as follows: the derivation of (1) was valid since the "last" term of series N is zero (that is, $\lim_{k \to \infty} a/(kc) = 0$), whereas the parallel derivation of (3) was invalid since the "last" term of series S is non-zero (because $\lim_{k \to \infty} (k+1)a/(kc) = a/c \neq 0$). In modern terms, he had correctly recognized that, regardless of the convergence or divergence of the series $\sum_{k=1}^{\infty} x_k$, the new series $\sum_{k=1}^{\infty} (x_k - x_{k+1})$ converges to x_1 *provided* $\lim_{k \to \infty} x_k = 0$. Thus, he not only explained the need for "caution" in his earlier discussion but also exhibited a fairly penetrating insight, by the standards of his day, into the general convergence/divergence issue.

Having thus established (2) to his satisfaction, Jakob addressed the harmonic series itself. Using his brother's analysis of the harmonic series, he proclaimed in Section XVI of *Tractatus*:

XVI. *Summa ſeriei inſinita harmonicè progreſſionalium* , $\frac{1}{1} + \frac{1}{2} + \frac{1}{3} + \frac{1}{4} + \frac{1}{5}$ *&c. eſt inſinita.*

He began the argument that "the sum of the infinite harmonic series

$$\frac{1}{1} + \frac{1}{2} + \frac{1}{3} + \frac{1}{4} + \frac{1}{5} \text{ etc.}$$

is infinite" by introducing

$$A = \frac{1}{2} + \frac{1}{3} + \frac{1}{4} + \frac{1}{5} + \frac{1}{6} + \frac{1}{7} + \cdots,$$

which "transformed into fractions whose numerators are 1, 2, 3, 4 etc" becomes

$$\frac{1}{2} + \frac{2}{6} + \frac{3}{12} + \frac{4}{20} + \frac{5}{30} + \frac{6}{42} + \cdots.$$

Using (2), Jakob next evaluated:

$$C = \frac{1}{2} + \frac{1}{6} + \frac{1}{12} + \frac{1}{20} + \cdots = 1$$

$$D = \quad \frac{1}{6} + \frac{1}{12} + \frac{1}{20} + \cdots = C \quad -\frac{1}{2} = 1 - \frac{1}{2} \quad = \frac{1}{2}$$

$$E = \quad\quad \frac{1}{12} + \frac{1}{20} + \cdots = D \quad -\frac{1}{6} = \frac{1}{2} - \frac{1}{6} \quad = \frac{1}{3}$$

$$F = \quad\quad\quad \frac{1}{20} + \cdots = E - \frac{1}{12} = \frac{1}{3} - \frac{1}{12} = \frac{1}{4}$$

$$\vdots \qquad\qquad \vdots \qquad \vdots \qquad\qquad \vdots$$

By adding this array columnwise, and again implicitly assuming that termwise addition of infinite series is permissible, he arrived at

$$C + D + E + F + \cdots = \frac{1}{2} + \left(\frac{1}{6} + \frac{1}{6}\right) + \left(\frac{1}{12} + \frac{1}{12} + \frac{1}{12}\right) + \cdots$$

$$= \frac{1}{2} + \frac{2}{6} + \frac{3}{12} + \frac{4}{20} + \cdots$$

$$= A.$$

On the other hand, upon separately summing the terms forming the extreme left and the extreme right of the arrayed equations above, he got

$$C + D + E + F + \cdots = 1 + \frac{1}{2} + \frac{1}{3} + \frac{1}{4} + \cdots = 1 + A.$$

Hence, $A = 1 + A$. In Jakob's words, "The whole" equals "the part"—that is, the harmonic series $1 + A$ equals its part A—which is impossible for a finite quantity. From this, he concluded that $1 + A$ is infinite.

XVI. *Summa seriei infinitæ harmonicè progressionalium*, $\frac{1}{1} + \frac{1}{2} + \frac{1}{3} + \frac{1}{4} + \frac{1}{5}$ *&c. est infinita.*

Id primus deprehendit Frater: inventa namque per præced. summa seriei $\frac{1}{2} + \frac{1}{6} + \frac{1}{12} + \frac{1}{25} + \frac{1}{30}$, &c. visurus porrò, quid emergeret ex ista serie, $\frac{2}{2} + \frac{3}{2} + \frac{4}{12} + \frac{4}{20} + \frac{5}{30}$, &c. si resolveretur methodo Prop. XIV. collegit propositionis veritatem ex absurditate manifesta, quæ sequeretur, si summa seriei harmonicæ finita statueretur. Animadvertit enim,

Seriem A, $\frac{1}{2} + \frac{1}{3} + \frac{1}{4} + \frac{1}{5} + \frac{1}{6} + \frac{1}{7}$, &c. ∞ (fractionibus singulis in alias, quarum numeratores sunt 1, 2, 3, 4, &c. transmutatis)

seriei B, $\frac{1}{2} + \frac{2}{6} + \frac{3}{12} + \frac{4}{20} + \frac{5}{30} + \frac{6}{42}$, &c. ∞ C+D+E+F, &c.

C. $\frac{1}{2} + \frac{1}{6} + \frac{1}{12} + \frac{1}{25} + \frac{1}{30} + \frac{1}{42}$, &c. ∞ per præc. $\frac{1}{1}$

D. .. $+\frac{1}{6} + \frac{1}{12} + \frac{1}{20} + \frac{1}{30} + \frac{1}{42}$, &c. ∞ C $-\frac{1}{2}$ ∞ $\frac{1}{2}$

E. ... $+\frac{1}{12} + \frac{1}{20} + \frac{1}{30} + \frac{1}{42}$, &c. ∞ D $-\frac{1}{6}$ ∞ $\frac{1}{3}$

F. ..... $+\frac{1}{20} + \frac{1}{30} + \frac{1}{42}$, &c. ∞ E $-\frac{1}{12}$ ∞ $\frac{1}{4}$

&c. ∞   &c.

∞ G; unde sequitur, seriem G ∞ A, totum parti, si summa finita esset.

Ego

Jakob Bernoulli was certainly convinced of the importance of his brother's deduction and emphasized its salient point when he wrote:

> The sum of an infinite series whose final term vanishes perhaps is finite, perhaps infinite.

Obviously, this proof features a naive treatment both of series manipulation and of the nature of "infinity." In addition, it attacks infinite series "holistically" as single entities, without recourse to the modern idea of partial sums. Before getting overly critical of its distinctly 17th-century flavor, however, we must acknowledge that Bernoulli devised this proof a century and a half before the appearance of a truly rigorous theory of series. Further, we can not deny the simplicity and cleverness of his reasoning nor the fact that, if bolstered by the necessary supports of modern analysis, it can serve as a suitable alternative to the standard proof.

Indeed, this argument provides us with an example of the history of mathematics at its best—paying homage to the past yet adding a note of freshness and ingenuity to the modern classroom. Perhaps, in contemplating this work, some of today's students might even come to share a bit of the enthusiasm and wonder that moved Jakob Bernoulli to close his *Tractatus* with the verse [7]

> So the soul of immensity dwells in minutia.
> And in narrowest limits no limits inhere.
> What joy to discern the minute in infinity!
> The vast to perceive in the small, what divinity!

Remark. Jakob Bernoulli, eager to examine other infinite series, soon turned his attention in section XVII of *Tractatus* to

$$1 + \frac{1}{4} + \frac{1}{9} + \frac{1}{16} + \cdots = \sum_{k=1}^{\infty} \frac{1}{k^2}, \tag{4}$$

the evaluation of which "is more difficult than one would expect" ("difficilior est quam quis expectaverit"), an observation that turned out to be quite an understatement. He correctly established the convergence of (4) by comparing it termwise with the greater, yet convergent series

$$1 + \frac{1}{3} + \frac{1}{6} + \frac{1}{10} + \cdots$$
$$= 2\left(\frac{1}{2} + \frac{1}{6} + \frac{1}{12} + \frac{1}{20} + \cdots\right) = 2(1) = 2.$$

But evaluating the sum in (4) was too much for Jakob, who noted rather plaintively

> If anyone finds and communicates to us that which up to now has eluded our efforts, great will be our gratitude.

The evaluation of (4), of course, resisted the attempts of another generation of mathematicians until 1734, when the incomparable Leonhard Euler devised an enormously clever argument to show that it summed to $\pi^2/6$. This result, which Jakob Bernoulli unfortunately did not live to see, surely ranks among the most unexpected and peculiar in all of mathematics. For the original proof, see ([4], pp. 83–85). A modern outline of Euler's reasoning can be found in ([2], pp. 486–487).

REFERENCES

1. Jakob Bernoulli, *Ars Conjectandi*, Basel, 1713.
2. Carl B. Boyer, *A History of Mathematics*, Princeton University Press, 1985.
3. C. H. Edwards, *The Historical Development of the Calculus*, Springer-Verlag, New York, 1979.
4. Leonhard Euler, *Opera Omnia* (1), Vol. 14 (C. Boehm and G. Faber, editors), Leipzig, 1925.
5. Morris Kline, *Mathematical Thought from Ancient to Modern Times*, Oxford University Press, New York, 1972.
6. D. J. Struik (editor), *A Source Book in Mathematics (1200–1800)*, Harvard University Press, 1969.
7. Translated from the Latin by Helen M. Walker, as noted in David E. Smith's *A Source Book in Mathematics*, Dover, New York, 1959, p. 271.

REARRANGING THE ALTERNATING HARMONIC SERIES

C. C. Cowen

Department of Mathematics, Purdue University, West Lafayette, IN 47906

K. R. Davidson

Department of Mathematics, University of Waterloo, Waterloo, Ontario N2L 3G1

R. P. Kaufman

Department of Mathematics, University of Illinois, Urbana, IL 61801

In most elementary courses that include a discussion of infinite series, the instructor states and sometimes proves Riemann's theorem, "A conditionally convergent series can be rearranged to sum to any real number." However, after presenting this theorem, even in an advanced calculus or elementary analysis course, the instructor may get the uneasy feeling that the point of the theorem has been lost, that the issues involved are too subtle for students still afraid of *infinite* series. Perhaps some examples would be helpful—that is the point of this note. We calculate, explicitly, the sums of rearranged alternating harmonic series for a large class of rearrangements. More important, we give an example of a rearrangement that can be summed for freshman calculus students and a technique for adding such series that can be the basis of a set of exercises for more advanced students. The results presented here are old (1883) results of Pringsheim ([3], or see [1, pp. 74–77], [2, pp. 96–98], or [5, p. 25]), but they are not as well known as they should be.

By the alternating harmonic series, we mean the series

$$\sum_{k=1}^{\infty} (-1)^{k+1} k^{-1} = 1 - \frac{1}{2} + \frac{1}{3} - \frac{1}{4} \cdots$$

whose sum is $\ln 2$. We say that a series is a *simple rearrangement* of an alternating series if it is a rearrangement of the series and the subsequence of positive terms and the subsequence of negative terms are in their original order. For example,

$$1 + \frac{1}{3} - \frac{1}{2} + \frac{1}{5} + \frac{1}{7} - \frac{1}{4} + \frac{1}{9} + \frac{1}{11} - \frac{1}{6} \cdots \tag{1}$$

is a simple rearrangement of the alternating harmonic series whereas

$$1 + \frac{1}{7} - \frac{1}{4} + \frac{1}{3} - \frac{1}{2} \cdots$$

is not. If $\sum_{k=1}^{\infty} a_k$ is a simple rearrangement of the alternating harmonic series, let p_n be the number of positive terms in $\{a_1, a_2, \ldots, a_n\}$ and let α denote the asymptotic density of the positive terms in the rearrangement. That is, $\alpha = \lim_{n \to \infty} p_n/n$, if the limit exists. Thus, $\alpha = \frac{1}{2}$ for the unrearranged alternating harmonic series and $\alpha = \frac{2}{3}$ for rearrangement (1) above.

THEOREM 1 [3]. *A simple rearrangement of the alternating harmonic series converges to an extended real number if and only if α, the asymptotic density of the positive terms in the rearrangement, exists. Moreover, the sum of a rearrangement with density α is* $\ln 2 + \frac{1}{2}\ln(\alpha(1 - \alpha)^{-1})$.

We shall give a short proof of this theorem below, but first we work out some special cases in a more naive way.

This paper was presented to the Indiana Section of the MAA on October 27, 1979, by C. C. Cowen.

The simplest example of adding a rearrangement is attributed by Manning to Laurent [2, page 98]. Laurent's rearrangement is

$$1 - \frac{1}{2} - \frac{1}{4} + \frac{1}{3} - \frac{1}{6} - \frac{1}{8} + \frac{1}{5} - \frac{1}{10} - \frac{1}{12} + \cdots \tag{2}$$

where $\alpha = \frac{1}{3}$. One easily justifies inserting parentheses to get

$$\left(1 - \frac{1}{2}\right) - \frac{1}{4} + \left(\frac{1}{3} - \frac{1}{6}\right) - \frac{1}{8} + \left(\frac{1}{5} - \frac{1}{10}\right) - \frac{1}{12} \cdots = \frac{1}{2}\left(1 - \frac{1}{2} + \frac{1}{3} - \frac{1}{4} \cdots\right),$$

so the sum of series (2) is $\frac{1}{2}\ln 2$. The trick of inserting parentheses makes this an attractive example, but generalizing it to find sums of other rearrangements is more difficult.

Rearrangement (1) can be handled using power series in the same way that one uses the Taylor series for $\ln(1+x)$ to show $\sum_{k=1}^{\infty}(-1)^{k+1}k^{-1} = \ln 2$. Let

$$f(x) = x + \frac{x^3}{3} - \frac{x^4}{2} + \frac{x^5}{5} + \frac{x^7}{7} - \frac{x^8}{4} \cdots.$$

Elementary estimates show that series (1) converges and we conclude by Abel's theorem [4, Theorem 8.2, page 160] that its sum is $\lim_{x\to 1^-} f(x)$. Since

$$f(x) = \frac{1}{2}\left[\ln(1+x) - \ln(1-x) + \ln(1-x^4)\right] = \frac{1}{2}\ln\left[(1+x)^2(1+x^2)\right],$$

the sum of rearrangement (1) is $\frac{3}{2}\ln 2$. More generally, this technique works for rearrangements in which blocks of n positive terms alternate with blocks of m negative terms. For this case one uses

$$f(x) = \frac{1}{2}\left[\ln(1+x^m) - \ln(1-x^m) + \ln(1-x^{2n})\right]$$

$$= \frac{1}{2}\ln\left[(1+x^m)(1+x^n)(1+x+\cdots+x^{n-1})(1+x+\cdots+x^{m-1})^{-1}\right].$$

Computations of this sort make interesting exercises because their rigorous analysis requires several standard techniques from the theory of power series. (In fact, it is possible to prove Theorem 1 from this by noting that the sum of a rearrangement is an increasing function of the asymptotic density α.)

Proof of Theorem 1. Suppose $\sum_{k=1}^{\infty} a_k$ is a simple rearrangement of the alternating harmonic series. Let p_n be as above and $q_n = n - p_n$ so that

$$\sum_{k=1}^{n} a_k = \sum_{j=1}^{p_n} (2j-1)^{-1} - \sum_{j=1}^{q_n} (2j)^{-1}.$$

For each positive integer n, let $E_n = (\sum_{k=1}^{n} k^{-1}) - \ln n$. The sequence $(E_n)_{n=1}^{\infty}$ is a decreasing sequence of positive numbers whose limit γ is called Euler's constant.
Now

$$\sum_{j=1}^{q_n} (2j)^{-1} = \frac{1}{2}\sum_{j=1}^{q_n} j^{-1} = \frac{1}{2}\ln q_n + \frac{1}{2}E_{q_n},$$

and

$$\sum_{j=1}^{p_n} (2j-1)^{-1} = \sum_{j=1}^{2p_n} j^{-1} - \sum_{j=1}^{p_n} (2j)^{-1} = \ln(2p_n) + E_{2p_n} - \frac{1}{2}\ln p_n - \frac{1}{2}E_{p_n}.$$

Thus

$$\lim_{n \to \infty} \sum_{k=1}^{n} a_k = \lim_{n \to \infty} \left[\ln 2p_n - \frac{1}{2}\ln p_n - \frac{1}{2}\ln q_n + E_{2p_n} - \frac{1}{2}E_{p_n} - \frac{1}{2}E_{q_n} \right]$$

$$= \ln 2 + \lim_{n \to \infty} \frac{1}{2}\ln(p_n q_n^{-1}) + \gamma - \frac{1}{2}\gamma - \frac{1}{2}\gamma,$$

$$= \ln 2 + \frac{1}{2}\ln\left(\lim_{n \to \infty} p_n q_n^{-1} \right).$$

That is, the series converges iff the limit on the right, which is $\ln 2 + \frac{1}{2}\ln(\alpha(1-\alpha)^{-1})$, exists.   ∎

We have seen that the sums of rearrangements of the alternating harmonic series depend only on the asymptotic density α. This behavior is in some sense specific to series like the harmonic series, as Theorem 2 indicates. Readers are invited to construct proofs for themselves or to consult Pringsheim's paper [3].

THEOREM 2 [3]. *Suppose* $\{a_n\}_{n=1}^{\infty}$ *is a sequence of real numbers such that* $|a_1| > |a_2| > |a_3| > \cdots$, $\lim_{n \to \infty} a_n = 0$, *and* $a_{2k-1} > 0 > a_{2k}$ *for* $k = 1, 2, 3, \ldots$.
(i) *If* $\lim_{n \to \infty} n|a_n| = \infty$, *and if* S *is a real number, there is a simple rearrangement of the series* $\sum_{k=1}^{\infty} a_k$ *with asymptotic density* $\frac{1}{2}$ *whose sum is* S.
(ii) *If* $\lim_{n \to \infty} na_n = 0$, *if* $\sum_{k=1}^{\infty} b_k$ *is a simple rearrangement of the series* $\sum_{k=1}^{\infty} a_k$ *for which the asymptotic density* α *exists, and if* $0 < \alpha < 1$, *then* $\sum_{k=1}^{\infty} b_k = \sum_{k=1}^{\infty} a_k$.

The authors gratefully acknowledge support from NSF (Cowen and Kaufman) and NSERC (Davidson).

References

1. T. J. Ia. Bromwich, An Introduction to the Theory of Infinite Series, 2nd ed., Macmillan, London, 1947.
2. H. P. Manning, Irrational Numbers, Wiley, New York, 1906.
3. A. Pringsheim, Über die Werthveränderungen bedingt convergierten Reihe und Producte, Mathematische Annalen, 22 (1883) 455–503.
4. W. Rudin, Principles of Mathematical Analysis, 2nd, ed., McGraw-Hill, New York, 1964.
5. E. T. Whittaker and G. N. Watson, A Course of Modern Analysis, Amer. ed., Cambridge University Press, 1943.

BIBLIOGRAPHIC ENTRIES: SERIES RELATED TO THE HARMONIC SERIES

1. *Monthly* Vol. 88, No. 1, pp. 33–40. Paul Schaefer, Sum-preserving rearrangements of infinite series.
2. *Math. Mag.* Vol. 54, No. 5, pp. 244–246. Richard Beigel, Rearranging terms in alternating series.
3. CMJ Vol. 16, No. 2, pp. 135–138. Fon Brown (student), L. O. Cannon, Joe Elich, and David G. Wright, On rearrangements of the alternating harmonic series.
4. CMJ Vol. 18, No. 1, p. 51. Frank Burk, $\pi/4$ and $\ln 2$ recursively.
5. CMJ Vol. 20, No. 3, pp. 194–200. Arthur C. Segal, Subharmonic series.
6. CMJ Vol. 20, No. 5, pp. 433–435. David P. Kraines, Vivian Y. Kraines, and David A. Smith, Sum the alternating harmonic series.

(c)

SUMS OF SPECIAL SERIES

A SIMPLE PROOF OF THE FORMULA $\sum\limits_{k=1}^{\infty} k^{-2} = \pi^2/6$

IOANNIS PAPADIMITRIOU, Athens, Greece

Start with the inequality $\sin x < x < \tan x$ for $0 < x < \pi/2$, take reciprocals, and square each member to obtain

$$\cot^2 x < 1/x^2 < 1 + \cot^2 x.$$

Now put $x = k\pi/(2m + 1)$ where k and m are integers, $1 \leq k \leq m$, and sum on k to obtain

$$(1) \qquad \sum_{k=1}^{m} \cot^2 \frac{k\pi}{2m+1} < \frac{(2m+1)^2}{\pi^2} \sum_{k=1}^{m} \frac{1}{k^2} < m + \sum_{k=1}^{m} \cot^2 \frac{k\pi}{2m+1}.$$

But since we have

$$(2) \qquad \sum_{k=1}^{m} \cot^2 \frac{k\pi}{2m+1} = \frac{m(2m-1)}{3},$$

(a proof of (2) is given below) relation (1) gives us

$$\frac{m(2m-1)}{3} < \frac{(2m+1)^2}{\pi^2} \sum_{k=1}^{m} \frac{1}{k^2} < m + \frac{m(2m-1)}{3}.$$

Multiply this relation by $\pi^2/(4m^2)$ and let $m \to \infty$ to obtain

$$\lim_{m \to \infty} \sum_{k=1}^{m} \frac{1}{k^2} = \frac{\pi^2}{6}.$$

Proof of (2). By equating imaginary parts in the formula

$$\cos n\theta + i \sin n\theta = (\cos \theta + i \sin \theta)^n = \sin^n\theta(\cot \theta + i)^n$$

$$= \sin^n\theta \sum_{k=0}^{n} \binom{n}{k} i^k \cot^{n-k}\theta,$$

428

we obtain the trigonometric identity

$$\sin n\theta = \sin^n\theta \left\{ \binom{n}{1}\cot^{n-1}\theta - \binom{n}{3}\cot^{n-3}\theta + \binom{n}{5}\cot^{n-5}\theta - + \cdots \right\}.$$

Take $n = 2m + 1$ and write this in the form

(3) $$\sin(2m + 1)\theta = \sin^{2m+1}\theta P_m(\cot^2\theta) \text{ with } 0 < \theta < \frac{\pi}{2},$$

where P_m is the polynomial of degree m given by

$$P_m(x) = \binom{2m + 1}{1}x^m - \binom{2m + 1}{3}x^{m-1} + \binom{2m + 1}{5}x^{m-2} - + \cdots .$$

Since $\sin\theta \neq 0$ for $0 < \theta < \pi/2$, equation (3) shows that $P_m(\cot^2\theta) = 0$ if and only if $(2m + 1)\theta = k\pi$ for some integer k. Therefore $P_m(x)$ vanishes at the m distinct points $x_k = \cot^2 \pi k/(2m + 1)$ for $k = 1, 2, \cdots, m$. These are all the zeros of $P_m(x)$ and their sum is

$$\sum_{k=1}^{m} \cot^2 \frac{\pi k}{2m + 1} = \binom{2m + 1}{3} \Big/ \binom{2m + 1}{1} = \frac{m(2m - 1)}{3},$$

which proves (2).

NOTE. This paper was translated from a Greek manuscript and communicated to the MONTHLY on behalf of the author by Tom M. Apostol, California Institute of Technology. After this paper was written it was learned that the same proof was discovered independently and published in Norwegian by Finn Holme in Nordisk Matematisk Tidskrift, vol. 18 (1970), pp. 91–92. See also A. M. Yaglom and I. M. Yaglom, *Challenging mathematical problems with elementary solutions,* vol. II, Holden-Day, San Francisco, 1967, problem 145.

APPLICATION OF A MEAN VALUE THEOREM FOR INTEGRALS
TO SERIES SUMMATION

Eberhard L. Stark

This note will give an answer to a query by M. Spivak [1, p. 237]. In his popular *Calculus*, Problem 13*29 reads: "Suppose that g is increasing on $[a,b]$ and that f is integrable on $[a,b]$. Prove that there is a number ξ in $[a,b]$ such that

$$\int_a^b f(t)g(t)\,dt = g(a)\int_a^\xi f(t)\,dt + g(b)\int_\xi^b f(t)\,dt."$$ (1)

It is accompanied by the remark: "The result of the problem, which I have cribbed from several older calculus texts, is called The Third Mean Value Theorem for Integrals. To be quite frank, I haven't the slightest idea what it is good for."

Indeed, (1) is good for a rather elementary proof of

$$\sum_{k=1}^\infty \frac{1}{k^2} = \frac{\pi^2}{6}$$ (2)

which would avoid, e.g., convergence theory for Fourier series ([2, p. 381, 410; 3, p. 531; 4, p. 480]). For the Dirichlet kernel the two representations

$$D_n(x) = \frac{1}{2} + \sum_{k=1}^n \cos kx = \frac{\sin(2n+1)(x/2)}{2\sin(x/2)} \qquad (x \in \mathbf{R}, n \in \mathbf{N})$$ (3)

are well known. Set $M_n = \int_0^\pi t D_n(t)\,dt$; by employing the polynomial representation of (3), partial integration results in

$$M_{2m-1} = 2\left\{ \frac{\pi^2}{8} - \sum_{k=1}^m \frac{1}{(2k-1)^2} \right\} \qquad (m \in \mathbf{N}).$$ (4)

On the other hand, the closed representation of (3) leads to

$$M_{2m-1} = \int_0^\pi \left(\frac{t/2}{\sin(t/2)} \right) \sin(4m-1)\frac{t}{2}\,dt$$

$$= 2\left\{ 1 + \left(\frac{\pi}{2} - 1 \right)\cos(4m-1)\frac{\xi}{2} \right\} \frac{1}{4m-1} \qquad (0 < \xi < \pi)$$

$$= O(1/m) \qquad (m \to \infty),$$ (5)

where, in the second step, (1) has been applied for the functions $f(x) = \sin((4m-1)x/2) \in C[0,\pi]$, $g(x) = (x/2)/\sin(x/2)$, with $g(0) = 1$, $g(\pi) = \pi/2$. A combination of (5) and (4) yields

$$\sum_{k=1}^\infty \frac{1}{(2k-1)^2} = \frac{\pi^2}{8};$$

the value of the sum (2) is now immediate.

For the first part of the proof, i.e., (3) implies (4), see [5]; however, the representation ([5, (2)]) for $D_n(x)$, though being "easily verified by induction and routine trigonometric identities," is wrong throughout that note. (For a simple proof, without induction, see, e.g., [2, p. 405; 3, p. 538; 4, p. 470].)

As a matter of fact, it should be mentioned that in textbooks (1) or slightly altered versions are generally called *The Second Mean Value Theorem* (e.g., [3, p. 130, 153; 4, p. 95]) or, more precisely, *The Weierstrass Form of Bonnet's Theorem* ([2, p. 163]). In addition, there are even sharper versions of Bonnet's theorem ([2, p. 163, Cor. 4.2; 3, p. 153, Ex. 26]), for which the first integral on the right-hand side of (1) may be cancelled, thus reducing the actual calculations in (5) to a minimum. Further useful references are, e.g., [6, p. 189, 217 (giving an application to the proof of Taylor's formula); 7, p. 304; 8, p. 214; 9, p. 325; and 10, p. 618, 666 (pointing out some historical sources)]. For a somewhat related proof, compare [11], and for further elementary proofs of (2), see [12] and the literature cited there.

Editor's note: The second mean value theorem is a bread-and-butter lemma in theories like trigonometric series, Laplace transforms, etc. See, for example, *Zygmund Trigonometric Series*, 2nd ed., vol. 1, p. 58 (Cambridge, 1959), Widder, *The Laplace Transform* (Princeton, 1941), p. 65.

References

1. M. Spivak, Calculus, Benjamin, New York–Amsterdam, 1967.
2. D. V. Widder, Advanced Calculus, Prentice-Hall, Englewood Cliffs, N.J., 1961.
3. J. M. H. Olmsted, Advanced Calculus, Appleton-Century-Crofts, New York, 1961.
4. W. Fulks, Advanced Calculus, Wiley, New York–London, 1961.
5. D. P. Giesy, Still another elementary proof that $\Sigma 1/k^2 = \pi^2/6$, Math. Mag., 45 (1972) 148–149.
6. R. R. Goldberg, Methods of Real Analysis, Wiley, New York–Toronto–London, 1964.
7. R. G. Bartle, The Elements of Real Analysis, Wiley, New York–London–Sydney, 1964.
8. T. M. Apostol, Mathematical Analysis, Addison-Wesley, Reading, Mass., 1957.
9. G. H. Hardy, A Course of Pure Mathematics, Cambridge Univ. Press, New York, 1963 (orig. ed., 1908).
10. E. W. Hobson, The Theory of Functions of a Real Variable, Cambridge Univ. Press, New York, 1957 (orig. ed., Cambridge, 1907).
11. E. L. Stark, Another proof of the formula $\Sigma 1/k^2 = \pi^2/6$, this MONTHLY, 76 (1969) 552–553.
12. T. M. Apostol, Another elementary proof of Euler's formula for $\zeta(2n)$, this MONTHLY, 80 (1973) 425–431.

RHEINISCH-WESTFÄLISCHE TECHNISCHE HOCHSCHULE AACHEN, LEHRSTUHL A FÜR MATHEMATIK, TEMPLERGRABEN 55, D 5100 AACHEN, BUNDESREPUBLIK DEUTSCHLAND.

SUMMING POWER SERIES WITH POLYNOMIAL COEFFICIENTS

JOHN KLIPPERT

*Department of Mathematics and Computer Science, James Madison University,
Harrisonburg, Virginia 22807*

Third semester calculus students quickly learn that

$$\sum_{n=0}^{\infty} x^n = \frac{1}{1-x} \quad \text{for} \quad |x| < 1$$

by showing that the sequence of partial sums

$$S_k = \sum_{n=0}^{k} x^n = \frac{1 - x^{k+1}}{1-x}, \quad x \neq 1.$$

But by rewriting a power series in terms of finite differences, we can almost as easily sum *any* power series with polynomial coefficients. Consider the series $f(x) = \sum_{n=0}^{\infty} a_n x^n$ with positive radius of convergence. Then

$$f(x) = a_0 + x\phi(x) \quad \text{where} \quad \phi(x) = \sum_{n=1}^{\infty} a_n x^{n-1} = \sum_{n=0}^{\infty} a_{n+1} x^n.$$

So

$$(1-x)\phi(x) = \sum_{n=0}^{\infty} a_{n+1} x^n - x\left(\sum_{n=1}^{\infty} a_n x^{n-1} \right)$$

$$= \sum_{n=0}^{\infty} a_{n+1} x^n - \sum_{n=1}^{\infty} a_n x^n = \sum_{n=0}^{\infty} a_{n+1} x^n - \sum_{n=0}^{\infty} a_n x^n + a_0$$

$$= a_0 + \sum_{n=0}^{\infty} (a_{n+1} - a_n) x^n = a_0 + \sum_{n=0}^{\infty} \Delta a_n x^n,$$

with $\Delta a_n = a_{n+1} - a_n, \quad n = 0, 1, 2, \ldots$.

Therefore if $x \neq 1$ and $x \in D$, the domain of f, we have

$$\phi(x) = \frac{a_0}{1-x} + \frac{1}{1-x} \sum_{n=0}^{\infty} \Delta a_n x^n.$$

Thus

$$f(x) = a_0 + x\phi(x) = a_0 + \frac{a_0 x}{1-x} + \frac{x}{1-x} \sum_{n=0}^{\infty} \Delta a_n x^n.$$

That is,

$$(1) \qquad \sum_{n=0}^{\infty} a_n x^n = \frac{a_0}{1-x} + \frac{x}{1-x} \sum_{n=0}^{\infty} \Delta a_n x^n, \quad x \neq 1, \quad x \in D.$$

Putting $\Delta^2 a_n = \Delta(\Delta a_n) = \Delta a_{n+1} - \Delta a_n$ (and inductively,

$$\Delta^p a_n = \Delta(\Delta^{p-1} a_n) = \Delta^{p-1} a_{n+1} - \Delta^{p-1} a_n, \quad p \in N)$$

and applying (1) to $\sum_{n=0}^{\infty} \Delta a_n x^n$, we have

$$\sum_{n=0}^{\infty} \Delta a_n x^n = \frac{\Delta a_0}{1-x} + \frac{x}{1-x} \sum_{n=0}^{\infty} \Delta^2 a_n x^n$$

432

and so

$$\sum_{n=0}^{\infty} a_n x^n = \frac{a_0}{1-x} + \frac{x}{1-x}\left(\frac{\Delta a_0}{1-x} + \frac{x}{1-x}\sum_{n=0}^{\infty}\Delta^2 a_n x^n\right)$$

$$= \frac{a_0}{1-x} + \frac{x\Delta a_0}{(1-x)^2} + \left(\frac{x}{1-x}\right)^2 \sum_{n=0}^{\infty}\Delta^2 a_n x^n.$$

Repeating p times in succession, we get

$$\sum_{n=0}^{\infty} a_n x^n = \frac{a_0}{1-x} + \frac{x\Delta a_0}{(1-x)^2} + \cdots + \frac{x^{p-1}\Delta^{p-1}a_0}{(1-x)^p} + \left(\frac{x}{1-x}\right)^p \sum_{n=0}^{\infty}\Delta^p a_n x^n.$$

That is,

$$(2) \qquad \sum_{n=0}^{\infty} a_n x^n = \sum_{k=0}^{p-1}\Delta^k a_0 \frac{x^k}{(1-x)^{k+1}} + \left(\frac{x}{1-x}\right)^p \sum_{n=0}^{\infty}\Delta^p a_n x^n, \quad x \neq 1, x \in D$$

where we define $\Delta^0 a_0 = a_0$.

Now if $a_n = P(n)$, a polynomial of degree $\leqslant p - 1$, we have

$$\Delta^p P(n) = 0, \quad n = 0, 1, 2, \ldots .$$

Observing that the radius of convergence of $\sum_{n=0}^{\infty} P(n)x^n$ is 1, we have according to (2) that

$$(3) \qquad \sum_{n=0}^{\infty} P(n)x^n = \sum_{k=0}^{p-1}\Delta^k P(0) \frac{x^k}{(1-x)^{k+1}}, \quad |x| < 1.$$

For example, if $P(n) = n^2 + n + 1$, then $\Delta a_n = 2n + 1$, $\Delta^2 a_n = 2$ and $\Delta^3 a_n = 0$. In particular, $P(0) = 1, \Delta P(0) = 2, \Delta^2 P(0) = 2$. Putting $p = 3$ into (3), we have

$$\sum_{n=0}^{\infty} P(n)x^n = \sum_{k=0}^{2}\Delta^k P(0) \frac{x^k}{(1-x)^{k+1}}$$

and so

$$\sum_{n=0}^{\infty} (n^2 + n + 1)x^n = P(0)\frac{1}{1-x} + \Delta P(0)\frac{x}{(1-x)^2} + \Delta^2 P(0)\frac{x^2}{(1-x)^3}$$

$$= \frac{1}{1-x} + \frac{2x}{(1-x)^2} + \frac{2x^2}{(1-x)^3}.$$

Finally, we remark that the transform (2) can also be used to accelerate the convergence of a series when $\Delta^p a_n$ converges to 0 faster than a_n (assuming that a_n *does* converge to 0).

Reference

1. Konrad Knopp, Theory and Application of Infinite Series, Blackie and Son Limited, London, 1966, pp. 230–273.

A GEOMETRIC PROOF OF THE FORMULA FOR ln 2

FRANK KOST, SUNY at Oneonta

Consider $f(x) = (1/x)$ defined on $[1, 2]$. The area under this curve and bounded by the lines $y = 0$, $x = 1$, $x = 2$ is $\int_1^2 (1/x)\,dx = \ln 2$.

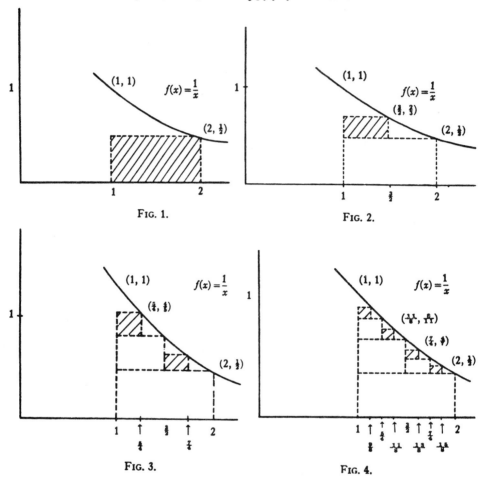

FIG. 1. FIG. 2.

FIG. 3. FIG. 4.

Our first approximation to this area is $A_1 = \frac{1}{2} = 1 - \frac{1}{2}$. See Figure 1. Next (Figure 2) we introduce the point $\frac{3}{2}$ and obtain a rectangle whose area $A_2 = [f(\frac{3}{2}) - f(2)]\frac{1}{2} = \frac{1}{3} - \frac{1}{4}$. In step 3 (Figure 3) we introduce the points $\frac{5}{4}$, $\frac{7}{4}$ and form two rectangles with total area

$$A_3 = \left[f\left(\frac{5}{4}\right) - f\left(\frac{3}{2}\right)\right]\frac{1}{4} + \left[f\left(\frac{7}{4}\right) - f(2)\right]\frac{1}{4} = \frac{1}{5} - \frac{1}{6} + \frac{1}{7} - \frac{1}{8}.$$

In step 4 (Figure 4) add points $\frac{9}{8}$, $\frac{11}{8}$, $\frac{13}{8}$, $\frac{15}{8}$ and obtain four rectangles with

434

combined area

$$A_4 = \left[f\left(\frac{9}{8}\right) - f\left(\frac{5}{4}\right) \right] \frac{1}{8} + \left[f\left(\frac{11}{8}\right) - f\left(\frac{3}{2}\right) \right] \frac{1}{8}$$

$$+ \left[f\left(\frac{13}{8}\right) - f\left(\frac{15}{8}\right) \right] \frac{1}{8} + \left[f\left(\frac{15}{8}\right) - f(2) \right] \frac{1}{8}$$

$$= \frac{1}{9} - \frac{1}{10} + \frac{1}{11} - \frac{1}{12} + \frac{1}{13} - \frac{1}{14} + \frac{1}{15} - \frac{1}{16}.$$

At the nth step $(n>1)$ 2^{n-2} points and 2^{n-2} rectangles are introduced with combined area

$$A_n = \frac{1}{2^{n-1}} \sum_{k=2^{n-1}+1}^{2^n} f\left(\frac{k}{2^{n-1}}\right)(-1)^{k+1} = \sum_{k=2^{n-1}+1}^{2^n} \frac{(-1)^{k+1}}{k}.$$

The rectangles are disjoint and their union is the region under the curve. Hence $A_1+A_2+ \cdots +A_n+ \cdots = \sum_{k=1}^{\infty}(-1)^{k+1}/k = \ln 2$.

On Sum-Guessing

Mangho Ahuja

Mangho Ahuja is Associate Professor of Mathematics at Southeast Missouri State University. He received his Ph.D. in Mathematics from the University of Colorado in 1971. His area of specialization is analysis.

Can you guess an approximate value of the sum

$$S = 1 + \frac{1}{\sqrt{2}} + \frac{1}{\sqrt{3}} + \cdots + \frac{1}{\sqrt{10^4}} \ ?$$

The actual value of S happens to be between 198 and 199. Of course this is a crude estimate, but it is far better than a wild guess. The trick, or the mathematics involved here, is an algebraic inequality which says the following.

Theorem 1. *For every positive integer n,*

$$2\sqrt{n+1} - 2\sqrt{n} < \frac{1}{\sqrt{n}} < 2\sqrt{n} - 2\sqrt{n-1} \ .$$

In this paper we will explore some results of inequalities which will help us in 'guessing' the sum. We will also see how the concepts of 'sum' and 'integral' are related in the case of positive, continuous, and monotonic functions.

First we will give an algebraic proof of Theorem 1. Later we will see that Theorem 1 is a special case of a more general theorem (Theorem 3).

Proof of Theorem 1. The inequality on the left side is obtained by noting that $4n(n+1) < (2n+1)^2$, taking the positive square root, and then dividing by $\sqrt{n}$. The inequality on the right side is obtained if we start with $4n(n-1) < (2n-1)^2$, and follow the same route.

Using Theorem 1 for $n = 1, 2, 3, \ldots, 10^4$,

$$
\begin{array}{lcl}
2(\sqrt{2} - 1) & < 1 < & 2 \qquad\qquad\qquad\qquad (1)\\[4pt]
2(\sqrt{3} - \sqrt{2}) & < \dfrac{1}{\sqrt{2}} < & 2(\sqrt{2} - 1) \\[10pt]
2(\sqrt{4} - \sqrt{3}) & < \dfrac{1}{\sqrt{3}} < & 2(\sqrt{3} - \sqrt{2}) \\[6pt]
\cdots & \cdots & \cdots \\[4pt]
2(\sqrt{10001} - \sqrt{10^4}) & < \dfrac{1}{\sqrt{10^4}} < & 2(\sqrt{10^4} - \sqrt{9999}).
\end{array}
$$

436

On adding we get $2(\sqrt{10001} - 1) < S < 2(\sqrt{10^4})$. However, the right side of (1) could be improved to

$$2(\sqrt{2} - 1) < 1 \leqslant 1. \tag{1'}$$

Using (1') instead of (1) in the summation we get,

$$2(\sqrt{10001} - 1) < S < 2\sqrt{10^4} - 1.$$

Thus $2(100 - 1) < S < 200 - 1$, or S lies between 198 and 199.

The above method can be easily generalized to a formula where the sum goes from $n = 1$ to $n = m^2$ for some positive integer m.

Corollary 1. *For every positive integer m,*

$$2m - 2 < \sum_{n=1}^{n=m^2} \frac{1}{\sqrt{n}} < 2m - 1.$$

Let T and U be the sums

$$T = 1 + \frac{1}{\sqrt[3]{2}} + \frac{1}{\sqrt[3]{3}} + \frac{1}{\sqrt[3]{4}} + \cdots + \frac{1}{\sqrt[3]{10^3}},$$

$$U = 1 + \frac{1}{\sqrt[4]{2}} + \frac{1}{\sqrt[4]{3}} + \frac{1}{\sqrt[4]{4}} + \cdots + \frac{1}{\sqrt[4]{10^4}}.$$

Can you guess the values of T and U? We realize that for each sum we need a different inequality to replace Theorem 1. What we need is a general approach to this problem. We need a method to derive an inequality (like the one in Theorem 1) for $1/\sqrt[3]{n}$, $1/\sqrt[4]{n}$, and in fact for $1/\sqrt[k]{n}$ for every positive integer $k \geqslant 2$. An approach to such a method is shown below. To be specific, we take $k = 3$. It will be easy to see that the method will work for any positive integer $k \geqslant 2$. In particular when $k = 2$, it provides us with an alternate proof of Theorem 1.

Theorem 2. *For every positive integer n,*

$$\frac{3}{2}\left[(n+1)^{\frac{2}{3}} - n^{\frac{2}{3}}\right] < \frac{1}{\sqrt[3]{n}} < \frac{3}{2}\left[n^{\frac{2}{3}} - (n-1)^{\frac{2}{3}}\right].$$

Proof. When $n = 1$, the inequality on the right side is pretty obvious. For all other inequalities, we consider the graph of the function $f(x) = x^{-\frac{1}{3}}$. (See Figure 1.) For each $i = n - 1$, n, $n + 1$, let P_i denote the point on the curve whose x-coordinate is i, and P_iQ_i be the perpendicular to X-axis. Let the horizontal line through P_n meet $P_{n-1}Q_{n-1}$ in R and $P_{n+1}Q_{n+1}$ in S. Area $P_nP_{n+1}Q_{n+1}Q_n$ under the

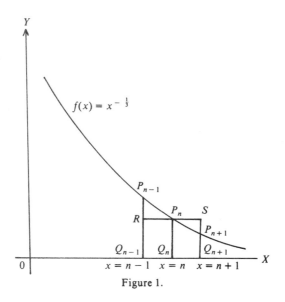

Figure 1.

curve < area of rectangle $P_n SQ_{n+1}Q_n$ = area of rectangle $RP_nQ_nQ_{n-1}$ < area $P_{n-1}P_nQ_nQ_{n-1}$ under the curve. Thus

$$\int_n^{n+1} x^{-\frac{1}{3}}\,dx < n^{-\frac{1}{3}} < \int_{n-1}^n x^{-\frac{1}{3}}\,dx,$$

or

$$\frac{3}{2}\left[(n+1)^{\frac{2}{3}} - n^{\frac{2}{3}}\right] < n^{-\frac{1}{3}} < \frac{3}{2}\left[n^{\frac{2}{3}} - (n-1)^{\frac{2}{3}}\right]$$

and the theorem is proved.

In the general case, for every positive integer $k \geqslant 2$, the function $f(x) = x^{-1/k}$ is monotonically decreasing, and a similar argument will establish the next theorem.

Theorem 3. *For every positive integer $k \geqslant 2$, and for every positive integer n, we have*

$$\frac{k}{k-1}\left[(n+1)^{(k-1)/k} - n^{(k-1)/k}\right] < n^{-1/k} < \frac{k}{k-1}\left[n^{(k-1)/k} - (n-1)^{(k-1)/k}\right].$$

We will now compute the value of sum T using Theorem 2. When $n = 1$,

$$\frac{3}{2}\left[2^{\frac{2}{3}} - 1\right] < 1 < \frac{3}{2}\left[1^{\frac{2}{3}} - 0\right]. \tag{2}$$

As before, the right side of (2) will be improved to

$$\frac{3}{2}\left[2^{\frac{2}{3}}-1\right]<1\leqslant 1. \tag{2'}$$

Now using $n = 2, 3, 4, \ldots, 10^3$ in Theorem 2 and adding, we obtain

$$\frac{3}{2}\left[(1001)^{\frac{2}{3}}-1\right]<T<\frac{3}{2}\left[(1000)^{\frac{2}{3}}-1\right]+1,$$

or

$$\frac{3}{2}\left[(1000)^{\frac{2}{3}}-1\right]<T<\frac{3}{2}\left[(1000)^{\frac{2}{3}}-1\right]+1.$$

Thus the value of T lies between 148.5 and 149.5. The estimation of the value of U is left to the reader.

A careful examination of the computation of S and T will show that we have approximated the 'sum' by two 'integrals'. In fact we can do this for any bounded, monotonic, and real valued function. The next theorem gives us a general view of the situation in the case of a positive monotonically decreasing continuous function.

Theorem 4. *If $f(x)$ is a positive, monotonically decreasing continuous function on* $(0, \infty)$, *then*

$$\int_1^{n+1} f(x)\,dx \leqslant \sum_1^n f(i) \leqslant \int_1^n f(x)\,dx + f(1). \tag{A}$$

Proof. In Figure 2 we see that each rectangle with base $[k, k + 1]$ for $k = 1, 2, 3, \ldots, n$, has height $f(k)$. Let $W = \sum_1^n f(k)$ denote the sum of the areas of these rectangles. Obviously W is greater than the area under the curve given by $\int_1^{n+1} f(x)\,dx$. Therefore $\int_1^{n+1} f(x)\,dx \leqslant \sum_1^n f(i)$. We also see that the value of $\int_1^n f(x)\,dx$ falls short of W by the shaded area. If we now translate the shaded regions moving them parallel to the X-axis till they all touch the Y-axis, we see that the total area of these shaded regions is less than the area of the rectangle on base $[0, 1]$, with height $f(1)$. Thus

$$\sum_1^n f(i) \leqslant \int_1^n f(x)\,dx + f(1).$$

From result (A) we can obtain

$$\sum_1^n f(i) - f(1) \leqslant \int_1^n f(x)\,dx \leqslant \sum_1^{n-1} f(i). \tag{B}$$

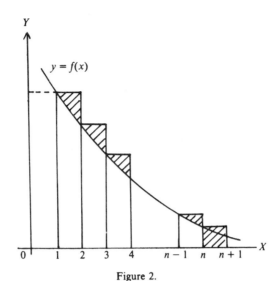

Figure 2.

Note that (A) could also be derived from (B). Both (A) and (B) show how the 'sum' and the 'integral' of a monotonic function are intertwined. In the case where the sum and the integral are thus related either one could be used to find the approximate value of the other. (Perhaps students would remember finding the approximate value of the 'integral' of a monotonic function by computing its 'sum'.) Also, using either (A) or (B), and by taking the limit as n tends to infinity, we observe that the series $\sum_1^\infty f(i)$ converges if and only if the improper integral $\int_1^\infty f(x)\,dx$ converges. Students will recognize this result as the well-known 'Integral Test' which is used to show the convergence of p-series, etc. Almost all text books on calculus use either (A) or (B) to prove the 'Integral Test'.

We hope that playing the game of 'guessing' the sum will motivate students to learn and derive simple inequalities which usually appear dull and meaningless. Results (A) and (B) are examples of the many nice properties which 'monotonic' functions possess. In this paper we have attempted to 'bring to life' the inequality in result (A), by making its repeated use in 'guessing' the sum.

The Sum is One

John H. Mathews, California State University Fullerton, Fullerton, CA 92634

Many calculus texts contain the following telescoping series.

$$\sum_{n=1}^{\infty} \frac{1}{n(n+1)} = \sum_{n=1}^{\infty} \left(\frac{1}{n} - \frac{1}{n+1} \right) = 1. \tag{1}$$

An interesting geometric realization of this result can be obtained from the fact that the family of curves $\{x^n\}$ partitions the unit square into an infinite number of regions with total area 1. First, find the area between the curves $y = x^{n-1}$ and $y = x^n$ over the interval $[0, 1]$:

$$\int_0^1 (x^{n-1} - x^n) \, dx = \int_0^1 x^{n-1} \, dx - \int_0^1 x^n \, dx = \frac{1}{n} - \frac{1}{n+1} = \frac{1}{n(n+1)}. \tag{2}$$

Summing each side of (2) produces the result (1).

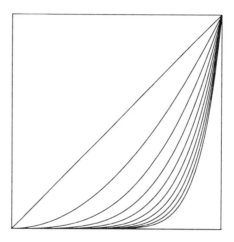

BIBLIOGRAPHIC ENTRIES: SUMS OF SPECIAL SERIES

1. *Monthly* Vol. 80, No. 4, pp. 425–431. Tom M. Apostol, Another elementary proof of Euler's formula for $\zeta(2n)$.

2. *Monthly* Vol. 86, No. 8, pp. 637–648. Richard Johnsonbaugh, Summing an alternating series.

3. *Monthly* Vol. 93, No. 5, pp. 387–389. M. D. Atkinson, How to compute the series expansions of sec x and tan x.

4. *Monthly* Vol. 94, No. 6, pp. 541–542. Mark Finkelstein, $\ln 2 = 1 - \frac{1}{2} + \frac{1}{3} - \frac{1}{4} + \cdots$ (proof by exhaustion).

> The same idea is used in an earlier paper, *Math. Mag.* Vol. 44, No. 1, pp. 37–38, reproduced above on pp. 434–435.

5. *Monthly* Vol. 94, No. 7, pp. 662–663. Boo Rim Choe, An elementary proof of $\sum_{n=1}^{\infty} 1/n^2 = \pi^2/6$

6. *Math. Mag.* Vol. 45, No. 3, pp. 148–149. Daniel P. Giesy, Still another elementary proof that $\sum_{k=1}^{\infty} 1/k^2 = \pi^2/6$.

> This is essentially the standard proof via Fourier series; what makes it 'elementary' is that it treats the nth partial sum of the Fourier series throughout, then lets $n \to \infty$ at the end.

7. *Math. Mag.* Vol. 47, No. 4, pp. 197–202. E. L. Stark, The series $\sum_{k=1}^{\infty} k^{-s}$, $s = 2, 3, 4, \ldots$ once more.

8. *Math. Mag.* Vol. 48, No. 3, pp. 148–154. Bruce C. Berndt, Elementary evaluation of $\zeta(2n)$.

9. *Math. Mag.* Vol. 63, No. 5, pp. 291–306. Ranjan Roy, The discovery of the series formula for π by Leibniz, Gregory and Nilakantha.

10. TYCMJ Vol. 12, No. 5, pp. 306–308. A. J. Ceasar, The Saint Petersburg paradox and some related series.

11. TYCMJ Vol. 14, No. 2, pp. 118–122. Arthur C. Segal, Closed-form formulas for quasigeometric series.

12. CMJ Vol. 15, No. 4, pp. 334–338. R. A. Mureika and R. D. Small, An almost correct series.

> A partial explanation of errors that result from truncating a series for $\pi/4$.

13. CMJ Vol. 18, No. 3, pp. 223–225. Norman Schaumberger, Extending the series for $\ln 2$.

14. CMJ Vol. 19, No. 2, pp. 149–153. James D. Harper, Estimating the sum of alternating series.

15. CMJ Vol. 19, No. 3, pp. 252–253. Leonard Gillman, More on the series for $\ln 2$.

16. CMJ Vol. 20, No. 4, pp. 329–331. Alan Gorfin, Evaluating the sum of the series $\sum_{k=1}^{\infty} k^j/M^k$.

12

SPECIAL NUMBERS
(a)

e

A DISCOVERY APPROACH TO *e*

J. P. TULL, The Ohio State University

While teaching the first year course in mathematics at the University of Zambia in 1970 and 1971, I came up with the following approach to exponential and logarithmic functions. It is perhaps novel in one small way. Namely, once we had defined the exponential and logarithmic functions to an arbitrary base, starting with the usual $a^{p/q} = \sqrt[q]{a^p}$, we eventually came to the question of the derivative. As usual we found

$$(a^{x+h} - a^x)/h = a^x(a^h - 1)/h$$

and so we needed only find the derivative at 0.

We noticed, by looking at graphs, that the derivative at 0 increases as a increases, and there are large values and small values. There ought to be one particular base, call it e, for which this derivative is 1. (Needless to say, we knew very little about continuous functions at this stage.)

For this base e, since the graph of log is the reflection in $y = x$ of the graph of exp, then $\log' 1 = 1$. This means that

(*) $(1/\delta) \log(1 + \delta) \to 1$

as $\delta \to 0$. We readily find that $\log' x = 1/x$ by the direct approach, using (*). But also from (*) we see that as $\delta \to 0$

$$e^{(1/\delta)\log(1 + \delta)} \to e^1 = e;$$

i.e., $(1 + \delta)^{1/\delta} \to e$ as $\delta \to 0$. Thus $e = \lim_{n \to \infty} (1 + 1/n)^n = \lim_{n \to \infty} (1 - 1/n)^{-n}$.

443

SIMPLE PROOFS OF TWO ESTIMATES FOR e

R. B. DARST, Colorado State University

Let $a_n = [1 + (1/n)]^{(n+1)}$, $b_n = [1 - (1/n)]^n$, and $c_n = [1 + (1/n)]^n$, $n = 1,2,\cdots$

In elementary calculus classes one frequently establishes (C): the sequence $\{c_n\}$ increases $(c_n < c_{n+1})$; sometimes one also shows (A): $\{a_n\}$ decreases, and (B): $\{b_n\}$ increases. Then $\lim a_n = \lim c_n$, and one can show that this common value is e.

We shall give simple proofs of (A), (B) and (C) based on the fact that $(1 + x)^{(n+1)} > 1 + (n + 1)x$ when $x > 0$ and n is a positive integer. Thus, to establish (A), notice that

$$(a_n/a_{n+1}) = [(n + 1)/n]^{(n+1)}/\{[(n + 2)/(n + 1)]^{(n+1)}[1 + 1/(n + 1)]\}$$

$$= [(n + 1)^2/(n^2 + 2n)]^{(n+1)}/[1 + 1/(n + 1)]$$

$$= [1 + 1/(n^2 + 2n)]^{(n+1)}/[1 + 1/(n + 1)]$$

$$> [1 + 1/(n + 1)^2]^{(n+1)}/[1 + 1/(n + 1)] > 1.$$

Since $b_{(n+1)} = 1/a_n$, (B) is also established. Finally,

$$(c_{n+1}/c_n) = [(n + 2)/(n + 1)]^{n+1}/\{[(n + 1)/n]^{n+1}[n/(n + 1)]\}$$

$$= [(n^2 + 2n)/(n + 1)^2]^{n+1}/[1 - 1/(n + 1)]$$

$$= [1 - 1/(n + 1)^2]^{n+1}/[1 - 1/(n + 1)]$$

$$= [b_{(n+1)^2}/b_{(n+1)}]^{1/(n+1)} > 1.$$

Which is Larger, e^π or π^e?

IVAN NIVEN

IVAN NIVEN is Professor of Mathematics at the University of Oregon, Eugene, Oregon, where he has been a faculty member since 1947. Before this he taught at the University of Illinois and at Purdue University. He has published numerous articles and six books, including *Numbers, Rational and Irrational, Mathematics of Choice*, and Calculus: An Introductory Approach, *which are related to the two-year college program.*

The answer is e^π, as will be evident from the discussion below. The question raised in the title is very special, but it clearly can be extended to other cases, such as 7^9 versus 9^7, or in general, which is larger, α^β or β^α, for given positive real numbers α and β. Although any such question can be settled in specific numerical cases by calculation, there are some general results that can be stated. These results can be established by a simple application of elementary calculus. All that one needs to know is that the derivative of log x is $1/x$. (We will write the natural logarithm as log x, whereas many calculus books use the notation ln x.)

The logarithmic function enters the question rather readily, because the inequality $\alpha^\beta > \beta^\alpha$ becomes, on taking logs, $\beta \log \alpha > \alpha \log \beta$, or dividing by the positive number $\alpha\beta$, $(\log \alpha)/\alpha > (\log \beta)/\beta$. This analysis suggests that we should look at the function $(\log x)/x$. It also suggests that we discuss the logic involved in taking logs in an inequality.

To deal with the latter question, we observe that for positive values of x the derivative of log x is $1/x$, which is also positive, so that log x increases with x. This implies that *if u and v are positive real numbers then $u > v$ implies* log $u >$ log v *and conversely*. Thus we are justified in taking logs in an inequality; to prove that $e^\pi > \pi^e$, for example, it suffices to prove that $\pi \log e > e \log \pi$.

Next it is noted that for positive values of x the function $(\log x)/x$ *increases with x in the interval $0 < x < e$, attains a maximum at $x = e$, and then decreases for $x > e$.* To establish this we calculate the derivative of $(\log x)/x$ to get $(1 - \log x)/x^2$. This derivative equals zero if log $x = 1$, i.e., if $x = e$. Also, the derivative is positive for positive values of $x < e$, but negative for $x > e$, and this proves the assertion made about $(\log x)/x$ by a simple application of calculus.

Now we are ready to look at the question of α^β versus β^α. To keep matters consistent we shall write all results with $\beta > \alpha$.

THEOREM 1. *If α and β are positive real numbers such that $\beta > \alpha \geq e$, then $\alpha^\beta > \beta^\alpha$.*

We notice that once this theorem is proved, we can settle the question in the title of this paper because $\pi > e$ and hence $e^\pi > \pi^e$. Also the question that was raised about 7^9 versus 9^7 is answered by the inequality $7^9 > 9^7$.

445

To prove the result we notice that since $(\log x)/x$ decreases from its maximum point at $x = e$, it follows that $(\log \alpha)/\alpha > (\log \beta)/\beta$ for $x > e$. Multiplying by the positive quantity $\alpha\beta$ we get

$$\beta \log \alpha > \alpha \log \beta \quad \text{or} \quad \log \alpha^\beta > \log \beta^\alpha.$$

But this implies that $\alpha^\beta > \beta^\alpha$.

THEOREM 2. *If α and β satisfy $e \geqslant \beta > \alpha > 0$, then $\beta^\alpha > \alpha^\beta$.*

The proof of this is similar to that of Theorem 1, because again from the behavior of the function $(\log x)/x$, we conclude that $(\log \beta)/\beta > (\log \alpha)/\alpha$, from which it follows that

$$\alpha \log \beta > \beta \log \alpha, \qquad \log \beta^\alpha > \log \alpha^\beta, \qquad \beta^\alpha > \alpha^\beta.$$

It may be noted that although the assumption $\beta > \alpha$ appears in both Theorems 1 and 2, the conclusions are reversed as inequalities. There is one other result that is easily proved.

THEOREM 3. *If $\beta \geqslant 1$ but $1 > \alpha > 0$, then $\beta^\alpha > \alpha^\beta$.*

This is obvious because $\beta^\alpha \geqslant 1$ whereas $\alpha^\beta < 1$. Now an examination of these three theorems shows that not all cases are covered. In particular, what can we say if $\beta > e$ but $1 < \alpha < e$? The answer is that no simple general rule can be stated. This point is illustrated by the inequalities

$$2^3 < 3^2, \qquad 2^4 = 4^2, \qquad 2^5 > 5^2.$$

However, our earlier analysis of $(\log x)/x$ can be used to draw some conclusions about the graph of $x^y = y^x$ for positive x and y, i.e., the graph in the first quadrant. The function $(\log x)/x$ is positive only if $x > 1$, and as x tends to infinity $(\log x)/x$ tends to zero. This latter assertion can be argued directly by writing x in the form e^u, so that $(\log x)/x$ becomes u/e^u. Replacing e^u by the first three terms of its power series expansion $1 + u + u^2/2$, we see that for positive values of u,

$$\frac{u}{e^u} < \frac{u}{1 + u + u^2/2} < \frac{u}{u + u^2/2} = \frac{2}{2 + u},$$

and $2/(2 + u)$ tends to zero as u tends to infinity.

Thus the graph of the function $f(x) = (\log x)/x$ rises from $f(1) = 0$ at $x = 1$ to $f(e) = e^{-1}$ at $x = e$, then falls as x increases beyond $x = e$ with $f(x)$ approaching 0 in value. Thus for any value of x, say $x = \alpha$, satisfying $1 < x < e$, there is a unique value of $y > e$, say $y = \beta$, such that

$$(\log \alpha)/\alpha = (\log \beta)/\beta, \qquad \beta \log \alpha = \alpha \log \beta,$$

$$\log \alpha^\beta = \log \beta^\alpha, \qquad \alpha^\beta = \beta^\alpha.$$

Also as α decreases over the open interval from e to 1, the corresponding β value increases from e through all positive real numbers greater than e.

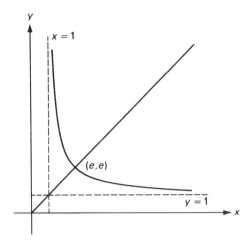

FIG. 1. Graph of $x^y = y^x$ in the first quadrant

Now the graph of $x^y = y^x$ in the first quadrant, shown in FIGURE 1, includes all points on the line $x = y$ in that quadrant. The rest of the graph is obviously symmetric about the line $x = y$, and by the above argument it is a curve through the point (e, e). This curve somewhat resembles the graph of the hyperbola $xy = 1$ in the first quadrant, except that (e, e) now plays the role of $(1, 1)$ on the hyperbola, and the curve is asymptotic to the vertical line $x = 1$ and the horizontal line $y = 1$, instead of $x = 0$ and $y = 0$ for the hyperbola.

Finally, it may be noted that throughout this paper the question of α^β versus β^α, and the graph of $x^y = y^x$, have been discussed only for positive values of α, β, x and y. The reason, of course, is that if either α or β is negative there may be no meaning to x^β or β^α. For example, if β is irrational, $(-2)^\beta$ has no meaning as a real number. Also, $(-2)^\beta$ has a real value for some but not all rational numbers β; it has a real value for $\beta = 1/3$, for example, but not for $\beta = 1/\alpha$.

Proof without Words:

$$\pi^e < e^\pi$$

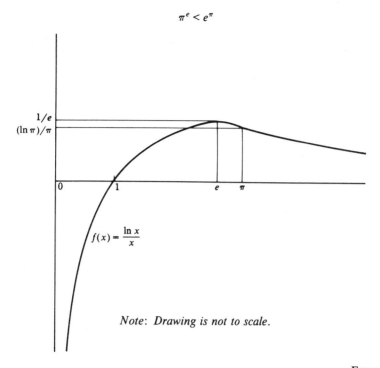

$$f(x) = \frac{\ln x}{x}$$

Note: *Drawing is not to scale.*

—FOUAD NAKHLI
American University of Beirut

Two More Proofs of a Familiar Inequality

Erwin Just
Norman Schaumberger

Norman Schaumberger has been teaching in the New York City area since 1951 and is presently Professor of Mathematics at Bronx Community College, CUNY, where he has been since 1959. He is Associate Editor of the New York State Mathematics Teachers Journal and recently completed a three-year term as a book review editor of the Mathematics Teacher.

Erwin Just is Professor and Chairman of the department of mathematics at Bronx Community College in New York City. He is the Editor of the "Problems and Solutions" Section in the Two-Year College Mathematics Journal.

A well-known approach used to demonstrate that $e^\pi > \pi^e$ is to observe that the function defined by $f(x) = \sqrt[x]{x}$ attains its maximum at $x = e$. It follows that $e^{1/e} > \pi^{1/\pi}$, which implies $e^\pi > \pi^e$. That $(e, e^{1/e})$ is the maximum point of f may be deduced by standard techniques involving the derivative of f or by use of the inequality $e^y > 1 + y$, $(y > 0)$. When y is replaced by $(x - e)/e$, the latter inequality implies $e^{(x-e)/e} > 1 + (x - e)/e$ which, after simplification, asserts that $e^{x/e} > x$ or, finally, $e^{1/e} > x^{1/x}$. This method for obtaining the maximum value of $\sqrt[x]{x}$ is attributed to the 19th century Swiss Mathematician, Jacob Steiner [1].

The inequality $e^\pi > \pi^e$ serves not only as a challenging problem for students who are interested in mathematics, but as an application of the methods of the calculus. It is in the spirit of the latter category that we offer two additional solutions that can be readily demonstrated to students in the first course in calculus.

We begin by observing that the area under the curve $y = \log x$ and bounded by $x = e$, $x = \pi$, and $y = 0$ is less than the rectangular area with height $\log \pi$ and base equal to $\pi - e$. Thus $(\pi - e) \log \pi > \int_e^\pi \log x \, dx$ which yields $(\pi - e) \log \pi > \pi \log \pi - \pi$ or $\pi > e \log \pi$. Therefore, $\log e^\pi > \log \pi^e$ which yields $e^\pi > \pi^e$. This completes our first proof.

The second solution makes use of the function defined by $f(x) = x - e \log x$, $x > 0$. It is easily found by routine methods of elementary calculus that f attains an absolute minimum at $x = e$. As a consequence, $\pi - e \log \pi > e - e \log e$, which implies $\pi > e \log \pi$, from which we deduce, as in the first proof, that $e^\pi > \pi^e$.

REFERENCES

1. H. Dorrie, 100 Great Problems of Elementary Mathematics, Dover, New York, 1965, 359.
2. I. Niven, Which is larger, e^π or π^e?, this Journal, Fall, 1972, 13–15.

An Alternate Classroom Proof of the Familiar Limit for e

Norman Schaumberger, *Bronx Community College*

The procedure usually followed by so many elementary calculus texts in presenting the logarithmic function is to begin by defining the natural logarithm $y = \log x$ for $x > 0$ as the integral

$$\int_1^x \frac{dt}{t}.$$

As a consequence of this definition the usual properties of the logarithm, including the differentiation formula, the addition theorem, and the existence of an inverse, are readily deduced. The inverse function $y = e^x$ is then investigated. Here e is defined as that number whose logarithm is 1.

Many textbooks that use this approach present either incorrect or quite messy proofs of the facts that

$$e = \lim_{n \to \infty} \left(1 + \frac{1}{n}\right)^n$$

and of the important inequalities

$$\left(1 + \frac{1}{n}\right)^{n+1} > e > \left(1 + \frac{1}{n}\right)^n.$$

Some of the books unfortunately omit any mention of either one or the other or even both of these relations. Although basic calculus can be developed without recourse to them, these relations are useful in giving sidelights to the theory. From among many well-known examples, we observe that they can be used to enrich discussions of both discrete and continuous compound interest, they are useful in testing the convergence and divergence of series of constants, they are often an essential ingredient in deriving significant inequalities, and the series expansion for e^x can be developed from the limit relation for e and the binomial theorem.

The following simple derivation of the above limit relation for e makes use of several important concepts encountered in elementary calculus and can be given immediately after the customary properties of $y = \log x$ have been presented. As a byproduct of our derivation, we obtain the inequalities

$$\left(1 + \frac{1}{n}\right)^{n+1} > e > \left(1 + \frac{1}{n}\right)^n$$

where n is any positive number.

We begin by considering the region bounded by the hyperbola $y = 1/x$, the x axis, and the lines $x = n$ and $x = n + 1$, as shown in Figure 1. The area of this shaded region is less than the area of rectangle $MNOP$ and greater than the area of rectangle $MQRP$.

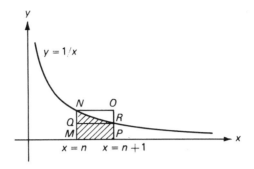

Fig. 1

Since $MNOP = 1/n \cdot 1$ and $MQRP = 1/(n + 1) \cdot 1$, we have

$$\frac{1}{n} > \int_n^{n+1} \frac{1}{x}\, dx > \frac{1}{n + 1}$$

or

$$\frac{1}{n} > \log(n + 1) - \log n > \frac{1}{n + 1} \tag{1}$$

and

$$\frac{1}{n} > \log\left(1 + \frac{1}{n}\right) > \frac{1}{n + 1}.$$

Multiplying the inequalities in (1) by n gives

$$1 > n \log\left(1 + \frac{1}{n}\right) > \frac{n}{n + 1},$$

which can be rewritten as

$$\log e > \log\left(1 + \frac{1}{n}\right)^n > \log e^{n/(n+1)}.$$

It follows that

$$e > \left(1 + \frac{1}{n}\right)^n > e^{n/(n+1)}. \tag{2}$$

Since the outside term on the right of (2) tends to the limit e as $n \to \infty$, we have

$$\lim_{n \to \infty} \left(1 + \frac{1}{n}\right)^n = e.$$

We can also multiply the inequalities in (1) by $n + 1$ to obtain

$$\frac{n + 1}{n} > (n + 1)\log\left(1 + \frac{1}{n}\right) > 1.$$

These inequalities can be rewritten in the form

$$e^{(n+1)/n} > \left(1 + \frac{1}{n}\right)^{n+1} > e. \tag{3}$$

The outside term on the left of (3) tends to e as $n \to \infty$ and thus, we have

$$\lim_{n \to \infty} \left(1 + \frac{1}{n}\right)^{n+1} = e.$$

Finally, from the left inequality in (2) and the right inequality in (3), we get

$$\left(1 + \frac{1}{n}\right)^{n+1} > e > \left(1 + \frac{1}{n}\right)^{n}$$

whenever $n > 0$.

BIBLIOGRAPHIC ENTRIES: e

1. *Monthly* Vol. 77, No. 9, pp. 968–974. C. D. Olds, The simple continued fraction expansion of e.

2. *Monthly* Vol. 81, No. 9, pp. 1011–1012, R. F. Johnsonbaugh, Another proof of an estimate for e.

  Similar to an old proof in Chrystal's *Algebra*, Vol. 2, p. 77 (Edinburgh, 1899).

3. *Monthly* Vol. 81, No. 9, pp. 1012–1013. H. Samelson, To e via convexity.

4. *Monthly* Vol. 93, No. 8, pp. 638–640. T. N. T. Goodman, Maximum products and $\lim(1 + 1/n)^n = e$.

5. *Monthly* Vol. 96, No. 4, pp. 354–355. Chung-Lie Wang, Simple inequalities and old limits.

6. *Math. Mag.* Vol. 62, No. 4, pp. 269–271. Harris S. Shultz and Bill Leonard, Unexpected occurrences of the number e.

  Six interesting examples, easy to understand.

7. TYCMJ Vol. 7, No. 1, pp. 11–12. Norman Schaumberger, A geometrical approach to a basic limit.

8. TYCMJ Vol. 7, No. 4, p. 46. John T. Varner III, Comparing a^b and b^a using elementary calculus.

9. TYCMJ Vol. 13, No. 5, pp. 331–332. Lee Badger. A nonlogarithmic proof that $(1 + 1/n)^n \to e$.

10. TYCMJ Vol. 14, No. 5, pp. 424–426. Robert R. Christian, Another way to introduce natural logarithms and e.

11. CMJ Vol. 20, No. 5, pp. 416–418. Norman Schaumberger, A generalization of $\lim_{n \to \infty}(n!)^{1/n}/n = 1/e$.

12. CMJ Vol. 15, No. 1, pp. 63–64. Sheldon P. Gordon, Evaluating e^x using limits.

Chapter 12. SPECIAL NUMBERS. Sec. (b) π

(No papers reproduced in this section)

BIBLIOGRAPHIC ENTRIES: π

1. *Monthly* vol. 92, No. 3, pp. 213–214. E. F. Assimus, Jr., Pi.
2. *Math. Mag.* Vol. 60, No. 3, pp. 141–150. G. A. Edgar, Pi: Difficult or easy?
3. *Math. Mag.* Vol. 61, No. 2, pp. 67–98. Dario Castellanos, The ubiquitous π, I.
4. *Math. Mag.* Vol. 61, No. 3, pp. 148–163. Dario Castellanos, The ubiquitous π, II.

Chapter 12. SPECIAL NUMBERS. Sec. (c) Euler's constant.

(No papers reproduced in this section)

BIBLIOGRAPHIC ENTRIES: EULER'S CONSTANT

1. *Monthly* Vol. 88, No. 9, pp. 696–698. Richard Johnsonbaugh, The trapezoidal rule, Stirling's formula, and Euler's constant.
2. *Monthly* Vol. 91, No. 7, pp. 428–430. C. W. Barnes, Euler's constant and e.
3. CMJ Vol. 16, No. 4, p. 279. Frank Burk, Euler's constant.

13

THE LIGHT TOUCH

THE DERIVATIVE SONG

Words by Tom Lehrer — Tune: "There'll be Some Changes Made"

You take a function of x and you call it y,
Take any x-nought that you care to try,
You make a little change and call it delta x,
The corresponding change in y is what you find nex',
And then you take the quotient and now carefully
Send delta x to zero, and I think you'll see
That what the limit gives us, if our work all checks,
Is what we call dy/dx,
It's just dy/dx.

THERE'S A DELTA FOR EVERY EPSILON (Calypso)

Words and Music by Tom Lehrer

There's a delta for every epsilon,
It's a fact that you can always count upon.
There's a delta for every epsilon
    And now and again,
    There's also an N.

But one condition I must give:
The epsilon must be positive
A lonely life all the others live,
    In no theorem
    A delta for them.

How sad, how cruel, how tragic,
How pitiful, and other adjec-
Tives that I might mention.
The matter merits our attention.
If an epsilon is a hero,
Just because it is greater than zero,
It must be mighty discouragin'
To lie to the left of the origin.

This rank discrimination is not for us,
We must fight for an enlightened calculus,
Where epsilons all, both minus and plus,
    Have deltas
    To call their own.

THE PROFESSOR'S SONG

Words by Tom Lehrer — Tune: "If You Give Me Your Attention"
from *Princess Ida* (Gilbert and Sullivan)

If you give me your attention, I will tell you what I am.
I'm a brilliant math'matician — also something of a ham.
I have tried for numerous degrees, in fact I've one of each;
Of course that makes me eminently qualified to teach.
I understand the subject matter thoroughly, it's true,
And I can't see why it isn't all as obvious to *you*.
Each lecture is a masterpiece, meticulously planned,
Yet everybody tells me that I'm hard to understand,
    And I can't think why.

My diagrams are models of true art, you must agree,
And my handwriting is famous for its legibility.
Take a word like "minimum" (to choose a random word), (*)
For anyone to say he cannot read that, is absurd.
The anecdotes I tell get more amusing every year,
Though frankly, what they go to prove is sometimes less than clear,
And all my explanations are quite lucid, I am sure,
Yet everybody tells me that my lectures are obscure,
    And I can't think why.

Consider, for example, just the force of gravity:
It's inversely proportional to something —— let me see ——
It's r^3 — no, r^2 — no, it's just r, I'll bet ——
The sign in front is plus — or is it minus, I forget ——
Well, anyway, there *is* a force, of that there is no doubt.
All these formulas are trivial if you only think them out.
Yet students tell me, "I have memorized the whole year through
Ev'rything you've told us, but the problems I can't do."
    And I can't think why!

Limerick

A teacher once, having some fun,
In presenting that two equals one,
    Remained quite aloof
    From his rigorous proof;
But his class was convinced and undone.

—Arthur White
Western Michigan University
Kalamazoo, MI 49008

(*) This was performed at a blackboard, and the professor wrote: ⋀⋁⋁⋁⋁⋁⋁⋁⋁⋀

The Versed of Boas

Ralph P. Boas

Prerequisites

How could you be a cowhand
    If you couldn't ride a horse?
If you yearn to cook for gourmets
    You'll need some food, of course.
You can master many subjects
    If you only have the will;
But how do you cope with calculus
    If your algebra is nil?

How could you sing in opera
    If you haven't any voice?
If music seems too difficult
    There is another choice.
Rewards in Math are plenty
    But this obstacle looms big:
How can you shine in calculus
    If you won't learn any trig?

Pereant

I've proved some theorems, once or twice,
And thought that they were rather nice.
My presentations were rejected
By referees who had detected
Those theorems that I thought my own
In journals I had never known,
And in a strange and knotty tongue.
Oh! For a world still fresh and young,
When fame was won by work alone,
Abstracting journals weren't known,
And (if report can be believed)
No information was retrieved,
Nor academic reputations
Achieved by counting up citations.

Aelius Donatus (fourth century) is quoted by his
student St. Jerome as saying "Pereant qui ante
nos nostra dixerunt." ("Damn the guys who
published our stuff—first.")

"That was on Sheba's obelisk
at the summer solstice!"

FALLACIES, FLAWS, AND FLIMFLAM

Edited by
Ed Barbeau

This column solicits mistakes, fallacies, howlers, anomalies and the like, that raise interesting mathematical issues and may be useful for teachers. Readers are invited to send submissions (which need not be original provided a full reference is given), details about sources, and comments on material already published to:

Ed Barbeau
Department of Mathematics
University of Toronto
Toronto, Ontario, Canada M5S 1A1

FFF #4   Area of an Ellipse

The area enclosed by the ellipse with parametric equations $x = 4\cos\theta$, $y = 3\sin\theta$ is 12π. However, one student determined the area in the following way. What is wrong?

Problem.   Determine the area enclosed by the ellipse with equations

$$x = 4\cos\theta \quad y = 3\sin\theta \quad (0 \le \theta < 2\pi).$$

Solution.   Using the formula $\int r^2(\theta)\,d\theta/2$ for area in polar coordinates, one finds the answer to be $\int_0^{2\pi}(4^2\cos^2\theta + 3^2\sin^2\theta)\,d\theta/2 = 25\pi/2$.

Comments on these will appear in the next issue. They appeared in the *Notes* of the Canadian Mathematical Society and are reproduced with the kind permission of the Publications Committee of the Society. FFF #4 was published as Quickie Q669 in *Mathematics Magazine* 55 (1982) 45, 49.

FFF #6   Cauchy's Negative Definite Integral

If the integrand is always nonnegative, then the definite integral should also be nonnegative. However, the following example from A. L. Cauchy, *Mémoire sur les intégrales définies* (Seconde partie, III, Exemple II, p. 405–406) Oeuvres (1) 1 (1882), 319–506 led Cauchy to reflect more deeply on the evaluation of the definite integral and to develop his calculus of residues.

Problem.   Evaluate

$$\int_0^{3\pi/4} \frac{\sin x}{1 + \cos^2 x}\,dx.$$

Solution.   Since the derivative of $\arctan(\sec x)$ is the integrand, the integral is equal to $\arctan(-\sqrt{2}) - \arctan 1 = -\arctan\sqrt{2} - \pi/4$, a negative quantity.

However, it is easy to see that the integrand is nonnegative on the closed interval $[0, 3\pi/4]$.

FFF #8. A Positive Vanishing Integral

M. Bencze of Brasov, Romania provides the following evaluation:

Problem. Evaluate

$$\int_{-1}^{1} (1 + x^2)^{-1} dx.$$

Solution. Since the derivative of

$$\frac{1}{2} \arccos \frac{1 - x^2}{1 + x^2}$$

is $(1 + x^2)^{-1}$, the given integral is equal to $\frac{1}{2}\arccos 0 - \frac{1}{2}\arccos 0 = 0$.

Thanks are due to the Publications Committee of the Canadian Mathematical Society for permission to reproduce the foregoing items which originally appeared in the *Notes* of the CMS.

FFF #17   cosh x = sinh x and 1 = 0

I am indebted to Robert Weinstock of the Oberlin College Physics Department in Oberlin, Ohio for the following argument:

Integrating by parts twice yields

$$\int e^x \sinh x \, dx = e^x \cosh x - \int e^x \cosh x \, dx$$

$$= e^x \cosh x - \left\{ e^x \sinh x - \int e^x \sinh x \, dx \right\},$$

whence

$$e^x(\cosh x - \sinh x) = 0.$$

An immediate corollary is that $\cosh x = \sinh x$ and $1 = e^x e^{-x} = 0$. If, however, one is more careful and inserts an arbitrary constant with each integration by parts, one readily concludes that $c = e^x(\cosh x - \sinh x) = 1$ for all real c.

FFF #20. A Power Series Representation

Problem. Expand about the origin $f(x) = (1 + x^2)/(1 - x^2)$.

Solution. By the quotient rule, we find that

$$f'(x) = 2\left[\frac{2x}{(1 - x^2)^2} \right] = 2\left[\frac{1}{1 - x^2} \right]',$$

whence $f(x) = 2(1 - x^2)^{-1} = 2(1 + x^2 + x^4 + x^6 + \dots)$ (so that, in particular, $f(0) = 2$). ∎

For other examples of the baleful effect of neglecting the integration constant, consult Phillip J. Sloan, The significance of the "insignificant" constants (Sharing teaching ideas), *Mathematics Teacher* 82 (1989) 186, 188. The example in FFF #19 can be generalized to anything of the form $\int g'(x)/g(x)\, dx$.

FFF #26.   Differentiating the Square of x

Analysis of the following argument is invited: At $x = c$, the function $y = (x - c)^2 = x^2 - 2cx + c^2$ has a minimum, so that $0 = Dy = D(x^2) - 2cD(x) = D(x^2) - 2c$. But c is arbitrary and $c = x$. Hence $D(x^2) = 2c = 2x$. (Submitted by A. W. Walker, Toronto, Ontario, Canada).

FFF #28 More Fun with Series

$$\frac{1}{1+t} = \frac{1}{1-t^2} - t\frac{1}{1-t^4} - t^3\frac{1}{1-t^4}$$

$$= 1 - t - t^3 + t^2 - t^5 - t^7 + t^4 - t^9 - t^{11} + \cdots$$

for $|t| < 1$. Integrating both sides between 0 and x yields

$$\log(1+x) = x - \frac{x^2}{2} - \frac{x^4}{4} + \frac{x^3}{3} - \frac{x^6}{6} - \frac{x^8}{8} + \cdots \quad (|x| < 1).$$

Taking the limit as x tends to 1 and invoking Abel's theorem, we obtain $\log 2 = 1 - \frac{1}{2} - \frac{1}{4} + \frac{1}{3} - \frac{1}{6} - \frac{1}{8} + \cdots$.

However, grouping terms in the series yields

$$\left(1 - \tfrac{1}{2}\right) - \tfrac{1}{4} + \left(\tfrac{1}{3} - \tfrac{1}{6}\right) - \tfrac{1}{8} + \cdots = \tfrac{1}{2} - \tfrac{1}{4} + \tfrac{1}{6} - \tfrac{1}{8} + \cdots$$

$$= \tfrac{1}{2}\left(1 - \tfrac{1}{2} + \tfrac{1}{3} - \tfrac{1}{4} + \cdots\right) = \tfrac{1}{2}\log 2.$$

Thus, $\log 2 = \frac{1}{2}\log 2$.

FFF #44.   A New Way to Obtain the Logarithm

Lewis Lum of the University of Portland in Oregon was inspired by FFF #18 (a howler involving $\sin x$) to send in an evaluation of $\int \dfrac{1}{x+1}\,dx$ on a calculus quiz:

$$\int \frac{1}{x+1}dx = \int\left(\frac{1}{x} + \frac{1}{1}\right)dx = \int\frac{1}{x}dx + \int\frac{1}{1}dx = \ln(x) + \ln(1)$$

$$= \ln(x + 1) + C.$$

FFF #45. All Powers of x Are Constant

From Alex Kuperman at the Israel Institute of Technology (Technion) in Haifa comes a proof by induction that the first derivatives of $x^0, x^1, x^2, \ldots, x^n, \ldots$ are all identically zero.

Observe that $(x^0)' = 0$. Assume that the derivative of x^n is zero for $n = 0, 1, 2, \ldots, k$. Then $(x^{k+1})' = (x \cdot x^k)' = x' \cdot x^k + x \cdot (x^k)'$ also is zero since $x' = (x^1)' = (x^k)' = 0$.

FFF #47. A Natural Way to Differentiate an Exponential

Gerry Myerson of Macquarie University in North Ryde, NSW, Australia sends an item which "looks like a howler, but isn't one." He writes:

> On a calculus exam several years ago, I asked for the derivative of $(\sin x)^{\log x}$, expecting the students to use logarithmic differentiation. Instead of that, one student reasoned as follows: if $\log x$ were constant, the answer would be $(\log x)(\sin x)^{\log x - 1}(\cos x)$; if $\sin x$ were constant, the answer would be $(\sin x)^{\log x}\log(\sin x) \cdot 1/x$ (using the formula $Dc^{f(x)} = c^{f(x)}\log c \cdot f'(x)$). Since neither $\log x$ nor $\sin x$ is constant, we have to use both formulas:
>
> $$D(\sin x)^{\log x} = (\log x)(\sin x)^{\log x - 1}(\cos x)$$
> $$+ (\sin x)^{\log x}\log(\sin x) \cdot \frac{1}{x}.$$

The reasoning may be suspect but the answer is absolutely correct. In fact, the technique is correct in that it gives the correct formula for $D(f(x))^{g(x)}$ for any differentiable f and g.

No bibliographic entries for this chapter.

SOURCES

Chapter 1. HISTORY

Monthly Vol. 90, No. 3, pp. 185–194, Judith V. Grabiner, Who gave you the epsilon? Cauchy and the origins of rigorous calculus.

Math. Mag. Vol. 53, No. 3, pp. 162–166. V. F. Rickey and P. M. Tuchinsky, An application of geography to mathematics: History of the integral of the secant.

Math. Mag. Vol. 56, No. 4, pp. 195–206. J. V. Grabiner, The changing concept of change: The derivative from Fermat to Weierstrass.

Chapter 2. PEDAGOGY

Monthly Vol. 78, No. 6, pp. 664–667. R. P. Boas, Jr., Calculus as an experimental science.

Monthly Vol. 82, No. 5, pp. 466–476. P. R. Halmos, E. E. Moise, George Piranian, The problem of learning to teach.

Monthly Vol. 90, No. 1, pp. 52–53. Dennis Wildfogel, A mock symposium for your calculus class.

TYCMJ Vol. 5, No. 4, pp. 49–53. Louise S. Grinstein, Calculus by mistake.

CMJ Vol. 16, No. 3, pp. 178–185. Jean Pedersen and Peter Ross, Testing understanding and understanding testing.

Chapter 3. FUNCTIONS. Sec. (a) Concepts

Monthly Vol. 78, No. 2, pp. 188–189. G. J. Minty, On the notion of "function."

Chapter 3. FUNCTIONS. Sec. (b) Trigonometric functions

Monthly Vol. 96, No. 3, p. 252. Donald Hartig, On the differentiation formula for sin θ.
Monthly Vol. 98, No. 4, pp. 346–349. Leonard Gillman, π and the limit of $(\sin \alpha)/\alpha$.
CMJ Vol. 21, No. 5, p. 403. Herb Silverman, Trigonometric identities through calculus.
CMJ Vol. 22, No. 5, p. 417. Daniel A. Moran, Graphs and derivatives of the inverse trig functions.

Chapter 3. FUNCTIONS. Sec. (c) Logarithmic functions

Monthly Vol. 79, No. 6, pp. 615–618. B. C. Carlson, The logarithmic mean.
Monthly Vol. 94, No. 5, p. 450. Henry C. Finlayson, The place of ln x among the powers of x.
TYCMJ Vol. 12, No. 1, pp. 20–23. B. L. McAllister and J. E. Whitesitt, Is ln the other shoe?

Chapter 3. FUNCTIONS. Sec. (d) Exponential and hyperbolic functions

Monthly Vol. 77, No. 3, pp. 294–297. R. W. Hamming, An elementary discussion of the transcendental nature of the elementary transcendental functions.

Monthly Vol. 81, No. 6, pp. 643–647. M. C. Mitchelmore, A matter of definition.

CMJ Vol. 19, No. 1, pp. 54–56. Roger B. Nelsen, The relationship between hyperbolic and exponential functions.

Chapter 4. LIMITS AND CONTINUITY

Math. Mag. Vol. 62, No. 3, pp. 176–184. Ray Redheffer, Some thoughts about limits.

Chapter 5. DIFFERENTIATION. Sec. (a) Theory

Monthly Vol. 77, No. 2, pp. 187–189. D. G. Herr, An introduction to differential calculus.

Monthly Vol. 82, No. 5, pp. 505–506. J. A. Eidswick, The differentiability of a^x.

Monthly Vol. 92, No. 8, pp. 589–590. Donald E. Richmond, An elementary proof of a theorem in calculus.

Monthly Vol. 97, No. 2, pp. 144–147. Ernst Snapper, Inverse functions and their derivatives.

TYCMJ Vol. 8, No. 1, pp. 10–11. David A. Birnbaum and Northrup Fowler III, An elementary result on derivatives.

TYCMJ Vol. 9, No. 2, pp. 67–72. Thomas J. Brieske, Mapping diagrams, continuous functions, and derivatives.

CMJ Vol. 16, No. 2, pp. 131–132. Peter A. Lindstrom, A self-contained derivation of the formula $d/dx(x^r) = rx^{r-1}$ for rational r.

CMJ Vol. 21, No. 4, pp. 312–313. Russell Jay Hendel, $(x^n)' = nx^{n-1}$: six proofs.

CMJ Vol. 17, No. 2, pp. 166–167. Russell Euler, A note on differentiation.

CMJ Vol. 20, No. 1, pp. 52–53. D. F. Bailey, Differentials and elementary calculus.

Chapter 5. DIFFERENTIATION. Sec. (b) Applications to geometry.

Monthly Vol. 97. No. 10, pp. 907–911. S. C. Althoen and M. F. Wyneken, The width of a rose petal.

See also Monthly 98 #2, p. 139 for perimeter, Ma. Mag. 43 #3, p. 156 for area. (Listed as items 5 and 6 in the Bibliography for Sec. 9(d) of this volume, p. 334).

TYCMJ Vol. 9, No. 1, pp. 47–48. Jay I. Miller, Differentiating area and volume.

CMJ Vol. 15, No. 1, pp. 37–41. Victor A. Belfi, Convexity in elementary calculus: some geometric equivalences.

CMJ Vol. 17, No. 4, p. 341. R. P. Boas, Does "holds water" hold water?

CMJ Vol. 18, No. 2, pp. 124–133. Jeanne L. Agnew and James R. Choike, Transitions.

TYCMJ Vol. 13, No. 1, pp. 59–61, Allan J. Kroopnick, A note on parallel curves.

This is a simplified treatment of an earlier article by Stein, v. 11, 239–246, listed as item 3 in the Bibliography for this section, p. 153.

Chapter 5. DIFFERENTIATION. Sec. (c) Applications to mechanics

Math. Mag. Vol. 50, No. 5, pp. 257–258. Gerald T. Cargo , Velocity averages.
TYCMJ Vol. 10, No. 2, pp. 82–88. R. P. Boas, Travelers' surprises.
TYCMJ Vol. 13, No. 3, pp. 195–196. William A. Leonard, Intuition out to sea.
CMJ Vol. 16, No. 3, pp. 186–189. S. C. Althoen and J. F. Weidner, Related rates and the speed of light.

Chapter 5. DIFFERENTIATION. Sec. (d) Differential equations.

Math. Mag. Vol. 44, No. 1, pp. 33–34, William C. Waterhouse, A fact about falling bodies.
TYCMJ Vol. 9, No. 3, pp. 141–145. David A. Smith, The homicide problem revisited.
CMJ Vol. 18, No. 1, pp. 44–45. Arthur C. Segal, A linear diet model.

Chapter 5. DIFFERENTIATION. Sec. (e) Partial Derivatives

Monthly Vol. 92, No. 2, pp. 144–145. Fred Helpern, Using the multivariable chain rule.

Chapter 6. MEAN VALUE THEOREM FOR DERIVATIVES, INDETERMINATE FORMS. Sec. (a) Mean Value Theorem for Derivatives

Monthly Vol. 79, No. 4, pp. 381–383. D. E. Sanderson, A versatile vector mean value theorem.
TYCMJ Vol. 12, No. 3, pp. 178–181. R. P. Boas, Who needs those mean-value theorems, anyway?

Chapter 6. MEAN VALUE THEOREM FOR DERIVATIVES, INDETERMINATE FORMS. Sec. (b) Indeterminate Forms

Monthly Vol. 76, No. 9, pp. 1051–1053. R. P. Boas, Jr., Lhospital's rule without mean value theorems.
Monthly Vol. 85, No. 6, pp. 484–486. John V. Baxley and Elmer K. Hayashi, Indeterminate forms of exponential type.
Monthly Vol. 93, No. 8, pp. 644–645. R. P. Boas, Counterexamples to L'Hôpital's rule.
Monthly Vol. 98, No. 2, pp. 156–157. Donald Hartig, L'Hôpital's rule via integration.
Math. Mag. Vol. 63, No. 3, pp. 155–159. R. P. Boas, Indeterminate forms revisited.
TYCMJ Vol. 10, No. 3, pp. 197–198. J. P. King, L'Hôpital's rule and the continuity of the derivative.
CMJ Vol. 15, No. 1, pp. 51–52. Robert J. Bumcrot, Some subtleties in L'Hôpital's rule.

Chapter 7. TAYLOR POLYNOMIALS, BERNOULLI POLYNOMIALS
AND SUMS OF POWERS OF INTEGERS. Sec. (a) Taylor polynomials.

Monthly Vol. 90, No. 2, p. 130. Ray Redheffer, From center of gravity to Bernstein's theorem.
Monthly Vol. 97, No. 9, p. 836. Deng Bo, A simple derivation of the Maclaurin series for sine and cosine.
Math. Mag. Vol. 49, No. 3, pp. 147–148. John Staib, Trigonometric power series.
CMJ Vol. 16, No. 2, pp. 103–107. Dan Kalman, Rediscovering Taylor's theorem.

Chapter 8. MAXIMA AND MINIMA

Monthly Vol. 82, No. 3, pp. 287–289. J. H. C. Creighton, A strong second derivative test.
Monthly Vol. 83, No. 5, pp. 361–365. L. H. Lange, Cutting certain minimum corners.
Monthly Vol. 96, No. 8, pp. 721–725. Richard Bassein, An optimization problem.
Monthly Vol. 97, No. 5, pp. 421–423. Mary Embry-Wardrop, An old max-min problem revisited.
TYCMJ Vol. 5, No. 1, pp. 22–24. L. H. Lange, Maximize $x(a - x)$.
TYCMJ Vol. 5, No. 2, pp. 12–14. Robert Owen Armstrong, Construction of an exercise involving minimum time.
TYCMJ Vol. 14, No. 1, pp. 57–60. W. L. Perry, A bifurcation problem in first semester calculus.
CMJ Vol. 15, No. 1, pp. 30–36. Kay Dundas, To build a better box.
CMJ Vol. 18, No. 3, pp. 225–229. Herbert Bailey, A surprising max-min result.
CMJ Vol. 21, No. 2, pp. 129–130. John W. Dawson, Jr., Hanging a bird feeder: Food for thought.
Monthly Vol. 83, No. 5, pp. 370–371. Cliff Long, Peaks, ridge, passes, valley, and pits. A slide study of $f(x, y) = Ax^2 + By^2$.
Math. Mag. Vol. 58, No. 3, pp. 147–149. J. M. Ash and H. Sexton, A surface with one local minimum.
Math. Mag. Vol. 58, No. 3, pp. 149–150. I. Rosenholtz and L. Smylie, "The only critical point in town" test.

Chapter 9. INTEGRATION. Sec. (a) Theory

Math. Mag. Vol. 64, No. 5, pp. 347–348. Michael W. Botsko, A fundamental theorem of calculus that applies to all Riemann integrable functions.
CMJ Vol. 15, No. 5, pp. 426–429. W. Vance Underhill, Finding bounds for definite integrals.
CMJ Vol. 16, No. 2, pp. 132–135. David E. Dobbs, Average values and linear functions.
CMJ Vol. 20, No. 3, p. 237. John H. Mathews and Harris S. Shultz, Riemann integral of cos x.
CMJ Vol. 21, No. 1, pp. 20–27. Gilbert Strang, Sums and differences vs. integrals and derivatives.

Chapter 9. INTEGRATION.B Sec. (b) Techniques of Integration

Monthly Vol. 76, No. 5, pp. 546–547. N. Schaumberger, The evaluation of $\int_a^b x^k \, dx$.

Monthly Vol. 81, No. 7, pp. 760–761. R. P. Boas, Jr. and M. B. Marcus, Inverse functions and integration by parts.

Math. Mag. Vol. 64, No. 2, p. 130. Roger B. Nelsen, Proof without words: Integration by parts.

TYCMJ Vol. 5, No. 2, pp. 1–7. S. K. Stein, Formal integration: dangers and suggestions.

TYCMJ Vol. 10, No. 5, pp. 353–354. John Staib and Howard Anton, A discovery approach to integration by parts.

Chapter 9. INTEGRATION. Sec. (d) Applications

Monthly Vol. 98, No. 2, pp. 154–156. Walter Carlip, Disks and shells revisited.

Monthly Vol. 81, No. 4, pp. 385–387. Richard T. Bumby, Upper bounds on arc length.

Math. Mag. Vol. 42, No. 3, pp. 132–133. John Kaucher, A theorem on arc length.

Math. Mag. Vol. 43, No. 1, p. 44. John T. White, A note on arc length.

Math. Mag. Vol. 50, No. 3. pp. 160–162. Sherman K. Stein, "Mean distance" in Kepler's third law.

TYCMJ Vol. 3, No. 1, pp. 72–75. William C. Stretton, Some problems of utmost gravity.

TYCMJ Vol. 4, No. 3, pp. 52–55. Bert K. Waits, Jerry L. Silver, A new look at an old work problem.

TYCMJ Vol. 6, No. 3, pp. 13–15. G. L. Alexanderson and L. F. Klosinski, Some surprising volumes of revolution.

TYCMJ Vol. 8, No. 4, pp. 207–211. Frieda Zames, Surface area and the cylinder area paradox.

TYCMJ Vol. 10, No. 3, pp. 179–181. Philip D. Straffin, Jr., Using integrals to evaluate voting power.

CMJ Vol. 16, No. 5, pp. 400–402. R. Rozen and A. Sofo, Area of a parabolic region.

Chapter 9. INTEGRATION. Sec. (e) Multiple integrals and line integrals

Math. Mag. Vol. 55, No. 1, pp. 3–11. Victor J. Katz, Change of variables in multiple integrals: Euler to Cartan.

Math. Mag. Vol. 59, No. 1, pp. 40–42. J. Chris Fisher and J. Shilleto, Three aspects of Fubini's theorem.

Math. Mag. Vol. 59, No. 3, pp. 170–171. J. Nunemacher, The largest unit ball in any Euclidean space.

Math. Mag. Vol. 64, No. 2, pp. 122–123. Paul B. Massell, Volumes of cones, paraboloids, and other "vertex solids."

TYCMJ Vol. 5, No. 3, pp. 20–21. Stewart Venit, Interchanging the order of integration.

Chapter 10. NUMERICAL, GRAPHICAL, AND MECHANICAL METHODS AND APPROXIMATIONS (including use of computers)

Monthly Vol. 76, No. 8, pp. 929–930. Anon, Note on Simpson's rule.

Monthly Vol. 82, No. 3, pp. 284–287. David A. Smith, Numerical differentiation for calculus students.

Monthly Vol. 83, No. 8, pp. 643–645. S. K. Stein, The error of the trapezoidal method for a concave curve.

Monthly Vol. 92, No. 6, pp. 425–426. Arthur Richert, A non-Simpsonian use of parabolas in numerical analysis.

Monthly Vol. 95, No. 8, pp. 754–757. George P. Richardson, Reconsidering area approximations.

TYCMJ Vol. 4, No. 1, pp. 36–38. Martin D. Landau and William R. Jones, An interpolation question resolved by calculus.

TYCMJ Vol. 13, No. 1, pp. 22–27. Sheldon P. Gordon, Generalized cycloids: Discovery via computer graphics.

CMJ Vol. 16, No. 1, p. 56. Frank Burk, Behold! The midpoint rule is better than the trapezoidal rule for concave functions.

CMJ Vol. 19, No. 2, pp. 166–168. Chris W. Avery and Frank P. Soler, Applications of transformations to numerical integration.

CMJ Vol. 21, No. 2, pp. 142–144. David P. Kraines, Vivian Y. Kraines, and David A. Smith, Circumference of a circle—the hard way.

Chapter 11. INFINITE SEQUENCES AND SERIES. Sec. (a) Theory

Monthly Vol. 78, No. 8, pp. 890–892. Eugene Schenkman, Interval of convergence of some power series.

Monthly Vol. 79, No. 6, pp. 634–635. G. J. Porter, An alternative to the integral test for infinite series.

Monthly Vol. 86, No. 8, pp. 679–681. J. R. Nurcombe, A sequence of convergence tests.

Monthly Vol. 88, No. 1, pp. 51–52. Mark D. Meyerson, Every power series is a Taylor series.

Monthly Vol. 91, No. 6, pp. 367–369. Wells Johnson, Power series without Taylor's theorem.

Monthly Vol. 92, No. 8, pp. 588–589. I. E. Leonard and James Duemmel, More- and Moore-power series without Taylor's theorem.

Monthly Vol. 93, No. 7, p. 561. Charles C., Mumma II, $N!$ and the root test.

Monthly Vol. 95, No. 10, p. 942. Jonathan Lewin and Myrtle Lewin, A simple test for the nth term of a series to approach zero.

Monthly Vol. 52, No. 3, p. 178. T. Cohen and W. J. Knight. Convergence and divergence of $\sum_{n=1}^{\infty} 1/n^p$.

Math. Mag. Vol. 57, No. 4, pp. 228–231. Y. S. Abu-Mostafa, A differentiation test for absolute convergence.

TYCMJ Vol. 9, No. 1, pp. 46–47, Louise S. Grinstein, A note on infinite series.

TYCMJ Vol. 13, No. 3, pp. 191–195. Ralph Boas, Power series for practical purposes.

Chapter 11. INFINITE SEQUENCES AND SERIES. Sec. (b) Series related to the harmonic series

Monthly Vol. 78, No. 3, pp. 272–273. Ivan Niven, A proof of the divergence of $\Sigma 1/p$.

Monthly Vol. 82, No. 9, pp. 931–933. A. D. Wadhwa, An interesting subseries of the harmonic series.

Monthly Vol. 87, No. 5, pp. 394–397. Charles Vanden Eynden, Proofs that $\Sigma 1/p$ diverges.

Monthly Vol. 87, No. 10, pp. 817–819. C. C. Cowen, K. R. Davidson and R. P. Kaufman, Rearranging the alternating harmonic series.

TYCMJ Vol. 10, No. 3, pp. 198–199. W. G. Leavitt, The sum of the reciprocals of the primes.

CMJ Vol. 18, No. 1, pp. 18–23. William Dunham, The Bernoullis and the harmonic series.

Chapter 11. INFINITE SERIES AND SEQUENCES. Sec. (c) Sums of special series

Monthly Vol. 80, No. 4, pp. 424–425. Ioannis Papadimitriou, A simple proof of the formula $\Sigma_{k=1}^{\infty} k^{-2} = \pi^2/6$.

Monthly Vol. 85, No. 6, pp. 481–483. Eberhard L. Stark, Application of a mean value theorem for integrals to series summation.

Monthly Vol. 90, No. 4, pp. 284–285. John Klippert, Summing power series with polynomial coefficients.

Math. Mag. Vol. 44, No. 1, pp. 37–38. Frank Kost, A geometric proof of the formula for ln 2.

TYCMJ Vol. 10, No. 2, pp. 95–99. Mangho Ahuja, On sum-guessing.

CMJ Vol. 22, No. 4, p. 322. John H. Mathews, The sum is one.

Chapter 12. SPECIAL NUMBERS. Sec. (a) *e*

Monthly Vol. 80, No. 2, pp. 193–194. J. P. Tull, A discovery approach to *e*.

Monthly Vol. 80, No. 2, p. 194. R. B. Darst, Simple proofs of two estimates for *e*.

TYCMJ Vol. 3, No. 2, pp. 13–15. Ivan Niven, Which is Larger, e^{π} or π^e?

Math. Mag. Vol. 60, No. 3, p. 165. Fouad Nakhli, Proof without words: $\pi^e < e^{\pi}$.

TYCMJ Vol. 3, No. 2, pp. 72–73. Norman Schaumberger, An alternative classroom proof of the familiar limit for *e*.

TYCMJ Vol. 6, No. 2, p. 45. Erwin Just and Norman Schaumberger, Two more proofs of a familiar inequality.

Chapter 13. THE LIGHT TOUCH.

Monthly Vol. 81, No. 5, p. 490. Tom Lehrer, The derivative song.
Monthly Vol. 81, No. 6, p. 612. Tom Lehrer, There's a delta for every epsilon (Calypso).
Monthly Vol. 81, No. 7, p. 745. Tom Lehrer, The professor's song.
Math. Mag. Vol. 64, No. 2, p. 91. Arthur White, Limerick.
TYCMJ Vol. 14, No. 4, p. 342. Ralph P. Boas, The Versed of Boas.
CMJ Vol. 20, No. 2, pp. 132–133. Ed Barbeau. Area of an ellipse.

> Explanation in CMJ Vol. 20, No. 3, p. 227.

CMJ Vol. 20, No. 3, p. 226. Ed Barbeau, Cauchy's negative definite integral.

> Explanation in CMJ Vol. 20, No. 4, p. 318.

CMJ Vol. 20, No. 4, p. 317. Ed Barbeau and M. Bencze, A positive vanishing integral.

> Explanation in CMJ Vol. 20, No. 5, p. 404.

CMJ Vol. 21, No. 2, p. 128. Ed Barbeau and Robert Weinstock, $\cosh x = \sinh x$ and $1 = 0$.
CMJ Vol. 21, No. 3, p. 217. Ed Barbeau, A power series representation.
CMJ Vol. 21, No. 4, p. 304. Ed Barbeau and A. W. Walker, Differentiating the square of x.
CMJ Vol. 21, No. 5, pp. 395–396. Ed Barbeau, More fun with series.

> Explanation in CMJ Vol. 23, No. 1, p. 38.

CMJ Vol. 22, No. 5, p. 403. Ed Barbeau and Lewis Lum, A new way to obtain the logarithm.
CMJ Vol. 22, No. 5, p. 403. Ed Barbeau and Alex Kuperman, All powers of x are constant.
CMJ Vol. 22, No. 5, p. 404. Ed Barbeau and Gerry Myerson, A natural way to differentiate an exponential.

AUTHOR INDEX

CONTENTS

Part I

3. FUNCTIONS

(a) CONCEPTS AND NOTATION

(b) TRIGONOMETRIC FUNCTIONS

(c) LOGARITHMIC FUNCTIONS

4. CONTINUITY, ϵ AND δ, DISCONTINUITIES

5. DIFFERENTIATION

6. MEAN VALUE THEOREM FOR DERIVATIVES, INDETERMINATE FORMS

(a) MEAN VALUE THEOREM

7. POLYNOMIALS AND POLYNOMIAL APPROXIMATIONS

(a) TAYLOR POLYNOMIALS

(b) OTHER POLYNOMIALS

8. MAXIMA AND MINIMA

9. INTEGRATION

(a) THEORY

(b) TECHNIQUES OF INTEGRATION

(c) SPECIAL INTEGRALS

(d) APPLICATIONS

480 CONTENTS, PART I

12. SPECIAL NUMBERS